Modeling and Simulation of Carbon Emission Related Issues

Modeling and Simulation of Carbon Emission Related Issues

Special Issue Editor

Wen-Hsien Tsai

MDPI • Basel • Beijing • Wuhan • Barcelona • Belgrade

Special Issue Editor
Wen-Hsien Tsai
National Central University
Taiwan

Editorial Office
MDPI
St. Alban-Anlage 66
4052 Basel, Switzerland

This is a reprint of articles from the Special Issue published online in the open access journal *Energies* (ISSN 1996-1073) from 2018 to 2019 (available at: https://www.mdpi.com/journal/energies/special_issues/carbon_emission)

For citation purposes, cite each article independently as indicated on the article page online and as indicated below:

LastName, A.A.; LastName, B.B.; LastName, C.C. Article Title. *Journal Name* **Year**, *Article Number*, Page Range.

ISBN 978-3-03921-311-5 (Pbk)
ISBN 978-3-03921-312-2 (PDF)

Cover image courtesy of Wen-Hsien Tsai.

Contents

About the Special Issue Editor

Wen-Hsien Tsai is a distinguished professor of accounting and information systems in the Department of Business Administration at National Central University, Taiwan. He has served as a guest editor for Special Issues of the journals *Energies* and *Sustainability* and as an associate editor of the journal *Decision Support Systems*. He is also a certified consultant of SAP financial modules. He received his Ph.D. degree in industrial management from the National Taiwan Science and Technology University. He received his MBA degree and his M.Sc. degree in industrial engineering from the National Taiwan University and National Tsing-Hwa University, respectively. His research interests include Industry 4.0, carbon emissions, carbon tax, activity-based costing (ABC), ERP implementation and auditing, green production and optimization decision, and the International Financial Reporting Standards (IFRS). He has published several papers in high-quality international journals, such as *Energies, Sustainability, Decision Support Systems, European Journal of Operational Research, Omega, Transportation Science, Industrial Marketing Management, Journal of the Operational Research Society, Computers and Operations Research, Journal of Cleaner Production, International Journal of Production Economics, Computers and Industrial Engineering, International Journal of Production Research*, etc.

Preface to "Modeling and Simulation of Carbon Emission Related Issues"

Carbon emissions reached an all-time high in 2018, when global carbon dioxide emissions from burning fossil fuels increased by about 2.7%, after a 1.6% increase in 2017. Thus, we need to pay special attention to carbon emissions and work out possible solutions if we still want to meet the targets of the Paris climate agreement. This Special Issue collects 16 carbon emissions-related papers (including 5 that are carbon tax-related) and 4 energy-related papers using various methods or models, such as the input–output model, decoupling analysis, life cycle impact analysis (LCIA), relational analysis model, generalized Divisia index model (GDIM), forecasting model, three-indicator allocation model, mathematical programming, real options model, multiple linear regression, etc. The research studies come from China, Taiwan, Brazil, Thailand, and United States. These researches involved various industries such as agricultural industry, transportation industry, power industry, tire industry, textile industry, wave energy industry, natural gas industry, and petroleum industry. Although this Special Issue does not fully solve our concerns, it still provides abundant material for implementing energy conservation and carbon emissions reduction. However, there are still many issues regarding the problems caused by global warming that require research. Finally, I am grateful to MDPI for the invitation to act as the Guest Editor of this Special Issue and I am indebted to the editorial office of *Energies* for the kind cooperation, patience, and committed engagement. I would like to thank the authors for submitting their excellent contributions to this Special Issue. My thanks are extended to the reviewers for evaluating the manuscripts and providing helpful suggestions. Sincere thanks also go to the editorial team of MDPI and *Energies* for providing the opportunity to publish this book and helping in all possible ways.

Wen-Hsien Tsai
Special Issue Editor

energies

MDPI

Editorial

Modeling and Simulation of Carbon Emission-Related Issues

Wen-Hsien Tsai

Department of Business Administration, National Central University, Jhongli, Taoyuan 32001, Taiwan; whtsai@mgt.ncu.edu.tw; Tel.: +886-3-426-7247

Received: 5 March 2019; Accepted: 27 June 2019; Published: 1 July 2019

1. Introduction

According to the Intergovernmental Panel on Climate Change (IPCC) Fifth Assessment Report in 2013 (IPCC, 2013) [1], global warming is mainly caused by several greenhouse gases, such as carbon dioxide (CO_2), methane, nitrous oxide, and ozone, which are emitted by human activities in a variety of ways. Baroness Anelay, the former UK Minister of State of the Foreign and Commonwealth Office, said: "The threat of climate change needs to be assessed in the same comprehensive way as nuclear weapons proliferation." [2]. In addition, both former Vice-President Al Gore and former President Barack Obama of United States deemed that climate change was a more dangerous threat to the world than international terrorism [3]. The Paris Agreement was signed by 195 nations in December 2015 to strengthen the global response to the threat of climate change, following the 1992 United Nations Framework Convention on Climate Change (UNFCC) and the 1997 Kyoto Protocol. In Article 2 of the Paris Agreement, the increase in the global average temperature is anticipated to be held to well below 2 °C above pre-industrial levels, and efforts are being employed to limit the temperature increase to 1.5 °C above pre-industrial levels [4].

It is estimated that about 72% of the totally emitted greenhouse gases is carbon dioxide (CO_2), 18% methane, and 9% nitrous oxide [5]. Therefore, carbon dioxide (CO_2) emission (or carbon emission) is the most important cause of global warming. The vast majority of anthropogenic carbon emissions come from the combustion of fossil fuels, principally coal, oil, and natural gas, with additional contributions coming from deforestation, changes in land use, soil erosion, and agriculture [6]. The United Nations had made possible efforts on greenhouse gas emissions mitigation. In Article 6 of the Paris Agreement, three cooperative approaches were presented that countries can take in attaining the goal of their carbon emission reduction, including direct bilateral cooperation, new sustainable development mechanisms, and non-market-based approaches [7].

For the carbon emission reduction, several related issues and practical technologies were proposed, such as carbon footprint, carbon tax, cap and trade, carbon right purchasing, carbon emission cost analysis, internal carbon pricing, and so on. Cap and trade is one method for regulating and ultimately reducing the amount of carbon emission [8]. The government sets a cap on carbon emission, limiting the amount of carbon dioxide that companies are allowed to release. Companies that can more efficiently reduce carbon emission can sell any extra permits in the emission market. Thus, the carbon trading markets were set up. Currently there are five trading in carbon allowances: the European Climate Exchange, NASDAQ OMX Commodities Europe, PowerNext, Commodity Exchange Bratislava, and the European Energy Exchange [9].

However, Harvey stated that, "A report released yesterday by a consortium of researchers known as the Global Carbon Project finds that global carbon dioxide emissions from burning fossil fuels are likely to have increased by about 2.7% in 2018, after a 1.6% increase in 2017" [10]. We need to pay special attention to carbon emissions and work out the possible solutions if we still want to meet the targets of the Paris climate agreement. In this urgent time for carbon emission reduction, this special issue collects 20 related papers concerning carbon emissions from household to various industries by using various models and methods.

2. Summary Information of 20 Papers in the Special Issue

Table 1 shows the summary information of 20 papers in this special issue, including Research Topic, Papers' Author, Method/Model, Research Object, and Industry/Field. From Table 1, we can see that this special issue has 16 carbon emissions-related papers (including five that are carbon tax-related) and four energy-related papers in various industries by using various methods or models. In carbon emissions, it explores household, transportation, and agricultural carbon emissions, carbon emissions reduction, carbon emissions forecasting, and quotas allocation for carbon emissions. In energy, it discusses renewable energy and energy consumption forecasting. These papers will be reviewed in the next section.

3. Review of the Special Issue

3.1. Carbon Emissions

3.1.1. Household Carbon Emissions

Wang et al. [11] used the carbon emissions coefficient method and Consumer Lifestyle Approach (CLA) to calculate the total carbon emissions of households in 30 provinces of China from 2006 to 2015, and adopted the extended Stochastic Impacts by Regression on Population, Affluence, and Technology (STIRPAT) model to analyze the factors influencing the total carbon emissions of households. The findings indicate that the energy and products' carbon emissions from China's households had demonstrated a rapid growth trend over the past 10 years and primarily derived from residents' high carbon emission categories: residences, food, transportation, and communications.

Huang et al. [12] analyzed the direct and indirect CO_2 emissions by urban and rural households in Beijing, Tianjin, Shanghai, and Chongqing. The results show that urban total household carbon emissions are larger than rural total household carbon emissions for the four megacities. Electricity and hot water production and supply is the largest contributor of indirect household carbon emissions for both rural and urban households. Besides, Beijing, Tianjin, Shanghai, and Chongqing outsource a large amount of indirect carbon emissions to their neighboring provinces.

3.1.2. Transportation Carbon Emissions

Transportation is an important source of carbon emissions in China. Wang et al. [13] analyzed the drivers of carbon emissions in China's transportation sector from 2000 to 2015 by using the Generalized Divisia Index Method (GDIM). The findings show that the added value of transportation, energy consumption, and per capita carbon emissions in transportation have always been major factors affecting China's carbon emissions from transportation. The carbon intensity of the added value and the energy intensity have a continuous effect on carbon emissions in transportation.

3.1.3. Agricultural Carbon Emissions

Zhang et al. [14] utilized decoupling analysis to construct a decoupling index based on carbon footprint and crop yield, and evaluated the relationship between crop production and greenhouse gas emissions using the most modern grain production base in China as a case study. The findings show that a weak but variable decoupling trend occurs from 2001 to 2015, and that there is a weak decoupling across the study period. Besides, rice production constituted 80% of the regional carbon footprint in a crop's life cycle.

Table 1. Summary information of 20 papers in this special issue.

Topic			Paper/Author	Method/Model	Research Object	Industry/Field
1. Carbon Emissions	1.1 Household Carbon Emissions		Wang, Yang, Dong, Cheng, Shang [11]	STIRPAT * Model	30 Provinces of China	Household
			Huang, Zhang, Liu [12]	Input–Output Model	Beijing, Tianjin, Shanghai, and Chongqing of China	Urban and Rural Household
	1.2 Transportation Carbon Emissions		Wang, Zhou, Zhu, Zhang, Zhang [13]	Generalized Divisia Index Method (GDIM).	China	Transportation Industry
	1.3 Agricultural Carbon Emissions		Zhang, Zou, Xu, Yang [14]	Decoupling Analysis	Heilongjiang Land Reclamation Area, China	Agricultural Industry
			Hu, Chen, Huang, Chung, Lan, Chen [15]	PAS 2050-based Life Cycle Analysis (LCA) approach; Life Cycle Inventory (LCI) Analysis; Life Cycle Impact Assessment (LCIA)	Taiwan	Agricultural Industry (Tea)
	1.4 Carbon Emissions Reduction	Carbon emissions reduction for departure aircraft pushback control	Zhu, Li, Sun, Zhang, Wang, Tsai [16]	A Time Prediction Model of aircraft departure taxiing time by using Multivariate Linear Regression	Shanghai Hongqiao Airport, China	Air Transportation Industry
		CO_2 emissions reduction evaluation of Li-ion batteries	Peralta, Vieira, Meunier, Vale, Salles, Carmo [17]	Optimization and simulations using HOMER (Hybrid Optimization Model for Multiple Energy Resources) software	Brazil	Ship Power Industry
	1.5 Carbon Emissions Forecasting		Sutthichaimethee, Kubaha [18]	Relational Analysis Model; VARIMAX-ECM ** Model	Thailand	Industrial Sector
			Zhu, He, Shang, Zhang, Ma [19]	Generalized Divisia Index Model (GDIM); Monte Carlo simulation	China	Power Industry
			Wang, Shang, He, Zhang, Liu [20]	Gray model (GM (1, 1)); Generalized Regression Neural Network (GRNN); Markov Forecasting model; Non-Linear programming	China	Government
	1.6 Quotas Allocation for Carbon Emissions		Meng, Wang, Chen [21]	A Hybrid Trend Forecasting model; A Three-Indicator Allocation model	30 provinces of China	Electric Power Industry

Table 1. *Cont.*

Topic			Paper/Author	Method/Model	Research Object	Industry/Field
2. Carbon Tax		Inventory routing problem with carbon tax	Wang, Tao, Shi [22]	Low-Carbon Inventory Routing Problem (LCIRP) model	China	Refined Oil Logistics
		Carbon tax	Fan, Gao, Chen, Wei, Xu, Lu, Wang, Zhang, Ren, Ulgiati, Dong [23]	3EAD-CGE *** model	China	Agricultural Industry
		Green quality management decision with carbon tax	Tsai [24]	Mathematical Programming	Taiwan	Tire Industry
		Green production planning and control with carbon tax under Industry 4.0	Tsai [25]	Mathematical Programming	Taiwan	Textile Industry
		Carbon taxes and carbon right costs analysis	Tsai [26]	Mathematical Programming	Taiwan	Tire Industry
3. Energy	3.1 Renewable Energy	Renewable energy investment decisions	Li, Wu, Li [27]	Real Options model	China	China Carbon Trading Market
		Life cycle assessment of Buoy–Rope–Drum wave energy converter	Zhai, Zhu, Lu [28]	Life Cycle Inventory (LCI) Analysis; Life Cycle Impact Analysis (LCIA)	China	Wave Energy Industry
		Ethylene economics and production forecasting because of the advent of shale gas	Foster [29]	Vector Error Correction Model (VECM); Multiple Linear Regression Model (MLR)	United States (U.S.)	Ethylene Industry; Natural Gas Industry
	3.2 Energy Forecasting	Energy consumption long-term forecasting	Sutthichaimethee, Kubaha [30]	LT-ARIMAXS Model ****	Thailand	Petroleum Industry

Note: * STIRPAT: Stochastic Impacts by Regression on Population, Affluence, and Technology; ** VARIMAX-ECM: Variance Maximization-Error Correction Model; *** 3EAD-CGE: Economy–Energy–Environmental–Agricultural–Dynamics Computable General Equilibrium; **** LT-ARIMAXS Model: The Long Term-Autoregressive Integrated Moving Average with Exogeneous variables and Error Correction Mechanism mode.

Hu et al. [15] evaluated the environmental impact and carbon footprint of Dongshan tea from Yilan County in Taiwan. The results indicate that climate change has the largest impact upon it, followed by human health, natural resources, and ecosystem quality. It is also found that the environmental impact of Taiwanese tea mainly came from fertilizer input during the raw material phase, electricity use during manufacturing, and electricity use during water boiling in the consumer-use phase.

3.1.4. Carbon Emissions Reduction

Zhu et al. selected Shanghai Hongqiao Airport to explore the control strategy for aircraft departure [16]. In this paper, the influence of the number of departure aircraft on the runway utilization rate, the takeoff rate, and the departure rate of flight departures under the conditions of airport runway capacity constraints are studied. A time prediction model of aircraft departure taxiing time is established in this study by using a multivariate linear regression equation, and the experimental results indicate that without reducing the utilization rate of the runway and the departure rate of flights, implementing a reasonable pushback number for the control of departing aircraft during busy hours can reduce the departure taxiing time of aircraft by about 32%, which will reduce the fuel consumption and pollutant emissions during taxiing on the airport surface.

Peralta et al. [17] analyzed the potential implementation of Li-ion batteries (lithium titanate or lithium iron phosphate) in a platform supply vessel system through simulations using HOMER software (Hybrid Optimization Model for Multiple Energy Resources). They also analyzed the potential emissions reduction for different parts of a mission to an offshore platform for different configurations of the ship power system.

3.1.5. Carbon Emissions Forecasting

Sutthichaimethee and Kubaha [18] used a Relational Analysis Model and VARIMAX-ECM Model to forecast carbon emissions in Thailand for the period between 2018–2029. The research results indicate that carbon emissions will continue to increase steadily by 14.68%, or 289.58 $MtCO_2eq$. by 2029, which is not in line with Thailand's carbon emissions reduction policy.

Zhu et al. [19] adopted Generalized Divisia Index Model (GDIM) and Monte Carlo simulation to explore the influencing factors and scenario forecasts of carbon emissions of the Chinese power industry. The results show that the output scale is the most important factor leading to an increase in carbon emissions in China's power industry from 2000 to 2015, followed by the energy consumption scale and population size. The results also indicate that China's power industry still has great potential to reduce carbon emissions, and the focus can be placed on the innovation and development of energy saving and emissions reduction technology.

Wang et al. [20] used the Gray model (GM (1, 1)), Generalized Regression Neural Network (GRNN), Markov forecasting model, and non-linear programming to evaluate whether China can achieve the 2020 and 2030 carbon intensity targets set by government through energy structure adjustment. The conclusions are that in 2020, the optimal energy structure will enable China to achieve its carbon intensity target under three scenarios. However, in 2030, the optimal energy structure cannot fully achieve China's carbon intensity target under any of the three scenarios.

3.1.6. Quotas Allocation for Carbon Emissions

The electric power industry is the first sector that was introduced into the Carbon Emissions Trading market, which is being constructed in China. Meng et al. [21] utilized a hybrid trend forecasting model and a three-indicator allocation model to propose a quota allocation scheme for carbon emissions in China's electric power industry in 30 provinces from 2016 to 2030. The research findings indicated that nine provinces are expected to be the buyers in the Carbon Emissions Trading market. These provinces are mostly located in eastern China, and account for approximately 63.65% of China's carbon emissions generated by the electric power sector.

3.2. Carbon Tax

Implementing a carbon tax is one method of carbon pricing to mitigate carbon emissions. Wang et al. [22] used Low-Carbon Inventory Routing Problem (LCIRP) model for the inventory routing problem in the distribution process of refined oil with the perspective of carbon tax, and proposed an improved adaptive genetic algorithm combined with greedy algorithm to solve the model. Fan et al. [23] utilized a 3EAD-CGE (economy–energy–environmental–agricultural–dynamics Computable General Equilibrium) model to analyze the degree of carbon tax on the macroenvironment, macroeconomy, and agricultural sectors during the period 2020–2050, in order to investigate whether carbon tax is suitable for China's agricultural-related sectors. This research provides detailed data that supports the views of most people against the imposition of a carbon tax on agricultural-related sectors.

Tsai [24] proposed a green quality management decision model with carbon tax under Activity-Based Costing (ABC) in the tire manufacturing industry. The optimal green quality production portfolio can be selected via a mathematical programming model. Activity-Based Costing (ABC) is used to assess green quality management and production cost. Tsai [9] also considered the environmental issues of carbon emissions, energy recycling, and waste reuse, and proposed a green production planning and control model with carbon tax. Tsai [25] used a mathematical programming model with Activity-Based Costing (ABC) and the Theory of Constraints (TOC) to achieve the optimal product mix to maximize profit under various resource, production, and sale-related constraints.

Cap and trade is one method for regulating and ultimately reducing the amount of carbon emissions. The government sets a cap on carbon emissions for the whole country, and then limited the amount of carbon dioxide that companies are allowed to release. A company that can more efficiently reduce carbon emission can sell any extra permits in the emission market to companies that cannot easily afford reducing carbon emissions. Tsai [26] combined mathematical programming, Theory of Constraints (TOC), and Activity-Based Costing (ABC) to formulate the green production decision model with carbon taxes and carbon right costs under the cap-and-trade scheme, in order to achieve the optimal product-mix decision under various constraints. This paper proposed three different scenario models with carbon taxes and carbon right, and used them to evaluate the effect on profit of changes in carbon tax rates.

3.3. Energy

3.3.1. Renewable Energy

Renewable energy is safe, abundant, and clean to use when compared to fossil fuels. However, many forms of renewable energy are location-specific and require storage capabilities. Even regarding this, renewable energy has great potential investment value. Li et al. [27] adopted a real option model considering carbon price fluctuation as a tool for renewable energy investment. Considering optimal investment timing and carbon price, the model introduces a carbon price fluctuation as part of the optimization paper.

Zhai et al. [28] applied Life Cycle Inventory (LCI) analysis and Life Cycle Impact Analysis (LCIA) to conduct a life cycle assessment (LCA) study for a buoy–rope–drum (BRD) wave energy converter (WEC).

Foster [29] utilized the Vector Error Correction model (VECM) and Multiple Linear Regression model (MLR) to projects the United States' (U.S.) future ethylene supply in the context of two megatrends: the natural gas surge and global climate change. The results indicated that the availability of shale gas in the U.S. and low-priced feedstocks from natural gas relative to crude oil were key factors influencing ethylene supply.

3.3.2. Energy Forecasting

Sutthichaimethee and Kubaha [30] applied LT-ARIMAXS (the Long Term-Autoregressive Integrated Moving Average with Exogeneous variables and Error Correction Mechanism model) to conduct energy consumption long-term forecasting for the petroleum industry in Thailand.

4. Concluding Remarks

Since carbon emissions reached an all-time high in 2018, where global carbon dioxide emissions from burning fossil fuels have increased by about 2.7% in 2018, after a 1.6% increase in 2017. We need to pay special attention to carbon emissions and work out the possible solutions if we still want to meet the targets of the Paris climate agreement. This special issue collects 16 carbon emissions-related papers (including five that were carbon tax-related) and four energy-related papers in various industries by using various methods or models. Although this special issue did not fully satisfy our needs, it still provides abundant related material for energy conservation and carbon emissions reduction. However, there still are many research topics waiting for our efforts to study to solve the problems of global warming.

Acknowledgments: Wen-Hsien Tsai is grateful to the MDPI Publisher for the invitation to act as guest editor of this special issue and is indebted to the editorial staff of "Energies" for the kind cooperation, patience and committed engagement, especially the senior editor, Miss Julyn Li. The guest editor would also like to thank the authors for submitting their excellent contributions to this special issue. Thanks are also extended to the reviewers for evaluating the manuscripts and providing helpful suggestions.

Conflicts of Interest: The author declares no conflict of interest.

References

1. IPCC (Intergovernmental Panel on Climate Change). *Climate Change 2013: The Physical Science Basis. Working Group I Contribution to the IPCC Fifth Assessment Report*; Cambridge University Press: Cambridge, UK, 2013; Available online: www.ipcc.ch/report/ar5/wg1 (accessed on 4 March 2019).
2. Carrington, D.; UK Minister: There's a Looming Threat That Should Be Taken as Seriously as Nuclear War. The Guardian. 14 July 2015. Available online: https://www.businessinsider.com/uk-minister-climate-change-must-be-taken-as-seriously-as-nuclear-war-2015-7 (accessed on 4 March 2019).
3. Voorhees, J. Obama Tells the Truth About Climate Change. Oops? *Slate*. 11 February 2015. Available online: https://slate.com/news-and-politics/2015/02/terrorism-vs-climate-change-obama-says-global-warming-impacts-more-people-than-terrorism-hes-right.html (accessed on 4 March 2019).
4. UNFCCC (United Nations Framework on Climate Change). Adoption of the Paris Agreement. In *Report of the Conference of the Parties on Its Twentyfirst Session (Held in Paris from 30 November to 13 December 2015)*; Addendum, Report No. FCCC/CP/2015/10/Add.1; United Nations, 29 January 2016; Available online: http://unfccc.int/resource/docs/2015/cop21/eng/10a01.pdf (accessed on 4 March 2019).
5. Juerg, R. CO_2—The Major Cause of Global Warming. Available online: https://timeforchange.org/CO2-cause-of-global-warming (accessed on 4 March 2019).
6. United States Environmental Protection Agency (EPA). Sources of Greenhouse Gas Emissions. Available online: https://www.epa.gov/ghgemissions/sources-greenhouse-gas-emissions (accessed on 4 March 2019).
7. Asian Development Bank. *Decoding Article 6 of the Paris Agreement*; Asian Development Bank: Manila, Philippines, April 2018; Available online: https://www.adb.org/sites/default/files/publication/418831/article6-paris-agreement.pdf (accessed on 4 March 2019).
8. European Commission. EU Emissions Trading System (EU ETS). Available online: https://ec.europa.eu/clima/policies/ets_en (accessed on 4 March 2019).
9. Redshaw, L. Carbon Trading. Available online: http://www.ecigl.com/investment-carbon-trading.shtml (accessed on 4 March 2019).
10. Harvey, C. CO_2 Emissions Reached an All-Time High in 2018. *E&E News*. 6 December 2018. Available online: https://www.scientificamerican.com/article/co2-emissions-reached-an-all-time-high-in-2018/ (accessed on 4 March 2019).

11. Wang, Y.; Yang, G.; Dong, Y.; Cheng, Y.; Shang, P. The Scale, Structure and Influencing Factors of Total Carbon Emissions from Households in 30 Provinces of China—Based on the Extended STIRPAT Model. *Energies* **2018**, *11*, 1125. [CrossRef]
12. Huang, R.; Zhang, S.; Liu, C. Comparing Urban and Rural Household CO_2 Emissions—Case from China's Four Megacities: Beijing, Tianjin, Shanghai, and Chongqing. *Energies* **2018**, *11*, 1257. [CrossRef]
13. Wang, Y.; Zhou, Y.; Zhu, L.; Zhang, F.; Zhang, Y. Influencing Factors and Decoupling Elasticity of China's Transportation Carbon Emissions. *Energies* **2018**, *11*, 1157. [CrossRef]
14. Zhang, Y.; Zou, X.; Xu, C.; Yang, Q. Decoupling Greenhouse Gas Emissions from Crop Production: A Case Study in the Heilongjiang Land Reclamation Area, China. *Energies* **2018**, *11*, 1480. [CrossRef]
15. Hu, A.; Chen, C.-H.; Huang, L.H.; Chung, M.-H.; Lan, Y.-C.; Chen, Z. Environmental Impact and Carbon Footprint Assessment of Taiwanese Agricultural Products: A Case Study on Taiwanese Dongshan Tea. *Energies* **2019**, *12*, 138. [CrossRef]
16. Zhu, X.; Li, N.; Sun, Y.; Zhang, H.; Wang, K.; Tsai, S.-B. A Study on the Strategy for Departure Aircraft Pushback Control from the Perspective of Reducing Carbon Emissions. *Energies* **2018**, *11*, 2473. [CrossRef]
17. Peralta, C.O.; Vieira, G.T.T.; Meunier, S.; Vale, R.J.; Salles, M.B.C.; Carmo, B.S. Evaluation of the CO_2 Emissions Reduction Potential of Li-ion Batteries in Ship Power Systems. *Energies* **2019**, *12*, 375. [CrossRef]
18. Sutthichaimethee, P.; Kubaha, K. A Relational Analysis Model of the Causal Factors Influencing CO_2 in Thailand's Industrial Sector under a Sustainability Policy Adapting the VARIMAX-ECM Model. *Energies* **2018**, *11*, 1704. [CrossRef]
19. Zhu, L.; He, L.; Shang, P.; Zhang, Y.; Ma, X. Influencing Factors and Scenario Forecasts of Carbon Emissions of the Chinese Power Industry: Based on a Generalized Divisia Index Model and Monte Carlo Simulation. *Energies* **2018**, *11*, 2398. [CrossRef]
20. Wang, Y.; Shang, P.; He, L.; Zhang, Y.; Liu, D. Can China Achieve the 2020 and 2030 Carbon Intensity Targets through Energy Structure Adjustment? *Energies* **2018**, *11*, 2721. [CrossRef]
21. Meng, M.; Wang, L.; Chen, Q. Quota Allocation for Carbon Emissions in China's Electric Power Industry Based Upon the Fairness Principle. *Energies* **2018**, *11*, 2256. [CrossRef]
22. Wang, S.; Tao, F.; Shi, Y. Optimization of Inventory Routing Problem in Refined Oil Logistics with the Perspective of Carbon Tax. *Energies* **2018**, *11*, 1437. [CrossRef]
23. Fan, W.; Gao, Z.; Chen, N.; Wei, H.; Xu, Z.; Lu, N.; Wang, X.; Zhang, P.; Ren, J.; Ulgiati, S.; et al. It is Worth Pondering Whether a Carbon Tax is Suitable for China's Agricultural-Related Sectors. *Energies* **2018**, *11*, 2296. [CrossRef]
24. Tsai, W.-H. A Green Quality Management Decision Model with Carbon Tax and Capacity Expansion under Activity-Based Costing (ABC)—A Case Study in the Tire Manufacturing Industry. *Energies* **2018**, *11*, 1858. [CrossRef]
25. Tsai, W.-H. Green Production Planning and Control for the Textile Industry by Using Mathematical Programming and Industry 4.0 Techniques. *Energies* **2018**, *11*, 2072. [CrossRef]
26. Tsai, W.H. Carbon Taxes and Carbon Right Costs Analysis for the Tire Industry. *Energies* **2018**, *11*, 2121. [CrossRef]
27. Li, Y.; Wu, M.; Li, Z. A Real Options Analysis for Renewable Energy Investment Decisions under China Carbon Trading Market. *Energies* **2018**, *11*, 1817. [CrossRef]
28. Zhai, Q.; Zhu, L.; Lu, S. Life Cycle Assessment of a Buoy-Rope-Drum Wave Energy Converter. *Energies* **2018**, *11*, 2432. [CrossRef]
29. Foster, G. Ethylene Supply in a Fluid Context: Implications of Shale Gas and Climate Change. *Energies* **2018**, *11*, 2967. [CrossRef]
30. Sutthichaimethee, P.; Kubaha, K. The Efficiency of Long-Term Forecasting Model on Final Energy Consumption in Thailand's Petroleum Industries Sector: Enriching the LT-ARIMAXS Model under a Sustainability Policy. *Energies* **2018**, *11*, 2063. [CrossRef]

energies

MDPI

Article

The Scale, Structure and Influencing Factors of Total Carbon Emissions from Households in 30 Provinces of China—Based on the Extended STIRPAT Model

Yong Wang [1,2], Guangchun Yang [1], Ying Dong [1], Yu Cheng [1] and Peipei Shang [3,*]

1 School of Statistics, Dongbei University of Finance and Economics, Dalian 116025, China; ywang@dufe.edu.cn (Y.W.); 18239908553@163.com (G.Y.); chengyu930706@163.com (Y.D.); dy123ing@163.com (Y.C.)
2 Postdoctoral Research Station, Dongbei University of Finance and Economics, Dalian 116025, China
3 Editorial Department, Dongbei University of Finance and Economics, Dalian 116025, China
* Correspondence: sjshangpeipei@163.com; Tel.: +86-150-411-65045

Received: 30 March 2018; Accepted: 1 May 2018; Published: 2 May 2018

Abstract: Household carbon emissions are important components of total carbon emissions. The consumer side of energy-saving emissions reduction is an essential factor in reducing carbon emissions. In this paper, the carbon emissions coefficient method and Consumer Lifestyle Approach (CLA) were used to calculate the total carbon emissions of households in 30 provinces of China from 2006 to 2015, and based on the extended Stochastic Impacts by Regression on Population, Affluence, and Technology (STIRPAT) model, the factors influencing the total carbon emissions of households were analyzed. The results indicated that, first, over the past ten years, the energy and products carbon emissions from China's households have demonstrated a rapid growth trend and that regional distributions present obvious differences. Second, China's energy carbon emissions due to household consumption primarily derived from the residents' consumption of electricity and coal; China's products household carbon emissions primarily derived from residents' consumption of the high carbon emission categories: residences, food, transportation and communications. Third, in terms of influencing factors, the number of households in China plays a significant role in the total carbon emissions of China's households. The ratio of children 0–14 years old and gender ratio (female = 100) are two factors that reflect the demographic structure, have significant effects on the total carbon emissions of China's households, and are all positive. Gross Domestic Product (GDP) per capita plays a role in boosting the total carbon emissions of China's households. The effect of the carbon emission intensity on total household carbon emissions is positive. The industrial structure (the proportion of secondary industries' added value to the regional GDP) has curbed the growth of total carbon emissions from China's household consumption. The results of this study provide data to support the assessment of the total carbon emissions of China's households and provide a reasonable reference that the government can use to formulate energy-saving and emission-reduction measures.

Keywords: household consumption; total carbon emissions; CLA Model; influence factor; STIRPAT model

1. Introduction

According to the International Energy Agency (IEA) [1], China surpassed the US as the world's largest emitter of carbon in 2007, and according to data on carbon dioxide emissions released by the BP World Energy Statistics Yearbook (2017) [2], China's global share of carbon emissions rose from 20.9% in 2005 to 27.5% in 2014 and remains on the rise. China, the largest developing country in the world, has undergone vigorous and rapid development over the past 40 years, and the increased large-scale, mechanized production, growing consumer demand, and improved living standards have caused China's carbon dioxide emissions to increase every year. To reduce carbon emissions, China has established a series of emissions reduction plans: at the World Climate Conference in Copenhagen in 2009, China vowed that by 2020 its carbon intensity would decrease by 40–45% from the 2005 amount. In 2014, China increased its goal of peak carbon emissions to approximately 2030. In 2015, at the Paris summit, China reaffirmed its commitment of peak carbon emissions to approximately 2030; China has also proposed that to approximately 2030, non-fossil energy consumption would account for about 20% of primary energy consumption, and carbon intensity would be 60 to 65% lower than 2005, and forest stock would be about 4.5 billion cubic meters more than in 2005. In December 2017, China officially launched the national carbon emission trading system [3], taking an important step on the path of reducing emissions. Therefore, how to achieve energy savings and emissions reduction and how to develop an appropriate path for China's low-carbon development have become timely issues for scholars.

Currently, most of our efforts to reduce emissions are committed to the field of industrial production. Recent years have seen diminishing marginal benefits of industrial emission reductions, and changing consumption patterns have been an effective manner by which to mitigate climate warming. All types of research on carbon emissions at the production level have gradually shifted to the consumption level. At present, household carbon emissions account for more than 40% [4] of China's total carbon emissions and increase yearly. In 2012, China's GDP growth rate was 7.7% [5], farewell to the past more than 30 years average of 10% per cent growth, indicating that China's economic growth phase has undergone a fundamental shift in the economic development of the "new normal". An important manifestation of the new normal of economic development is that the contribution rate of consumption to economic growth is obviously increased. In 2010, consumption contributed 61.9% [6] to China's economic growth, surpassing investment for the first time since 2006. In 2016, the contribution of consumption to China's economic growth increased to 64.6% [6]. Therefore, expanding domestic demand and promoting consumption are important pillars for economic growth. Changes in the size and structure of residents' consumption will have significant impacts on China's carbon emission. It is of great practical significance to study the carbon emissions of households and their influencing factors. To determine whether residents directly generate carbon emissions when they consume products and services, we can divide household consumption carbon emissions into energy carbon emissions of households and products carbon emissions of households. Energy carbon emissions of households are produced by residents' direct consumption of energy products. Products carbon emissions of households are generated by the residents' consumption of products and services that consume energy in all aspects of production and sales. The combination of energy and products carbon emissions of household consumption is called total household carbon emissions. This paper calculates the energy and products carbon emissions of household consumption and comprehensively analyses the present situation and influencing factors on total carbon emissions of China's household consumption.

2. Literature Review

At present, the research on the carbon emissions of households is divided into two areas: scale calculation and influential factor analysis. The measurement of scale is divided into the calculation of the energy carbon emissions of household consumption and calculation of the products carbon emissions of household consumption, and research on the influencing factors of carbon emissions from households primarily concentrates on the population, per capita income and technical levels. Currently, studies have focused on several areas, which are discussed in the following sections.

2.1. Measurement of the Size of Carbon Emissions of Household Consumption

Existing research primarily uses the carbon emission coefficient method to calculate the energy carbon emissions of household consumption. There are three methods for calculating products carbon emissions from households: Input-Output Analysis (IOA), Life-Cycle Analysis (LCA) and the Consumer Lifestyle Approach (CLA). ① The input-output analysis method, proposed by Professor Leon (Wassily Leontief) in 1936, was applied to the quantitative analysis of the American economic system in the same year [7], and the theoretical setting and empirical application were further analyzed in 1937 [8]. Herendeen [9] first applied it to the analysis and calculation of energy consumption in the United States in 1976. Druckman [10] established a quasi-multi-region-input-output model to study the carbon emissions produced by British residents from consumer products and services from 1990 to 2004; the results indicated that more than one-fourth of the UK household consumption carbon emissions in 2004 came from recreational and leisure consumption. Peng et al. [11] developed a non-competitive input-output model to calculate the carbon emissions caused by Chinese residents' consumption and their sectoral distribution from 1992 to 2007. Their results indicated that during the period of inspection, residents' consumption and carbon emissions indicated a rapid growth trend, which is an important component of China's total carbon emissions. Based on the statistical data from Shanghai from 1997 to 2010, Wu, Guo et al. [12] used the input-output model to calculate the products carbon emissions from Shanghai residents' consumption, and the study observed that the products carbon emissions of Shanghai residents' consumption was increasing, which was the primary source of the total carbon emissions of residents' consumption. Tian et al. [13] used the input-output model to calculate the products carbon emissions of residents in Liaoning Province in 1997, 2002 and 2007. The study determined that the products carbon emissions from urban residents' consumption were the primary component of products carbon emissions of residents' consumption. ② Life-Cycle Analysis examines the effect of a product or service on the environment throughout its life cycle. Liu et al. [14] used the life-cycle analysis method to construct the sustainable consumption evaluation model, and those authors analyzed the ecological influence of the Chinese urban household consumption behavior in 2000 and compared the contribution of different consumption behaviors to the ecological effect. Yiao et al. [15] accounted for the total amount of products carbon dioxide emitted by Chinese residents from 1997 to 2007 using the comprehensive life-cycle analysis method and determined that the products carbon emissions from urban residents was the primary component of household consumption products emissions. ③ The Consumer Lifestyle Approach (CLA), based on household consumer goods, calculates the carbon emissions from each type of consumption activity according to the expenditure of each category of consumer goods. Bin et al. [16] analyzed the relation between the energy consumption and carbon emissions of American residents using the consumer lifestyle approach and presented a detailed calculation method of the energy and products carbon emissions of residents' consumption. Wei et al. [17] studied carbon emissions from the end energy consumption of rural and urban residents in China from 1999 to 2002 using the consumer lifestyle approach and determined that the total carbon emissions of residents' consumption accounted for approximately 30% of the total carbon emissions. Fan and Wang [18] adopted the consumer lifestyle method to measure the products carbon emissions of Chinese residents from 1993 to 2007 as well as the trend in consumption carbon emissions of urban and rural residents per capita.

2.2. Study of the Factors Influencing the Consumption Carbon Emissions of Households

The research methods of the factors influencing household consumption carbon emissions primarily include two types. The first type is decomposition analysis, including Index Decomposition Analysis (IDA) and Structural Decomposition Analysis (SDA). The other type is the environmental effect (I) = population (P) × affluence (A) × technology (T) (IPAT) equation or STIRPAT model. Greenring et al. [19] used Di's index method to analyse 10 countries in the OECD regarding their carbon emissions from residential terminal services and private transport sectors from 1970 to 1993. It was determined that the effects of the energy structure of terminal consumption, fuel composition and energy intensity on the reduction of the intensity of carbon emissions were different. Using the LMDI model, Chai [20] constructed a complete decomposition model of carbon emissions of urban residents and evaluated the factors influencing the daily carbon emissions of Chinese urban residents from the perspectives of the consumption pattern, income and family size. Du [21] calculated and analyzed the influence of the urban-rural structure, consumption carbon intensity, consumption level and consumption structure on the products carbon emissions of Chinese residents from 2000 to 2015 using the LMDI model. Wang and Xia [22] used the SDA Model to analyse the factors influencing consumption carbon emissions of Chinese residents from 1995 to 2009, and the study determined that the carbon emissions generated by residents' consumption were generally increasing during the study period. However, from the perspective of structure and influence factors, Chinese residents' consumption remains in the area of low-carbon development (low-carbon development is a sustainable development model characterized by low energy consumption, low pollution and low emission, which is of great significance to the sustainable development of economy and society.). Based on the IPAT equation, Hubacek et al. [23] analyzed the influence factors of environmental change and carbon emissions in China and India from 1960 to 2000, and the results indicated that the influence of wealth factors on carbon emissions increased and that the influence of technical factors on carbon emissions decreased. Hubacek et al. [24] used the IPAT model to analyse the influence factors of China's carbon emissions from 1978 to 2008. The article reported that the increase in economic levels rendered residents inclined to consume products that generate more carbon emissions in the production process, leading to increased carbon emissions. Fu et al. [25] analyzed the factors influencing the products carbon emissions of Chinese residents from 1996 to 2011 using the STIRPAT model and determined that per capita output, the energy intensity and energy structure were the primary factors influencing the products carbon emissions of residents' consumption. Tang et al. [26] calculated the energy carbon emissions of Chinese residents from 1990 to 2014 and used the STIRPAT model to quantify the effects of variables such as the population size, energy structure, household consumption level and urbanization rate on China's consumption of energy carbon emissions. Based on China's provincial data from 2003 to 2012, Ji et al. [27] constructed the dynamic panel data model of the influencing factors of energy carbon emissions from households using the extended STIRPAT model. Based on the extended STIRPAT model, Richard [28] analyzed the influencing factors of energy consumption in 14 countries in the EU from 1960 to 2000. The study found that the total population, urbanization rate, per capita GDP, and the proportion of people over 65 years of age had catalytic effects on energy consumption, and the impact of GDP per capita on energy consumption was consistent with the Kuznets hypothesis. To examine the determinants of nitrogen oxides (NOx) emissions in the Community of Madrid in Spain, which is one of the most densely populated regions in Europe, Tiziana et al. [29] used an extended STIRPAT model to analyse the effects of the total population, income level, the population over 64 years of age and the proportion of male population aged 22–55 to the emission of NOx. Rosalia et al. [30] extended the classic Pressure-State-Response model. And the effects of GDP, urbanization rate, 0–14-year-old children's ratio on the emission of carbon monoxide (CO), nitrogen oxides (NOx) and volatile organic compounds (VOC) in European Union countries in 1995–2005 were analyzed. Brantley [31] summarized the literatures, which used the STIRPAT model to study the impacts of total population, age structure, family size and urbanization rate on carbon emissions. This paper summarizes the methods, conclusions, and limitations of existing studies as presented in Table 1.

On the whole, the majority of the literature concerns the total carbon emissions of household consumption and their influencing factors at the national level or in the single area. There are few studies on the provincial level, which cannot fully reflect the consumption carbon emissions of Chinese residents from the spatial scope. In the selection of influencing factors, most scholars use a single population to examine the effect of demographic changes on the total carbon emissions of household consumption without analyzing the demographic structure. With the development of human society, population structure has become an important component of the population problem and should be included in the field of investigation.

Table 1. Summary of household consumption carbon emission research.

Research Topics	Author	Research Methods	Research Process and Primary Conclusions	Research Limitations
Scale calculation of household consumption carbon emissions	Druckman et al. [10]	Quasi-Multi-Region-Input-Output (QMRIO)	The 2004 British household consumption carbon emissions were calculated based on the quasi-multi-region-input-output model. The study determined that the primary source of household carbon emissions in Britain was consumption of recreational and leisure goods.	Only accounted for data from 2004; it is not possible to examine the trend of household consumption carbon emissions in the UK vertically.
	Shuijun Peng et al. [11]	Non-Competitive-Input-Output Analysis	Based on the non-competitive input-output model, the consumption carbon emissions and their sectoral distribution of Chinese residents from 1992 to 2007 were estimated. The results indicated that the total carbon emissions of Chinese residents during the study period demonstrated a rapid growth trend, which is an important component of total carbon emissions. Products carbon emissions from residents are the primary component of total carbon emissions of residents, and the majority of them come from the consumption activities of urban residents.	Using interval data to account for four years of household consumption carbon emissions, the analysis year was inconsistent and less.
	Xu Tian et al. [13]	Input-Output Analysis	The carbon emissions coefficient method and the input-output model method were used to calculate the total carbon emissions of households in Liaoning Province in 1997, 2002 and 2007. The time change and urban-rural difference of total carbon emissions of households in Liaoning Province were analyzed. It was determined that the total carbon emissions of households in Liaoning province was increasing during the study period, the products carbon emissions from households was the primary source of total carbon emissions, and the difference between urban and rural residents was significant.	The analysis with interval data cannot continuously reflect the change and development of household consumption carbon emissions. In this single area of Liaoning, there was no comparative analysis between regions.
	Jingru Liu et al. [14]	Life-Cycle Analysis (LCA)	Based on the LCA model, this paper constructed a sustainable consumption evaluation model to calculate the products carbon emissions of Chinese urban household consumption in 2000 and compared the ecological effects of consumption behavior. The study determined that the most important effect on household consumption products carbon emissions was traffic-related consumption behavior, followed by food and housing.	The traditional LCA model was used to study only the environmental effect within the system boundary, ignoring the environmental effect outside the system boundary, and the evaluation scope was incomplete.
	Liang Yao et al. [15]	Comprehensive Life-Cycle Analysis	The LCA method and the input-output table were combined to calculate the products carbon emissions of Chinese residents from 1997–2007. The study determined that the products carbon emissions of household consumption in 2007 were 1.61 times that of household consumption in 1997; the average annual growth rate was 4.89%.	Although Comprehensive Life Cycle Analysis was adopted, the data were not available. The accounting year was too small and inconsistent and cannot fully reflect the change rule of products carbon emissions of residents' consumption. The input-output table was a value type, the environmental effect was a physical type, and the combination of the two also resulted in inaccurate accounting.

Table 1. *Cont.*

Research Topics	Author	Research Methods	Research Process and Primary Conclusions	Research Limitations
	Yiming Wei et al. [17]	Consumer Lifestyle Approach (CLA)	The Consumer Lifestyle Approach (CLA) was used to account for China's urban and rural residents' consumption of carbon emissions from 1999–2002. It determined that approximately 26% of energy consumption and 30% of carbon emissions came from residents' consumption activities.	The lack of consideration of industrial relevance in CLA method may underestimate the products carbon emissions of households.
Influential factor analysis of household consumption carbon emissions	Greenring et al. [19]	Di's index method	The carbon emissions of ten countries in the OECD were analyzed for their carbon emissions from residential terminal services and private transport sectors from 1970–1993 using Di's Index method. It was determined that the effects of the energy structure of terminal consumption, fuel composition and energy intensity on the reduction of carbon emission intensity differed.	Although the use of Di's index method can reflect the influence of various factors on the direction and extent of household consumption carbon emissions, there is an inability to establish a relation between the appropriate predictions.
	Hubacek et al. [23]	IPAT model	Based on the IPAT equation, the influence factors of environmental change and carbon emissions in China and India from 1960–2000 were analyzed. The results indicated that the influence of wealth factors on carbon emissions increased and the influence of technical factors on carbon emission decreased.	Using the IPAT equation, we can change one of the factors by fixing other factors to analyse the problem, assuming that the influence of each factor on the dependent variable is equal to the actual situation. However, in real life, the changes in various factors are not equal.
	Zhiying Ji et al. [27]	Extended STIRPAT and Kaya model	Based on the calculation of the energy consumption carbon emissions of Chinese households from 2003–2012, the influence factors of energy carbon emissions from residents' consumption were analyzed using the extended STIRPAT model and KAYA model. It was determined that the population size, consumption level, energy consumption structure, carbon emission intensity, energy consumption intensity and urbanization significantly affected residents' energy carbon emissions, and there were differences between urban and rural factors.	This article examined only the household consumption energy carbon emissions. There was no analysis of household consumption products carbon emissions. Therefore, the article cannot fully reflect China's household consumption carbon emissions.

This paper measured the total carbon emissions of China's households from 2006 to 2015 at the provincial level. Therefore, this paper can provide support for a comprehensive forecast of the future development trend of China's household consumption total carbon emissions and guidance on the government's policy for formulating targeted emission reduction policies according to the differences in the total carbon emissions of households in different regions. This article introduced the ratio of children 0–14 and gender ratio. These two factors can be used to study the effect of demographic changes on the total carbon emissions of household consumption. China is now vigorously pursuing a comprehensive fetus policy, and China's current gender ratio is seriously unbalanced (in 2015, it was 105.02). In this context, it is possible to provide theoretical guidance for China to specify a reasonable population policy to promote sustainable low-carbon development by studying the 0–14-year-old children's ratio and gender ratio.

The following sections of this article are arranged as follows: the third section introduces the calculation method of the total carbon emissions of household consumption, presents the model construction of the factors affecting the total carbon emissions of household consumption and provides a data explanation. The fourth section analyses the results of the total carbon emissions of household consumption and the results of the model construction. The fifth section, the discussion section, compares the similarities and differences between this article and previous research and discusses the limitations of this article. The sixth section includes the conclusions and suggestions for future research.

3. Research Methods and Data Explanation

3.1. Calculation Model of Total Carbon Emissions from Households

Total carbon emissions from households C_T are expressed as Equation (1):

$$C_T = C_D + C_{IN} \tag{1}$$

In Equation (1), C_D and C_{IN} indicate energy consumer carbon emissions and products consumer carbon emissions, respectively.

The method of calculating energy carbon emissions of household consumption adopts the method of the carbon emission coefficient, and C_D is expressed as Equation (2):

$$C_D = \sum_{i=1}^{n} F_i E_i \tag{2}$$

In Equation (2), F_i is the carbon emission factor for various energy sources. E_i indicates the population's consumption of various energy sources. Data were from the regional energy balance table in the China Energy Statistics Yearbook (2007–2016) [32]. $i = 1, 2, \ldots, 5$ were used to indicate the five types of energy that residents consume in their lives: coal, petroleum, natural gas, electricity and heat. Choosing these five kinds of energy products to calculate the energy carbon emission of Chinese residents' consumption was mainly based on the previous research [12] and consideration of the residents' demand for energy products in real life. The direct consumption of coal by residents is primarily used for cooking and heating. Especially in rural China, the use of coal is more common; the direct consumption of petroleum by residents is mainly used for providing fuel (petrol) to vehicles and cooking (liquefied petroleum gas). It is necessary to emphasize that there are various kinds of petroleum products. This paper mainly refers to gasoline and liquefied petroleum gas (LPG); natural gas is mainly used by residents for cooking. There is a greater demand for natural gas in urban areas; electricity and heat are essential civil energies in residents' life. The carbon emission coefficients of the three primary energy sources of coal, petroleum and natural gas are presented in Table 2. Electricity and heat are two energy sources that do not directly produce carbon emissions in the process of consumption but produce carbon emissions during production. Thus, the carbon emission coefficients of electricity and heat (Table 3) must be computed in conjunction with the carbon emissions generated by electricity and heat in the production process and their yields, and the formula for calculating the carbon emission coefficients of electricity is expressed as Equation (3):

$$F_{electricity} = \frac{\sum_{i}^{3} F_i A_i}{I_{electricity}} \quad (3)$$

In Formula (3), $F_{Electricity}$ represents the carbon emission factor of the electricity F_i represents the carbon emission factor of coal, petroleum and natural gas. A_i represents the amount of coal, petroleum and natural gas consumed by electricity generation. These data were derived from the regional energy balance table in the China Energy Statistics Yearbook (2007–2016) [32], $i = 1, 2, 3$. $I_{Electricity}$ represents the total supply of electricity.

Similarly, the formula for calculating the carbon emission coefficient of heat is expressed as Equation (4):

$$F_{heat} = \frac{\sum_{j}^{3} F_j A_j}{I_{heat}} \quad (4)$$

In Equation (4), F_{heat} indicates the carbon emission coefficient of heat. F_j represents the carbon emission factor of coal, petroleum and natural gas. A_j indicates the amount of coal, petroleum and natural gas consumed by heating, $j = 1, 2, 3$. I_{heat} represents the total supply of heat.

Table 2. Carbon emission coefficients for all types of primary energy (ton carbon/ ton tce).

Data Sources	Carbon Emission Coefficient of Coal	Carbon Emission Coefficient of Petroleum	Carbon Emission Coefficient of Natural Gas
DOE [33]/EIA [34]	0.70	0.48	0.39
ORNL [35]	0.72	0.59	0.40
IPCC [36]	0.76	0.59	0.45
National Science and Technology Commission Climate Change Project [37]	0.73	0.58	0.41
National Development and Reform Commission Energy Research Institute [38]	0.75	0.58	0.44
Average	0.73	0.56	0.42

Table 3. Carbon emission coefficients of electricity and heat.

	2006	2007	2008	2009	2010	2011	2012	2013	2014	2015
Carbon emission coefficient of electricity (million t carbon/billion kW·h)	2.6584	2.4873	2.5100	2.5525	2.4424	2.4236	2.4937	2.4301	2.2887	2.2255
Carbon emission coefficient of heat (million t carbon/10 ^ 10 kJ)	0.0326	0.0329	0.0322	0.0318	0.0329	0.0335	0.0382	0.0343	0.0332	0.0333

Note: The electricity carbon emission coefficient is calculated by Equation (3), and the heat carbon emission coefficient is calculated by Equation (4).

The calculation of products carbon emissions of household consumption primarily refers to the research method of Wei [17]. Using the CLA method to calculate the products carbon emissions of household consumption, the formula is expressed as Equation (5):

$$C_{IN} = \sum_{i=1}^{n} (Q_i P_i) \times L \quad (5)$$

In Equation (5), Q_i indicates that residents buy consumer goods per capita of consumption expenditure (data are from the China Statistical Yearbook (2007–2016) [39]). P_i indicates the carbon emission factor for each consumer product (Table 4). L is the total population (data are from the China Statistical Yearbook (2007–2016) [39]). Finally, $i = 1, 2, \ldots, 8$ indicates the type of the eight major categories of consumer goods: food, clothing, residences, household appliances and supplies, transportation and communications, culture and entertainment, health care and others. The official statistics of China divide the household consumption into eight categories, and the eight categories are subdivided into many small categories. Considering the availability of data, the paper no longer accounts for the products carbon emissions of residents according to the small subdivided categories.

Table 4. Consumption of consumer goods carbon emission factor table [40] (unit: ton carbon/million).

Consumption Type	Carbon emission Factor (Ton Carbon/Million)	Consumption Type	Carbon Emission Factor (Ton Carbon/Million)
Food	0.66	Transportation and communications	1.1
Clothing	1.21	Culture and entertainment	0.6
Residences	2.21	Health care	0.9
Household appliances and supplies	0.52	Others	0.3

3.2. Analysis Model of Factors Affecting Total Carbon Emissions from Households

3.2.1. STIRPAT Model

The IPAT model, can be used to study the effects of demographic, economic and technological factors on environmental pressure. The IPAT model has been widely used since it was introduced in the 1970s. However, the IPAT model has limitations, by changing only one of the factors to analyse the problem, the effect of various factors on the dependent variable is of equal proportions. To overcome this limitation, Dietz & Rosa (1994) [41] expanded the model into a random form, the STIRPAT model, the expression of which is Equation (6):

$$I = aP^b A^c T^d \varepsilon \tag{6}$$

In Equation (6), I represents the environmental effect. P, A and T represent the population, wealth and technology, respectively; a is constant, and b, c and d are indices that can be used to analyse the nonproportional effects of the changes of various factors on the environment. ε is a random perturbation term.

In the empirical analysis, we generally take the logarithm of the two sides and obtain the following models:

$$\ln I = \ln a + b \ln P + c \ln A + d \ln T + \ln \varepsilon \tag{7}$$

In Equation (7), the logarithm of the three exponents b, c and d in the model can be estimated as parameters, and the various influencing factors can be decomposed and analyzed appropriately, which provides a theoretical basis for studying the effects of various factors on the environment.

The studies of Du [21], Hubacek [23] and Ji [27] found that population size, living standards, carbon intensity and industrial structure had significant impacts on consumer carbon emissions. In addition to these factors, Richard [28] and Rosalia [30] introduced age structures in the study of carbon emissions. Considering the existing research and the actual situation in China, this paper adds the variable of gender ratio. It is primarily due to the imbalance of gender ratio at present in China, which leads to differences in consumption habits and consumption structure. Thus, this paper constructs an extended STIRPAT model of the influence factors of total carbon emission in residents' consumption, which is expressed as Equation (8):

$$\begin{aligned} \ln C_{it} = \beta_0 + \beta_1 \ln PSIZE_{it} + \beta_2 \ln CHI_{it} + \beta_3 \ln SEX_{it} + \\ \beta_4 \ln RGDP_{it} + \beta_5 \ln GDPC_{it} + \beta_6 \ln INST_{it} + \alpha_i + \mu_{it} \end{aligned} \tag{8}$$

In Equation (8), (i = 1, 2, L, 30) represents 30 provinces in mainland China except Tibet. In this formula, t(t = 2006, 2007, L, 2015) is the sample observation period. C is the total carbon emissions of residents' consumption. $PSIZE$ is the size of the population. CHI is the 0–14-year-old children's ratio. *SEX* is the gender ratio (female = 100). *RGDP* is GDP per capita. *GDPC* is the carbon emissions intensity. *INST* is the industrial structure. The data of the population size, 0–14-year-old children's ratio, gender ratio, GDP per capita, carbon intensity and industrial structure were all derived from the Statistical Yearbook of China (2007–2016) [39], and some transformations and calculations have been made.

3.2.2. Variable Description

1. Population sizes (*PSIZE*). This paper used the number of households to represent the population size factor. Related research indicates that as the household numbers increase, two factors change in residents' consumption: one is the increase in the basic unit of consumption, which inevitably leads to the expansion of the consumption scale and an increase in households' total carbon emissions. The second factor is that as the number of households increases, the households shrink and consumption of some common consumer goods increases, which can also lead to an increase in the total carbon emissions of household consumption. Therefore, in theory, the increase in household numbers leads to an increase in the total carbon emissions of household consumption.
2. Zero-to-fourteen-year-old children's ratio (*CHI*). In this paper, the proportion of the 0–14-year-old children of the total population is represented by *CHI*. This indicator is a factor that reflects the demographic structure. These children are not part of the working age population, their consumption type and consumption psychology have particularity, and their proportion changes inevitably lead to changes in residents' consumption. The introduction of this variable is of practical significance because of the widespread implementation of the fetus policy.
3. Gender ratio (*SEX*). The gender ratio in this article is calculated by the female population = 100. A higher gender ratio indicates a larger male population. Males and females have different consumption propensities, and a change in the gender ratio leads to changes in the total carbon emissions of household consumption. Considering the reality of the gender imbalance in China, it is also of practical significance to introduce this variable.
4. Per capita income level (*RGDP*). This paper used GDP per capita to express *RGDP*. Based on the 2006-year base period, the GDP per capita was reduced by the gross domestic product index, excluding the effect of the price factor on GDP per capita. If people's per capita income level is different, then they have different requirements for environmental and material consumption, which leads to differences in the total carbon emissions of household consumption. This paper presents the index as a wealth factor in the STIRPAT model.
5. Carbon emissions intensity (*GDPC*). Carbon intensity refers to the amount of carbon emitted by the one million GDP, or "carbon emissions per unit of GDP". This index can be used as a technical factor in the STIRPAT model. If this indicator is low, it will inhibit the total carbon emissions of household consumption.
6. Industrial structure (*INST*). In this paper, the proportion of the second industrial added value to the GDP was used to express the *INST*. Production determines consumption; differences in industrial structures lead to different consumption structures, and differences in consumption structures affect the total household carbon emissions. Therefore, in addition to the above variables, which are used to embody the three factors of population, wealth and technology in the model, this paper introduces the industrial structure, which can reflect changes in the production structure.

The independent variables mentioned above are presents in Table 5 in original units.

Table 5. Summary statistics for pooled data (n = 300).

		2006	2007	2008	2009	2010	2011	2012	2013	2014	2015
C_T	Minimum	3,272,615.9	3,389,096.7	3,810,304.3	4,222,105.6	5,038,779.2	5,665,363.6	6,216,264.8	7,042,161.3	9,300,614.5	9,977,888.0
	Mean	28,713,827.3	32,423,317.7	38,269,260.8	41,007,200.9	47,447,563.6	53,344,770.2	59,916,367.3	65,164,069.3	82,424,758.7	89,278,260.9
	Maximum	94,772,323.6	109,511,861.9	121,897,595.2	132,347,908.1	15,8443,649.0	1,75,628,659.8	194,937,327.9	211,315,200.7	250,312,464.5	274,917,288.0
PSIZE	Minimum	149.3	151.2	151.0	155.7	162.9	162.8	168.6	171.5	182.8	169.9
	Mean	1354.7	1364.5	1376.1	1393.6	1431.9	1477.7	1487.8	1517.0	1528.4	1470.9
	Maximum	3166.3	3186.1	3181.5	3178.0	3262.8	3323.1	3328.2	3476.2	3446.8	3419.1
CHI	Minimum	8.1	8.0	7.9	7.6	8.6	8.3	8.5	9.4	10.1	9.3
	Mean	18.4	17.8	17.3	16.8	16.5	16.3	16.3	16.2	16.2	16.3
	Maximum	27.3	27.2	26.0	24.8	25.2	24.2	23.0	22.1	22.1	22.6
SEX	Minimum	94.9	95.6	95.1	97.3	101.5	95.8	98.6	98.9	98.2	100.5
	Mean	102.8	103.0	103.3	103.7	105.7	105.1	105.1	105.4	105.1	105.7
	Maximum	109.9	110.0	109.8	114.2	114.5	115.2	113.6	111.7	118.6	120.4
RGDP	Minimum	0.6	0.7	0.8	0.9	1.1	1.2	1.4	1.6	1.7	1.9
	Mean	1.9	2.2	2.4	2.6	2.9	3.2	3.5	3.8	4.1	4.4
	Maximum	5.8	6.6	7.1	7.5	6.9	7.3	7.8	8.3	8.9	9.5
GDPC	Minimum	0.3	0.3	0.3	0.3	0.3	0.3	0.3	0.3	0.3	0.3
	Mean	0.4	0.4	0.4	0.4	0.4	0.4	0.4	0.4	0.5	0.5
	Maximum	0.7	0.7	0.7	0.7	0.7	0.6	0.7	0.6	0.7	0.7
INST	Minimum	27.4	26.8	25.7	23.5	24.0	23.1	22.7	22.3	21.3	19.7
	Mean	47.7	48.3	49.4	47.5	49.1	49.7	48.6	47.9	46.0	43.3
	Maximum	57.8	60.0	61.5	56.5	57.3	59.0	57.7	57.3	54.1	50.5

The method model and research concept of this paper are presented in Figure 1.

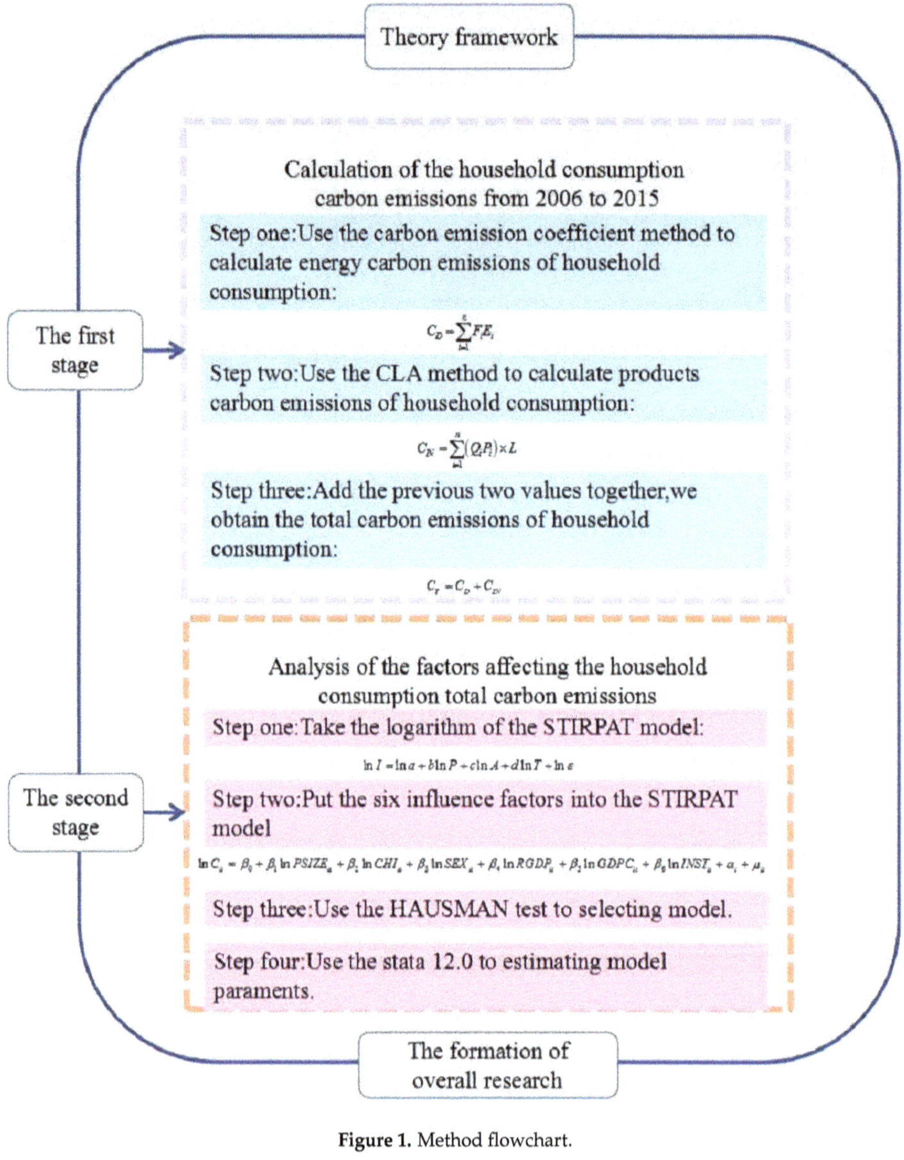

Figure 1. Method flowchart.

4. Results and Analysis

4.1. Calculation and Analysis of the Total Carbon Emissions of China's Household Consumption

4.1.1. Analysis of The Scale and Composition of the Total Carbon Emissions of China's Household Consumption

According to Equation (2), we can estimate the energy carbon emissions of China's household consumption from 2006 to 2015, and according to Equation (5), we can estimate the products' carbon emissions of China's household consumption from 2006 to 2015; the total carbon emissions of China's households can be obtained by adding the two together. The changes in the size and composition of the total carbon emissions of China's households from 2006 to 2015 are presented in Figure 2.

According to the general trend of change, the energy carbon emissions of China's household consumption, products carbon emissions of China's household consumption and total carbon emissions of China's household consumption from 2006 to 2015 are increasing. In 2006, the energy carbon emissions from China's households were 190.74 million tons; in 2010, the energy carbon emissions reached 258.72 million tons, which reflected an average annual growth rate of 7.92%. In 2015, the energy carbon emissions of China's household consumption grew to 333.44 million tons, with an average annual growth rate of 5.21% from 2010 to 2015.

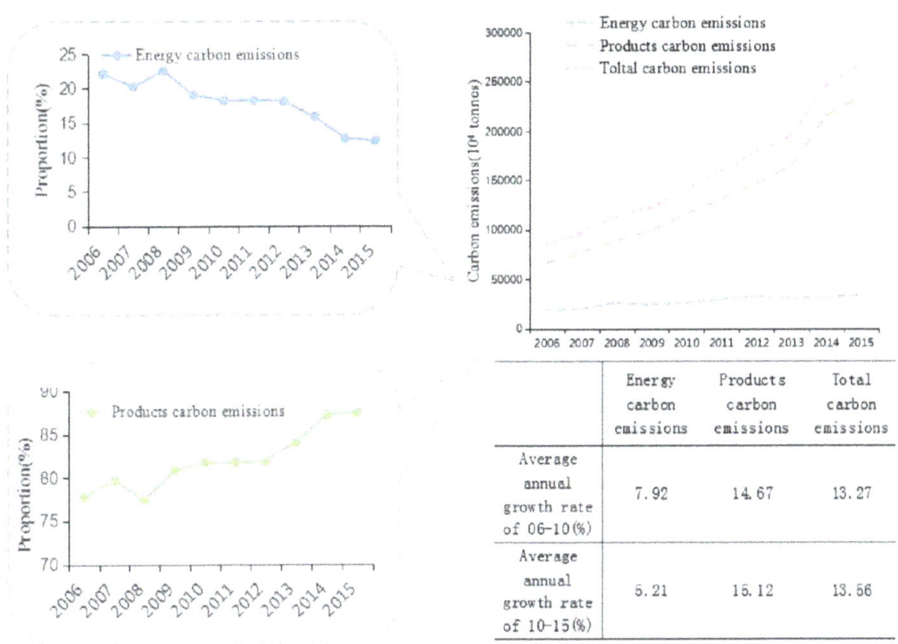

	Energy carbon emissions	Products carbon emissions	Total carbon emissions
Average annual growth rate of 06-10(%)	7.92	14.67	13.27
Average annual growth rate of 10-15(%)	5.21	15.12	13.56

Figure 2. The scale and structural change of the total carbon emissions of China's household consumption from 2006 to 2015.

In 2006, the products carbon emissions from China's households were 670.68 million tons; by 2010, the emissions had increased to 1159.52 million tons, an average annual growth rate of 14.67%. By 2015, the products carbon emissions of China's household consumption had increased further to 2344.90 million tons; the annual average growth rate from 2010 to 2015 was 15.12%. In 2006, the total carbon emissions from China's households were 861.41 million tons, reaching 1418.24 million tons in 2010, with an average annual growth rate of 13.27%. In 2015, the total carbon emissions of China's household consumption grew to 2678.35 million tons, with an average annual growth rate of 13.56% from 2010 to 2015. Thus, the rate of the products carbon emissions of China's household consumption grew faster than that of the energy carbon emissions of China's household consumption. In the future, the products carbon emissions of China's household consumption will increase further, which is the primary force stimulating the total carbon emissions from China's households. In terms of structural changes, the products carbon emissions of China's household consumption have accounted for more than 70% of the total carbon emissions from China's households from 2006 to 2015. In 2006, it was 77.86% and grew to 87.55% by 2015. The proportion of energy carbon emissions of China's household consumption has hovered at approximately 20%, and there has been a downward trend in recent years. With the rapid development of China's economy and the improvement of living standards, the people's desire to consume is strengthened. Thus, the products carbon emissions of China's household consumption and total carbon emissions of China's household consumption are bound to demonstrate rapid growth. Simultaneously, progress in science and technology and the promotion of clean energy, such as wind and water, have slowed the growth of the energy carbon emissions of China's household consumption.

4.1.2. Analysis of the Regional Characteristics of Total Carbon Emissions from China's Household Consumption

Because of the vastness of the Chinese territory, the region's economic development status, living habits, population distribution and geographical locations have extremely large differences. Thus, China's regional residents' consumption and consumption structure differ greatly, leading to regional differences in total carbon emissions from China's household consumption. Figure 3 presents the spatial distribution of total carbon emissions from China's household consumption in 2015. Figure 3 indicates clear regional differences in total carbon emissions from China's household consumption. There are ten provinces with a total carbon footprint of more than 1 billion tons, whereas households in the lower provinces consume approximately 10 million tons of total carbon emissions. Rounding out the top five are Guangdong (274.92 million tons), Jiangsu (196.05 million tons), Shandong (176.82 million tons), Zhejiang (167.89 million tons) and Henan (136.93 million tons). The common characteristics of these provinces are their vast areas and large populations. Although Beijing and Shanghai do not have the advantages of area and population, they do have developed economies and population flow; their household consumption total carbon emissions reached 99.87 million tons and 111.59 million tons, which were located in the forefront of 30 provinces. The bottom five were Qinghai (9.98 million tons), Ningxia (10.98 million tons), Hainan (14.29 million tons), Gansu (35.22 million tons) and Xinjiang (42.19 million tons); the majority of these provinces are located in the midwest, in which there is less population and economic development is relatively backward.

Because products' carbon emissions accounted for approximately 80% of the total carbon emissions from China's household consumption, they are a major component of China's household consumption carbon emissions. The geographical distribution of products carbon emissions is identical to that of the total carbon emissions from China's households. The same provinces are at the top and bottom of the list of China's household consumption products carbon emissions and China's household consumption total carbon emissions. The geographical distribution of the energy carbon emissions of China's households is quite different from that of the former two. The regional distribution of the energy carbon emissions of China's households indicates a significant difference between the north and south, and the energy carbon emissions of households in northern areas are higher than those in the south, which indicates that the total carbon emissions of China's households and the products carbon emissions of China's households are greatly influenced by the economic development level and population factors, whereas the regional distribution of the energy carbon emissions of China's households is relatively more vulnerable to the geographical location. Heating in the north consumes more fossil energy in winter, leading to more energy carbon emissions from households in the north than in the south as a whole. Liaoning Province, a typical area, ranked thirteenth in the products carbon emissions of households (81.15 million tons) and fourth in the energy carbon emissions of household consumption (17.82 million tons).

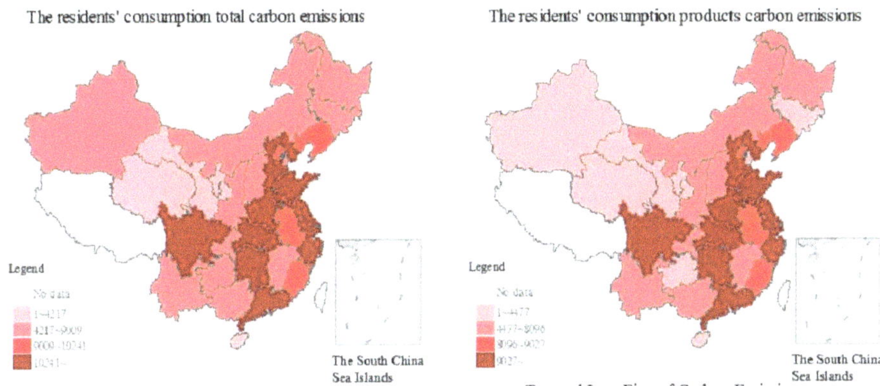

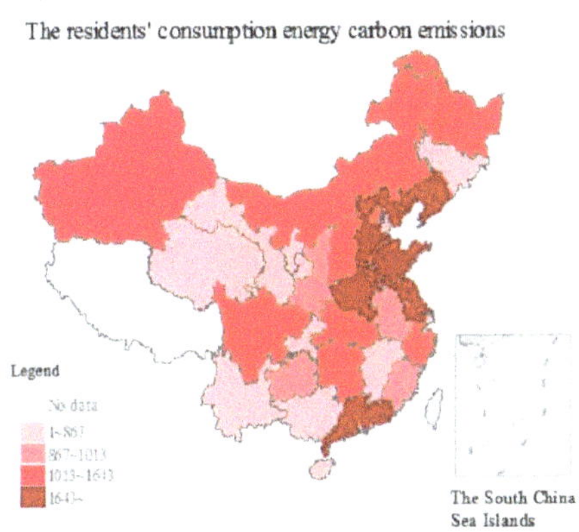

Top and Last Five of Carbon Emissions

No.	Total	Products	Energy
1	Guangdong	Guangdong	Guangdong
2	Jiangsu	Jiangsu	Hebei
3	Shandong	Shandong	Shandong
4	Zhejiang	Zhejiang	Liaoning
5	Henan	Henan	Henan
26	Xinjiang	Xinjiang	Chongqing
27	Gansu	Gansu	Gansu
28	Hainan	Hainan	Qinghai
29	Ningxia	Ningxia	Hainan
30	Qinghai	Qinghai	Ningxia

Figure 3. The spatial distribution of the total carbon emissions of China's household consumption in 2015.

4.1.3. Analysis of the Sources of Energy Carbon Emissions from China's Household Consumption

The energy carbon emissions of household consumption refer to the carbon emissions produced by residents' consumption of energy products in their daily lives. The energy products here refer to coal, petroleum, natural gas, electricity and heat. Because different energy products have different levels of cleanliness, different energy consumption combinations affect carbon emissions.

Figure 4 is the contribution of five energy sources to the energy carbon emissions of China's household consumption. As seen in Figure 4, from 2006 to 2015, the carbon emissions generated by the electricity supply were the primary source of the energy carbon emissions of China's household consumption; the proportion of the change was not small, between 40% and 50%. The second-highest contribution to the energy carbon emissions of China's household consumption was carbon emissions from coal consumption; however, the proportion was in a downward trend from 2006 to 2015 and was reduced to 15.73% by 2015. Carbon emissions from coal consumption are already lower than those generated by petroleum and heat consumption. Of the three primary sources of energy, coal has the highest carbon emission coefficient. The reduction of coal consumption is due to the optimization of the energy consumption structure, which has a certain inhibitory effect on the energy carbon emissions of China's household consumption. The total carbon emissions generated by these three types of energy, petroleum, natural gas and heat, account for approximately 30% of the energy carbon emissions of China's household consumption. Although the proportion is not large, it is increasing. In general, the carbon emissions generated by residents' consumption of various energy sources continue to grow, and the pressure on the environment cannot be ignored.

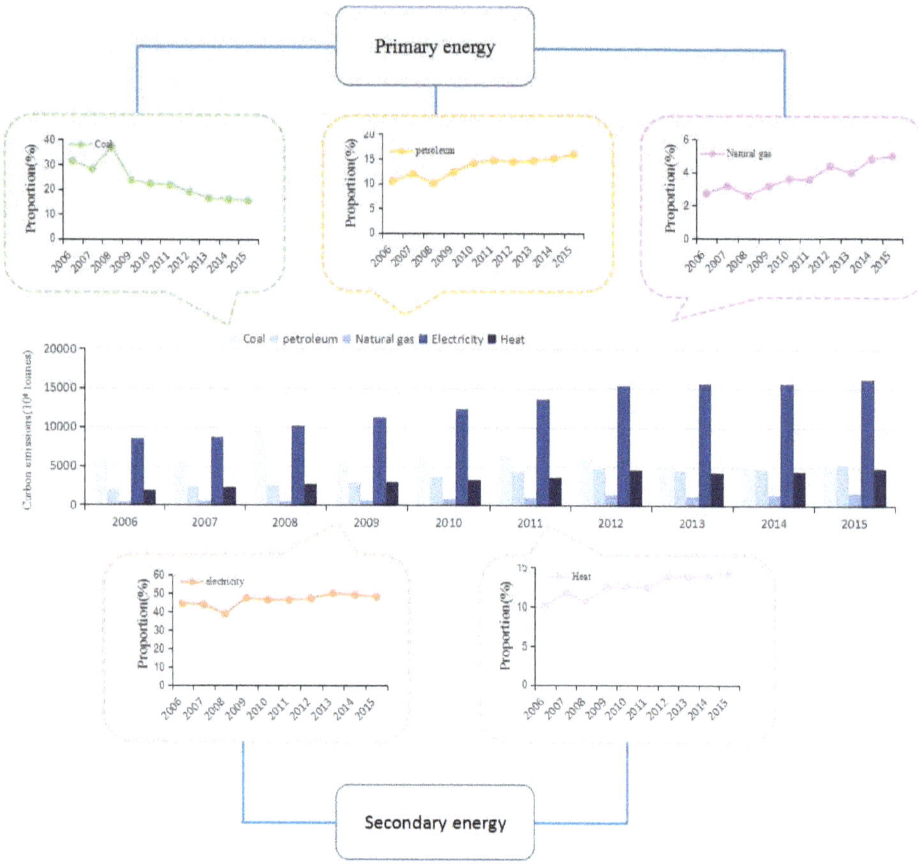

Figure 4. The contribution of five energy sources to the energy carbon emissions of China's household consumption.

4.1.4. Analysis of the Sources of Products Carbon Emissions from China's Household Consumption

The products carbon emissions of household consumption refer to the carbon emissions generated by the consumption of non-energy products in daily life. The products carbon emissions of household consumption come from household consumption of food, clothing, residences, household appliances and supplies, transportation and communications, culture and entertainment, health care and other types of consumer goods.

In this paper, the products carbon emissions of eight types of consumer goods were divided into a high carbon group and a low carbon group. The high carbon group included residences, food, transportation and communications. The low carbon group included clothing, culture and entertainment, health care, household appliances and supplies. Figure 5 presents the change in products carbon emissions of eight consumer goods and indicates clear differences in products carbon emissions from different client categories. The largest average amount of carbon emissions comes from the residence category, and its annual products carbon emissions is 448,060 tons, which is increasing yearly. In 2006, the products carbon emissions of residences was 1904.3 million tons, which increased to 1034.56 million tons in 2015, with an average annual growth rate of 20.69%. The residence rate was followed by food (316.49 million tons) and transportation and communications (19,849 million tons). Products carbon emissions from the first three high carbon sequestrations accounted for 71.78% of the products carbon emissions of all eight types of consumer goods. Products carbon emissions from other and household appliances and supplies were low, accounting for only 4.21 percent of the total products carbon emissions of household consumption. Clearly, the products carbon emissions of China's household consumption are characterized by a high concentration in the consumption category. In general, the products carbon emissions of eight consumer products are increasing yearly, which places increasingly more pressure on the environment.

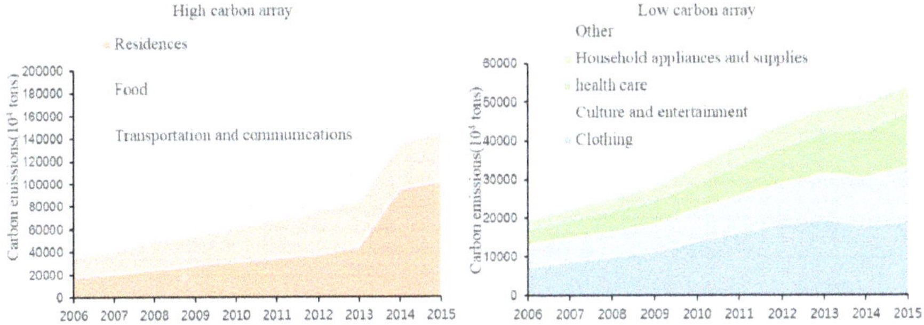

	High carbon array			Low carbon array				
	residences	Food	Traffic and communications	Clothing	Cultural and entertainment	Health care	Household appliances and supplies	Other
Annual average carbon emissions (Million tons)	44806.15	31649.03	19849.29	14513.06	9156.45	8555.59	4369.69	1272.71
Annual average growth rate(%)	60.36	27.20	35.25	27.84	28.28	35.87	37.86	26.23
Proportion(%)	33.39	23.59	14.79	10.82	6.82	6.38	3.26	0.95

Figure 5. The change in products carbon emissions of eight consumer goods.

Summary: The total carbon emissions of China's household consumption increased during the study period, and the regional differences are clear. The consumption mix of different energy and non-energy products significantly affected the total carbon emissions of China's household consumption.

4.2. *Analyses of the Factors Influencing the Total Carbon Emissions of China's Household Consumption*

4.2.1. Parameter Estimation Results of Panel Data Model

Based on panel data of 30 provinces in China from 2006 to 2015, this paper used the developed STIRPAT model to analyse the factors influencing the total carbon emissions of China's household consumption using Stata12.0 software.

In order to determine whether the impact of GDP per capita on China's consumption of total carbon emissions was consistent with the Kuznets hypothesis, the square of GDP per capita was introduced as an explanatory variable in model (1) [25]. If the coefficient of the square of GDP per capita was negative and the t-test under a given condition was adopted, it indicates that the impact of GDP per capita on the total carbon emission of Chinese residents was consistent with the Kuznets hypothesis. The Hausman test results of model (1) and model (2) were shown in Table 6, and both model (1) and model (2) chose fixed effect model. The estimation results of the fixed effect model were presented in Table 7. In the model (1), the coefficient of the square of GDP per capita did not pass the t-test, which showed that the impact of GDP per capita on the consumption of carbon emissions by Chinese residents did not conform to the Kuznets hypothesis. Further, the square of GDP per capita was removed from the model (1) to obtain the model (2). From the estimation results of model (2), regression models adjusted R^2 is close to 1, and the overall significance test of the model of the F test indicated that the p value was less than 0.05; thus, the overall model fitting effect was good. All of the coefficients of the explanatory variables passed the t test under the given condition, and the fitting effect of each coefficient was good.

Table 6. Hausman Test.

Inspection	Chi square Statistic	p Value
Model (1)	93.69	0.0000
Model (2)	79.99	0.0000

Table 7. Estimated results of fixed effect model.

Statistics	Model (1)	Model (2)
ln*PSIZE*	0.7735 *** (8.46)	0.7903 *** (9.05)
ln*CHI*	0.5572 *** (6.28)	0.5729 *** (6.74)
ln*SEX*	0.3097 * (1.66)	0.3282 * (1.79)
ln*RGDP*	1.1009 *** (27.75)	1.1215 *** (50.48)
$(\ln RGDP)^2$	0.0132 (0.63)	
ln*GDPC*	0.0839 *** (4.08)	0.08389 *** (4.09)
ln*INST*	−0.5685 *** (−8.09)	−0.5831 *** (−8.81)
β_0 (constant)	9.6374 *** (9.00)	9.4422 *** (9.22)
Individual effects	yes	yes
Time effect	no	no
N	300	300
Adjusted R^2	0.9649	0.9649
F	1033.5500	1208.6400
P	0.0000	0.0000

Note: *** and * indicate that the statistical quantity is significant at the significance level of 1% and 10%, respectively.

4.2.2. Analysis of Influencing Factors

1. The influence coefficient of the number of households is 0.7903, which indicates that the increase in the number of households will lead to an increase in total carbon emissions of China's household consumption. First, the increase in the number of households indicates the increase in the number of basic units of consumption, which will lead to the expansion of the consumption scale and hence to an increase in total household carbon emissions. Second, the increase in the number of households represents, to a certain extent, shrinking family units and some increase in the public consumption of consumer goods, which will lead to an increase in the total carbon emissions from China's households.
2. The influence factor of the age structure factor of the 0–14-year-old children's ratio was 0.5729. Considering other invariable factors, each additional unit will lead to a 0.5729 increase in total carbon emissions from China's households, with a significant effect on the total carbon emissions of China's household consumption. This is because education expenditure accounts for about 7% [6] of household consumption expenditure, which has a great impact on family consumption tendency. The effect of increasing the ratio of total carbon emissions from household consumption is more complicated [42]: in the short term, education expenses will stimulate the growth of household consumption and thus lead to an increase in the total carbon emissions from household consumption. In the long run, the increase in the number of children will increase social and

economic burdens, such as parents meeting their children's consumption needs, which may inhibit the fulfilment of the parents' own needs. At the national level, society as a whole must devote more human, physical and financial resources to protecting and raising children and adolescents, which will certainly reduce investment in other areas. Thus, the increase in the number of children in the long term will inhibit economic growth, thereby indirectly reducing total carbon emissions from households. At present, the role of Chinese children in the total carbon emissions from household consumption is primarily reflected in the short-term effects (the influence coefficient is positive); thus, with the implementation of the fetus policy, the total carbon emissions of China's household consumption will be further increased.

3. The influence coefficient of the gender ratio was 0.3282, indicating that with other factors unchanged, the total carbon emissions of China's household consumption will increase by 0.3282 units per unit of increase in the gender ratio. Compared with the age factor, the gender ratio has a weaker effect on the carbon emissions from China's household consumption. People of different genders have different needs for consumer goods, which leads to the difference of consumption carbon emissions. According to a previous article, of the eight consumer products of residents, the high-carbon group includes residences, food and transportation and communications, accounting for 71.78% of the products carbon emissions of China's household consumption. Among these three types of consumer goods, the consumption of residences is mostly family public consumption, and the gender difference does not have much influence in this area. In terms of food consumption, for physiological reasons, men consume more food than women and are more likely to consume meat, leading to higher products carbon emissions from food consumption than women. In the area of transportation and communications, the majority of people who buy private cars are male, which leads men to consume more than women in this area. Products carbon emissions from men's consumption in transportation and communications are higher than women's. With the implementation of the fetus policy, the imbalance of the gender ratio will improve; in the long run, the development trend of China's gender ratio is conducive to reducing the total carbon emissions of household consumption.
4. The absolute value of the effect coefficient of the per capita GDP is the largest (1.1215) and has the most significant effect on the total carbon emissions of China's household consumption. First, the per capita GDP is higher, indicating that the living standards of the residents in the region are higher and that they are more inclined to high-energy consumption and high-carbon emission products. Second, GDP per capita reflects the overall level of economic development in the region. Higher levels of economic development lead to greater production and consumption demands supported by the entire region's economic development and the inevitable increase in total carbon emissions from household consumption.
5. The influence coefficient of the technical factor of the carbon emission intensity is 0.0839, which trends in the same direction as the total carbon emissions of China's household consumption. Therefore, the improvement of technology can promote the improvement of energy use efficiency, thereby inhibiting the total carbon emissions of household consumption, which is an effective manner in which to reduce the total carbon emissions of household consumption.
6. The influence coefficient of the industrial structure on the total carbon emissions of household consumption is -0.6236. In recent years, with the development of China's economy and the upgrading of the industrial structure, tertiary industry has developed rapidly. The improvement in people's living standards has caused people to no longer be satisfied with the basic necessities of food and clothing and has increased the consumption demand for entertainment and enjoyment, which are mostly in tertiary industries. Therefore, the development trend of China's industrial structure and the total carbon emissions of China's household consumption are moving in the reverse direction.

Summary: The number of households, 0–14-year-old children's ratio, gender ratio, GDP per capita and carbon emissions intensity are moving in the same direction as the total carbon emissions of

China's household consumption. The industrial structure and carbon emissions of China's household consumption are moving in the opposite direction.

5. Discussion

5.1. Comparisons between this Paper and Previous Studies

This paper measures the total carbon emissions from household consumption in 30 provinces in China from 2006 to 2015 and analyses their influencing factors based on the STIRPAT model. The research results of this paper have similarities with and differences from previous studies.

With regard to the scale and composition of total carbon emissions of China's household consumption, this paper and the studies of Peng [11], Wu [12], Wei [17], and Tang [26] observed that the total carbon emissions of China's household consumption increased during the study period. Products carbon emissions from households are the primary component of the total carbon emissions from households. This is because as the standard of living of Chinese residents increases, people's consumption increases and they tend to consume products that produce more carbon emissions in the production process; therefore, the total carbon emissions of household consumption is increasing. However, this article observed that carbon emissions from residents' consumption of electricity are the focal components of energy carbon emissions from China's households. In products carbon emissions of China's household consumption, products carbon emissions from residents' consumption of transportation and communications occupy three of the eight categories of consumer goods. Products carbon emissions from residents' consumption of culture and entertainment are ranked fifth. The research results of Wei et al. [17] indicated that the energy carbon emissions of China's household consumption were primarily from residents' consumption of coal. In products carbon emissions of China's household consumption, the residents' products carbon emissions from culture and entertainment consumption were higher than the residents' products carbon emissions from transportation and communications consumption. This is because the research period of Wei et al. [17] was from 1999 to 2002, and the research period of this paper was from 2006 to 2015. As time passed, residents' consumption of energy products diversified; residents tended to consume more clean energy and consume less coal. In recent years, traffic communication products have rapidly been popularized, and private cars and mobile phones have caused residents' consumption of transportation and communications to increase significantly.

The analysis of the factors influencing the total carbon emissions of China's household consumption, consistent with Du [21], Fu [25], and Ji [27] et al., determined that the population size, per capita GDP and carbon emission intensity have a driving effect on China's household consumption total carbon emissions. However, this paper argues that the effect of the per capita GDP on China's household consumption total carbon emissions has exceeded the population size factor, which differs from previous research. The effects of economic development on total carbon emissions from household consumption extends beyond the demographic factor: this gap will widen further in the future, and the effect of economic factors on the total carbon emissions of China's household consumption will be increasingly more significant.

5.2. Significance of This Study in Other Countries and the Limitations of This Study

The methods and the conclusions of this paper not only apply to China but also to other countries. In recent years, the birth rate in developing countries has remained high. Residents' consumption presents a new and changing characteristic, which will lead to further aggravation of carbon emissions from household consumption. Richard et al. [28] found that the increase in total population was the main factor that led to the increase of energy consumption in EU countries. This is consistent with the conclusion of this paper, which shows that the impact of total population on energy consumption and carbon emissions is the same in different countries. Yeh et al. [43] found that the increase in population and the proportion of the working-age population will lead to an increase in energy consumption in Taiwan. This is consistent with the findings of this study, which show that changes in population and population structure have significant effects on energy consumption and carbon emissions. Because of the diminishing efficiency of industrial emissions reduction, it is obviously more effective and reasonable to reduce carbon emissions from the consumer side. To measure the products carbon emissions of China's household consumption in consecutive years, this paper adopted the Consumer Lifestyle Approach (CLA), which measures the products carbon emissions from terminal consumption. Because there is no combination of an input-output table and a lack of dependence on industry, products carbon emissions from household consumption may be underestimated.

6. Conclusions and Suggestions

6.1. Conclusions

In this paper, the carbon emissions coefficient method and Consumer Lifestyle Approach (CLA) were used to measure the total carbon emissions of China's household consumption from 2006 to 2015. This paper analyzed the spatio-temporal characteristics of total carbon emissions from China's households and the sources of total carbon emissions from China's households. On this basis, using the STIRPAT model as the theoretical basis, this paper expanded the model by introducing factors such as the number of households, 0–14-year-old children's ratio, gender ratio, per capita GDP, carbon emissions intensity and industrial structure; constructed the panel data model of the total carbon emissions of China's household consumption; and analyzed the effects of six factors on the total carbon emissions of China's household consumption. Main conclusions of this paper are shown in Table 8. The research of this paper resulted in the following conclusions:

First, in terms of the scale and structural changes of total carbon emissions from China's households, energy and products carbon emissions and total carbon emissions from China's households have been increasing in recent decades. Products carbon emissions from China's households are the primary source and primary driving force of the total carbon emissions of China's household consumption. With the further development of the economy, the total carbon emissions of China's household consumption will continue to increase over a long period of time.

Second, in terms of the spatial distribution of the total carbon emissions from China's households, the results indicated that the high-carbon platoon is primarily distributed in the east and the middle region and that the carbon emissions level is low in the western region, which is primarily influenced by regional economic development levels. The regional distribution of energy carbon emissions from China's households is also affected by geographical factors, such as the geographical location; its regional distribution demonstrates clear north-south differences, and the energy carbon emissions from households is higher in the north than in the south.

Third, with regard to the source of total carbon emissions from China's households, the proportion of carbon emissions generated by electricity and coal consumption in energy carbon emissions from China's household consumption is greater, representing 47% and 22% of energy carbon emissions from China's household consumption in the past decade, respectively. For products carbon emissions of China's household consumption, residences, food, and transportation and communications belong to the high carbon emissions group. Products carbon emissions of these three types of high-carbon emission products account for more than 70% of the total products carbon emissions.

Finally, with regard to the effects of the variables on the total carbon emissions from China's households, the number of households, 0–14-year-old children's ratio, gender ratio, GDP per capita and carbon emissions intensity were positively correlated with the total carbon emissions of China's household consumption. The industrial structure was negatively correlated with the total carbon emissions of China's household consumption.

Table 8. Main conclusions of this paper.

		Trend of Change	Main Component	Regional Distribution
Scale calculation of household consumption carbon emissions	Total carbon emissions	The total carbon emissions of households increased from 861.41 million tons to 2678.35 million tons in 2006–2015, with an average annual growth rate of 13.43%.	About 80% of the total carbon emissions came from products carbon emissions.	High carbon areas were mainly located in the eastern and central regions.
	Products carbon emissions	The products carbon emissions of households increased from 670.68 million tons to 2344.90 million tons in 2006–2015, with an average annual growth rate of 14.92%.	About 33.39% of the products carbon emissions came from the consumption of residences.	High carbon areas were mainly located in the eastern and central regions.
	Energy carbon emissions	The energy carbon emissions of households increased from 190.74 million tons to 333.44 million tons in 2006–2015, with an average annual growth rate of 6.40%.	About 40% of the energy carbon emissions came from the consumption of electricity.	High carbon areas were mainly in the North.
Influential factor analysis of household consumption carbon emissions		**Demographic factors**	**Economic factor**	**Technical factor**
	Coefficients of factors	$PSIZE = 0.7903$	$RGDP = 1.1215$	$GDPC = 0.0839$
		$CHI = 0.5729$		
		$SEX = 0.3282$		

6.2. Suggestions

According to the study, consumption carbon emission of Chinese residents was still on the rise, and the regional distribution was obviously different. Demographic, economic and technological factors had significant impacts on Chinese residents' consumption of carbon emissions. Therefore, in terms of policy design, the first step should be to formulate targeted policies to guide and encourage residents to adopt low-carbon consumption patterns, and secondly, to improve energy use efficiency and reduce carbon intensity by promoting technological innovation.

In response to a series of emission reduction policies and plans currently formulated by the Chinese government, combined with the existing research, this article specifically put forward the following recommendations:

(1) Many of China's current administrative policies have helped curb the increase in consumer carbon emissions, but further improvements are needed. The carbon intensity regulation policy can promote the production sector to optimize the energy supply structure, and increase the supply of the energy with lower carbon emission factor. So the policy plays a guiding role in the consumption of residents. Such a policy would also increase the price of energy products with higher carbon emissions, leading residents to voluntarily choose clean energy. But at present, China mainly promotes the implementation of carbon intensity control by administrative means, lack of legislative support. Therefore the government should strengthen the formulation of supporting legal system in the future; low-carbon labeling policies can help consumers identify low-carbon products, but such policies need to be complemented by promotional and price incentives. Therefore, the government should increase the low carbon propaganda, expand low-carbon propaganda channels. Advocacy through the popular mobile phone app is a good choice. The Government should also increase the subsidy for the residents to buy low energy products; Step electricity price policy can promote the residents to form a low-carbon lifestyle, and can be extended to the heating aspect; China's two-child policy and the prohibition of illegal gender identification policy in the long run are conducive to curb the increase in consumer carbon emissions. So these policies should be continued to be implemented, and supplemented by public propaganda.

(2) The policies of energy saving and emission reduction based on market regulation mainly includes establishing carbon trading market and levying carbon tax. China has set up a nationwide carbon trading market, but has not yet implemented a carbon tax policy. As with carbon intensity regulation, carbon trading markets also indirectly regulate consumer carbon emissions through the production sector. The establishment of a carbon trading market can promote the optimal distribution of carbon emission rights, which will undoubtedly reduce the cost of China's emission reduction. But China's carbon trading market has been established late, a series of regulatory and regulatory policies are not yet standardized, the Chinese government needs improve it based on the actual situation and draw on the experience of developed countries; Carbon tax policy is also an effective way to promote the early completion of carbon peak in China, but the levy of the carbon tax needs to consider the issue of fairness and public acceptance. It is a good choice to carry out a progressive tax rate and return the carbon tax to companies and residents.

Author Contributions: Yong Wang and Guangchun Yang were mainly responsible for the writing of the full text. Peipei Shang conceived and designed the study. Ying Dong and Yu Cheng built the models of the paper.

Acknowledgments: This paper is supported by Liaoning Economic and Social Development Research Project (2018lslktzd-010), National Natural Science Foundation of China (71573034), Liaoning Social Science Fund (L17CTJ001, L17BJY042), China Postdoctoral Science Fund (2016M601318, 2017T100180) and Research Project of Dongbei University of Finance and Economics (DUFE2017Q16).

Conflicts of Interest: The authors declare no conflict of interest.

References

1. International Energy Agency. Available online: http://www.iea.org/ (accessed on 20 April 2018).
2. BP World Energy Statistics Yearbook. 2017. Available online: https://www.bp.com/zh_cn/china/reports-and-publications/_bp_2017-_.html (accessed on 20 April 2018).
3. National Development and Reform Commission. Available online: http://qhs.ndrc.gov.cn/qjfzjz/201712/W020171221535050998016.pdf (accessed on 20 April 2018).
4. Liu, L.; Wu, G.; Wang, J.; Wei, Y. China's carbon emissions from urban and rural households during 1992–2007. *J. Clean. Prod.* **2011**, *19*, 1754–1762. [CrossRef]
5. National Bureau of Statistics of the People's Republic of China. Available online: http://www.stats.gov.cn/tjsj/zxfb/201602/t20160229_1323991.html (accessed on 20 April 2018).
6. National Bureau of Statistics of the People's Republic of China. Available online: http://www.stats.gov.cn/tjsj/ndsj/2017/indexch.htm (accessed on 20 April 2018).
7. Leontief, W.W. Quantitative Input and Output Relations in the Economic Systems of the United States. *Rev. Econ. Stat.* **1936**, *18*, 105–125. [CrossRef]
8. Leontief, W.W. Interrelation of Prices, Output, Savings, and Investment. *Rev. Econ. Stat.* **1937**, *19*, 109–132. [CrossRef]
9. Herendeen, R. Total energy cost of household consumption in Norway, 1973. *Energy* **1978**, *3*, 615–630. [CrossRef]
10. Druckman, A.; Jackson, T. The carbon footprint of UK households 1990–2004: A socio-economically disaggregated, quasi-multi-regional input–output model. *Ecol. Econ.* **2009**, *68*, 2066–2077. [CrossRef]
11. Peng, S.J.; Zhang, W.C. The empirical analysis of carbon emission trend and its influencing factors in Chinese residents' consumption. *World Econ.* **2013**, *36*, 124–142.
12. Wu, K.Y.; Guo, X.; Wang, W.X.; Zhang, H. An empirical analysis of consumption carbon emissions of Shanghai residents. *Resour. Environ. Yangtze River Basin* **2013**, *22*, 535–543.
13. Tian, X.; Geng, Y.; Dong, H.; Dong, L.; Fujita, T.; Wang, Y.; Zhao, H.; Wu, R.; Liu, Z.; Sun, L. Regional household carbon footprint in China: A case of Liaoning province. *J. Clean. Prod.* **2015**, *114*, 401–411. [CrossRef]
14. Liu, J.R.; Wang, R.S.; Yang, J.W. Ecological impact assessment of household consumption based on life cycle analysis. *Urban Environ. Urban Ecol.* **2005**, *1*, 15–17.
15. Yao, L.; Liu, J.R.; Wang, R.S. Comparative analysis of carbon emissions implied by urban and rural residents in China. *Chin. Popul.·Resour. Environ.* **2011**, *21*, 25–29.
16. Shui, B.; Dowlatabadi, H. Consumer lifestyle approach to US energy use and the related CO emissions. *Energy Policy* **2005**, *33*, 197–208.
17. Wei, Y.M.; Liu, L.C.; Fan, Y.; Wu, G. The impact of lifestyle on energy use and CO_2, emission: An empirical analysis of China's residents. *Energy Policy* **2007**, *35*, 247–257. [CrossRef]
18. Fan, L.; Wang, D. Calculation and analysis of carbon emission from indirect energy consumption of Chinese residents. *Eco-Economy* **2014**, *30*, 28–32.
19. Greening, L.A. Effects of human behavior on aggregate carbon intensity of personal transportation: Comparison of 10 OECD countries for the period 1970–1993. *Energy Econ.* **2004**, *26*, 1–30. [CrossRef]
20. Chai, S.G. An empirical study on the driving factors of carbon emission from urban residents. *Stat. Decis.* **2016**, *32*, 95–99.
21. Du, W. Analysis of indirect carbon emission characteristics of Chinese residents' life based on LMDI decomposition model. *China's Popul. Resour. Environ.* **2016**, *26*, 5–9.
22. Wang, H.J.; Xia, Y. Analysis on the influencing factors and development path of Chinese residents' consumption carbon emission. *China Manag. Sci.* **2017**, *25*, 1–10.
23. Hubacek, K.; Guan, D.; Barua, A. Changing lifestyles and consumption patterns in developing countries: A scenario analysis for China and India. *Futures* **2007**, *39*, 1084–1096. [CrossRef]
24. Hubacek, K.; Feng, K.; Chen, B. Changing Lifestyles towards a Low Carbon Economy: An IPAT Analysis for China. *Energies* **2011**, *5*, 22–31. [CrossRef]
25. Fu, J.Y.; Li, C.L. Calculation and driving factors of indirect energy-use carbon emission from Chinese residents' consumption—Panel data analysis based on STIRPAT model. *Consum. Econ.* **2015**, *31*, 92–97.

26. Tang, Y.Y.; Li, S.; Xia, Q. Analysis on the factors affecting the carbon emission of Chinese residents 'living energy—Based on STIRPAT model. *Eco-Economy* **2017**, *33*, 42–47.
27. Ji, Z.Y.; Lai, X.F.; Jia, L.J. Carbon emissions from domestic energy consumption in the household sector: A study of measurement and driving factors. *China's Popul. Resour. Environ.* **2016**, *26*, 64–72.
28. York, R. Demographic trends and energy consumption in European Union Nations, 1960–2025. *Soc. Sci. Res.* **2007**, *36*, 855–872. [CrossRef]
29. Laureti, T.; Montero, J.M.; Fernández-Avilés, G. A local scale analysis on influencing factors of NOx, emissions: Evidence from the Community of Madrid, Spain. *Energy Policy* **2014**, *74*, 557–568. [CrossRef]
30. Castellano, R.; Laurete, T.; Rego, A. Estimating the effects of road transportation on environmental quality. *Environ. Eng. Manag. J.* **2010**, *9*, 1151–1160.
31. Liddle, B. Impact of population, age structure, and urbanization on carbon emissions/energy consumption: Evidence from macro-level, cross-country analyses. *Popul. Environ.* **2014**, *35*, 286–304. [CrossRef]
32. The China Energy Statistics Yearbook. Available online: http://epub.cnki.net/KNS/brief/result.aspx?dbPrefix=CYFD (accessed on 8 December 2017).
33. ENERGY.GOV. Available online: https://www.energy.gov/ (accessed on 12 November 2017).
34. EIA. Available online: https://www.eia.gov/ (accessed on 12 November 2017).
35. Oak Ridge National Laboratory. Available online: https://www.ornl.gov/ (accessed on 10 November 2017).
36. Intergovernmental Panel on Climate Change. Available online: https://www.ipcc.ch/ (accessed on 12 November 2017).
37. Ministry of Science and Technology of the People's Republic of China. Available online: http://www.most.gov.cn/index.htm (accessed on 15 November 2017).
38. Energy Research Institute National Development and Reform Commission. Available online: http://www.eri.org.cn/ (accessed on 15 November 2017).
39. National Bureau of Statistics of the People's Republic of China. Available online: http://www.stats.gov.cn/tjsj/ndsj/ (accessed on 15 November 2017).
40. Fu, Y.P.; Ma, S.C.; Song, B.Y. The difference of consumption carbon emission between urban and rural residents in China and its influencing factors—Empirical analysis based on panel data. *Explor. Econ. Probl.* **2016**, *37*, 43–50.
41. Dietz, T.; Rosa, E.A. Rethinking the environmental impacts of population, affluence and technology. *Hum. Ecol. Rev.* **1994**, *2*, 277–300.
42. CNKI. Available online: http://kns.cnki.net/KCMS/detail/detail.aspx?dbcode=CJFQ&dbname=CJFDLAST2016&filename=ZRZY201512001&uid=WEEvREcwSlJHSldRa1FhcTdWajFtaW1FbDByRHVsTWNzU0JrMFFNR3VsYz0=$9A4hF_YAuvQ5obgVAqNKPCYcEjKensW4ggI8Fm4gTkoUKaID8j8gFw!!&v=MDc1MjNMdXhZUzdEaDFUM3FUcldNMUZyQ1VSTEtmWXVadEZ5M2xVcjdJUHovUmQ3RzRIOVROclk5RlpZUjhlWDE= (accessed on 20 April 2018).
43. Yeh, J.C.; Liao, C.H. Impact of population and economic growth on carbon emissions in Taiwan using an analytic tool STIRPAT. *Sustain. Environ. Res.* **2017**, *27*, 41–48. [CrossRef]

energies

MDPI

Article

Comparing Urban and Rural Household CO_2 Emissions—Case from China's Four Megacities: Beijing, Tianjin, Shanghai, and Chongqing

Rui Huang [1,2,*], Shaohui Zhang [3,4,*] and Changxin Liu [5,*]

1 Key Laboratory of Virtual Geographic Environment for the Ministry of Education, Nanjing Normal University, Nanjing 210023, China
2 Jiangsu Center for Collaborative Innovation in Geographical Information Resource Development and Application, Nanjing 210023, China
3 School of Economics & Management, Beihang University, Beijing 100191, China
4 International Institute for Applied Systems Analysis, Schossplatz 1, A-2361 Laxenburg, Austria
5 Institutes of Science and Developments, Chinese Academy of Sciences, Beijing 100190, China
* Correspondence: huangrui4420@163.com (R.H.); shaohui.zhang@iiasa.ac.at (S.Z.); liuchangxin@casipm.ac.cn (C.L.); Tel.: +86-152-6186-5906 (R.H.)

Received: 3 May 2018; Accepted: 10 May 2018; Published: 15 May 2018

Abstract: CO_2 emissions caused by household consumption have become one of the main sources of greenhouse gas emissions. Studying household CO_2 emissions (HCEs) is of great significance to energy conservation and emissions reduction. In this study, we quantitatively analyzed the direct and indirect CO_2 emissions by urban and rural households in Beijing, Tianjin, Shanghai, and Chongqing. The results show that urban total HCEs are larger than rural total HCEs for the four megacities. Urban total per capita household CO_2 emissions (PHCEs) are larger than rural total PHCEs in Beijing, Tianjin, and Chongqing, while rural total PHCEs in Shanghai are larger than urban total PHCEs. Electricity and hot water production and supply was the largest contributor of indirect HCEs for both rural and urban households. Beijing, Tianjin, Shanghai, and Chongqing outsourced a large amount of indirect CO_2 emissions to their neighboring provinces.

Keywords: household CO_2 emissions (HCEs); per capita household CO_2 emissions (PHCEs); input–output model

1. Introduction

CO_2 is increasing rapidly due to human activities. Cities are related to about 70–80% of the global carbon emissions: as the main locus of human economic activities and energy consumption, cities play an important role in implementing carbon reduction policies [1–3]. Inhabitants of cities are a key driving force of greenhouse gas (GHG) emissions due to global urbanization development [4]. Biesiot and Noorman [5] proposed that "most of the environmental load in an economy can be allocated to households". The consumption of goods and services in households plays a key role for energy use and CO_2 emissions, especially for developing countries [6]. The activities of consumers (i.e., personal transportation, personal services, and homes) accounts for 45–55% of total energy consumption [7]. Among the key determinants of household energy requirements are socio-economic, demographic, geographic and residential factors [8,9]. Therefore, the consumption patterns of households differ widely within countries, because household characteristics vary (e.g., personal income, household size and related age, the level of education). These factors usually indicate variance in rural and urban areas, meaning that the trajectory of energy consumption in these areas is different [10]. As such, it is significant to study urban and rural energy consumption and CO_2 emissions at a city scale.

China has promised to achieve peak CO_2 emissions around 2030 and to make their best efforts to achieve this goal earlier (National Development & Reform Commission of China, 2015). Given that China's regions have different resource endowments, energy structures, and economic development levels, China has delegated emissions reduction targets to the lower administrative units [11,12]. Tackling global climate change needs to be integrated into city management [13]. Beijing, Tianjin, Shanghai, and Chongqing, as the four municipalities of China, are the economic leaders for other provinces and cities. Thus, these four metropolitan areas' household CO_2 emissions (HCEs) and per capita household CO_2 emissions (PHCEs) need to be studied as examples for other provinces to make policies about energy conservation and emission reduction. On the other hand, the existing research on HCEs at a micro level are mostly based on survey data [14], which provides useful and detailed information for community and households. However, the indirect CO_2 emissions caused by consuming goods and services have not been considered. Park and Heo [15] quantified the direct and indirect energy use of Korean households from 1980 to 2000 and found that the share of indirect household energy consumption accounts for above 60% of the total energy consumption. Markaki et al. [16] found that indirect emissions of Greek households accounted for more than 70% of the total carbon footprint. Therefore, it is essential to evaluate the indirect CO_2 emissions when making policies for household emission reduction. In addition, due to the characteristics of survey data, the results have great uncertainties. It may be difficult for city planners and policy-makers to establish and implement united environmental practices. In light of the above, we adopted the data from the National Bureau of Statistics and an input–output table in this study to estimate direct and indirect CO_2 emissions of urban and rural households in Beijing, Tianjin, Shanghai, and Chongqing.

Household energy consumption is a subject that has attracted considerable scholarly interest. Frequently, studies of household energy consumption, household carbon/CO_2 emissions, and household carbon footprints have been springing up. Some scholars made cross-national comparative studies. For example, Reinders et al. [17] investigated both the direct and indirect energy use of households in 11 EU member countries. Sommer and Kratena [18], and Ivanova et al. [19] calculated the household carbon footprint in the EU27. Lenzen et al. [20] comparatively analyzed the energy requirements of the household sector in Australia, Brazil, Denmark, India, and Japan. Maraseni et al. [21] compared the household carbon emissions between China, Canada, and the UK. Kerkhof et al. [6] examined the household CO_2 emissions of Netherlands, UK, Sweden, and Norway. Brizga et al. [22] estimated the household CO_2 emissions for the three Baltic States (Estonia, Latvia, and Lithuania). Their results show that per capita household CO_2 emissions (PHCEs) in developing countries were much lower than developed countries, while the indirect energy consumption in the sectors of housing, food, beverages, and tobacco, and recreation and culture, and hotel, cafes and restaurants vary significantly per country.

Some research based on a national scale has also been widely studied [23–32]. For instance, Baiocchi et al. [33] pointed out that private households accounted for 75% of the total UK CO_2 emissions, whereas China's household energy consumption was about 25% of the total final energy consumption [34]. With the economic development and improvement of peoples' living standards, the share of household CO_2 emissions is supposed to increase; for example, carbon footprint per household in Norwegian increased by 26% between 1999 and 2012 [35].

There are some household CO_2 emissions studies at the micro scale, such as Sydney, Australia [36], Melbourne, Australia [1], Xiamen, China [37], Tianjin, China [38], and Noakhali, Bangladesh [10]. In China, due to the regional differences between economic structure, resource endowment, industry structure, consumption structures and patterns, urban household CO_2 emissions in eastern regions were much larger, while the provinces in undeveloped western regions had the smallest carbon footprint [39,40].

The analysis of social structures and their evolution trends could inform the government planners and households [41]. In order to find out the impacts of socio-economic factors on household CO_2 emissions, many variables, such as population, affluence, energy intensity, the urbanization level, employment rate, and the share of the tertiary industry, are considered. A large amount of research has shown that household energy requirements, carbon emissions and carbon footprint are closely related to income [42], level of education [43], age [36], gender [38], occupation [14], household size [44], urbanization [45], car ownership [43], urban density [46,47], consumption patterns [48,49], and imports [50]. Different methods, such as index decomposition analysis (IDA) [51], logarithmic mean Divisia index (LMDI) [52], and Stochastic Impacts by Regression on Population, Affluence, and Technology (STIRPAT)model [53,54] were adopted. More discussions can be seen in the review by Zhang et al. [2]. However, the similarities and differences of the direct and indirect HCEs between the urban and rural households are the focus in this study.

2. Materials and Methods

2.1. Household CO_2 Emissions

Household CO_2 emissions include both direct and indirect components of energy consumption. Direct energy consumption refers to the end use of energy, such as for lighting and space heating. Indirect energy, also referred to as "embodied energy," is the amount of energy use throughout the production of goods and services used by households [55,56]. The framework of household CO_2 emissions accounting is shown in Figure 1.

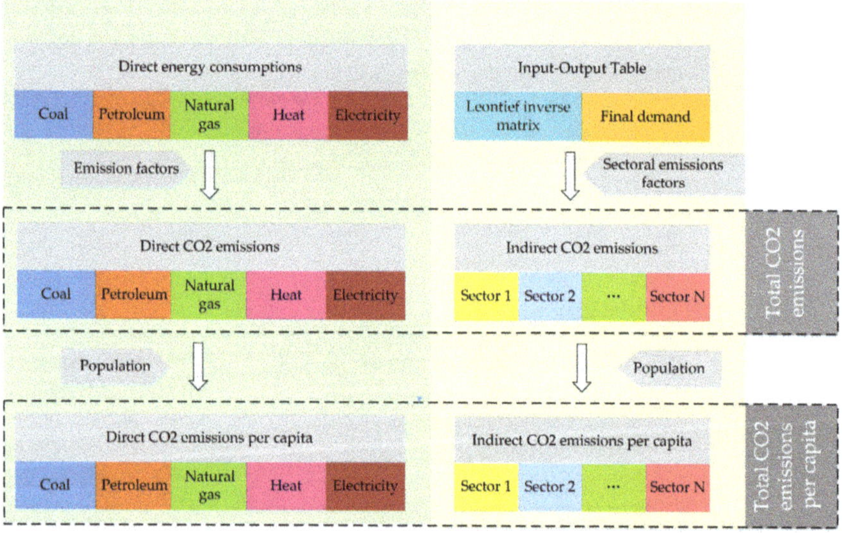

Figure 1. The framework of household CO_2 emissions accounting.

2.1.1. Direct CO_2 Emissions

For direct energy consumption in Beijing's households, we mainly consider coal, oil, natural gas, electricity, and heat. In order to calculate CO_2 emissions for a given energy type, we multiplied its use by a carbon emission coefficient and then added up the results. Expressed mathematically, the procedure is as follows:

$$DC = \sum_i EC_i \bullet Coef_i \quad (1)$$

where DC represents the direct CO_2 emissions and EC_i denotes direct energy consumption of each energy variety i. $Coef_i$ is the CO_2 coefficient for each energy variety i. According to Equation (1), we can calculate the direct CO_2 emissions of urban and rural households, respectively.

2.1.2. Indirect CO_2 Emissions

Based on the input–output model, a region's indirect CO_2 emissions can be obtained by

$$IndC = InCoef \bullet (I - A)^{-1} \bullet Y \tag{2}$$

where $IndC$ denotes the indirect CO_2 emissions, $InCoef$ is the CO_2 coefficient of each sector, I is the identity matrix, A is the intermediate consumption coefficients, and Y is the household final demand.

2.1.3. Total CO_2 Emissions

Total CO_2 emissions are obtained by summing the direct CO_2 emissions and the indirect CO_2 emissions, as shown in Equation (3). TC represents the total CO_2 emissions for urban or rural households. We calculated both urban and rural households' CO_2 emissions in this study.

$$TC = DC + IndC \tag{3}$$

2.1.4. Total CO_2 Emissions Per Capita

Total CO_2 emissions per capita are obtained by total CO_2 emissions divided by the population:

$$PC = TC/P \tag{4}$$

where PC and P denote the PHCEs and population, respectively.

2.2. Data

In this paper, energy consumption data are obtained from the China Energy Statistical Yearbook [57] compiled by the Department of Energy Statistics, National Bureau of Statistics (2008–2016). Direct CO_2 coefficients are obtained from the IPCC report as shown in Table 1. Heat value is adjusted according to principles for calculation of total production energy consumption in 2008 in China. The China Multi-Regional Input–Output Table 2007 [58] and 2012 [59] are used to calculate indirect CO_2 emissions, including 30 sectors. The indirect CO_2 emissions of each province at a sectoral level are obtained from China Emission Account and Datasets (CEADs, http://www.ceads.net/). Population data are from the Beijing Statistical Yearbook (2016) [60], Tianjin Statistical Yearbook (2016) [61], Shanghai Statistical Yearbook (2016) [62], and Chongqing Statistical Yearbook (2016) [63], as shown in Table 2. Due to the lack of data regarding Shanghai's urban and rural population, its rural population is represented by agricultural population and urban population is obtained by total population minus its agricultural population. Although Beijing and Shanghai municipal governments have adopted the strictest household registration system to control their population, the population still increased to a large extent. For example, Beijing's urban population increased by 32.6% from 2007 to 2012, while rural population increased by 12.8%.

Table 1. Direct CO_2 emissions coefficients.

Fuel	Unit	Heat Value	Carbon Content	Oxidation Rata	CO_2 Emission Factor Unit (Kg/GJ)
Coal	GJ/t	20.91	27.4	94%	94.44
Oil	GJ/t	41.82	20.1	98%	72.73
Natural gas	GJ/ 10^4 Nm^3	38.93	15.3	99%	55.54
Heat	-	-	-	-	110
Electricity	-	-	-	-	873

Table 2. Population data (10,000 person).

	Urban Population				Rural Population			
	Beijing	Tianjin	Shanghai	Chongqing	Beijing	Tianjin	Shanghai	Chongqing
2007	1416	851	1882	1361	260	264	182	1455
2008	1504	908	1966	1419	267	268	174	1420
2009	1581	958	2046	1475	279	270	165	1384
2010	1686	1034	2145	1530	276	266	157	1355
2011	1741	1090	2196	1606	278	264	152	1313
2012	1784	1152	2234	1678	286	261	146	1267
2013	1825	1207	2272	1733	290	265	143	1237
2014	1859	1248	2286	1783	293	269	139	1208
2015	1878	1278	2280	1838	293	269	136	1178

3. Results

3.1. Urban and Rural Direct HCEs

3.1.1. Direct HCEs

Direct household CO_2 emissions (HCEs) of Beijing, Tianjin, Shanghai, and Chongqing are shown in Figure 2. Beijing's total direct HCEs increased by approximately 60% from 49.1 Mt in 2007 to 78 Mt in 2015. Shanghai's total direct HCEs increased by approximately 47.7% from 48.7 Mt in 2007 to 71.9 Mt in 2015. The total direct HCEs in Tianjin and Chongqing were smaller than that of Beijing and Shanghai; for example, Tianjin's total direct HCEs were around 59% of that of Beijing in 2015, and Chongqing's total direct HCEs were about 73% of that in Shanghai in 2015. However, total direct HCEs of Tianjin and Chongqing increased by 89.8% and 84.2% from 2007 to 2015, respectively.

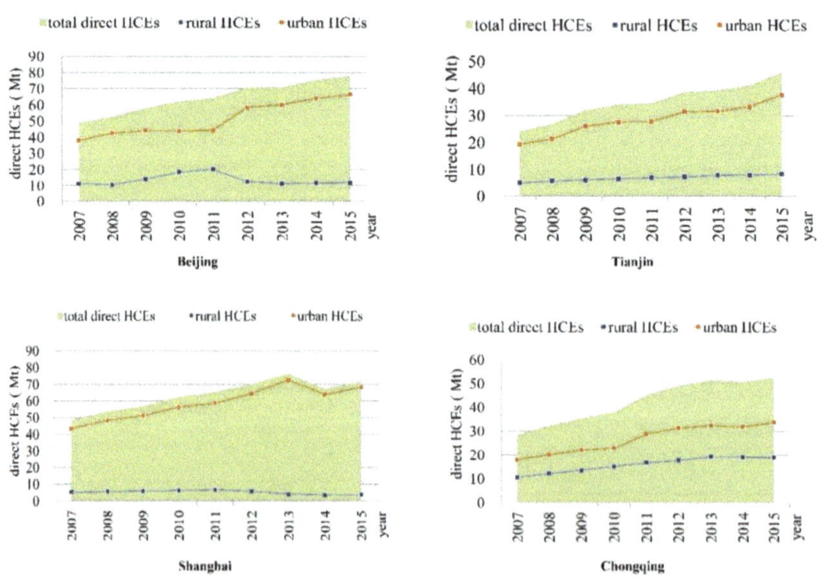

Figure 2. Direct household CO_2 emissions (HCEs).

Urban direct HCEs were much larger than rural direct HCEs for the four megacities; for instance, Shanghai's urban direct HCEs were more than 18 times larger than rural direct HCEs in 2015, which accounted for about 95% of its total direct HCEs. Beijing's rural and urban HCEs show different trends.

These can be divided into two phases. The first phase is from 2007 to 2011. During this phase, both rural and urban direct HCEs kept a similar increasing trend. However, they have showed different trends since 2012. Urban direct HCEs increased sharply in 2012. After that, they kept increasing steadily. On the contrary, rural direct HCEs declined significantly in 2012, then remained about the same. Tianjin's urban direct HCEs increased rapidly during 2007–2015 with an annual increase rate of 9%, while the annual increase rate of rural direct HCEs were 7%, whereas Chongqing's urban and rural direct HCEs kept the same annual increase rate, which was 8%.

3.1.2. Direct Energy Consumption Structure

Energy consumption structure for direct HCEs are shown in Figure 3. The energy consumption structure of Beijing's urban households remained stable from 2007 to 2015. By contrast, rural households' energy consumption structure had a large fluctuation during 2008–2011. Due to the global financial crisis, coal prices rose sharply [64]. The coal consumption of rural households dropped significantly. In 2011, the share of coal was only 20.6%. After the financial crisis, coal consumption rose and stayed stable with a relatively lower coal price. Heat consumption in Tianjin's urban households accounted for 26–29% of their total direct energy consumption, which was much higher than Beijing. It is unexpected to find that the oil consumption of Shanghai's rural households accounted for about one third of their total direct energy consumption. After the financial crisis, the share increased to more than 60%. By contrast, the household energy consumption structure in Chongqing was cleaner.

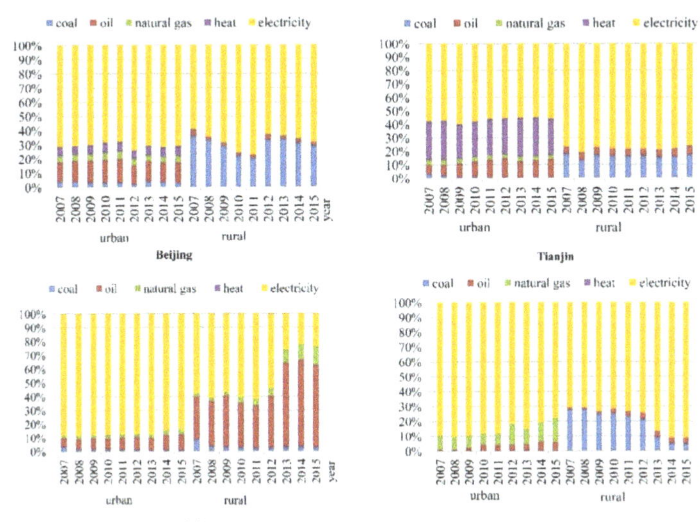

Figure 3. Energy structure of direct HCEs.

3.1.3. Direct PHCEs

Direct PHCEs in Beijing, Tianjin, Shanghai, and Chongqing from 2007 to 2015 are shown in Figure 4. It is interesting to find that the direct PHCEs of rural and urban households were getting close in the last three years for the four cities. For example, direct PHCEs of Beijing's rural households were larger than that of urban households. In 2011, the former was 2.85 times larger than the latter. Since 2012, PHCEs of urban and rural household were about 1 ton of CO_2 (tC) per person, which is the smallest. PHCEs of urban and rural households in Tianjin and Shanghai were approximately three times that of Beijing.

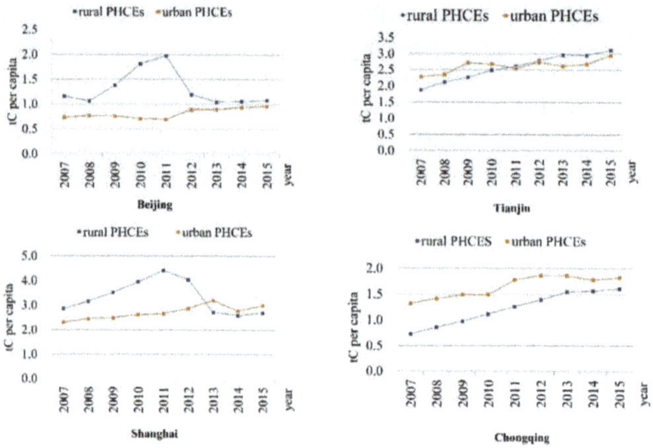

Figure 4. Direct per capita HCEs (PHCEs).

3.2. *Urban and Rural Indirect HCEs*

3.2.1. Indirect HCEs and PHCEs

By adding urban and rural indirect HCEs, we can obtain the total indirect HCEs of each city. Total indirect HCEs of Beijing, Tianjin, and Shanghai, respectively, decreased by 2.96%, 27.54%, and 16.67% from 2007 to 2012, while Chongqing's total indirect HCEs increased by 32.36%. Urban and rural indirect HCEs and PHCEs are shown in Figure 5. We can see that urban indirect HCEs were much larger than that of rural households. For example, Beijing's urban indirect HCEs were more than 13 times those of rural households in 2015. Chongqing's urban indirect HCEs were more than four times that of rural households in 2015.

From the perspective of per capita, urban and rural indirect PHCEs of Beijing and Tianjin decreased from 2007 to 2012, while urban and rural indirect PHCEs of Chongqing increased. Urban indirect PHCEs of Shanghai were two times that of rural indirect PHCEs in 2007. However, they were about the same in 2012.

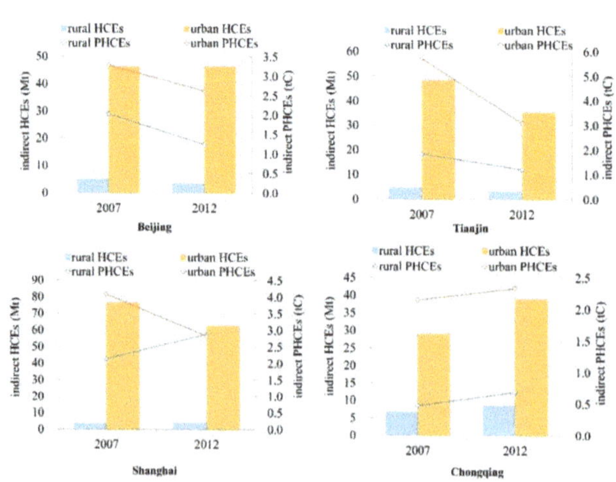

Figure 5. Indirect HCEs and PHCEs.

3.2.2. Sectoral Indirect HCEs

Sectoral abbreviation and indirect HCEs are shown in Table A1. Indirect HCEs from electricity and hot water production and supply were much larger than other sectors for all the four cities. For instance, rural and urban indirect HCEs from electricity and hot water production and supply in Tianjin accounted for 63.3% and 69.4% in 2012, respectively. Thus, to better express the indirect HCEs at sectoral level, we give the percentage-stacked bar chart of indirect HCEs from all the sectors except electricity and hot water production and supply, as shown in Figure 6.

Figure 6. Sectoral indirect HCEs.

For Beijing, Tianjin, Shanghai, and Chongqing, the indirect HCEs from agriculture, coal mining, food processing and tobacco, petroleum refining, coking, etc., chemical industry, nonmetal products, metallurgy, construction, transport and storage increased. The share of indirect HCEs from agriculture were relatively large and increased from 2007 to 2012 for both urban and rural residents in Chongqing. The share of indirect HCEs from coal mining decreased from 2007 to 2012 in Shanghai, Tianjin, and Chongqing; however, the share of indirect HCEs from petroleum refining, coking, etc. increased. For Beijing, Shanghai, and Chongqing, the share of indirect HCEs from transport and storage increased from 2007 to 2012, but the share decreased by 5.9% and 6.7% for rural and urban residents in Tianjin, respectively. However, the share of indirect HCEs from metallurgy respectively increased by 4.5% and 3% for rural and urban residents in Tianjin.

3.2.3. Outsourced Indirect HCEs

Due to the difference of regional resource endowment and industrial structure, the four cities outsourced large amounts of CO_2 emissions to other provinces to meet their own demands for products and services through inter-regional trade. For example, outsourced indirect HCEs accounted for 73.7%

for Beijing in 2007, and the share increased to 87.6% in 2012. Similarly, the share of outsourced indirect HCEs in Chongqing increased from 43.9% in 2007 to 59.7% in 2012. On the contrary, the share of outsourced indirect HCEs in Shanghai and Tianjin decreased by 6.9% and 8.7%, respectively. However, the outsourced indirect HCEs in Shanghai and Tianjin still accounted for more than 60%.

The outsourced indirect HCEs of Beijing, Tianjin, Shanghai, and Chongqing in 2012 are shown in Figure 7. Beijing, Tianjin, Shanghai, and Chongqing respectively outsourced 142 Mt, 127.1 Mt, 108.6 Mt, and 130.6 Mt indirect HCEs to other provinces in 2012, most of which were neighboring provinces with rich resources and less developed economic structure. For example, Inner Mongolia, Hebei, and Shanxi were the top three contributors to Beijing's outsourced indirect HCEs; the shares were 17.8%, 17.4%, and 8.6%, respectively. 26.8% of Chongqing's outsourced indirect HCEs were from Guizhou, Yunnan, and Sichuan.

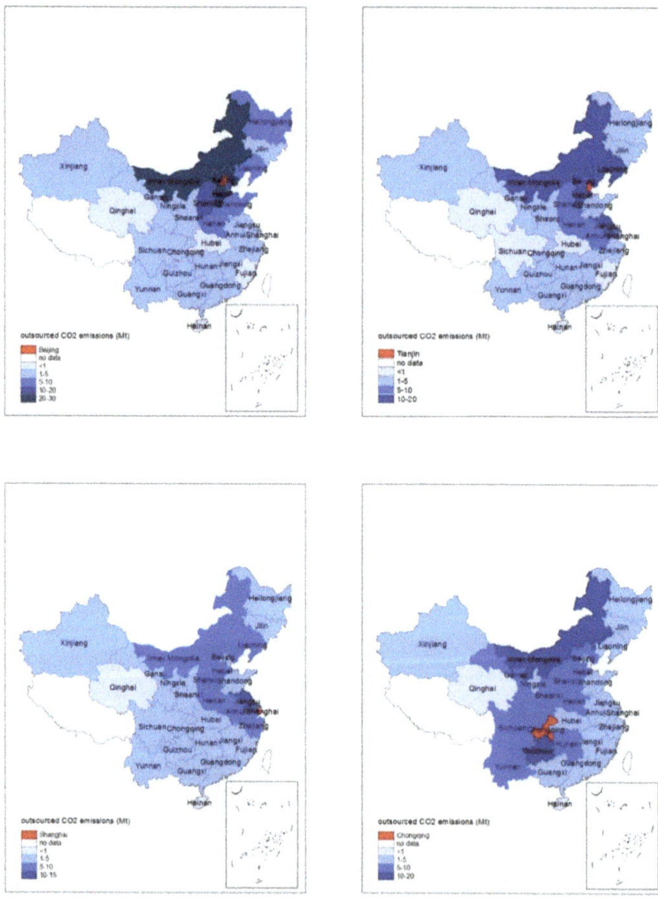

Figure 7. Outsourced indirect HCEs.

3.3. Urban and Rural Total HCEs and PHCEs

The total CO_2 emissions can be obtained by summing up urban and rural households' direct and indirect CO_2 emissions. Chongqing's total CO_2 emissions increased significantly with the increase rate of 49.71% from 64.63 Mt in 2007 to 96.76 Mt in 2012. Beijing's total CO_2 emissions increased by 20.2% from 100.79 Mt in 2007 to 121.15 Mt in 2012. Shanghai's total CO_2 emissions increased by 6.21% from

129.45 Mt in 2007 to 137.49 Mt in 2012, whereas Tianjin's total CO_2 emissions decreased slightly from 77.75 Mt in 2007 to 77.53 Mt in 2012.

Rural and urban households' HCEs and PHCEs are shown in Figure 8. The urban–rural total HCEs gap in Shanghai is the largest, followed by Beijing and Tianjin. Chongqing's urban–rural total HCEs gap is the smallest. From the amount of total HCEs, Chongqing has the largest rural HCEs and the smallest urban HCEs. On the contrary, Shanghai has the smallest rural HCEs and the largest urban HCEs.

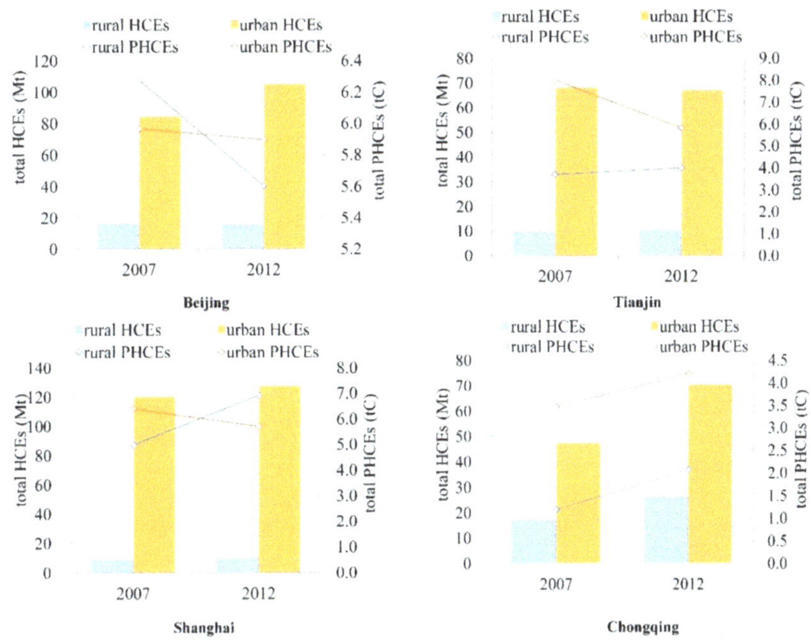

Figure 8. Total HCEs and total PHCEs.

From the perspective of total PHCEs, Chongqing's rural and urban PHCEs increased by 73.59% and 21.01%, respectively. Beijing's rural and urban PHCEs decreased by 10.69% and 1.19%. Rural PHCEs in both Tianjin and Shanghai respectively increased by 8.03% and 38.72%, while urban PHCEs decreased by 27.10% and 10.89%, respectively.

PHCEs in our study and other studies are compared in Table 3. PHCEs in Beijing, Tianjin, Shanghai, and Chongqing were larger than the national average household footprint shown by Wiedenhofer et al. [65], Fan et al. [66], and Qu et al. [67], but much smaller than the U.S. [68] and European countries [18,69,70]. Compared to the results of Tian et al. [71] and Fry et al. [72], Beijing's total PHCEs in our results were 31.56% and 29.02% smaller, respectively, due to different research methods and data sources. Shanghai's total PHCEs in our results were close to other cities in the Yangtze River delta region [14].

Table 3. Results comparison (ton of CO_2).

Sources	Study Area	Carbon Footprints	Study Period
This study	Beijing	5.75	2012
	Tianjin	4.91	2012
	Shanghai	6.31	2012
	Chongqing	3.14	2012
Wiedenhofer et al. [65]	China	1.7	2012
Fan et al. [66]	China	2	2005
Qu et al. [67]	China	1.75	2011
Jones and Kammen [68]	US	20	2005
Isaksen et al. [69]	Norway	12.2	2007
	West Germany	19.8	
Weber and Perrels [70]	Netherlands	18.7	1990
	France	12.9	
Sommer and Kratena [18]	EU27	15.7	-
Tian et al. [71]	Jingjin region	8.4	2007
Fry et al. [72]	Beijing	8.1	2011
Xu et al. [14]	Nanjing, Ningbo, and Changzhou	6.0	2010
Lin et al. [73]	Xiamen, China	3.9	2009
Tian et al. [74]	Liaoning	3.5	2007
Qu et al. [75]	Northwestern China arid-alpine regions	1.4	2008

4. Discussion

In this study, we considered both direct and indirect emissions caused by rural and urban household consumption (as shown in Figure 1). Total emissions are obtained by summing direct CO_2 emissions and indirect CO_2 emissions [56]. The direct CO_2 emissions mainly refer to the consumption of coal, oil, gas, electricity, and heat from China energy statistical yearbook, while the indirect CO_2 emissions are caused by the consumption of products and services, which is also named embodied emissions [40,72].

Urban direct HCEs were much larger than rural direct HCEs. There are several reasons for this: (1) in terms of both quantity and variety, urban residents have more household equipment than rural residents; (2) urban citizens have more cars, which not only brings about severe traffic problems, but also consumes lots of gasoline and produces more emissions; and (3) the population of urban areas is larger than that of rural areas. With rapid urbanization, more and more people flood into the city. For example, Beijing's urban population was six times larger than the rural population in 2014.

For both urban and rural households in Beijing, Tianjin, and Chongqing in China, CO_2 emissions caused by electricity consumption accounted for the largest proportion of their direct CO_2 emissions: the most carbon-intensive categories were electricity and hot water production and supply. For instance, the shares of direct HCEs from electricity in Beijing were 71.3% and 58.2% in 2007 for urban and rural household, respectively, and increased to 73.7% and 62.3% in 2012, respectively. An increased level of income or consumption increased the probability of the use of electricity [76,77]. Thus, the result reflects the improvement of the income and living standard of urban and rural household and the widespread use of household electrical appliances with the rapid development of economy.

For rural households in Beijing and Shanghai, direct HCEs from coal and oil consumption occupied a larger relative proportion. This is related to the large amounts of coal use for heating and cooking in rural areas of Beijing. Oil is the main energy consumption in rural areas of Shanghai, and the share of direct HCEs from oil consumption was approximately 60% in 2015. Affected by the financial crisis and post-crisis, the coal and oil price rose dramatically and the consumption of coal and oil of rural household declined, thus direct HCEs decreased significantly in 2012. Increasing the price of coal and oil may be an effective way to control fossil energy use and reduce CO_2 emissions, such as the through implementation of a carbon tax or environmental tax [78]. However, to avoid the economic loss and urban–rural household welfare losses caused by carbon tax, the optimal carbon tax rate should be formulated carefully.

Large amounts of CO_2 emissions are outsourced to other provinces to meet the demand of local residents. For example, about 68.5% of Beijing's household emissions were outsourced to other provinces in 2007, which is consistent with Feng et al. [79]. The share increased to 81.7% in 2012. The Chinese government has taken active measures to improve the capacity of key areas to adapt to climate change and mitigate the adverse effects of climate change on economic and social development and people's livelihood. The National Development and Reform Commission (NDRC) started the pilot work of carbon emissions trading in Beijing, Tianjin, Shanghai, Chongqing, Hubei, Guangdong, and Shenzhen in 2011. The completion of the reduction of carbon dioxide emission intensity is included in the comprehensive evaluation system of economic and social development in various regions and the system of cadre performance assessment [80]. To reduce Beijing's CO_2 emissions and environmental pressure, Beijing adjusted its industrial structure: heavy industries were moved to its neighboring provinces, such as Hebei, Inner Mongolia, and Shanxi. Through interregional trade, products and services are imported to meet the demands of local household. Government should pay more attention to interprovincial carbon leakage to make an equitable and effective regional emissions reduction scheme. To reduce China's total CO_2 emissions, energy efficiency improvement and clean energy development are significant.

Urban total HCEs increased to a large extent with the increase of urban population. For example, urban population increased by 23.27% in Chongqing from 2007 to 2012, while its urban total HCEs increased by 49.16%. In our study, urban households contributed 72.81–92.65% of total HCEs in 2012. Yang et al. [81] find that urban households contribute 92.6% of the particulate matter 2.5 (PM 2.5) footprint of Beijing's households. Therefore, it is urgent to control urban population. City planners should promote economic development and increase the job opportunities in rural areas and the rural–urban fringe zone to reduce the migrants who move to the city and seek jobs. For example, on 1 April 2017, the State Council of China has decided to build Xiongan New Area, which is a new area of national significance after Shenzhen Special Economic Zone and Pudong New Area of Shanghai. It is expected to relieve the stress of Beijing's population and environment.

5. Conclusions

We examined the direct and indirect CO_2 emissions of urban and rural households in Beijing, Tianjin, Shanghai, and Chongqing in this study. The results showed that total PHCEs were larger than the national average level, but much smaller compared to developed countries such as the US and EU countries [82]. Direct HCEs caused by electricity consumption account for a large proportion of emissions. Despite the urban/rural differential for both groups, the most carbon-intensive categories were electricity and hot water production and supply, agriculture, coal mining, food processing and tobacco, petroleum refining, coking, etc., chemical industry, nonmetal products, metallurgy, construction, transport and storage.

Most household CO_2 emissions are contributed by urban HCEs in Beijing, Tianjin, Shanghai, and Chongqing. Chongqing's total HCEs are approximately 70–80% of Beijing and Shanghai in 2012; however, this increased by about 50% from 2007 to 2012. With the acceleration of urbanization, this is supposed to increase in future. Therefore, it is important to advocate low carbon consumption patterns to control household CO_2 emissions.

Measuring and understanding energy consumption helps in forming a proper policy to motivate the citizens of metropolitan areas to become "greener" consumers and promote renewable energy development. This "greener" character needs to be achieved, as urban cities are environmentally compromised regions because of their metropolitan character [83]. Therefore, the following suggestions are proposed for city planners and policy makers: (1) continue to promote low-carbon green lifestyles and encourage residents to use low-carbon and renewable energy to save energy with the aid of the media; (2) control cities' populations: promote the development of neighbouring districts, create more jobs and opportunities in the neighbouring districts, and divert migrant workers; (3) in the process of urbanization, encourage the development of low-carbon infrastructure, along with the use of materials that improve building quality and sustainability; and (4) judge government performance on the basis not only of GDP, but also of energy efficiency and technical progress.

Author Contributions: R.H. designed the research, R.H., S.Z. and C.L. discussed the results and contributed to writing the paper. We would like to thank Klaus Hubacek from University of Maryland and the reviewers' suggestions, which helps to improve our paper.

Acknowledgments: This work was supported by Chinese National Natural Science Foundation (41701615, 71690245), Jiangsu Provincial Natural Science Foundation (BK20171038), China Postdoctoral Science Foundation (2016M600429), and Natural science fund for colleges and universities in Jiangsu Province (16KJB170003).

Conflicts of Interest: The authors declare no conflict of interest.

Appendix A

Table A1. Indirect CO_2 emissions of urban and rural household in 2012 (10,000 tons).

		Rural				Urban			
	Abbreviation	**Beijing**	**Tianjin**	**Shanghai**	**Chongqing**	**Beijing**	**Tianjin**	**Shanghai**	**Chongqing**
Agriculture	Agri	15.80	14.59	17.73	121.28	217.41	136.95	314.11	410.59
Coal mining	Coal	10.01	5.74	13.00	49.64	126.23	55.35	187.89	186.31
Petroleum and gas	Petr	1.64	3.83	4.20	3.37	22.81	38.56	69.87	23.54
Metal mining	Meta	0.23	0.22	0.20	0.79	3.23	2.20	3.33	3.70
Nonmetal mining	Nonm	0.25	0.20	0.20	0.68	3.49	1.75	3.30	2.74
Food processing and tobacco	Food	14.63	7.39	10.76	22.08	183.13	75.33	197.05	82.99
Textile	Text	0.57	1.87	1.59	2.21	10.32	18.92	31.31	18.38
Clothing, leather, fur, etc.	Clot	0.46	0.33	0.62	0.25	9.04	3.38	15.10	2.02
Wood processing and furnishing	Wood	0.17	0.19	0.18	0.16	2.45	2.76	3.99	0.73
Paper making, printing, stationery, etc.	Pape	1.08	1.10	1.53	5.63	15.32	11.03	34.15	24.53
Petroleum refining, coking, etc.	Perc	10.32	12.96	15.04	39.75	143.91	127.27	275.21	148.75
Chemical industry	Chem	9.58	7.04	8.14	30.89	124.74	64.56	147.35	171.91
Nonmetal products	Npro	9.59	9.15	9.99	48.13	131.25	65.25	180.55	144.45
Metallurgy	Melu	17.78	18.62	19.50	42.22	255.95	188.25	331.14	194.87
Metal products	Mpro	0.29	0.44	0.26	0.70	4.21	4.03	4.77	2.72
General and specialist machinery	Gene	0.37	0.37	0.36	0.71	5.12	3.63	6.46	3.44
Transport equipment	Tran	0.29	0.33	0.50	0.59	4.50	3.63	7.62	3.01
Electrical equipment	Ecal	0.15	0.23	0.18	0.40	1.98	2.11	2.91	1.92
Electronic equipment	Enic	0.08	0.08	0.09	0.07	1.08	0.98	1.79	0.29
Instrument and meter	Inst	0.01	0.01	0.01	0.09	0.13	0.06	0.15	0.60
Other manufacturing	Oman	0.14	0.06	0.17	0.39	1.89	0.53	3.07	2.00
Electricity and hot water production and supply	Ehwp	214.24	199.05	268.23	384.03	2739.07	2467.62	3562.09	1988.70
Gas and water production and supply	Gasw	0.69	0.33	2.73	0.63	7.89	3.47	29.38	3.79
Construction	Cons	0.31	0.22	0.37	0.32	5.25	3.45	7.70	1.65
Transport and storage	Tras	24.53	14.72	26.30	74.21	362.53	145.55	579.24	324.73
Wholesale and retailing	Whol	4.99	3.94	4.99	13.66	69.66	40.47	96.68	59.56
Hotel and restaurant	Hote	3.17	3.04	1.61	8.69	56.30	32.69	37.62	54.32
Leasing and commercial services	Leas	1.22	0.63	1.37	0.85	25.05	6.55	35.43	6.07
Scientific research	Scie	0.20	0.06	0.08	0.13	3.11	0.52	1.39	0.60
Other services	Oser	9.78	7.63	10.10	8.83	131.26	50.39	141.01	41.87

References

1. Wiedmann, T.; Chen, G.W.; Barrett, J. The concept of city carbon maps, a case study of Melbourne, Australia. *J. Ind. Ecol.* **2016**, *20*, 676–691. [CrossRef]
2. Zhang, X.L.; Luo, L.Z.; Skitmore, M. Household carbon emission research: An analytical review of measurement, influencing factors and mitigation prospects. *J. Clean. Prod.* **2015**, *103*, 873–883. [CrossRef]
3. Kennedy, C.; Steinberger, J.; Gasson, B.; Hansen, Y.; Hillman, T.; Havránek, M.; Pataki, D.; Phdungsilp, A.; Ramaswami, A.; Mendez, G.V. Methodology for inventorying greenhouse gas emissions from global cities. *Energy Policy* **2010**, *38*, 4828–4837. [CrossRef]
4. Satterthwaite, D. Cities' contribution to global warnings: Notes on the allocation of greenhouse gas emissions. *Environ. Urban.* **2008**, *20*, 539–549. [CrossRef]
5. Biesiot, W.; Noorman, K.J. Energy requirements of household consumption: A case study of the Netherlands. *Ecol. Econ.* **1999**, *28*, 367–383. [CrossRef]
6. Kerkhof, A.C.; Benders, R.M.J.; Moll, H.C. Determinants of variation in household CO_2 emissions between and within countries. *Energy Policy* **2009**, *37*, 1509–1517. [CrossRef]
7. Schipper, L.; Bartlett, S.; Hawk, D.; Vine, E. Linking life-styles and energy use: A matter of time? *Annu. Rev. Energy* **1989**, *14*, 271–320. [CrossRef]
8. Pachauri, S. An analysis of cross-sectional variations in total household energy requirements in India using micro survey data. *Energy Policy* **2004**, *32*, 1723–1735. [CrossRef]
9. Rao, M.N.; Reddy, B.S. Variations in energy use by Indian households: An analysis of micro level data. *Energy* **2007**, *32*, 143–152.
10. Miah, M.D.; Foysal, M.A.; Koike, M.; Kobayashi, H. Domestic energy-use pattern by the households: A comparison between rural and semi-urban areas of Noakhali in Bangladesh. *Energy Policy* **2011**, *39*, 3757–3765. [CrossRef]
11. Liu, L.C.; Liang, Q.M.; Wang, Q. Accounting for China's regional carbon emissions in 2002 and 2007: Production-based versus consumption-based principles. *J. Clean. Prod.* **2015**, *103*, 384–392. [CrossRef]
12. Shan, Y.L.; Liu, J.H.; Liu, Z.; Xu, X.W.H.; Shao, S.; Wang, P.; Guan, D.B. New provincial CO_2 emission inventories in China based on apparent energy consumption data and updated emission factors. *Appl. Energy* **2016**, *184*, 742–750. [CrossRef]
13. Bai, X. Integrating global environmental concerns into urban management: The scale and readiness arguments. *J. Ind. Ecol.* **2007**, *11*, 15–29. [CrossRef]
14. Xu, X.; Tan, Y.; Chen, S.; Yang, G.; Su, W. Urban household carbon emission and contributing factors in the Yangtze River Delta, China. *PLoS ONE* **2015**, *10*, e0121604. [CrossRef] [PubMed]
15. Park, H.C.; Heo, E. The direct and indirect household energy requirements in the Republic of Korea from 1980 to 2000-An input-output analysis. *Energy Policy* **2007**, *35*, 2839–2851. [CrossRef]
16. Markaki, M.; Belegri-Roboli, A.; Sarafidis, Y.; Mirasgedis, S. The carbon footprint of Greek households (1995–2012). *Energy Policy* **2017**, *100*, 206–215. [CrossRef]
17. Reinders, A.H.M.E.; Vringer, K.; Blok, K. The direct and indirect energy requirement of households in the European Union. *Energy Policy* **2003**, *31*, 139–153. [CrossRef]
18. Sommer, M.; Kratena, K. The carbon footprint of European households and income distribution. *Ecol. Econ.* **2017**, *136*, 62–72. [CrossRef]
19. Ivanova, D.; Vita, G.; Steen-Olsen, K.; Stadler, K.; Melo, P.C.; Wood, R.; Hertwich, E.G. Mapping the carbon footprint of EU regions. *Environ. Res. Lett.* **2017**, *12*, 54013. [CrossRef]
20. Lenzen, M.; Wier, M.; Cohen, C. A comparative multivariate analysis of household energy requirements in Australia, Brazil, Denmark, India and Japan. *Energy* **2006**, *31*, 181–207. [CrossRef]
21. Maraseni, T.N.; Qu, J.S.; Zeng, J.J. A comparison of trends and magnitudes of household carbon emissions between China, Canada and UK. *Environ. Dev.* **2015**, *15*, 103–119. [CrossRef]
22. Brizga, J.; Feng, K.; Hubacek, K. Household carbon footprints in the Baltic States: A global multi-regional input–output analysis from 1995 to 2011. *Appl. Energy* **2017**, *189*, 780–788. [CrossRef]
23. Bin, S.; Dowlatabadi, H. Consumer lifestyle approach to US energy use and the related CO_2 emissions. *Energy Policy* **2005**, *33*, 197–208. [CrossRef]

24. Kok, R.; Benders, R.M.J.; Moll, H.C. Measuring the environmental load of household consumption using some methods based on input-output energy analysis: A comparison of methods and a discussion of results. *Energy Policy* **2006**, *34*, 2744–2761. [CrossRef]
25. Wei, Y.M.; Liu, L.C.; Fan, Y.; Wu, G. The impact of lifestyle on energy use and CO_2 emission: An empirical analysis of China's residents. *Energy Policy* **2007**, *35*, 247–257. [CrossRef]
26. Weber, C.L.; Matthews, H.S. Quantifying the global and distributional aspects of American household carbon footprint. *Ecol. Econ.* **2008**, *66*, 379–391. [CrossRef]
27. Dhakal, S. Urban energy use and carbon emissions from cities in China and policy implications. *Energy Policy* **2009**, *37*, 4208–4219. [CrossRef]
28. Feng, Z.H.; Zou, L.L.; Wei, Y.M. The impact of household consumption on energy use and CO_2 emissions in China. *Energy* **2011**, *36*, 656–670. [CrossRef]
29. Wang, Z.H.; Yang, L. Indirect carbon emissions in household consumption: Evidence from the urban and rural area in China. *J. Clean. Prod.* **2014**, *78*, 94–103. [CrossRef]
30. Zhu, Q.; Peng, X.Z.; Wu, K.Y. Calculation and decomposition of indirect carbon emissions from residential consumption in China based on the input-output model. *Energy Policy* **2012**, *48*, 618–626. [CrossRef]
31. Teubler, J.; Buhl, J.; Lettenmeier, M.; Greiff, K.; Liedike, C. A household's burden- the embodied resource use of household equipment in Germany. *Ecol. Econ.* **2018**, *146*, 96–105. [CrossRef]
32. Pachauri, S.; Spreng, D. Direct and indirect energy requirements of households in India. *Energy Policy* **2002**, *30*, 511–553. [CrossRef]
33. Baiocchi, G.; Minx, J.; Hubacek, K. The impact of social factors and consumer behavior on carbon dioxide emissions in the United Kingdom. *J. Ind. Ecol.* **2010**, *14*, 50–72. [CrossRef]
34. Ding, Q.; Cai, W.; Wang, C.; Sanwal, M. The relationships between household consumption activities and energy consumption in China-an input-output analysis from the lifestyle perspective. *Appl. Energy* **2017**, *207*, 520–532. [CrossRef]
35. Steen-Olsen, K.; Wood, R.; Hertwich, E.G. The carbon footprint of Norwegian household consumption 1999–2012. *J. Ind. Ecol.* **2016**, *20*, 582–592. [CrossRef]
36. Lenzen, M.; Dey, C.; Foran, B. Energy requirements of Sydney households. *Ecol. Econ.* **2004**, *49*, 375–399. [CrossRef]
37. Yang, D.; Gao, L.; Xiao, L.; Wang, R. Cross-boundary environmental effects of urban household metabolism based on an urban spatial conceptual framework: A comparative case of Xiamen. *J. Clean. Prod.* **2012**, *27*, 1–10. [CrossRef]
38. Bai, Y.; Liu, Y. An exploration of residents' low-carbon awareness and behavior in Tianjin, China. *Energy Policy* **2013**, *61*, 1261–1270. [CrossRef]
39. Wang, Z.H.; Yang, Y.T.; Wang, B. Carbon footprints and embodied CO_2 transfers among provinces in China. *Renew. Sustain. Energy Rev.* **2018**, *82*, 1068–1078. [CrossRef]
40. Feng, K.S.; Davis, S.J.; Sun, L.X.; Li, X.; Guan, D.B.; Liu, W.D.; Liu, Z.; Hubacek, K. Outsourcing CO_2 within China. *Proc. Natl. Acad. Sci. USA* **2013**, *110*, 11654–11659. [CrossRef] [PubMed]
41. Duchin, F.; Hubacek, K. Linking social expenditures to household lifestyles. *Futures* **2003**, *35*, 61–74. [CrossRef]
42. López, L.A.; Arce, G.; Morenate, M.; Zafrilla, J.E. How does income redistribution affect households' material footprint? *J. Clean. Prod.* **2017**, *153*, 515–527. [CrossRef]
43. Minx, J.; Baiocchi, G.; Wiedmann, T.; Barrett, J.; Cueutzig, F.; Feng, K.S.; Forster, M.; Pichler, P.; Weisz, H.; Hubacek, K. Carbon footprints of cities and other human settlements in the UK. *Environ. Res. Lett.* **2013**, *8*, 35039. [CrossRef]
44. Fremstad, A.; Underwood, A.; Zahram, S. The environmental impact of sharing: Household and urban economics in CO_2 emissions. *Ecol. Econ.* **2018**, *145*, 137–147. [CrossRef]
45. Zhang, C.; Cao, X.Y.; Ramaswami, A. A novel analysis of consumption-based carbon footprints in China: Unpacking the effects of urban settlement and rural-to-urban migration. *Glob. Environ. Chang.* **2016**, *39*, 285–293. [CrossRef]
46. Jones, C.; Kammen, D.M. Spatial distribution of U.S. household carbon footprints reveals suburbanization undermines greenhouse gas benefits of urban population density. *Environ. Sci. Technol.* **2014**, *48*, 895–902. [CrossRef] [PubMed]

47. Gill, B.; Moeller, S. GHG Emissions and the Rural-Urban Divide. A Carbon Footprint Analysis Based on the German Official Income and Expenditure Survey. *Ecol. Econ.* **2018**, *145*, 160–169. [CrossRef]
48. Duarte, R.; Mainar, A.; Sánchez-Chóliz, J. The impact of household consumption patterns on emissions in Spain. *Energy Econ.* **2010**, *32*, 176–185. [CrossRef]
49. Dai, H.C.; Musui, T.; Matsuoka, Y.; Fujimori, S. The impacts of China's household consumption expenditure patterns on energy demand and carbon emissions towards 2050. *Energy Policy* **2012**, *50*, 736–750. [CrossRef]
50. López, L.A.; Arce, G.; Morenate, M.; Monsalve, F. Assessing the inequality of Spanish households through the carbon footprint: The 21st century great recession effect. *J. Ind. Ecol.* **2016**, *20*, 571–581. [CrossRef]
51. Zha, D.L.; Zhou, D.Q.; Zhou, P. Driving forces of residential CO_2 emissions in urban and rural China: An index decomposition analysis. *Energy Policy* **2010**, *38*, 3377–3383.
52. Zhao, X.L.; Li, N.; Ma, C.B. Residential energy consumption in urban China: A decomposition analysis. *Energy Policy* **2012**, *41*, 644–653. [CrossRef]
53. Zhang, C.G.; Lin, Y. Panel estimation for urbanization, energy consumption and CO_2 emissions: A regional analysis in China. *Energy Policy* **2012**, *49*, 488–498. [CrossRef]
54. Zhao, C.S.; Niu, S.W.; Zhang, X. Effects of household energy consumption on environment and its influence factors in rural and urban areas. *Energy Procedia* **2012**, *14*, 805–811.
55. Miller, R.E.; Blair, P.D. *Input-Output Analysis: Foundations and Extensions*, 2nd ed.; Cambridge University Press: New York, NY, USA, 2009.
56. Feng, K.S.; Hubacek, K. Carbon implications of China's urbanization. *Energy Ecol. Environ.* **2006**, *1*, 39–44. [CrossRef]
57. Department of Energy Statistics, National Bureau of Statistics, People's Republic of China. *China Energy Statistical Yearbook (2008–2016)*; China Statistics Press: Beijing, China, 2015.
58. Liu, W.D.; Chen, J.; Tang, Z.P.; Liu, H.G.; Han, D.; Li, F.Y. *Theories and Practice of Constructing China's Interregional Input Output Tables between 30 Provinces in 2007*; China Statistics Press: Beijing, China, 2012.
59. Mi, Z.F.; Meng, J.; Guan, D.B.; Shan, Y.L.; Song, M.L.; Wei, Y.M.; Liu, Z.; Hubacek, K. Chinese CO_2 emission flows have reversed since the global financial crisis. *Nat. Commun.* **2017**, *8*, 1712. [CrossRef] [PubMed]
60. Beijing Municipal Bureau of Statistics. *Beijing Statistical Yearbook 2016*; China Statistics Press: Beijing, China, 2017.
61. Tianjin Municipal Bureau of Statistics. *Tianjin Statistical Yearbook 2016*; China Statistics Press: Beijing, China, 2017.
62. Shanghai Municipal Bureau of Statistics. *Shanghai Statistical Yearbook 2016*; China Statistics Press: Beijing, China, 2017.
63. Chongqing Municipal Bureau of Statistics. *Chongqing Statistical Yearbook 2016*; 2017.
64. Guo, J.; Zheng, X.Y.; Chen, Z.M. How does coal price drive up inflation? Reexamining the relationship between coal price and general price level in China. *Energy Econ.* **2016**, *57*, 265–276. [CrossRef]
65. Wiedenhofer, D.; Guan, D.; Liu, Z.; Meng, J.; Zhang, N.; Wei, Y.M. Unequal household carbon footprints in China. *Nat. Clim. Chang.* **2016**, *7*, 75–80. [CrossRef]
66. Fan, J.; Guo, X.M.; Marinova, D.; Wu, Y.R.; Zhao, D.T. Embedded carbon footprint of Chinese urban households: Structure and changes. *J. Clean. Prod.* **2012**, *33*, 50–59. [CrossRef]
67. Qu, J.S.; Maraseni, T.; Liu, L.N.; Zhang, Z.Q.; Yusaf, T. A comparsion of household carbon emission patterns of urban and rural China over the 17 years perild (1995–2011). *Energies* **2015**, *5*, 10537–10557. [CrossRef]
68. Jones, C.M.; Kammen, D.M. Quantifying carbon footprint reduction opportunities for U.S. households and communities. *Environ. Sci. Technol.* **2011**, *45*, 4088–4095. [CrossRef] [PubMed]
69. Isaksen, E.T.; Narbel, P.A. A carbon footprint proportional to expenditure- a case for Norway? *Ecol. Econ.* **2017**, *131*, 152–165. [CrossRef]
70. Weber, C.; Perrels, A. Modelling lifestyle effects on energy demand and related emissions. *Energy Policy* **2000**, *28*, 549–566. [CrossRef]
71. Tian, X.; Chang, M.; Lin, C.; Tanikawa, H. China's carbon footprint: A regional perspective on the effect of transitions in consumption and production patterns. *Appl. Energy* **2014**, *123*, 19–28. [CrossRef]
72. Fry, J.; Lenzen, M.; Jin, Y.T.; Wakiyama, T.; Baynes, T.; Wiedmann, T.; Malik, A.; Chen, G.W.; Wang, Y.F.; Geschke, A.; et al. Assessing carbon footprints of cities under limited information. *J. Clean. Prod.* **2017**, *176*, 1254–1270. [CrossRef]
73. Lin, T.; Yu, Y.; Bai, X.; Feng, L.; Wang, J. Greenhouse gas emissions accounting of urban residential consumption: A household survey based approach. *PLoS ONE* **2013**, *8*, e55642. [CrossRef] [PubMed]

74. Tian, X.; Geng, Y.; Dong, H.J.; Dong, L.; Fujita, T.; Wang, Y.T.; Zhao, H.Y.; Wu, R.; Liu, Z.; Sun, L. Regional household carbon footprint in China: A case of Liaoning province. *J. Clean. Prod.* **2016**, *114*, 401–411. [CrossRef]
75. Qu, J.; Zeng, J.; Li, Y.; Wang, Q.; Maraseni, T.; Zhang, L.H.; Zhang, Z.Q.; Clarke-Sather, A. Household carbon dioxide emissions from peasants and herdsmen in northwestern arid-alpine regions, China. *Energy Policy* **2013**, *57*, 133–140. [CrossRef]
76. Zhang, C.; Zhou, K.L.; Yang, S.L.; Shao, Z. On electricity consumption and economic growth in China. *Renew. Sustain. Energy Rev.* **2017**, *76*, 353–368. [CrossRef]
77. Bridge, B.A.; Adhikari, D.; Fontenla, M. Household-level effects of electricity on income. *Energy Econ.* **2016**, *58*, 222–228. [CrossRef]
78. Dong, H.J.; Dai, H.C.; Geng, Y.; Fujita, T.; Liu, Z.; Xie, Y.; Wu, R.; Fujii, M.; Masui, T.; Tang, L. Exploring impact of carbon tax on China's CO_2 reductions and provincial disparities. *Renew. Sustain. Energy Rev.* **2017**, *77*, 596–603. [CrossRef]
79. Feng, K.S.; Hubacek, K.; Sun, L.X.; Liu, Z. Consumption-based CO_2 accounting of China's megacities: The case of Beijing, Tianjin, Shanghai and Chongqing. *Ecol. Indic.* **2014**, *47*, 26–31. [CrossRef]
80. Su, M.; Liang, C.; Chen, B.; Yang, Z.F. Low-Carbon Development Patterns: Observations of Typical Chinese Cities. *Energies* **2012**, *5*, 1796–1803. [CrossRef]
81. Yang, S.Y.; Chen, B.; Wakeel, M.; Hayat, T.; Alasedi, A.; Ahmad, B. $PM_{2.5}$ footprint of household energy consumption. *Appl. Energy* **2017**. [CrossRef]
82. Ntanos, S.; Chalikias, M. Countries clustering with respect to carbon dioxide emissions by using the IEA database. In Proceedings of the 7th International Conference on Communication Technologies in Agriculture, Food and Envitronment (HAICTA 2015), Kavala, Greece, 17–20 September 2015.
83. Ntanos, S.; Kyriakopoulos, G.; Chalikias, M.; Arabatzis, G.; Skordoulis, M.; Galatsidas, S.; Drosos, D. A social assessment of the usage of renewable energy sources and its contribution to life quality: The case of an Attica urban area in Greece. *Sustainability* **2018**, *10*, 1414. [CrossRef]

energies

MDPI

Article

Influencing Factors and Decoupling Elasticity of China's Transportation Carbon Emissions

Yong Wang [1,2], Yu Zhou [1], Lin Zhu [1], Fei Zhang [1] and Yingchun Zhang [3,*]

1 School of Statistics, Dongbei University of Finance and Economics, Dalian 116025, China; ywang@dufe.edu.cn (Y.W.); zhouyu940919@163.com (Y.Z.); 18042687217@163.com (L.Z.); 15541145920@163.com (F.Z.)
2 Postdoctoral Research Station, Dongbei University of Finance and Economics, Dalian 116025, China
3 School of Economics, Qingdao University, Qingdao 266071, China
* Correspondence: zhangyc@qdu.edu.cn; Tel.: +86-138-642-66799

Received: 8 April 2018; Accepted: 2 May 2018; Published: 5 May 2018

Abstract: Transportation is an important source of carbon emissions in China. Reduction in carbon emissions in the transportation sector plays a key role in the success of China's energy conservation and emissions reduction. This paper, for the first time, analyzes the drivers of carbon emissions in China's transportation sector from 2000 to 2015 using the Generalized Divisia Index Method (GDIM). Based on this analysis, we use the improved Tapio model to estimate the decoupling elasticity between the development of China's transportation industry and carbon emissions. The results show that: (1) the added value of transportation, energy consumption and per capita carbon emissions in transportation have always been major contributors to China's carbon emissions from transportation. Energy carbon emission intensity is a key factor in reducing carbon emissions in transportation. The carbon intensity of the added value and the energy intensity have a continuous effect on carbon emissions in transportation; (2) compared with the increasing factors, the decreasing factors have a limited effect on inhibiting the increase in carbon emissions in China's transportation industry; (3) compared with the total carbon emissions decoupling state, the per capita decoupling state can more accurately reflect the relationship between transportation and carbon emissions in China. The state of decoupling between the development of the transportation industry and carbon emissions in China is relatively poor, with a worsening trend after a short period of improvement; (4) the decoupling of transportation and carbon emissions has made energy-saving elasticity more important than the per capita emissions reduction elasticity effect. Based on the conclusions of this study, this paper puts forward some policy suggestions for reducing carbon emissions in the transportation industry.

Keywords: carbon emissions; influencing factors; decoupling elasticity; Generalized Divisia Index; Tapio's model

1. Introduction

Since the beginning of this century, the concentration of greenhouse gases in the atmosphere, represented by carbon dioxide, has been steadily increasing, leading to global warming and more frequent natural disasters. Climate change has become one of the greatest challenges to mankind in the 21st century. Currently, all countries in the world are constantly seeking solutions and striving to achieve the goal of lower carbon emissions: the United Nations (UN) has held many international negotiations on climate change and formulated the "United Nations Framework Convention on Climate Change" [1] and the "Kyoto Protocol" [2]; in 2016, leaders from more than 170 countries jointly signed the Paris Agreement [3] focusing on climate change issues at the UN Headquarters. As the world's largest developing country, with its rapid economic development, China's carbon emissions remain high. China's carbon emissions account for about one-third of the world's total carbon emissions and rank first in the world in carbon emissions [4]. The research data of the Global Carbon Project [5], an international carbon emission research institution, shows that in 2015, China's carbon emissions accounted for 28.65% of the world's total carbon emissions, far exceeding the second highest in the United States (14.93%) and the third largest in the European Union (9.68%). To this end, the Chinese government has actively shouldered its responsibility and obligation to reduce emissions, and successively formulated numerous energy-saving and emission reduction policies [6–9].

The transportation industry is an indispensable key link in daily life and social production and a basic industry for economic development in China. Along with the increasing demand for transportation in China, the transportation industry, while promoting economic growth and facilitating human life, has also caused a great deal of energy consumption and carbon emissions. The energy consumption of China's transportation industry increased from 114,470,000 tons of coal equivalent in 2000 to 383,180,000 tons of coal equivalent in 2015. At the same time, China's transportation sector accounts for approximately one-quarter of the carbon emissions, second only to the third largest carbon sector in the energy and industrial sectors [10]. Moreover, with the continuous progress of urbanization in China and the increasing number of motor vehicles, carbon emissions in the transportation industry are still on the rise. At present, China's transportation industry is at a critical period of rapid development and transition. By accurately analyzing the drivers of changes in carbon emissions in the transportation sector, the relationship between transportation development and carbon emissions can be explored. This is of great and far-reaching practical significance for the early realization of low-carbon transportation.

Factor decomposition and decoupling are important parts of research on carbon emission. There are deficiencies in the methods used to decompose and decouple the carbon emission in the transportation industry, and indicators in the decoupling—causal chain need to be improved. The deficiencies of the existing research are described in detail in Section 2.3. Therefore, this paper innovates the research methods of carbon emission factor decomposing and the selection of decoupling—causal chain indicators in transportation industry, and provides a new research perspective for accurately analyzing the relationship between the development of transportation industry and carbon emissions. Specific innovations and improvements in this article are detailed in Section 2.3.

The remainder of this article is arranged as follows: Section 2 presents a literature review; Section 3 introduces the measurement of carbon emissions; Section 4 builds the index decomposition model of the historical evolution of transportation carbon emissions and the decoupling model of carbon emissions from transportation development; Section 5 analyzes the driving factors and the decoupling situation; Section 6 presents the discussion and analysis and finally Section 7 presents the conclusions and suggestions.

2. Literature Review

This paper studies the influencing factors and decoupling elasticity of China's transportation carbon emissions. The factor decomposition is used to analyze the specific driving effect of factors that have impacts on carbon emissions. The decoupling is used to explore the correlation between the development of the transportation industry and carbon emissions. Therefore, based on the purpose of this paper, the literature review contains two aspects, one is the influencing factors of carbon emissions, and the other is the decoupling of carbon emissions and economic development.

2.1. Literature Review of the Influencing Factors of Carbon Emissions

An analysis of influencing factors is an important part of the research on carbon emissions in the transportation industry. Wang et al. [11] investigated the influence mechanism of people's activity travel scheduling on transportation energy consumption and emissions on holidays in China. Hao et al. [12] comprehensively measured energy efficiency in China's transportation sector and identifies the opportunities for further energy efficiency improvements. Wei et al. [13] discussed carbon dioxide abatement for 29 provinces in China, and concluded that industry composition, energy mix, openness degree affected carbon dioxide abatement potential. Lugauer et al. [14] estimated the impact of age distribution on carbon emissions by exploiting demographic variation in a panel of 46 countries. Yang et al. [15] analyzed the allocation of carbon intensity reduction target by 2020 among industrial sectors in China. The factor decomposition method is gradually being used in research on carbon emissions addressing an impact mechanism that was originally applied primarily to examine energy consumption [16]. Factor decomposition of changes in carbon emissions includes structural decomposition and index decomposition. The structural decomposition method requires an input-output model as its basis so it's not convenient to analyze in practice. Compared with the structural decomposition method, the index decomposition method is widely used in environmental economics because it is suitable for decomposing time series data and models with fewer factors [16].

In general, the index decomposition methods mainly include the Laspeyres decomposition method, Arithmetic Mean Divisia Index (AMDI) method and Logarithmic Mean Divisia (LMDI) method. Sun [17] proposed a complete decomposition model, also known as the improved Laspeyres decomposition method, and used it to analyze the factor for the change of energy intensity and energy consumption in the world. Rüstemoğlu et al. [18] applied the refined Laspeyres index model to analyze the impact of four main factors, such as economic activity, employment, energy intensity, and carbon intensity in carbon emissions changes in Brazil and Russia. Hatzigeorgiou et al. [19] dealt with the decomposition analysis of energy-related carbon emissions in Greece using the AMDI technique, and concluded that the income effect was the most important factor contributing to the increase of carbon emissions. Although the Laspeyres decomposition method has been improved to solve the original residual problem well, the calculation process becomes very complicated when the influencing factor of the decomposition is more than 3 [20]. AMDI has residual problems and does not apply to cases in which there is a value of 0 in the data. Compared with the above two exponential decomposition methods, LMDI solves their existing problems well, and the model is simple. Therefore, it has been widely used in academia to study the influencing factors of carbon emissions in transportation and other fields. Gambhir et al. [21] used the LMDI decomposition method to determine the main factors of China's road transportation sector carbon emissions and set different scenarios to estimate changes in costs and carbon emissions. M'raihi et al. [22] investigated the effects of the main driving factors of carbon emissions changes from road freight transportation in Tunisia using decomposition analysis, mainly the LMDI, and the results showed that economic growth and average petroleum emissions were the main driving factors.

Shi et al. [23] took four Chinese megacities (Beijing, Tianjin, Shanghai, and Chongqing) as case studies, and decomposed per capita urban carbon emissions into manufacturing, transportation and construction sectors using LMDI method. Du et al. [24] used LMDI model to analyze the change of carbon emissions in China's metallurgy industry, and the empirical results showed that main factors were labor productivity, energy intensity and industry size. Ma et al. [25] built an LMDI method with a higher technical resolution and applied it to decompose the growth of energy-related carbon emissions in China. Wang et al. [26] used LMDI method based on the extended Kaya identity to explore the main driving factors for energy-related carbon emissions in Guangdong province annually. De Freitas et al. [27] conducted LMDI decomposition of carbon emissions change from energy consumption in Brazil. The results demonstrated that economic activity and demographic pressure were the leading forces explaining emission increase. On the other hand, carbon intensity reductions and diversification of energy mix towards cleaner sources were the main factors contributing to emission mitigation. Kharbach et al. [28] used LMDI to analyze the drivers for carbon emissions' increase in Moroccan road transportation sector, and found that population growth and increase in vehicles ownership were the main causes. Zhu et al. [29] applied LMDI decomposition method combined with Tapio decoupling model to study the driving factors and decoupling effects of the transportation sector's carbon emissions in the Beijing-Tianjin-Hebei area, China. Zhang et al. [30]

conducted an empirical analysis of urban traffic energy consumption in Beijing, Shanghai and Guangzhou using the LMDI method. The main factors affecting the energy consumption of Beijing's passenger traffic are the number of daily trips per capita followed by the proportion of motorized trips. The main factor in Shanghai is the urban population, followed by the average single trip distance. The main factor affecting Guangzhou is the urban population, followed by the number of daily trips per capita. Shen et al. [10] used the LMDI method to study the drivers of carbon emissions in China's transportation sector and found that the main positive drivers were urbanization, the tertiary industry's share of the total secondary and tertiary industries, and the total population. The main negative drivers are the traffic volume per unit of GDP, the energy consumption per unit of traffic volume, the total secondary and tertiary industries per urban resident population, and the contribution of tertiary industry to GDP. Zhou et al. [31] carried out LMDI decomposition of transportation carbon emissions and considered that economic growth is the most influential factor. Traffic energy conservation is the second most important factor contributing to the reduction of transportation carbon emissions, but the driving effect is not stable enough. However, the effect of traffic emissions reduction is the most limited.

With the growing academic research on index decomposition, the defects of LMDI have also gradually appeared. Vaninsky [32] noted problems that exist in index decomposition methods such as LMDI: the model cannot contain multiple relative and absolute factors simultaneously, and the result of decomposition depends on the interdependence of factors and may lead to results that conflict with economic common sense. To solve these shortcomings, Vaninsky [32] proposed a new index decomposition model, that is, Generalized Divisia Index Decomposition (GDIM). The GDIM model can simultaneously contain the impact of multiple absolute and relative variables on carbon emissions, solve the problems of other index decomposition methods, and more accurately and objectively analyze the contribution of each driving force to changes in carbon emissions. Currently, Shao Shuai et al. have used GDIM to disaggregate the historical evolution of China's manufacturing [33] and mining [34] carbon emissions and, respectively, have reached the following conclusions: (1) the main driver of the increase in manufacturing carbon emissions is the investment scale, and the key factors in the reduction of manufacturing emissions are the output carbon intensity and the carbon intensity of investment; (2) the primary factor of the promotion of the increase in the mining carbon emissions is the scale of output with the conclusion that the carbon intensity of the output contributes the most to the mining carbon emissions reduction.

2.2. Literature Review of Carbon Emissions Decoupling

Factor decomposition of the evolution of the trend of carbon emissions can determine the impacts of various factors on carbon emission changes. The decoupling analysis is helpful to explore the relationship between economic growth and carbon emissions to provide a practical solution for the realization of low-carbon development. According to the Environmental Kuznets Curve (EKC) hypothesis, economic growth generally leads to increased environmental pressures and resource consumption. However, when effective policies and new technologies are adopted, the result may be lower environmental pressures and resource consumption in return for the same or even faster economic growth; this process is called decoupling. The decoupling of carbon emissions in the transportation industry is the idealized process of continuously weakening or even causing the disappearance of the relationship between transportation development and carbon emissions, that is, realizing the growth of the transportation industry while gradually reducing energy consumption. To date, scholars have conducted a great deal of research on the decoupling of carbon emissions from the economy.

Li et al. [35] conducted a decoupling analysis of the added value of Shanxi's industrial sector and energy consumption and carbon emissions. They concluded that the province's industrial GDP, energy consumption and carbon emissions showed an expansive coupling status. Yang et al. [36] used the decoupling theory to analyze the characteristics of carbon emissions and economic growth in the western region and concluded that the major factor contributing to the increase in carbon emissions is the rapid growth of the economy, while the impacts of industrial structure, energy intensity and energy structure on carbon emissions are not the same. Wu et al. [37] discussed decoupling trends in world economic growth and carbon emissions based on decoupling theories. Lin et al. [38] calculated the decoupling trend of carbon emissions from China's heavy industry by presenting a theoretical framework for decoupling. Diakoulaki et al. [39] analyzed the decoupling relationship between industrial economic growth and carbon emissions in 14 EU countries and the efforts and achievements made by various countries. Li et al. [40] analyzed the decoupling state between the carbon emissions of the construction industry in Jiangsu Province and the province's economic growth based on the Tapio decoupling model. Román et al. [41] used the decoupling elasticity approach to analyze the importance of energy consumption changes in relation to the GDP changes in Colombia. The results showed that current decoupling-oriented measures were steps in the right direction but efforts made to achieve decoupling energy consumption from economic growth were not at all effective. Freitas et al. [42] examined the occurrence of a decoupling between economic activity and carbon emissions from energy consumption in Brazil. Grand [43] introduced different types of decoupling models in detail and used them to analyze the decoupling of economic activities and carbon emissions in Argentina. Roinioti et al. [44] analyzed the decoupling relationship between carbon emissions and economic growth in Greece with the use of the decoupling index. The results indicated that the decoupling progress achieved was intercepted during the years of intense recession. Wang et al. [45] studied China's industrial carbon emissions based on decoupling elasticity and the Tapio decoupling model. Hu et al. [46] used the non-competitive I-O model and the Tapio decoupling model to comprehensively analyze the decoupling relationship between the output of the product sector in China and its embodied carbon emissions under trade openness. Wan et al. [47] studied the decoupling relationship between carbon emissions and economic growth of the equipment manufacturing industry in China. As energy and the environment are facing increasingly serious forms, some domestic and foreign scholars have shifted their attention from decoupling the relationship between the total carbon emissions and economic growth to the transportation industry, a high-energy-consuming and high-emission industry, and they have researched decoupling economic growth from carbon emissions in transportation.

Tapio [48] studied the relationships among transportation volume, greenhouse gases and transportation economic growth in Europe from 1970 to 2001 by constructing a decoupling model. Gray et al. [49] studied the decoupling of traffic, carbon emissions and economic growth in Scotland. Loo et al. [50] explored the potential and the reality of decoupling transportation from economic growth in 15 major countries such as China, Russia and Canada. Zhao et al. [51] analyzed the relationship between transportation growth and carbon emissions associating the decomposition technique with the decoupling elasticity in Guangdong province, China. Using the Tapio elasticity analysis method, Liu et al. [52] explored the decoupling between transportation development and economic growth. The results show that there is a clear interaction between the economy and traffic volume, and they are mutually dependent. Zhou et al. [31] constructed causal chains by introducing intermediate variables and analyzed the decoupling relationship between economic growth and carbon emissions based on Tapio's decoupling model from the perspectives of the decoupling elasticity of traffic emissions reduction, decoupling elasticity of traffic energy-savings and decoupling elasticity of traffic development. Table 1 summarizes representative studies on the influencing factors of carbon emissions and the decoupling from economic development.

Table 1. Representative studies on the influencing factors of carbon emissions and the decoupling from economic development.

Research Topics	Study Authors	Research Content	Research Methods	Research Process	Insufficient Research
Carbon emission factor analysis	Timilsina et al. [53]	Factors affecting carbon emissions in the transportation sector in some countries	LMDI decomposition method	Studying the main influencing factors of carbon emissions in Latin America and the Caribbean countries, such as population growth, energy intensity, and energy structure and so on.	Taking some developing countries as their research subjects cannot completely explain the impact of urbanization and other factors on the transportation industry and has little reference to other countries.
	Wu et al. [54]	Evolution of carbon emissions in China and its driving factors	Three complete decomposition methods	A study is conducted on China's energy-related changes in carbon emissions from 1996–1999 and their influencing factors. The results show that energy intensity, the number of motor vehicles and the average distance traveled are the driving forces behind the growth of carbon emissions in transportation.	Consideration of the impact factors of transportation carbon emissions is relatively shallow, does not explain the impact of population growth and economic growth, and cannot truly reflect the drivers of transportation carbon emissions.
	M'raihi et al.	The driving factors of carbon emissions changes from road freight transportation in Tunisia	LMDI decomposition method	This paper decomposes the carbon emissions changes from road freight transportation in Tunisia into five kinds of factors from 1990 to 2006. The results show that economic growth and average petroleum emissions are the main driving factors.	Due to the inherent flaw in the LMDI decomposition method used, it is limited in the selection of factors and only includes the absolute GDP, which leads to some important factors not being considered.
The decoupling relationship between carbon emissions and economic development	Guo Shiyi [55]	The decoupling of carbon emissions from industrial economic growth	IGT decoupling index model	Based on the IGT decoupling index, the paper studies the relationship between industrial economic growth and carbon emissions in China from 2000 to 2014. The results show that China's industry is in a period of transition from "relative decoupling" to "absolute decoupling".	Although the unique advantage of the IGT model is the ability to quantitatively analyze the relationship between economic growth and environmental loading, the Tapio decoupling model provides a more detailed breakdown of decoupling types and is more appropriate for the current context of decoupling carbon emissions from industrial economies.

Table 1. *Cont.*

Research Topics	Study Authors	Research Content	Research Methods	Research Process	Insufficient Research
	Tang et al. [56]	Decoupling indicators of carbon emissions from the tourism industry in China	Tapio decoupling model	This paper explores the decoupling effects between tourism-related carbon emissions and the tourism economy from 1990 to 2012. The results show that the decoupling state is very complicated and alternates between negative decoupling and weak decoupling.	The decoupling elasticity is not further decomposed, leading to a shallow analysis of the decoupling situation, and the study fails to analyze in more detail the specific reasons for the decoupling of tourism carbon emissions.
	Zhou Yinxiang [57]	The decoupling and coupling relationship between traffic carbon emissions and industrial economic growth	Tapio decoupling model	This paper measures the decoupling relationship between China's transportation carbon emissions and industrial economic growth from 1990 to 2013 and the trends of their evolution, and it constructs the causal chain to further analyze the influencing factors and mechanism of action. The results show that the decoupling status is not ideal and that the decoupling depends mainly on the traffic energy-saving elasticity.	In the decoupling model, the total amount of carbon emissions in the transportation industry is taken into account without taking into account the differences among individuals, making the analysis not meticulous.

2.3. The Deficiencies of the Existing Studies and the Innovations of this Article

Currently, scholars have laid a solid foundation for future exploration with their studies of the factor decomposition of carbon emissions in transportation and economic development and the decoupling elasticity. However, taken together, the existing research has the following deficiencies:

(1) Decomposition methods of influencing factors of transportation carbon emissions are flawed. Most of the studies focused on index decomposition of the influencing factors of carbon emissions in the transportation sector use the Logarithmic Mean Divisia Index (LMDI) method. The method itself has some defects that cause inaccuracies in the decomposition result. Vaninsky [32] noted that there are two problems with the existing index decomposition methods, including LMDI. First, the methods decompose the carbon emissions into the product of various factors according to the Kaya identity. For example, carbon emissions are usually decomposed into the product of energy carbon emission intensity, energy intensity, GDP per capita and population, while at most one absolute factor (such as population) can be considered in the decomposition. When other absolute factors change, such as when energy consumption increases, while other factors remain unchanged, the model does not show an increase in carbon emissions; it simply reduces the energy carbon intensity and increases the energy intensity. This is contradictory. Second, due to the interdependence of various factors, when different factor decomposition models are used, the methods may lead to different decomposition results.

(2) The choice of influencing factors of transportation carbon emissions is not comprehensive enough. In analyzing the changes that affect the carbon emissions in the transportation sector, the limitations inherent in the selected decomposition method make the factors of the study less comprehensive. Some studies have expanded the basic decomposition model in order to analyze more selected factors. Although this can increase the research factors, the selection of factors is still subject to the Kaya identity; this cannot arbitrarily select the relative factor and the absolute factor, and at most, it can only study the impact of an absolute factor [32]. This leads to typical factors, such as the energy consumption and the value added of the sector, not being taken into account, leaving the final conclusion insufficiently comprehensive.

(3) The causal chain model of decoupling in transportation is insufficient. Existing research mainly analyzes the decoupling situation of carbon emissions from economic growth in the transportation industry by constructing a causal chain. The Tapio model divides the decoupling into eight types according to the magnitude of the decoupling elasticity. This fine division means that the calculation results of the decoupling elasticity will be classified into different categories with little accuracy. In addition, although the existing research can analyze the reasons for the decoupling of carbon emissions from the energy-saving elasticity and emissions-reducing elasticity of transportation, the total amount of carbon emissions in the model is too macroscopic and does not take into account the differences among individuals. It also makes the calculation results of the decoupling elasticity not accurate enough and cannot respond sensitively to changes in the decoupling state, which may lead to deviation in the final analysis.

Based on the status of existing research, this paper first uses GDIM to factor the changes in carbon emissions in China's transportation industry, then analyzes the situation of decoupling transportation development from carbon emissions. The improvements and innovations in this article are as follows:

(1) In terms of research methods, this paper is the first to analyze the driving factors of changes in carbon emissions in China's transportation industry using the Generalized Divisia Index Method (GDIM) proposed by Vaninsky [32]. Based on the Kaya identity, GDIM constructs a decomposition model that contains multiple absolute and relative variables by deforming Kaya identities. There are three advantages to analyzing the driving factors of carbon emissions changes in the transportation industry based on GDIM: ① GDIM avoids the inherent flaws of other index decomposition methods and breaks through limitations when selecting factors. Furthermore,

GDIM expands the analysis scope of Kaya's identity, breaks the formal interdependence of various influencing factors, can reveal the impact of more than one absolute factor of change on carbon emissions in particular, and fully considers the factors that have an impact on carbon emissions that are implicit in the decomposition process; ② GDIM overcomes the contradictions in the existing index decomposition methods, doesn't produce results that are inconsistent with economic common sense and breaks the current situation of interdependence when selecting factors. Moreover, GDIM avoids the factor decomposition results depending on the selection of influencing factors, and makes it impossible to produce paradoxical conclusions when using different factor decomposition forms; ③ Using GDIM can examine the implicit environmental impact factors, and its decomposition result can distinguish the correlation of all the factors without any double counting problem. Therefore, this paper examines a full range of eight key factors such as carbon intensity of added value, energy intensity and per capita carbon emissions in the transportation industry. In particular, it examines three absolute factors that the current literature fails to pay attention to at the same time, but cannot be ignored: population size, energy consumption, and added value of transportation industry. This paper uses GDIM to accurately and comprehensively quantify the actual contribution of the different factors in transportation carbon emissions, and draws more reasonable decomposition results.

In order to more intuitively demonstrate the reasons for selecting the GDIM model in this paper and the advantages of the GDIM model, the GDIM model is compared with other typical factor decomposition methods, as shown in Table 2.

Table 2. Comparison of typical factor decomposition methods.

Methods	Features
Structural decomposition method	It is based on an input-output model, which has higher requirements on data, and the decomposition result can only be an additive form. So it's not convenient to analyze in practice [20].
Laspeyres decomposition method	When the number of influencing factors exceeds 3, the calculation process becomes complicated.
AMDI method	The model contains residuals; it cannot be used when there are zero values in the data.
LMDI method	The result of decomposition depends on the interdependence of factors and the choice of factors is limited.
GDIM	It breaks the formal interdependence of various influencing factors and makes the results more comprehensive and accurate.

(2) In the selection of indicators, this paper adopts per capita carbon emissions in the transportation industry in the causal chain of decoupling elasticity for the first time. This paper doesn't use the indicator of total carbon emissions in the decoupling causal chain model of the transportation industry, but instead innovatively decomposes per capita decoupling elasticity into per capita emissions reduction elasticity and energy-saving elasticity. This can further explore the key reasons for the decoupling of the development of the transportation industry from carbon emissions, fully take into account the differences among individuals, and reflect the individual's role and value. In addition, it can more truly reflect the actual situation of carbon emissions in China's transportation industry and the evolution trend of decoupling state over time than the previous decoupling causal chain model, avoid the misjudgment of the decoupling situation caused by the mistake of the calculation result, and make up for the deficiency of the existing decoupling model.

3. Estimation of Carbon Emissions in China's Transportation Industry

Currently, China's official statistical agencies have not released data on carbon emissions in the transportation sector and need to calculate them. According to China's current statistical standards and the "Industrial classification for national economic activities of 2017" [58] published by the National Bureau of Statistics, the transportation industry in this paper includes transportation, storage and postal services. Among them, transportation includes railway transportation, road transportation, water transportation, air transportation, pipeline transportation, multimodal transportation and transportation agency industry. It should be pointed out that some people still have misunderstandings about the scope of the transportation industry and believe that the production of transportation goods, such as vehicles, fuels and infrastructure, also belong to the transportation industry. However, in the "Industrial classification for national economic activities of 2017", production of transportation goods belongs to the manufacturing industry, and the use of vehicles belongs to the transportation industry. This article uses the official classification, and the definition of the transportation industry in this paper is also widely used in other studies, such as Zhao et al. [51], Wang et al. [59], and Hao et al. [12].

The carbon emissions from human activities mainly come from the consumption of fossil fuels, while most of the carbon emissions data are calculated indirectly through energy consumption. The fossil fuel energy consumed by the transportation industry mainly includes raw coal, coke, gasoline, diesel oil, and natural gas. Based on fossil energy consumption data, this paper used the calculation method of carbon emissions from energy consumption described in the IPCC GHG Inventories Guide, and combined various coefficients published by the Chinese government to measure carbon emissions from the transportation industry. The carbon emission calculation method used in this paper is currently widely used by academics, such as Wei et al. [13], Freitas et al. [42], Zhou et al. [31]. The specific formula is (1):

$$CO_2 = \sum_i E_i \times CV_i \times CCF_i \times (1 - CS_i) \times O_i \times (44/12) \tag{1}$$

In Equation (1): $i = 1, 2, \cdots, 8$ indicates the type of energy. To ensure the accuracy of the estimation results, this paper fully considers eight kinds of fossil fuels, including raw coal, coke, crude oil, fuel oil, gasoline, kerosene, diesel oil and natural gas; CO_2 is the total carbon emissions from transportation energy consumption in units of 10^4 tons; E_i is the consumption of fossil fuels in units of 10^4 tons or one hundred million cubic meters; CV_i is the average low calorific value in units of kJ/kg or kJ/m^3; CCF_i is the unit of carbon content of calorific value in tons/TJ; CS_i is the carbon fixation rate; O_i is the rate of carbon oxidation; 44 is the molecular weight of CO_2; and 12 is the atomic weight of C.

The carbon emissions of China's transportation industry are calculated according to Equation (1). The values of the coefficients are derived from the literature [31] and reference "China Energy Statistical Yearbook" [60] (2000–2015), "A Study on City Greenhouse Gas Emissions Inventory" [61] and "2006 IPCC Guidelines for National Greenhouse Gas Inventories" [62]. The sequence length of each fossil energy consumption in transportation is 2000–2015, and the data are from "China Energy Statistical Yearbook". Considering that the electricity does not directly produce CO_2, in order to avoid double counting, the electricity consumed by the transportation industry has not been included in our calculation.

4. Methods

4.1. Decomposition Model of Influencing Factors of Carbon Emissions in the Transportation Industry—GDIM

We have considered the defects of the decomposition methods such as LMDI. Thus, in order to comprehensively investigate the driving factors of changes in carbon emissions, this paper uses the GDIM proposed by Vaninsky [32].

(1) Creating an expression of carbon emissions and related factors. The index Z is a function of the influencing factors $X_1, X_2, \ldots, X_n$, and X_i varies with time t: $X_i = X_i(t)$; they are related to each other by Equation (2):

$$\begin{aligned} Z &= f(X) = f(X_1, \ldots, X_n), \\ \Phi_j(X_1, \ldots, X_n) &= 0, j = 1, \ldots, k \end{aligned} \tag{2}$$

In this paper, let $Z = CO_2$, $X_1 = GDP_t$, $X_2 = (CO_2/GDP_t)$, $X_3 = E$, $X_4 = (CO_2/E)$, $X_5 = P$, $X_6 = (CO_2/P)$, $X_7 = (GDP_t/P)$, and $X_8 = (E/GDP_t)$, where CO_2 is the transportation carbon emissions (10^4 tons), GDP_t is the transportation added value (100 million yuan), CO_2/GDP_t is the carbon intensity of added value (tons/10^4 yuan), E is the energy consumption (10^4 tons of coal equivalent), CO_2/E is the energy carbon emission intensity (ton/ton of coal equivalent), P is the population size (10^4 persons), CO_2/P is the per capita carbon emissions in the transportation industry (ton/person), GDP_t/P is the per capita added value of transportation (10^4 yuan/person), and E/GDP_t is the energy intensity (tons of coal equivalent/10^4 yuan). The transportation carbon emissions and related factors are expressed by Equations (3)–(5):

$$Z = X_1X_2 = X_3X_4 = X_5X_6 \tag{3}$$

$$X_7 = X_6/X_2 \tag{4}$$

$$X_8 = X_2/X_4 \tag{5}$$

Further, Equations (3)–(5) are transformed into the form of the system of Equations (2); these are Equations (6)–(10):

$$Z = X_1X_2 \tag{6}$$

$$X_1X_2 - X_3X_4 = 0 \tag{7}$$

$$X_1X_2 - X_5X_6 = 0 \tag{8}$$

$$X_1 - X_5X_7 = 0 \tag{9}$$

$$X_3 - X_1X_8 = 0 \tag{10}$$

(2) Constructing the Jacobian matrix of influential factors. For factor X, let $Z(X)$ denote its contribution to the change in carbon emissions. Constructing a Jacobian matrix Φ_X composed of various influencing factors from Equations (6)–(10), where $(\Phi_X)_{ij} = \frac{\partial \Phi_j}{\partial X_i}$, $i = 1,2, \ldots, 8$; $j = 1,2, \ldots, 4$:

$$\Phi_X = \begin{pmatrix} X_2 & X_1 & -X_4 & -X_3 & 0 & 0 & 0 & 0 \\ X_2 & X_1 & 0 & 0 & -X_6 & -X_5 & 0 & 0 \\ 1 & 0 & 0 & 0 & -X_7 & 0 & -X_5 & 0 \\ -X_8 & 0 & 1 & 0 & 0 & 0 & 0 & -X_1 \end{pmatrix}^T \tag{11}$$

(3) Discovering the changes in carbon emissions and the contribution of various influencing factors. According to the principle of GDIM, the variation in carbon emissions ΔZ is decomposed into the following sum of contributions of each factor of Equation (12):

$$\Delta Z[X|\Phi] = \int_L \nabla Z^T (I - \Phi_X \Phi_X^+) dX \tag{12}$$

In Equation (12), L refers to the time span; $\nabla Z = (f_1', \ldots, f_n')^T$, $f_i' = \frac{\partial Z}{\partial X_i}$, i.e., $\nabla Z = (X_2, X_1, 0, 0, 0, 0, 0, 0)$; I is the unit matrix; $dX = diag(X_1', X_2', \ldots, X_n')dt$; "+" refers to the generalized inverse matrix; $\Phi_X^+ = \left(\Phi_X^T \Phi_X\right)^{-1} \Phi_X^T$ if the columns in the Jacobian matrix Φ_X are linearly independent, and the coordinates of the row vector $\Delta Z[X|\Phi]$ are the decomposition ΔZ_{X_i} of each factor.

From Equation (12), the variation ΔZ of carbon emissions is divided into the sum of eight effects: ΔZ_{X_1}, ΔZ_{X_2}, ΔZ_{X_3}, ΔZ_{X_4}, ΔZ_{X_5}, ΔZ_{X_6}, ΔZ_{X_7}, and ΔZ_{X_8}. Among them, the three absolute factors ΔZ_{X_1},

ΔZ_{X_3} and ΔZ_{X_5} represent the impact of changes in the added value of transportation, changes in energy consumption and changes in the population size on carbon emissions in the transportation sector, respectively. In terms of relative factors, ΔZ_{X_2} represents the impact of low-carbon changes in transportation development on changes in carbon emissions; ΔZ_{X_4} represents the impact of energy structure adjustment on changes in carbon emissions; and ΔZ_{X_6} reflects changes in carbon emissions caused by low-carbon changes in people's travel; ΔZ_{X_7} reflects the impact of changes in per capita added value of transportation on changes in carbon emissions; and ΔZ_{X_8} shows the impact of changes in energy efficiency on changes in carbon emissions.

The paper uses the data of China's transportation industry from 2000 to 2015. The added value of transportation and population size come from the "China Statistical Yearbook" [63] (2000–2015), and the energy consumption comes from the "China Energy Statistical Yearbook" [60] (2000–2015). To maintain the comparability of data, the added value of the transportation sector is adjusted at constant 2000 prices to eliminate the impact of price fluctuations.

4.2. The Decoupling Model of Carbon Emissions from Transportation Development—Tapio

"Decoupling" is the phenomenon of a breakdown in the coupling between economic development and environmental pressure [64]. If economic development did not lead to increased environmental pressure and even less environmental pressure, then "decoupling" could be found. Based on the Organization for Economic Co-operation and Development (OECD) decoupling model, Tapio introduced the concept of "decoupling elasticity" to construct the decoupling index, which solves the difficulty of the OECD decoupling model in the selection of the base period. It comprehensively and carefully divides the decoupling state and is the most widely used decoupling relationship research method.

According to Tapio's theory of decoupling, this paper constructs a decoupling model between the economic growth of transportation and carbon emissions, and in order to further explore the reasons for the decoupling between the two, a causal chain is constructed for the decoupling model:

$$\frac{\%\Delta CO_2}{\%\Delta GDP_t} = \frac{\%\Delta CO_2}{\%\Delta E} \times \frac{\%\Delta E}{\%\Delta GDP_t} \quad (13)$$

In Equation (13), $\%\Delta CO_2$, $\%\Delta GDP_t$ and $\%\Delta E$ represent the percentages of changes in carbon emissions, value added and energy consumption in transportation, respectively; $\frac{\%\Delta CO_2}{\%\Delta GDP_t}$ represents the decoupling elasticity between economic growth and carbon emissions in transportation; $\frac{\%\Delta CO_2}{\%\Delta E}$ represents the elasticity of emission reduction between transportation energy consumption and carbon emissions; and $\frac{\%E}{\%\Delta GDP_t}$ represents the energy-saving elasticity between transportation economic growth and energy consumption.

Most of the studies on the decoupling of transportation industry refer to Equation (13). To overcome the shortcomings of the model, the total carbon emissions in Equation (13) are replaced by per capita carbon emissions in the transportation industry:

$$\frac{\%\Delta(CO_2/P)}{\%\Delta GDP_t} = \frac{\%\Delta(CO_2/P)}{\%\Delta E} \times \frac{\%\Delta E}{\%\Delta GDP_t} \quad (14)$$

In Equation (14), $\%\Delta(CO_2/P)$ represents the percentage change in carbon emissions per capita in the transportation industry; $\frac{\%\Delta(CO_2/P)}{\%\Delta GDP_t}$ represents per capita decoupling elasticity between transportation economic growth and carbon emissions; and $\frac{\%\Delta(CO_2/P)}{\%\Delta E}$ represents per capita emissions reduction elasticity between transportation energy consumption and carbon emissions.

According to the Tapio decoupling elastic value of the division, the decoupling situation of development of transportation from carbon emissions is subdivided into eight kinds of states [48]. This division standard is also widely used in other studies at present, such as Román et al. [41], Zhao et al. [51] and Jiang et al. [65]. The specific divisions and their meanings are shown in Table 3.

Table 3. Division of the decoupling of transportation development from carbon emissions.

Decoupling State	$\%\Delta(CO_2/P)$	$\%\Delta GDP_t$	Per Capita Decoupling Elasticity	Meaning
Strong decoupling	-	+	$(-\infty,0)$	Economic growth, decreasing carbon emissions
Weak decoupling	+	+	[0,0.8)	Economic growth, increased carbon emissions, carbon emissions increased less than the rate of economic growth
Recessive decoupling	-	-	$(1.2, +\infty)$	Economic recession, carbon emissions decreasing, but the decrease in carbon emissions is greater than the economic recession
Strong negative decoupling	+	-	$(-\infty,0)$	Economic recession, carbon emissions increase
Weak negative decoupling	-	-	[0,0.8)	Economic recession, carbon emissions decrease, but the decrease in carbon emissions is less than the magnitude of economic recession
Expansive negative decoupling	+	+	$(1.2, +\infty)$	Economic growth, increased carbon emissions, carbon emissions increased more than economic growth
Expansive coupling	+	+	[0.8,1.2]	Economic growth, increased carbon emissions, the increase in carbon emissions is greater than or equal to the economic growth rate
Recessive coupling	-	-	[0.8,1.2]	Economic recession, carbon emissions decrease, carbon emissions are less than or equal to the magnitude of economic recession

4.3. Model Summary

The research methods and models of the influencing factors and the decoupling of carbon emissions from transportation development are summarized in Figure 1.

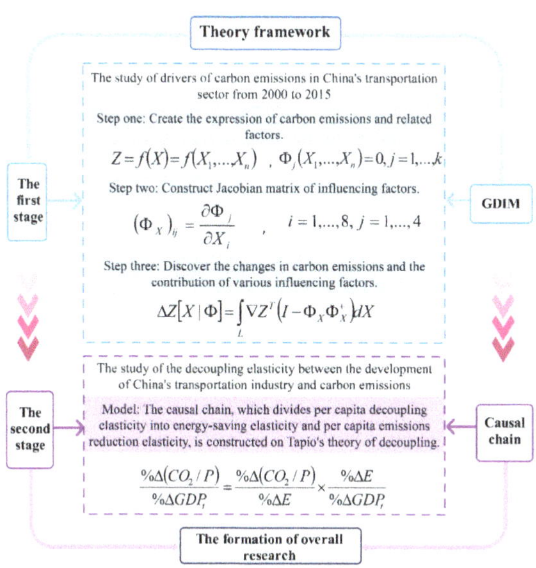

Figure 1. The summary of research methods and models.

5. Analysis of Results

5.1. Analysis of the Factors Affecting Carbon Emissions in the Transportation Industry

5.1.1. Concrete Analysis of the Factors Affecting Carbon Emissions in the Transportation Industry

The time range of this study is from 2000 to 2015. To facilitate the analysis of the results, it is divided into three sub-stages: 2000–2005, 2005–2010 and 2010–2015. Using the R software (R i386 3.4.1) [32] to decompose the driving force of carbon emissions in transportation by the generalized Divisia index method, the result of factorization can be calculated according to Equations (8) and (12) as shown in Figure 2.

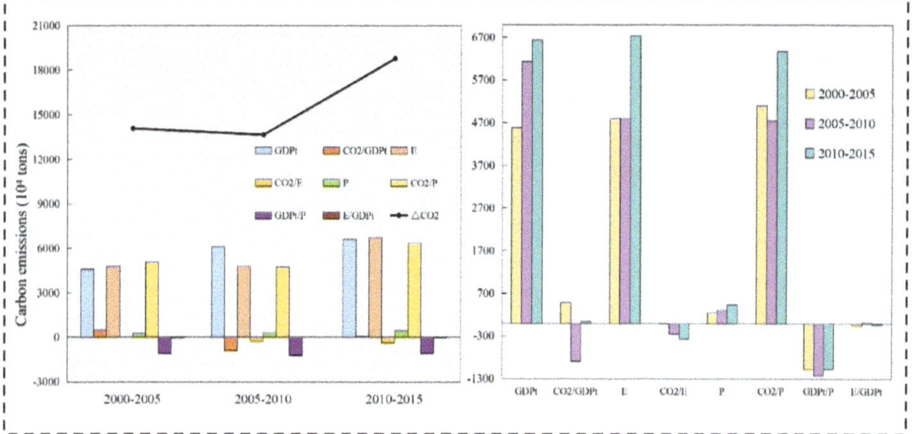

Figure 2. Staged factor decomposition results of the evolution of transportation carbon emissions.

As seen in Figure 2, among the eight factors affecting transportation carbon emissions, the added value of transportation, energy consumption, population size and per capita carbon emissions in the transportation industry have all shown an increasing effect on carbon emissions. However, the energy carbon emission intensity and per capita added value in the transportation industry show a declining effect on carbon emissions. The increasing and decreasing effects of carbon intensity on the added value and energy intensity have emerged:

(1) The driving factors for the growth of carbon emissions in the transportation industry are strong. Among the various factors that promote growth in carbon emissions, the growth-enhancing effect of the added value of transportation industry is constantly increasing. In 2000–2005, 2005–2010 and 2010–2015, the added value of transportation contributed to 45,883,200 tons, 61,226,800 tons and 66,330,900 tons of carbon emissions, respectively. This is mainly due to the rapid growth of China's economy, which has led to the ever-increasing length of its transportation routes. The average annual growth rate of regular-service airline routes in the transportation routes has even reached 8.79%. At the same time, the transportation equipment is also growing, resulting in a corresponding increase in carbon emissions.

The effect of energy consumption on increased transportation carbon emissions also shows an increasing trend, reflecting the continuous growth of China's high-energy transportation modes and the high share of all transportation modes. For instance, the proportion of passenger traffic of civil aviation increased from 0.45% in 2000 to 2.24% in 2015, and the proportion of passenger traffic on railways increased from 7.11% in 2000 to 13.04% in 2015. Although the proportion of passenger traffic on the road has decreased, it has still retained approximately 80% of the share in recent years.

At the same time, with the improvement in people's living standards and quality of life, people's demand for transportation is increasing. The number of motor vehicles such as private cars is rapidly increasing, and the courier and take-out industries are developing rapidly, which greatly increases the consumption of transportation energy so that the corresponding amount of carbon emissions continues to increase.

The evolutionary trend of the effect of population size on the promotion of carbon emissions in the transportation industry is identical to the above two factors, from 2.4918 million tons in 2000–2005 to 4.3635 million tons in 2010–2015. The reason for this is that the absolute number of the Chinese population is constantly increasing and is accompanied by an increase in the rate of urbanization. This has increased the rigid demand for transportation and has led to a continuous increase in carbon emissions in transportation. Quantitatively speaking, the effect of population size on the increase in carbon emissions is relatively small. This is due to China's implementation of the family planning policy, which limits the natural increase in the population.

The per capita carbon emissions in the transportation industry show greater volatility in promoting the growth of transportation carbon emissions, decreasing from 50.9032 million tons in 2000–2005 to 47.4734 million tons in 2005–2010 and then increasing to 63.7025 million tons in 2010–2015. This is because under the impact of the 2008 international financial crisis, the economic downturn caused a decrease in people's willingness to spend, leading to a drop in consumption-based travel, such as tourism, in approximately 2008 and resulting in a decrease in the increasing effect of per capita carbon emissions.

(2) The effect of each reduction factor on the suppression of carbon emissions growth is weaker. Among all the factors contributing to the decrease, the declining effect of energy carbon emission intensity has been steadily increasing from 80,200 tons in 2000–2005 to 2,611,700 tons in 2005–2010, then increasing to 3,760,800 tons in 2010–2015. This shows that the optimization of the energy structure in China's transportation industry shows obvious results. Among them, the energy structure showed a significant low-carbon adjustment in 2005–2010 thanks to China's goal of optimizing energy structure proposed during The Eleventh Five-Year Plan period. The proportions of coal and petroleum dropped by 3.0 and 0.5 percentage points, respectively; natural gas and other renewable energy sources increased by 2.5 and 0.3 percentage points, respectively. In recent years, China's high-speed rail construction has developed rapidly, and the electrification rate has thus been upgraded. As a result, this objectively optimizes the energy structure of the transportation industry and promotes a decrease in carbon emissions.

The per capita added value of transportation has a more stable effect on reducing carbon emissions. In 2000–2005, 2005–2010 and 2010–2015, the per capita added value of the transportation industry led to decreases of 10,886,100 tons, 12,160,400 tons and 10,700,800 tons of carbon emissions, respectively. It seems counterintuitive that the per capita added value of the transportation industry has a negative effect on carbon emissions, and the absolute value is small relative to other factors. Vaninsky [32] stated that per capita added value in the transportation sector is a relative quantity factor and includes two indicators that have an impact on carbon emissions: the added value of transportation and population size. Changes in these indicators affect their carbonization, and they are also energy-related. As the per capita added value of the transportation industry is correlated with some indicators, its change affects all through Equations (6)–(10). In this way, changes in per capita added value in the transportation industry are allocated to all of these indicators. Only part of its own change is due to changes in per capita added value and is calculated by Equation (12) in the impact on changes in carbon emissions. The remaining part is included in the impact of other indicators and accordingly adjusts the response level of the resulting indicator, Z. Therefore, even if the per capita added value of the transportation industry increases the carbon emissions, if the value is not large enough, it may show a negative value. On the other hand, with the rapid economic growth in China, the state gradually extends the welfare of the people from the most basic medical care to transportation and other fields. For instance, the government subsidizes transportation to impoverished laborers working across provinces. The negative impact of per capita added value of transportation on carbon emissions shows that the motivation of people's welfare lags behind the development of the national transportation economy [32].

The carbon intensity of added value contributed to the increases of 4,879,900 tons and 460,000 tons of carbon emissions in 2000–2005 and 2010–2015, respectively, while the effect of declining in 2005–2010 is very obvious, with 8,956,500 tons. This is because for the first time in China's "The Eleventh Five-Year Plan", energy conservation and emissions reduction were binding targets, and an energy-savings and emissions reduction indicator system, a testing system, an assessment system and a target responsibility system were established to make the transportation industry's energy conservation and emissions reduction efforts continue to strengthen and effectively enhance the carbon productivity of the transportation industry and raise the low-carbon level of its development. In the period of 2010–2015, due to a lack of overall planning and promotion of standards and policies related to transportation, the carbon intensity of added value shows a weak increasing effect.

In the period of 2000–2005, the effect of the decrease in energy intensity was 467,400 tons. In 2005–2010, it increased the carbon emissions by 186,900 tons. Finally, carbon emissions were restrained by 229,600 tons in 2010–2015. This shows that the energy efficiency of China's transportation industry has improved in recent years. This can be attributed to the fact that China attached great importance to energy development during The Twelfth Five-Year Plan period and set a target of 16% reduction in energy intensity per unit of GDP by 2015 to guide the transportation industry to continuously improve energy efficiency, avoid the unnecessary waste of energy, and extend the duration of energy use under a given supply.

5.1.2. Cumulative Contribution Analysis of Factors Affecting Carbon Emissions in the Transportation Industry

To more clearly and comprehensively reflect the dynamic impacts of the above eight factors on the changes in carbon emissions from 2000 to 2015, the contribution of each factor to carbon emissions was accumulated year by year. Based on 2000, the cumulative effect of each factor was calculated, as shown in Figure 3.

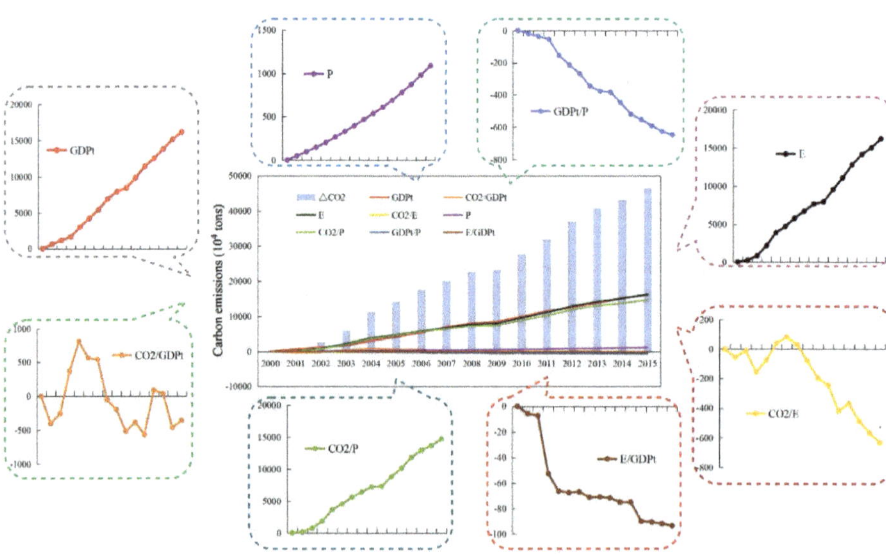

Figure 3. Cumulative contribution of drivers of changes in carbon emissions in transportation.

(1) The cumulative growth of transportation carbon emissions is larger. Figure 3 shows that during the period of 2000–2015, the cumulative carbon emissions from transportation increased by 464,478,100 tons, and the cumulative growth after 2000 was both positive and increasing. This is because after China acceded to the World Trade Organization in 2001, the transportation industry enjoyed many opportunities for development, resulting in the expansion of the transportation market. The industry, including warehousing, concentrated transportation and other industries related to transportation, was open to the outside world, and international trade was frequent. At the same time, the process of urbanization in China entered a phase of an all-round promotion since 2002. The population and area of cities constantly expanded, and the demand for transportation rapidly increased. Together, these factors contributed to the rapid development of the transportation sector and consumed a large amount of energy, resulting in a substantial increase in the accumulated carbon emissions of the transportation industry.

(2) The cumulative contribution of each factor to carbon emissions is different in size and trend. Figure 3 shows that the added value of transportation and energy consumption are the primary factors driving the increase in carbon emissions. Carbon emissions from transportation added value increased from 6,137,100 tons in 2001 to 161,496,400 tons in 2015, an average annual growth rate of 26.31%. Total energy consumption increased by 162,401,800 tons of carbon emissions in 2000–2015. Per capita carbon emissions in the transportation sector are also an important factor in promoting the growth of carbon emissions. However, the growth-boosting effect of per capita carbon emissions was surpassed by the added value of the transportation industry in 2007 and maintained its rapid growth at an average annual rate of 39.78%. The effect of population size on carbon emissions growth was relatively weak. By 2015, its cumulative result was 10,892,500 tons of carbon emissions. Per capita added value of transportation and energy carbon emission intensity are the main factors of carbon emissions reduction. Among them, the effect of energy carbon emission intensity gradually emerged after 2008, and the effect of its reduction increased rapidly from 2008 to 2015 at an average annual rate of 34.81%. The carbon intensity of added value has generally reduced carbon emissions in 2000–2015, reducing a total of 3.5895 million tons of carbon emissions, but its volatility was greater. Energy intensity had a small effect on reducing carbon emissions, and its growth rate was relatively slow; it reduced 934,300 tons of carbon emissions cumulatively by 2015. From the above results, we can see that the added value of transportation and energy consumption still play a significant role in the growth of carbon emissions. The adjustment of energy structure and the improvement in energy efficiency in the transportation sector, which are highly valued by China, have achieved initial success in their contribution to carbon emissions reduction but are still far from the expected targets and still have much room for improvement. Since economic development is the driving force of national rejuvenation and the guarantee of people's livelihood, the countermeasures to reduce carbon emissions by sacrificing the economic growth rate are not in keeping with the fact that China is still in a 'developing country's' situation; however, this is not conducive to achieving energy conservation, emissions reduction and sustainable development in the transportation industry. Therefore, based on the above results, in the future, China's transportation industry should focus on improving energy efficiency, increase the proportion of clean energy in the energy mix and actively implement a carbon reduction policy that focuses on low-carbon and energy-saving development.

Summary: The eight factors studied have different effects on carbon emissions, and the driving effect of the increasing factors are more obvious. The decreasing factors still have a lot of room for improvement in curbing carbon emissions in China's transportation industry.

5.2. Analysis of the Decoupling Elasticity between Transportation Development and Carbon Emissions

By decomposing the evolving trend of carbon emissions in the transportation sector, the present study investigated the actual contribution of each factor to the change in carbon emissions in the transportation industry. Decoupling can help to explore the interaction between carbon emissions and economic development. In the following section, we further analyze the decoupling situation between transportation development and carbon emissions based on factor decomposition.

Based on Equation (14), we calculated the per capita decoupling elasticity, per capita emissions reduction elasticity and energy-saving elasticity of transportation development and carbon emissions from 2001 to 2015. To more clearly reflect the size of each decoupling indicator and the relationships among them, we created a trend graph, as shown in Figure 4.

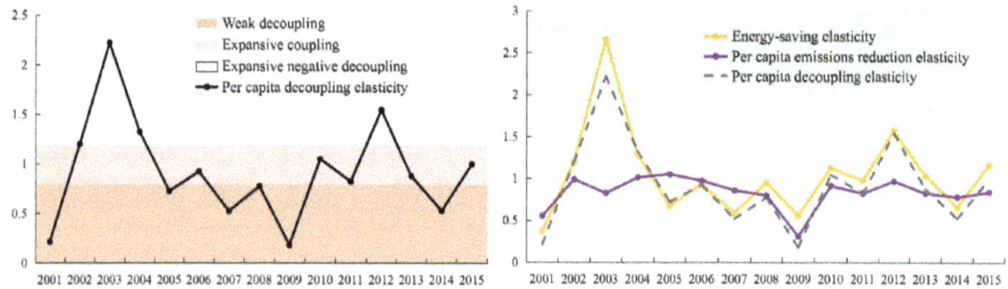

Years	2001	2002	2003	2004	2005	2006	2007	2008	2009	2010	2011	2012	2013	2014	2015
Energy-saving elasticity	0.38	1.21	2.67	1.30	0.68	0.94	0.60	0.96	0.57	1.14	0.99	1.59	1.05	0.67	1.18
Per capita emissions reduction elasticity	0.56	0.99	0.83	1.02	1.06	0.98	0.86	0.81	0.32	0.92	0.83	0.98	0.83	0.79	0.85
Per capita decoupling elasticity	0.21	1.20	2.22	1.32	0.72	0.92	0.52	0.77	0.18	1.05	0.82	1.55	0.88	0.53	1.00
Decoupling state	Weak decoupli-ng	Expansi-ve coupling	Expansi-ve negative decoupli-ng	Expansi-ve negative decoupli-ng	Weak decoupli-ng	Expansi-ve coupling	Weak decoupli-ng	Weak decoupli-ng	Weak decoupli-ng	Expansi-ve coupling	Expansi-ve coupling	Expansi-ve negative decoupli-ng	Expansi-ve coupling	Weak decoupli-ng	Expansi-ve coupling

Figure 4. The decoupling of carbon emissions from transportation development and the trend of decoupling elasticity.

As can be seen in Figure 4, the decoupling of the development of China's transportation industry and carbon emissions has fluctuated greatly. This is similar to the existing studies, such as the study by Zhao et al. [51] and Zhou et al. [31] on the decoupling of the development of transportation industry and carbon emissions in Guangdong Province and the whole country. In Figure 4, we can see that during the period of 2001–2015, there are three types of decoupling between the development of transportation industry and carbon emissions: "weak decoupling", "expansive coupling" and "expansive negative decoupling". Among them, more than half of the years experience "expansive coupling" and "expansive negative decoupling", indicating that the decoupling of carbon emissions from transportation development is poor, and the economic growth of the transportation industry is accompanied by an increase in carbon emissions.

Overall, during the period of 2001–2015, the decoupling of carbon emissions from transportation development shows a tendency to deteriorate first, then to improve, and then to slightly deteriorate again. The years 2001–2004 were a period of deterioration in decoupling; 2005–2009 was a period of improvement, and 2010–2015 was a period of slight deterioration in decoupling. ① From 2001 to 2004, the period was dominated by expansive negative decoupling, namely, the increase in carbon emissions was greater than the economic growth of the transportation sector. This is mainly due to the improvement in the global economic situation and the expansion of the economic scale of China's transportation industry. For the purpose of boosting domestic demand and increasing investment, China implemented a proactive macro-economic policy and launched a large number of high-energy-consuming and repetitive infrastructure projects [31]. In addition, online sales, such as

Taobao, developed rapidly during this period. However, these basic projects and online sales led to a sharp increase in demand and the consumption of energy while promoting the development of the transportation industry [31], but the state lacked effective emissions reduction measures to control this. ② During 2005–2009, in 2006 only, the decoupling state was expansive coupling. During the rest of the years, there was weak decoupling, meaning that the growth rate of the transportation carbon emissions was less than the rate of economic growth. This is mainly because during the period when China developed its transportation industry, China increased its emphasis on carbon emissions, set a hard target for energy conservation and emissions reduction, and promulgated a number of laws and regulations such as the "Renewable Energy Law". At the same time, China's overall improvement in the level of opening up provided ample opportunities for high and new technology industries to speed up the development of high and new technology industries, which have the characteristics of high added value and low transportation density [66], thus optimizing the decoupling of transportation development from carbon emissions and transforming the economic growth mode from extensive to intensive. ③ During the period of 2010–2015, the transportation carbon emissions showed obvious signs of increases compared with the previous period. The decoupling state was mainly dominated by expansive coupling and even showed expansive negative decoupling in 2012. This shows that the transportation industry's energy-saving emissions reduction efforts were insufficient at this stage and did not reach the desired state. Under the background of the problem that traffic congestion is difficult to solve and there has been a rapid increase in the number of motor vehicles, the road toward energy conservation and emissions reduction in China's transportation industry is still full of bumps.

After replacing total carbon emissions with per capita carbon emissions in the transportation sector, the decoupling status in 2002 and 2008 is different from the previous calculation. In 2002, the decoupling state of total carbon emissions was expansive negative decoupling, while the per capita decoupling state was expansive coupling. In 2008, the decoupling state of total carbon emissions was expansive coupling, while the per capita decoupling state was weak decoupling. China formally joined the World Trade Organization in late 2001, causing an increase in the traffic demand in 2002 and an increase in carbon emissions. At the same time, the state analyzed in depth the opportunities and challenges brought by the accession to the World Trade Organization for energy development in The Tenth Five-year Plan for Energy Development and formulated energy development strategies that focused on optimizing the energy structure and making efforts to improve energy efficiency. It guided the transportation industry to gradually pursue energy-saving emissions reduction and to some extent slowed the increase in carbon emissions. Therefore, the per capita decoupling state was changed from weak decoupling in 2001 to expansive coupling in 2002. With the continuous opening up of China, the transportation industry faced increasingly fierce price competition, especially in the impact of road transportation. Therefore, compared with the growth rate of the transportation economy, the increase in its carbon emissions was larger, leading to a worsening per capita decoupling status in the next two years, from expansive coupling in 2002 to expansive negative decoupling. In 2008, the global economic crisis caused a great impact on China's transportation industry. Coupled with the continuous progress of energy conservation and emissions reduction, the per capita decoupling status showed weak decoupling. It can be seen from this that the per capita decoupling state can more accurately and subtly reflect the relationship between the development of the transportation industry and carbon emissions compared with the decoupling state calculated through the transportation of total carbon emissions, so we can conduct a better and more comprehensive analysis and provide a reference for the formulation of energy conservation and emissions reduction policies by the state.

There is a relatively large gap between the trend of the energy-saving elasticity and per capita emissions reduction elasticity. As seen in Figure 4, the volatility of energy-saving elasticity is relatively large, especially in the sharp increase in energy-saving elasticity in 2001–2002. After a period of fluctuation, the energy-saving elasticity has also picked up in recent years, indicating that the transportation industry is more dependent on energy consumption and that energy efficiency needs to be improved. Per capita emissions reduction elasticity fluctuated relatively more in 2008–2010 and almost overlapped with per capita decoupling elasticity, which shows that per capita emissions reduction elasticity, which is the energy consumption structure, has a more important impact on decoupling the development of transportation from carbon emissions during this period. However, per capita emissions reduction elasticity changed little in general, tended to be stable, and had little effect on the decoupling status. Within the entire period of 2001–2015, the trends of energy-saving elasticity and per capita decoupling elasticity were almost the same, both rising and falling, indicating that compared with per capita emissions reduction elasticity, energy-saving elasticity played a more crucial role in decoupling the development of transportation from carbon emissions. This is consistent with the analysis of energy intensity in the factor decomposition above. Energy intensity showed a declining effect on carbon emissions in 2001, 2005, 2007, 2009 and 2014. Although energy intensity showed a stimulus-increasing effect in 2008, its value was relatively small, only 2770 tons. However, the decoupling of transportation development from carbon emissions has shown weak decoupling in the past few years.

Summary: The causal chain model of the per capita decoupling elasticity accurately reflects that the decoupling state of carbon emissions from the development of China's transportation industry is relatively poor. And energy-saving elasticity plays a more important role in decoupling than the per capita emissions reduction elasticity.

6. Discussion and Analysis

6.1. Discussion on the Comparison between the Results of This Paper and the Existing Research Results

This study explores the factors influencing carbon emissions in China's transportation industry and the decoupling relationship between carbon emissions and the development of the transportation industry. Many scholars have done similar research on these two aspects. The following two representative articles were selected to compare with the results of this study.

(1) Comparing driving factors of the evolution of China's transportation carbon emissions. Zhou et al. [31] conducted an LMDI decomposition of China's transportation carbon emissions from 1995 to 2012, extending the Kaya identity and discussing the impact of five factors on carbon emissions. Three of them are the same as the research factors in this paper, namely, energy carbon emission intensity, energy intensity and population size. The comparison of the results is shown in Table 4.

Table 4. Comparison of the results of driving factors. Unit: 10^4 tons

Years	Energy Carbon Emission Intensity Effect		Energy Intensity Effect		Population Size Effect	
	Comparative Literature	This Article	Comparative Literature	This Article	Comparative Literature	This Article
2001	−163.28	−54.03	−1124.04	−5.76	150.57	48.38
2002	157.67	44.81	−233.32	−1.60	146.24	47.57
2003	−505.51	−146.47	1916.89	−45.10	149.20	49.37
2004	131.51	83.39	843.15	−13.73	168.30	55.50
2005	449.20	111.62	−206.91	−1.18	194.37	64.37
2006	3.16	43.59	108.76	0.63	193.57	63.09
2007	17.09	−51.12	−1291.05	−4.26	207.09	66.69
2008	−412.6	−110.16	−1186.17	0.28	215.37	70.43
2009	−213.71	−119.30	−335.39	−0.75	212.77	70.68
2010	−112.07	−48.33	101.38	−3.37	222.65	72.27
2011	−387.76	−172.46	−187.74	0.17	243.08	79.12
2012	258.27	50.14	1793.08	−15.28	275.97	89.92

As seen in Table 4, ① the direction of the effect of energy carbon emission intensity on the carbon emissions in the transportation sector is almost the same in the two studies, with the opposite direction in 2007 only. The comparative study shows the promotion effect, and this paper shows the inhibitory effect. Considering that the two studies are basically consistent with the overall change trend of energy carbon emission intensity in terms of transportation carbon emissions and that the gap before and after is one year only in the turning year in which the energy carbon emission intensity plays a role, this is acceptable. ② In both papers, there is a big gap between the magnitude of the effect of energy intensity on transportation carbon emissions. In recent years, China has focused on energy development and has guided the transportation industry to improve the efficiency of energy utilization so that energy intensity has a negative effect on carbon emissions in most years. However, although the adjustment of energy intensity is effective, it does not reach the expected level, making the effect of reducing energy intensity weaker and in some years even showing a slight increase in the effect. Therefore, the maximum declining effect of the energy intensity in this paper is only 451,000 tons in Table 4, and the increasing effect does not exceed 10,000 tons. ③ The population size in both papers contributed to carbon emissions. This is due to the huge population base in China at present, but the sizes of the two studies are quite different. China's family planning policy, which has been implemented for several decades, has greatly limited the increase in the natural population growth rate and causes the population scale to promote carbon emissions, but its value is not great.

From the above comparative analysis, it can be seen that compared with the paper using the LMDI decomposition method used in previous studies to study the influencing factors, using GDIM to explore the drivers of transportation carbon emissions can not only allow more key factors to be chosen but also shows more reasonable decomposition results and helps to develop practical policy measures.

At present, there are studies on the factors of carbon emissions in the transportation industry, such as Tunisia and Morocco. M'raihi et al. [22] used LMDI to decompose Tunisia's carbon emissions from road freight from 1990 to 2006 into average oil emissions, oil consumption share of road freight, oil consumption intensity of road fright, road freight intensity, and GDP. The results show that economic growth is a major factor in the increase of carbon emissions in the transportation industry, which is the same as the study in this paper. In addition, the paper also analyzes the cumulative contributions of various influencing factors and can more clearly reflect the dynamic impact of economic growth on carbon emissions changes. Kharbach et al. [28] used LMDI to decompose the carbon emissions from the road transportation industry in Morocco into the average fleet emission factor, energy use, motor vehicle ownership, and population. The study found that population growth and the increase in motor vehicle ownership have a catalytic effect on carbon emissions, which is the same as this paper. In addition, the paper also studied the carbon emissions per capita and found that per capita carbon emissions contributed significantly to the growth of carbon emissions.

(2) Comparing the decoupling between the development of China's transportation industry and carbon emissions. Based on the Tapio decoupling model, Zhou Yinxiang [57] studied the decoupling relationship between the development of the transportation industry and carbon emissions in 1990–2013 and their evolution and constructed the causal chain to explore the influencing factors of decoupling. In the same manner as in this present paper, Zhou Yinxiang concluded that whether the development of the transportation industry can be decoupled from carbon emissions depends more on energy-saving elasticity, but the decoupling status in individual years is different. The comparison of the results is shown in Table 5.

As seen in Table 5, the decoupling status is the same for most years, but the change in the decoupling status in this study is slightly greater than that in the reference study, mainly due to the per-capita carbon emission indicator that allocates the transportation carbon emissions to each individual, specifically considers the changes in individual carbon emissions, and truly reflects the actual situation of China's carbon emissions. Thus, per capita decoupling elasticity can more accurately reflect the relationship between transportation development and carbon emissions and capture changes in the decoupling status more sensitively and subtly.

Table 5. Comparison of research results of the decoupling state.

Years	Decoupling State	
	Comparative Literature	This Article
2001	Weak decoupling	Weak decoupling
2002	Expansive coupling	Expansive coupling
2003	Expansive negative decoupling	Expansive negative decoupling
2004	Expansive negative decoupling	Expansive negative decoupling
2005	Expansive coupling	Weak decoupling
2006	Expansive coupling	Expansive coupling
2007	Weak decoupling	Weak decoupling
2008	Weak decoupling	Weak decoupling
2009	Weak decoupling	Weak decoupling
2010	Expansive coupling	Expansive coupling
2011	Weak decoupling	Expansive coupling
2012	Expansive negative decoupling	Expansive negative decoupling
2013	Expansive negative decoupling	Expansive coupling

At present, there are few studies on the decoupling of transportation development and carbon emissions in other countries. Loo et al. [50] studied the decoupling of transportation and economic growth in 15 countries, including Russia and Canada. The results showed that decarbonisation of the transportation sector were more difficult to achieve than the reduction in the levels of transportation-related fatalities. Alises et al. [67] compared the decoupling of road freight transportation between the United Kingdom and Spain. Studies have shown that Spain's decoupling rate is much lower than that of the United Kingdom driven by economic structure changes. It can thus be seen that the decoupling of the transportation industry is not particularly optimistic in China and other countries.

6.2. The Reference Meaning of This Study

(1) The research in this present paper can provide some reference to other countries. All the selected indices in this paper are widely applicable indices, such as the added value of transportation and total energy consumption. All of these factors play a vital role in the transportation carbon emissions both in developed and developing countries. An accurate understanding of the situation can help to develop a realistic carbon reduction policy for one or more of these factors. In today's world, all countries are actively developing the transportation industry. With the increasing number of motor vehicles and the ever-increasing number of trips and distances traveled by residents, the rapid growth of carbon emissions in the transportation industry has become an important issue that cannot be ignored. If we can grasp the decoupling relationship between the development of transportation and carbon emissions and block the coupling between the two, this will significantly improve the status of the continued increase in global carbon emissions.

(2) The research of this paper can provide a reference to other industries. The industrial sector, represented by electricity and machinery, has far more carbon emissions than transportation and is an important area for China's carbon emissions. Its development is accompanied by an increase in carbon emissions. Using the research methods of this paper, we can determine the factors that have a significant impact on carbon emissions, understand the decoupling state between carbon emissions and industrial development, and take effective measures to save energy and reduce emissions.

6.3. The Deficiency of This Research and Directions of Improvement

Due to the lack of data on various energy consumption in the transportation sector before 2000, this paper only studies the situation of carbon emissions in the transportation industry from 2000 to 2015 and fails to expand the study period, which is not conducive to a more accurate and comprehensive grasp of the law of changes of China's transportation carbon emissions. In the study of the drivers of carbon emissions in the transportation industry, the selected indicators are the most common key factors, such as the added value of the industry. In the future, we can try to make innovations in terms of the factors studied.

7. Conclusions and Suggestions

7.1. Main Conclusions

The transportation industry is a large source of energy consumption and carbon emissions in China. The transportation industry must shoulder the heavy responsibility of reducing carbon emissions and strive to find practical ways to save energy and reduce emissions. This is, to a large extent, related to whether China's carbon emission reduction targets and low-carbon sustainable development vision can be successfully achieved as soon as possible. In this paper, GDIM is first used to decompose the evolution of China's carbon emissions from 2000 to 2015 in the transportation industry, and then, the decoupling of carbon emissions from transportation development is analyzed based on the Tapio decoupling theory. Then, the interaction between carbon emissions and economic development is further explored. The main conclusions of this study are as follows:

(1) The added value of transportation, energy consumption and per capita carbon emissions in the transportation industry are the major factors leading to the increase in carbon emissions. The energy carbon emission intensity is the key factor leading to a reduction in carbon emissions. Among other factors, population size has a positive effect on carbon emissions, while per capita added value of transportation, energy intensity and carbon intensity of added value have a decreasing effect on carbon emissions.

(2) The decreasing factors have a limited effect on the suppression of carbon emissions in the transportation industry, and this effect is far less than the contribution of the increasing factors to the increase in carbon emissions.

(3) The decoupling state between the development of China's transportation industry and the carbon emission is poor, and it gradually shows a deteriorating trend after a short period of improvement.

(4) Per capita decoupling elasticity can reflect the decoupling status between transportation development and carbon emissions more accurately and subtly than the decoupling elasticity of total carbon emissions.

(5) Compared with the effect of per capita emissions reduction elasticity, energy-saving elasticity plays a more crucial role in decoupling the development of transportation from carbon emissions.

7.2. Policy Suggestions

Based on the above research results, we propose the following policy suggestions:

(1) Optimize the traffic structure. From the analysis of Figures 2–4, the added value of transportation industry is the main factor that promotes increases in carbon emissions, and the expansion of the economic scale also leads to deterioration of the decoupling state. The continuous economic growth has become the bottleneck of low-carbon development in China's transportation industry, but at the same time it is also the goal of sustainable development of our country and the guarantee of people's material well-being. Transportation carbon emissions reduction at the expense of economic development is not desirable, but the state can adjust the structure of the transportation industry, optimize the combination of various modes of transportation and the proportion of investment, gradually develop high-tech industries, and rationally plan the construction of low-carbon transportation infrastructure.

(2) Optimize the energy structure of the transportation industry. From the analysis of Figures 2 and 3, the energy carbon emission intensity is a key factor in reducing carbon emissions. Therefore, the government should focus on the following: optimizing the energy structure of the transportation industry in the future; reducing the consumption of traditional energy, such as

petrol and diesel, by adjusting prices and taxes; phasing out the high-energy-consumption and high-emissions transportation vehicles; increasing the investment in the use of clean energy, such as solar energy; and encouraging and promoting the development of new energy sources for transportation.

(3) Improve the energy efficiency of the transportation industry. Through the analysis of Figures 2–4, it can be seen that energy intensity has a decreasing effect on carbon emissions, and energy-saving elasticity plays an important role in decoupling the development of transportation from carbon emissions. Therefore, energy efficiency must be taken seriously. At present, the enhancement of energy efficiency in the transportation industry has a very limited effect on curbing carbon emissions and has not reached a satisfactory level. In the future, there will still be much room for improvement. Increasing energy efficiency plays an important role in decoupling the carbon emissions from the development of transportation and can achieve the effect of reducing carbon emissions in a relatively short period of time rather than optimizing the energy structure. In the future, the government should increase investment in and development of energy-saving technologies, actively develop and promote low-carbon transportation technologies, optimize the transportation system, and enhance the intelligence of transportation to reduce energy consumption.

(4) Increase public transportation system construction. Through the analysis of Figures 2–4, it can be seen that energy consumption is the main factor that promotes carbon emissions, and the large increase in energy consumption also hinders the decoupling process of China's transportation industry. With the advancing urbanization in China, the problems of the increase in the number of private cars, traffic congestion, slow driving and so on are becoming increasingly prominent. As a result, the demand and consumption of energy continues to increase, leading to more serious carbon emissions in the transportation sector. However, promoting public transportation construction is a fast and effective way to solve this problem. Therefore, the government should give priority to the development of public transportation, accelerate the construction of urban rail transit such as subways and skyrails, and rationally plan bus lanes and bus routes to facilitate citizens' travel and transfer.

(5) Enhance citizens' low-carbon traffic awareness. From Figures 2 and 3, it can be seen that the population size has a positive effect on carbon emissions and the per capita carbon emissions is the main factor that lead to an increase in carbon emissions. Therefore, the population cannot be ignored in the carbon emissions reduction in transportation industry. Although China implements the family planning policy and controls the natural population growth rate, the huge population base still causes the population size to promote an increase in carbon emissions, making citizens' choices of modes of transportation especially important for reducing carbon emissions. Therefore, the government should step up the publicity of low-carbon traffic, conduct lectures on the theme of low-carbon transportation, hold related art performances, carry out bicycle riding and other activities in schools, work units and communities, broadcast more of these public service advertisements and formulate corresponding incentive measures for citizens' environmental protection behaviors in different degrees to raise citizens' awareness of environmental protection and promote citizens' green travel.

The main conclusions of this paper on the influencing factors and decoupling elasticity of China's transportation carbon emissions are shown in Table 6.

Table 6. Main conclusions of this paper.

Research Topics	Factors	Contribution to Carbon Emissions (10^4 tons)			Cumulative Contribution to Carbon Emissions (10^4 tons)	Character
		2000–2005	2005–2010	2010–2015		
Decomposition of influencing factors—GDIM	Added value of transportation industry	4588.32	6122.68	6633.09	Increasing from 6,137,100 tons in 2001 to 161,496,400 tons in 2015	The primary increasing factor
	Carbon intensity of added value	487.99	−895.65	46.00	Reducing a total of 3.5895 million tons of carbon emissions in 2000-2015	The more volatile decreasing factor
	Energy consumption	4786.97	4803.62	6738.01	Increasing from 2,336,700 tons in 2001 to 162,401,800 tons in 2015	The primary increasing factor
	Energy carbon emission intensity	−8.02	−261.17	−376.08	The decreasing effect growing rapidly from 783,600 tons in 2008 to 6.3401 million tons in 2015	The main decreasing factors
	Population size	249.18	312.57	436.35	Cumulative resulting in 10,892,500 tons of carbon emissions by 2015	The relatively weak increasing factor
	Per capita carbon emissions in the transportation industry	5090.32	4747.34	6370.25	Maintaining its rapid growth at an average annual rate of 39.78%	The important increasing factor
	Per capita added value of transportation	−1088.61	−1216.04	−1070.08	The decreasing effect increasing from 200,100 tons in 2001 to 6.4745 million tons in 2015	The main decreasing factors
	Energy intensity	−46.74	18.69	−22.96	Reducing 934,300 tons of carbon emissions cumulatively by 2015	The small decreasing factor
		Situation				
Decoupling elasticity—Improved Tapio causal Chain	Decoupling state	The years 2001-2004 were a period of deterioration; 2005-2009 was a period of improvement, and 2010-2015 was a period of slight deterioration.				
	Per capita decoupling elasticity	It increased sharply from 0.21 in 2001 to 2.22 in 2003. After a period of fluctuation, it reached 1.00 in 2015.				
	Per capita emissions reduction elasticity	It fluctuated relatively more in 2008-2010, but it tended to be stable in general.				
	Energy-saving elasticity	Its volatility is relatively large, the trends of it and per capita decoupling elasticity are almost the same, and it plays a more crucial role in decoupling the development of transportation from carbon emissions.				

Author Contributions: Yingchun Zhang was mainly responsible for the writing of the full text; Yong Wang and Yu Zhou conceived and designed the study; Lin Zhu and Fei Zhang built the models of the paper.

Acknowledgments: This paper is supported by Liaoning Economic and Social Development Research Project (2018lslktzd-010), National Natural Science Foundation of China (71573034), Liaoning Social Science Fund (L17CTJ001, L17BJY042), China Postdoctoral Science Foundation special fund (2017T100180), China Postdoctoral Science Foundation(2016M601318) and Research Project of Dongbei University of Finance and Economics (DUFE2017Q16).

Conflicts of Interest: The authors declare no conflict of interest.

References

1. United Nations Framework Convention on Climate Change. Available online: https://wenku.baidu.com/view/a8aa555271fe910ef02df896.html (accessed on 16 April 2018).
2. Full Text of the Kyoto Protocol. Available online: https://wenku.baidu.com/view/98fdd2fb941ea76e58fa0402.html?from=search (accessed on 16 April 2018).
3. The Chinese Version of the Paris Agreement. Available online: https://wenku.baidu.com/view/f9492ef783c4bb4cf6ecd1a9.html (accessed on 15 April 2018).
4. Wu, Y.; Tan, W.; Shi, Y. China Ranked First in the World in Terms of Carbon Emissions in 2013 or Accelerated the Adjustment of Energy Structure. 2014. Available online: http://www.thepaper.cn/newsDetail_forward_1268328 (accessed on 2 December 2017).
5. Global Carbon Project. Available online: http://www.globalcarbonatlas.org/en/CO2-emissions (accessed on 10 May 2017).
6. Notice of the State Council Issuing "Work Plan on Controlling Greenhouse Gas Emission in the Twelfth Five-Year Plan". Available online: http://www.gov.cn/zwgk/2012-01/13/content_2043645.htm (accessed on 12 November 2017).
7. State Council Issues "Work Plan on Controlling Greenhouse Gas Emission in the Thirteenth Five-year Plan". *People's Daily*, 5 November 2016.
8. Notice of the State Council Issuing "Integrated Work Plan on Energy Saving and Emission Reduction in the Thirteenth Five-Year Plan. Available online: http://www.gov.cn/gongbao/content/2017/content_5163448.htm (accessed on 15 November 2017).
9. Notice of the State Council Issuing "Integrated Work Plan on Energy Saving and Emission Reduction in the Twelfth Five-Year Plan". Available online: http://www.gov.cn/zwgk/2011-09/07/content_1941731.htm (accessed on 15 November 2017).
10. Shen, M.; Chi, X. Analysis on the Driving Factors of the Growth of Carbon Emissions in China's Transportation Sector. *Jianghuai Trib.* **2012**, 31–38. [CrossRef]
11. Wang, B.; Shao, C.; Ji, X. Influencing Mechanism Analysis of Holiday Activity–travel Patterns on Transportation Energy Consumption and Emissions in China. *Energies* **2017**, *10*, 897. [CrossRef]
12. Hao, H.; Liu, F.; Liu, Z.; Zhao, F. Measuring Energy Efficiency in China's Transport Sector. *Energies* **2017**, *10*, 660. [CrossRef]
13. Wei, C.; Ni, J.; Du, L. Regional Allocation of Carbon Dioxide Abatement in China. *China Econ. Rev.* **2012**, *23*, 552–565. [CrossRef]
14. Lugauer, S.; Jensen, R.; Sadler, C. An Estimate of the Age Distribution's Effect on Carbon Dioxide Emissions. *Econ. Inq.* **2014**, *52*, 914–929. [CrossRef]
15. Yang, B.; Liu, C.; Su, Y.; Jing, X. The Allocation of Carbon Intensity Reduction Target by 2020 among Industrial Sectors in China. *Sustainability* **2017**, *9*, 148. [CrossRef]
16. Zhou, Y. Responsive Relationship between Transportation Carbon Emissions and Traffic Economic Growth—A Practical Study Based on Decoupling & Recoupling Theory and LMDI Decomposition. *Collect. Essays Financ. Econ.* **2014**, *11*, 764–765.
17. Sun, J.W. Changes in Energy Consumption and Energy Intensity: A Complete Decomposition Model. *Energy Econ.* **2004**, *20*, 85–100. [CrossRef]
18. Rüstemoğlu, H.; Andrés, A.R. Determinants of CO_2 Emissions in Brazil and Russia between 1992 and 2011: A Decomposition Analysis. *Environ. Sci. Policy* **2016**, *58*, 95–106. [CrossRef]

19. Hatzigeorgiou, E.; Polatidis, H.; Haralambopoulos, D. CO_2 Emissions in Greece for 1990–2002: A Decomposition Analysis and Comparison of Results Using the Arithmetic Mean Divisia Index and Logarithmic Mean Divisia Index Techniques. *Energy* **2008**, *33*, 492–499. [CrossRef]
20. Ang, B.; Zhang, F. A Survey of Index Decomposition Analysis in Energy and Environmental Studies. *Energy* **2000**, *25*, 1149–1176. [CrossRef]
21. Gambhir, A.; Tse, L.K.C.; Tong, D.; Martinez-Botas, R. Reducing China's Road Transport Sector CO_2 Emissions to 2050: Technologies, Costs and Decomposition Analysis. *Appl. Energy* **2015**, *157*, 905–917. [CrossRef]
22. M'raihi, R.; Mraihi, T.; Harizi, R.; Bouzidi, M.T. Carbon Emissions Growth and Road Freight: Analysis of the Influencing Factors in Tunisia. *Transp. Policy* **2015**, *42*, 121–129. [CrossRef]
23. Shi, L.; Sun, J.; Lin, J.; Zhao, Y. Factor Decomposition of Carbon Emissions in Chinese Megacities. *J. Environ. Sci.* **2018**. [CrossRef]
24. Du, Z.; Lin, B. Analysis of Carbon Emissions Reduction of China's Metallurgical Industry. *J. Clean. Prod.* **2017**, *176*, 1177–1184. [CrossRef]
25. Ma, L.; Chong, C.; Zhang, X.; Liu, P.; Li, W.; Li, Z.; Ni, W. LMDI Decomposition of Energy-related CO_2 Emissions Based on Energy and CO_2 Allocation Sankey Diagrams: The Method and an Application to China. *Sustainability* **2018**, *10*, 344. [CrossRef]
26. Wang, F.; Wang, C.; Su, Y.; Jin, L.; Wang, Y.; Zhang, X. Decomposition Analysis of Carbon Emission Factors from Energy Consumption in Guangdong Province from 1990 to 2014. *Sustainability* **2017**, *9*, 274. [CrossRef]
27. De freitas, L.C.; Kaneko, S. Decomposition of CO_2 Emissions Change from Energy Consumption in Brazil: Challenges and Policy Implications. *Energy Policy* **2011**, *39*, 1495–1504. [CrossRef]
28. Kharbach, M.; Chfadi, T. CO_2 Emissions in Moroccan Road Transport Sector: Divisia, Cointegration, and EKC Analyses. *Sustain. Cities Soc.* **2017**, *35*, 396–401. [CrossRef]
29. Zhu, X.; Li, R. An Analysis of Decoupling and Influencing Factors of Carbon Emissions from the Transportation Sector in the Beijing-Tianjin-Hebei Area, China. *Sustainability* **2017**, *9*, 722. [CrossRef]
30. Zhang, M.; Mu, H. Research on the Change of Motorized Passenger Transport Energy Consumption in Selected Chinese Cities Based on LMDI Method. *Acta Sci. Nat. Univ. Pekin.* **2010**, *46*, 483–486.
31. Zhou, Y.; Liu, W. Decoupling Elasticity and Driving Factors Analysis between the Relationship of Transportation Carbon Emissions and Economic Growth. *J. Guangxi Univ. Financ. Econ.* **2015**, *28*, 28–35.
32. Vaninsky, A. Factorial Decomposition of CO_2 Emissions: A Generalized Divisia Index Approach. *Energy Econ.* **2014**, *45*, 389–400. [CrossRef]
33. Shao, S.; Zhang, X.; Zhao, X. Empirical Decomposition and Peaking Pathway of Carbon Dioxide Emissions of China's Manufacturing Sector—Generalized Divisia Index Method and Dynamic Scenario Analysis. *Chin. Ind. Econ.* **2017**, *34*, 44–63.
34. Shao, S.; Liu, J.; Geng, Y.; Yang, Y. Uncovering Driving Factors of Carbon Emissions from China's Mining Sector. *Appl. Energy* **2016**, *166*, 220–238. [CrossRef]
35. Li, Z.; Qing, D. An Empirical Study of Decoupling Economic Growth from Carbon Dioxide—A Case Study of Shanxi Province. *Fujian Trib.* **2010**, *30*, 67–72.
36. Yang, R.; Chang, X. Decoupling and Driving Factors Analysis between the Relationship of Carbon Emissions and Economic Growth in Western Region. *Econ. Geogr.* **2012**, *32*, 34–39.
37. Wu, Y.; Zhu, Q.; Zhu, B. Decoupling Analysis of World Economic Growth and CO_2 Emissions: A Study Comparing Developed and Developing Countries. *J. Clean. Prod.* **2018**, *190*, 94–103. [CrossRef]
38. Lin, B.; Liu, K. Using LMDI to Analyze the Decoupling of Carbon Dioxide Emissions from China's Heavy Industry. *Sustainability* **2017**, *9*, 1198.
39. Diakoulaki, D.; Mandaraka, M. Decomposition Analysis for Assessing the Progress in Decoupling Industrial Growth from CO_2 Emissions in the Eu Manufacturing Sector. *Energy Econ.* **2007**, *29*, 636–664. [CrossRef]
40. Li, R.; Jiang, R. Moving Low-carbon Construction Industry in Jiangsu Province: Evidence from Decomposition and Decoupling Models. *Sustainability* **2017**, *9*, 1013.
41. Román, R.; Cansino, J.M.; Botia, C. How Far Is Colombia from Decoupling? Two-level Decomposition Analysis of Energy Consumption Changes. *Energy* **2018**, *148*, 687–700. [CrossRef]
42. Freitas, L.C.D.; Kaneko, S. Decomposing the Decoupling of CO_2 Emissions and Economic Growth in Brazil. *Ecol. Econ.* **2011**, *70*, 1459–1469. [CrossRef]

43. Grand, M.C. Carbon Emission Targets and Decoupling Indicators. *Ecol. Indic.* **2016**, *67*, 649–656. [CrossRef]
44. Roinioti, A.; Koroneos, C. The Decomposition of CO_2 Emissions from Energy Use in Greece Before and During the Economic Crisis and Their Decoupling from Economic Growth. *Renew. Sustain. Energy Rev.* **2017**, *76*, 448–459. [CrossRef]
45. Wang, Q.; Li, R.; Jiang, R. Decoupling and Decomposition Analysis of Carbon Emissions from Industry: A Case Study from China. *Sustainability* **2016**, *8*, 1059. [CrossRef]
46. Hu, J.; Gui, S.; Zhang, W. Decoupling Analysis of China's Product Sector Output and Its Embodied Carbon Emissions—An Empirical Study Based on Non-competitive I-O and Tapio Decoupling Model. *Sustainability* **2017**, *9*, 815. [CrossRef]
47. Wan, L.; Wang, Z.L.; Ng, J. Measurement Research on the Decoupling Effect of Industries' Carbon Emissions—Based on the Equipment Manufacturing Industry in China. *Energies* **2016**, *9*, 921. [CrossRef]
48. Tapio, P. Towards a Theory of Decoupling: Degrees of Decoupling in the Eu and the Case of Road Traffic in Finland between 1970 and 2001. *Transp. Policy* **2005**, *12*, 137–151. [CrossRef]
49. Gray, D.; Anable, J.; Illingworth, L.; Graham, W. Decoupling the Link between Economic Growth, Transport Growth and Carbon Emissions in Scotland. Available online: http://www.gov.scot/Resource/Doc/935/0042647.pdf (accessed on 7 October 2017).
50. Loo, B.P.Y.; Banister, D. Decoupling Transport from Economic Growth: Extending the Debate to Include Environmental and Social Externalities. *J. Transp. Geogr.* **2016**, *57*, 134–144. [CrossRef]
51. Zhao, Y.; Kuang, Y.; Huang, N. Decomposition Analysis in Decoupling Transport Output from Carbon Emissions in Guangdong Province, China. *Energies* **2016**, *9*, 295. [CrossRef]
52. Liu, X.; Sun, L.; Sun, Q. Decoupling Between China and Provincial Economic Growth and Carbon Emissions from Transport. *J. Shandong Univ. Sci. Technol.* **2016**, *18*, 47–54.
53. Timilsina, G.R.; Shrestha, A. Factors Affecting Transport Sector CO_2 Emissions Growth in Latin American and Caribbean Countries: A LMDI Decomposition Analysis. *Int. J. Energy Res.* **2009**, *33*, 396–414. [CrossRef]
54. Wu, L.; Kaneko, S.; Matsuoka, S. Driving Forces behind the Stagnancy of China's Energy-related CO_2 Emissions from 1996 to 1999: The Relative Importance of Structural Change, Intensity Change and Scale Change. *Energy Policy* **2005**, *33*, 319–335. [CrossRef]
55. Guo, S. Study on Decoupling and Emission Peak of China Industry. *Ind. Econ. Rev.* **2017**, *4*, 1–13.
56. Tang, Z.; Shang, J.; Shi, C.; Liu, Z.; Bi, K. Decoupling Indicators of CO_2 Emissions from the Tourism Industry in China: 1990–2012. *Ecol. Indic.* **2014**, *46*, 390–397. [CrossRef]
57. Zhou, Y. Study on the Decoupling and Coupling Relationship between Transportation Carbon Emissions and Industrial Economic Growth: Based on Tapio Decoupling Model and Cointegration Theory. *Inquiry Econ. Issues* **2016**, *37*, 41–48.
58. Industrial Classification for National Economic Activities of 2017. Available online: http://www.stats.gov.cn/tjsj/tjbz/hyflbz/201710/t20171012_1541679.html (accessed on 9 October 2017).
59. Wang, Y.F.; Li, K.P.; Xu, X.M.; Zhang, Y.R. Transport Energy Consumption and Saving in China. *Renew. Sustain. Energy Rev.* **2014**, *29*, 641–655. [CrossRef]
60. Department of Energy Statitics, National Bureau of Statitics, People's Republic of China. China Energy Statistical Yearbook. 2000–2015. Available online: http://www.doc88.com/p-0911392153730.html (accessed on 13 November 2017).
61. Cai, B. A Study on City Greenhouse Gas Emissions Inventory. *Adv. Clim. Chang. Res.* **2011**, *7*, 23–28.
62. Intergovernmental Panel on Climate Change (IPCC). *2006 IPCC Guidelines for National Greenhouse Gas Inventories*; Institute for Global Environmental Strategy: Kanagawa, Japan, 2006.
63. National Bureau of Statistics of People's Republic of China. China Statistical Yearbook. 2000–2015. Available online: http://www.stats.gov.cn/tjsj/ndsj/2016/indexch.htm (accessed on 13 November 2017).
64. Li, H.; Wang, L.; Zhang, F. Core Conception of the Modern Circular Economy in China: The Notion of Decoupling and Its Measurement. *Chin. Soft Sci.* **2007**, 43–48. [CrossRef]
65. Jiang, R.; Zhou, Y.; Li, R. Moving to a Low-carbon Economy in China: Decoupling and Decomposition Analysis of Emission and Economy from a Sector Perspective. *Sustainability* **2018**, *10*, 978. [CrossRef]

66. Wu, W.; Cao, Y.; Liang, S.; Xiao, C. Study on Dynamics of Freight Volume in the Yangtze River Delta since the Mid-1980s. *Resour. Environ. Yangtze Basin* **2011**, *20*, 843–847.
67. Alises, A.; Vassallo, J.M.; Guzmán, A.F. Road Freight Transport Decoupling: A Comparative Analysis between the United Kingdom and Spain. *Transp. Policy* **2014**, *32*, 186–193. [CrossRef]

energies

MDPI

Article

Decoupling Greenhouse Gas Emissions from Crop Production: A Case Study in the Heilongjiang Land Reclamation Area, China

Yu Zhang, Xiaojiao Zou, Caifen Xu and Qingshan Yang *

School of Geographical Science, Northeast Normal University, Changchun 130024, China; zhangy221@nenu.edu.cn (Y.Z.); zouxj437@nenu.edu.cn (X.Z.); xucf 2016@163.com (C.X.)

* Correspondence: yangqs027@163.com; Tel.: +86-431-8509-9550

Received: 28 April 2018; Accepted: 4 June 2018; Published: 6 June 2018

Abstract: Modern agriculture contributes significantly to greenhouse gas emissions in several ways. From the perspective of sustainability assessment, it is not enough to evaluate mitigation measures that rely only on emissions reductions. In this article, we use the method of decoupling analysis to construct a decoupling index based on carbon footprint and crop yield and evaluate the relationship between crop production and greenhouse gas emissions using the most modern grain production base in China as a case study. The results indicate that a weak but variable decoupling trend occurred from 2001 to 2015 and that each branch achieved on average a weak decoupling across the study period. In addition, rice production constituted 80% of the regional carbon footprint in a crop's life cycle. The results of our analysis of rice production show that weak decoupling was the most common outcome but was not consistent because a weak coupling occurred in 2015. Each branch on average achieved a weak decoupling except for the SH branch. Our research indicates that high agricultural material inputs with low utilization efficiency contributed to the poor relationship between crop production and greenhouse gas emissions in the study area. Fertilizer, especially N fertilizer, was an important contributor to the total greenhouse gas emissions of crop production. As a supplement to carbon footprint assessment, this decoupling analysis helps local decision-makers diagnose the level of green growth, identify key options to mitigate greenhouse gas emissions from agriculture, and adopt more targeted interventions towards sustainable agriculture.

Keywords: decoupling analysis; greenhouse gas emissions; carbon footprint; low-carbon agriculture

1. Introduction

The relationship between economic growth and environmental pressure has been intensively discussed [1–3]. In recent years, green economic growth has attracted worldwide attention as a way to maintain rapid economic development while limiting environmental degradation. Like the term "green economy," "decoupling" refers to the ability of an economy to grow without a corresponding increase in environmental pressure [4]. Today, decoupling environmental impacts from human well-being has been widely acknowledged by policy-makers, industry leaders, and civil society as a key issue to address in meeting sustainable development goals [5]. In the field of sustainability studies, following the environmental Kuznets curve (EKC) hypothesis [6], decoupling analysis has become increasingly popular, and there is a growing body of literature on the decoupling method and indicators of decoupling [7–12]. Indeed, as a policy goal, decoupling environmental impact from economic growth has been adopted by the European Union (EU) and the Green Economy Initiative of the United Nations Environment Program (UNEP) [13].

Climate change is one of the greatest challenges to mankind today. The increases in anthropogenic greenhouse gases (GHGs), including carbon dioxide (CO_2), methane (CH_4), and nitrous oxide (N_2O), have important effects on global warming [14,15]. Many studies have empirically assessed the potential impact of human activities or production sectors on global warming by quantifying the carbon footprint (CF) [16–18]. Modern agriculture is usually accompanied by high material inputs, high energy consumption, and high release of pollutants, which all play an important role in GHG emissions [19]. Extensive studies have evaluated agricultural CFs associated with material inputs or based on life cycle assessment (LCA). LCA is a commonly used environmental management tool to assess a product or service from "cradle to grave" [20]. The literature on evaluating the CFs of crop production generally quantified the GHG emissions from sowing to harvest, including the indirect emissions from agricultural material inputs and the direct emissions from energy consumption for farm mechanical operations, N_2O from N fertilizer use or the CH_4 emissions from rice paddies [21–24]. Studies of CFs for a diverse range of crops have been performed at different geographical scales using national statistical data or farm survey data [17,25–28], and other studies have described a certain crop's CF in more detail, such as for rice [29–31], spring barley [32], and wheat [33]. In addition, GHG emissions under different cropping systems and farm management practices have been addressed in detail [21–23,34,35], and the CFs of crop production have also been compared across countries [27]. All these studies have helped to further explore measures to mitigate agricultural GHG emissions and have put forward potential solutions to develop low-carbon agriculture.

China is a major agricultural producer, and GHG emissions in the agricultural sector account for 17% of the national total [36]. According to previous studies, the CFs of crop production in China [37] were higher than those in the USA [27] and the UK [17] based on national statistical data. From the perspective of sustainability assessment, it is not enough to evaluate mitigation measures depending only on emissions reduction; similarly, it remains difficult to examine if one farming region has taken effective measures to reduce the carbon intensity of agriculture. The Heilongjiang land reclamation area (HLRA) is both the most modern grain production base and the largest green grain production base in China, the application of chemical fertilizers and pesticides in the HLRA is far below the national standard, and its crop yield per unit area exceeds that of the US [38]. However, it is not clear if its high crop yield occurs at the expense of high GHG emissions. In this study, we use this area as an example to estimate the extent to which GHG emissions are decoupled from crop production.

The objectives of this study are, first, to quantify the CFs of crop production (including rice, maize, soybeans, and wheat) using the LCA approach in the HLRA during 2001–2015; second, to determine the relationship between crop production and GHG emissions based on a decoupling index; and, last, to analyze the composition of the CFs of crop production and further provide targeted suggestions for decision-making for low-carbon agriculture.

The flowchart for the decoupling analysis is shown in Figure 1.

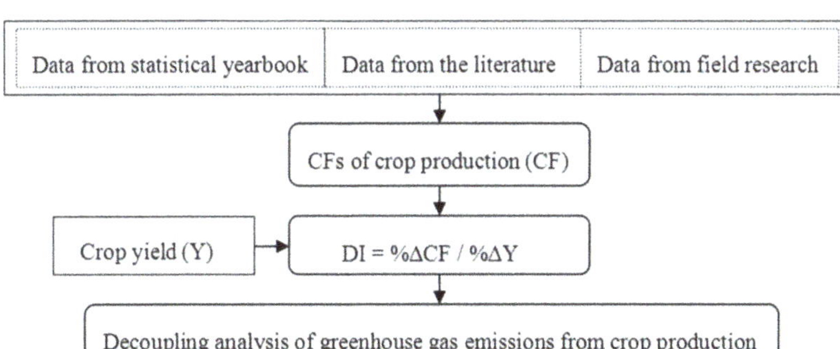

Figure 1. Steps in the decoupling analysis.

2. Materials and Methods

2.1. Carbon Footprint Calculation

The carbon footprint of crop production was expressed in this study in CO_2 equivalents (CE) following the LCA approach. GHG emissions included the direct and indirect emissions from crop production. The indirect emissions were attributed to the manufacture of agricultural material inputs (e.g., fertilizers, pesticides and plastic films) and electricity used for rice irrigation; the direct emissions were attributed to energy consumption from farm mechanical operations including seeding, tillage, transportation and harvesting as well as N_2O from N fertilizer use and CH_4 emissions from rice paddies [25].

GHG emissions from agricultural material inputs or sources were expressed as CF_i (in tCE) using Equation (1):

$$CF_i = \sum (I_i \times EF_i), \tag{1}$$

where I_i is the amount of each agricultural input or source i, including fertilizers (in t), pesticides (in t), seed (in t), plastic films (in t), electricity for rice irrigation (in kWh) and diesel for machinery (in t), and EF_i is the GHG emission factor in this study (Table 1).

The direct N_2O emissions from fertilizer N use were expressed as CF_{N2O} (in tCE) using Equation (2):

$$CF_{N_2O} = I_N \times EF_{N_2O} \times \frac{44}{28} \times 298, \tag{2}$$

where I_N represents the amount of N fertilizer used (in t), EF_{N2O} is the emission factor for N_2O emissions caused by N fertilizer used (in $tN_2O\text{-}Nt^{-1}$) [39,40], 44/28 is the ratio of molecular weights of N_2O to N_2, and 298 is the net global warming potential of N_2O over a 100-year period [40].

The CH_4 emissions from a submerged rice paddy in a single season were expressed as CF_{CH4} (in tCE) using Equation (3):

$$CF_{CH_4} = EF_d \times T \times A \times 25, \tag{3}$$

where EF_d is a daily emission factor (in $tCEha^{-1}day^{-1}$), T is the rice growing period (in days), A is the planting area (in ha), and 25 is the relative molecular warming forcing of CH_4 in a 100-year period [40].

Here, EF_d was estimated by Equation (4) due to the restricted condition of data:

$$EF_d = EF_c \times SF_w \times SF_p \times SF_m, \tag{4}$$

where EF_c is the basic emission factor for fields flooded without organic amendment; SF_w and SF_p are scaling factors for different hydrological conditions over the rice growing period and before rice transplanting, respectively; and SF_m is a scaling factor for quantifying organic amendment used for rice production [41]. All of the above emission factors for agricultural inputs or sources are shown in Table 1.

The total carbon footprint CF_t (in tCE) was calculated for rice production and for dry crop production (maize, soybeans, and wheat) by Equations (5) and (6), respectively:

$$CF_t = CF_i + CF_{N_2O} + CF_{CH_4} \tag{5}$$

$$CF_t = CF_i + CF_{N_2O}. \tag{6}$$

Based on the estimated CF_t, carbon intensity in crop yield, CF_Y (in $tCEt^{-1}$), and the carbon intensity in crop area, CF_A (in $tCEha^{-1}$), were calculated using Equations (7) and (8), respectively, in terms of crop yield (Y, in t) and crop planting area (A, in ha).

$$CF_Y = \frac{CF_t}{Y} \tag{7}$$

$$CF_A = \frac{CF_t}{A} \quad (8)$$

Table 1. Emission factors for agricultural inputs and sources.

Emission Source	Abbreviation	Emission Factor	Reference
Fertilizer	EFf	1.53 tCEt−1 (N fertilizer); 1.63 $tCEt-^{1}$ (P fertilizer); 0.66 $tCEt-^{1}$ (K fertilizer)	[30]
Pesticide	EF_p	0.20 $tCEt-^{1}$ (Herbicide); 16.60 $tCEt^{-1}$ (Insecticide)	[30]
Plastic film	EFpf	22.70 tCEt−1	[30]
Seed	EF_s	0.58 tCEt−1	[26]
Electricity for irrigation	EF_e	1.23 × 10−3 $tCEkWh-^{1}$	[26]
Diesel for machinery	EF_d	0.89 tCEt−1	[30]
N fertilizer-induced N_2O	EF_{N2O}	0.01 $tN_2O\text{-}Nt-^{1}$ (Dry cropland); 0.0073 $tN_2O\text{-}Nt^{-1}$ (Rice paddy)	[39,40]
CH_4 emissions from rice field	EFc	1.30 × 10−3 tCH4 ha−1day−1	[41]
	SFw	0.52	[41]
	SFP	0.68	[41]
	SFm	1	[41]

2.2. Decoupling Index

In this article, the decoupling index (*DI*) is used to indicate the degree of decoupling of GHG emissions from crop production, following Equation (9):

$$DI = \%\Delta CF/\%\Delta Y = (CF_j/CF_{j-1} - 1)/(Y_j/Y_{j-1} - 1), \quad (9)$$

where $\%\Delta CF$ is the percentage change in GHG emissions from crop production, and CF_j and CF_{j-1} denote GHG emissions in a target year *j* and the base year $j - 1$; $\%\Delta Y$ is the percentage change in crop yield, and Y_j and Y_{j-1} denote the crop yield in a target year *j* and the base year $j - 1$, respectively. Six decoupling index values are shown in Table 2.

Table 2. Degrees of decoupling GHG emissions from crop production.

Decoupling Degree	Relationship between GHG Emissions and Crop Production
Strong decoupling	$\Delta Y > 0$, $\Delta CF \leq 0$, $DI \leq 0$
Weak decoupling	$\Delta Y > 0$, $\Delta CF > 0$, $0 < DI < 1$
Recessive decoupling	$\Delta Y < 0$, $\Delta CF < 0$, $DI \geq 1$
Expansive coupling	$\Delta Y > 0$, $\Delta CF > 0$, $DI \geq 1$
Weak coupling	$\Delta Y < 0$, $\Delta CF < 0$, $0 < DI < 1$
Strong coupling	$\Delta Y < 0$, $\Delta CF \geq 0$, $DI \leq 0$

2.3. Study Area and Data Sources

The HLRA is located in northeast China and includes nine branches with an area of 57,600 km^2 (Figure 2). There are four main grain crops in the study area: rice, maize, soybeans, and wheat. Rice and maize are main crops generally grown in the humid eastern branches, whereas soybeans and wheat are the main crops grown in the semi-humid western branches. These four crops accounted for 97% of the total output in 2015. The comprehensive utilization rate of agricultural mechanization is over 94%, and these commodity grains achieve 91% of their annual crop yield in the HLRA, which has improved national food security.

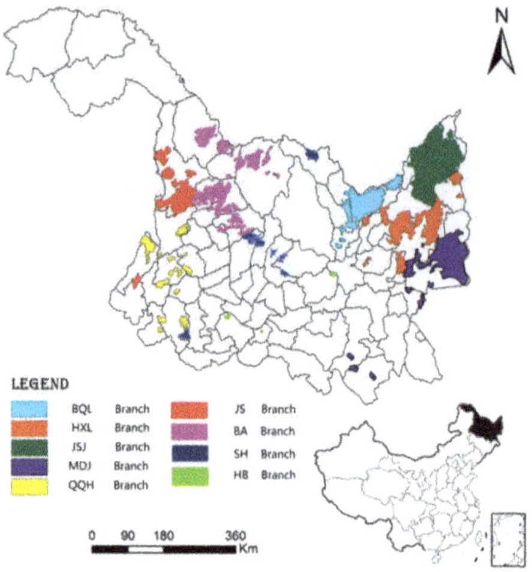

Figure 2. Location of the study area.

Data for quantifying GHG emissions from agricultural inputs or sources were collected from the National Cost-Benefit Survey for Agricultural Products (2001–2015). Crop yield and planting area data were collected from the Statistical Yearbook of the Heilongjiang Land Reclamation Area (2002–2016), and data for quantifying CH_4 emissions from the rice paddy, including the rice cultivation and growing periods, were obtained from field research and existing literature [26,30,39–41].

3. Results and Analysis

3.1. Relationship between Crop Yield and Carbon Footprint

The changes in the crop yields and the CFs of four crops in the HLRA (2001–2015) are shown in Figure 3; the correlation coefficient, R, between crop yield and the CF was 0.994 at a significance level of 0.01. Using crop yield as the independent variable, x, and the CF as the dependent variable, y, the best-fit linear equation relating these two variables was y = 0.2227x + 72.383. The R^2 and adjusted R^2 values of this equation were 0.988 and 0.987, respectively, which indicated a close relationship between GHG emissions and crop production in the HLRA.

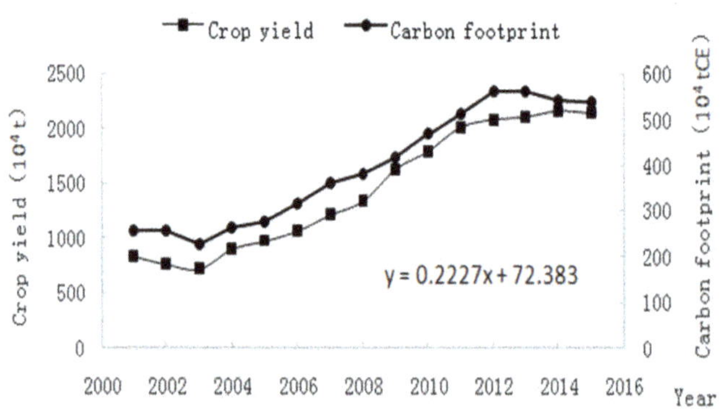

Figure 3. Relationship between carbon footprint and crop yield in the HLRA (2001–2015).

3.2. Decoupling GHG Emissions from Crop Production

According to Table 2, the results of decoupling GHG emissions from crop production during 2001–2015 in the HLRA are shown in Table 3, and the results based on the average value in the period 2001–2015 are shown in Table 4.

Table 3. Decoupling GHG emissions from crop production in the HLRA.

Year	Crop Yield (10^4 t)	Growth Rate of Crop Yield (%)	CF (10^4 tCE)	Growth Rate of CF (%)	DI	Decoupling Degree
2001	832.17	-	255.72	-	-	
2002	761.31	−8.52	254.98	−0.29	0.03	Weak coupling
2003	717.41	−5.77	226.22	−11.28	1.95	Recessive decoupling
2004	901.22	25.62	262.11	15.87	0.62	Weak decoupling
2005	973.10	7.98	275.83	5.23	0.66	Weak decoupling
2006	1065.11	9.46	316.12	14.61	1.54	Expansive coupling
2007	1210.07	13.61	360.04	13.89	1.02	Expansive coupling
2008	1337.51	10.53	381.27	5.9	0.56	Weak decoupling
2009	1631.90	22.01	415.27	8.92	0.41	Weak decoupling
2010	1794.78	9.98	467.96	12.69	1.27	Expansive coupling
2011	2014.16	12.22	512.72	9.56	0.78	Weak decoupling
2012	2085.14	3.52	560.7	9.36	2.66	Expansive coupling
2013	2109.67	1.18	559.8	−0.16	−0.14	Strong decoupling
2014	2165.06	2.63	542.11	−3.16	−1.2	Strong decoupling
2015	2146.15	−0.87	537.46	−0.86	0.99	Weak coupling

Table 4. Decoupling GHG emissions from crop production in the HLRA (average of the years 2001–2015).

Branch	Crop Yield (10^4 t)	Growth Rate of Crop Yield (%)	CF (10^4 tCE)	Growth Rate of CF (%)	DI	Decoupling Degree
BQL	220.33	6.56	57.72	5.24	0.8	Weak decoupling
HXL	283.32	7.13	70.3	3.67	0.51	Weak decoupling
JSJ	438.37	9.58	125.94	8.72	0.91	Weak decoupling
MDJ	277.5	5.33	73.87	3.71	0.7	Weak decoupling
BA	122.37	11.79	21.5	3.91	0.33	Weak decoupling
JS	93.06	11.15	12.81	1.39	0.12	Weak decoupling
QQH	67.357	9.82	15.56	6.21	0.63	Weak decoupling
SH	41.46	6.72	11.98	3.84	0.57	Weak decoupling
HB	12.68	11.93	4.84	8.37	0.7	Weak decoupling

According to Table 3, during 2001–2015, strong decoupling occurred for two years, weak decoupling occurred for five years, and recessive decoupling occurred for one year, which indicated that changes in carbon intensity were variable during this period and largely composed of weak decoupling. GHG emissions of crop production did not increase in proportion with crop yield in the HLRA for 2013 and 2014. Expansive coupling occurred for four years, and weak coupling occurred for two years. One of these weakly increasing years, 2015, followed two years of strong decoupling, which suggests that HLRA continues to face both challenges and opportunities as low-carbon agriculture continues to develop.

According to Table 4, from the perspective of the branch scale, all branches experienced weak decoupling between crop production and GHG emissions when considering the mean change from 2001–2015. This analysis revealed the potential for the HLRA to experience strong decoupling with continued progress.

3.3. Example: Decoupling GHG Emissions from Rice Production

Rice in the HLRA is economically and environmentally important to China, both as the largest green food base and because of its high-quality rice varieties. On average, rice acreage was 48% of the total grain-planting area, and rice accounted for 62% of the total grain yield in the HLRA during

2001–2015. During these years, GHG emissions from rice production accounted for 80% of the total HLRA CF, with maize, soybeans, and wheat contributing 11%, 7%, and 2% to the total CF, respectively. Further results of decoupling GHG emissions from rice production with regard to the whole HLRA and the branch scale over 2001–2015 are shown in Tables 5 and 6.

Table 5. Decoupling GHG emissions from rice production in the HLRA.

Year	Rice Yield (104 t)	Growth Rate of Rice Yield (%)	CF of Rice (10^4 tCE)	Growth Rate of CF of Rice (%)	DI	Decoupling Degree
2001	527.42	-	194.14	-	-	
2002	452.77	−14.15	197.92	1.94	−0.14	Strong coupling
2003	424.16	−6.32	165.73	−16.26	2.57	Recessive decoupling
2004	528.62	24.63	195.99	18.26	0.74	Weak decoupling
2005	573.43	8.48	209.58	6.93	0.82	Weak decoupling
2006	682.5	19.02	250.45	19.5	1.03	Expansive coupling
2007	798.07	16.93	288.25	15.09	0.89	Weak decoupling
2008	842.18	5.53	307.51	6.68	1.21	Expansive coupling
2009	927.32	10.11	322.21	4.78	0.47	Weak decoupling
2010	1094.39	18.02	373.24	15.84	0.88	Weak decoupling
2011	1278.91	16.86	422.52	13.2	0.78	Weak decoupling
2012	1370.42	7.16	464.38	9.91	1.38	Expansive coupling
2013	1385.67	1.11	464.6	0.05	0.05	Weak decoupling
2014	1329.35	−4.06	443.6	−4.52	1.11	Recessive decoupling
2015	1291.51	−2.85	435.93	−1.73	0.61	Weak coupling

Table 6. Decoupling GHG emissions from rice production at the branch scale of the HLRA (average of the years 2001–2015).

Branch	Rice Yield (10^4 t)	Growth Rate of Rice Yield (%)	CF of Rice (10^4 tCE)	Growth Rate of CF of Rice (%)	DI	Decoupling Degree
BQL	127.57	7.68	45.46	6.79	0.88	Weak decoupling
HXL	151.36	4.45	54	3.24	0.73	Weak decoupling
JSJ	384.88	11.03	117.57	9.95	0.9	Weak decoupling
MDJ	205.39	3.91	64.17	3.39	0.87	Weak decoupling
BA	3.7	57.74	4.62	5.38	0.09	Weak decoupling
JS	3.54	41.43	4.26	10.15	0.24	Weak decoupling
QQH	40.09	10.23	13.01	6.74	0.66	Weak decoupling
SH	16.18	1.3	7.17	2.36	1.82	Expansive coupling
HB	6.44	8.35	3.50	7.85	0.94	Weak decoupling

According to Table 5, weak decoupling was the most common outcome, observed for seven years of the study period, whereas recessive decoupling was observed for two years. No strong decoupling was observed, but expansive coupling was observed for three years, and both strong coupling and weak coupling were observed for one year each. In 2003 and 2014, when the growth rate of CF of rice production decreased by −16.26% and −4.52%, respectively, rice yield decreased accordingly by −6.32% and −4.06% compared with the previous year, respectively. The desired decoupling between GHG emissions and rice production was therefore not observed during these years. Worse than that, strong coupling was observed in 2002, when the CF of rice production increased by 1.94%, despite a decrease in rice yield of 14.15%. Both increases and decreases in carbon intensity in rice production were observed; even weak coupling occurred, most recently in 2015, and there was no clear trend across the time series.

Seen from the branch scale over 2001–2015 (Table 6), on average, each branch except the SH branch achieved a weak decoupling of GHG emissions from rice production. However, the rate of rice yield growth (1.3%) failed to exceed that of the rice CF growth rate (2.36%) in the SH branch, which led to the degree of expansive coupling.

4. Discussion

Generally, we can evaluate GHG emissions based on the reduction in the carbon footprint; however, some indeterminacy remains when diagnosing the effective quantity of emissions reduction, and we thus need to link it with the economic development process. As a supplementary method, the frame of decoupling focuses on the relationship between economic growth and environmental pressure, which helps to create a better understanding of the nature of green growth, further remove barriers to decoupling, and encourage policies towards decoupling [5]. Recently, many studies have used a panel VAR approach or a log mean division index (LMDI) decomposition method to analyze the factors that affect GHG emissions in the manufacturing or transport sector [12,42–45]; however, these methods are addressed less in the decoupling of GHG emissions from the agricultural sector. In this discussion section, we compare the carbon footprint of crop production in the HLRA within different countries, analyze the composition of carbon footprint, and further focus on rice production.

4.1. Comparative Analysis of Carbon Footprint

According to Equation (7) for carbon intensity in crop yield, the CF_Y in the HLRA varied by crop. On average, rice production possessed the highest CF_Y (0.36 $tCEt^{-1}$), maize production possessed the lowest CF_Y (0.12 $tCEt^{-1}$), and the CF_Y for soybean and wheat production showed intermediate values of 0.19 $tCEt^{-1}$ and 0.21 $tCEt^{-1}$, respectively. Compared with existing research results (Table 7), most CF_Y values of crop production in the HLRA were lower than the average value in China, except for soybeans (0.10 $tCEt^{-1}$), and the CF_Y of soybean and wheat production in the HLRA was close to that of the USA.

Table 7. Comparison of international carbon intensity in crop yield.

Country/Region	Crop	CF_Y ($tCEt^{-1}$)	Reference
HLRA	Rice	0.36	
	Maize	0.12	
	Soybeans	0.19	
	Wheat	0.21	
China	Rice	0.80	[25]
	Maize	0.33	[25]
	Soybeans	0.10	[46]
	Wheat	0.65	[25]
USA	Maize	0.12–0.25	[47]
	Wheat	0.25–0.35	[47]
Canada	Wheat	0.27–0.50	[48]
India	Rice	1.2–1.5	[49]
	Wheat	0.12	[49]

We observe a better relationship between crop production and GHG emissions in the HLRA than in other regions of the world. However, as various values of the decoupling index were observed, it did not appear that the carbon intensity of agriculture in the HLRA steadily decreased.

4.2. Composition Analysis of Carbon Footprint

The compositions of the CFs for the four major crops in the HLRA during the period 2001–2015 are shown in Figure 4. On average, CH_4 was the biggest contributor (41%) to the total CF, which indicated that rice production was the main source of GHG emissions in the HLRA. Direct N_2O emissions and indirect emissions from N fertilizer input together represented the second-biggest contributor (25%), with electricity for irrigation (11%) representing the third-largest contributor to the CF. All other sources, including P fertilizer (4%), seed (4%), plastic films (6%), diesel (7%), and K fertilizer and pesticides (1%), were minor contributors to the HLRA CF.

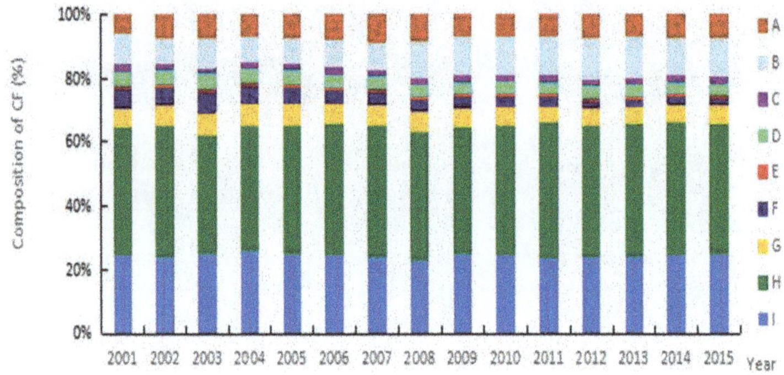

Figure 4. Composition of CFs for crop production in the HLRA (2001–2015). A: Plastic film; B: Electricity; C: Pesticide; D: P fertilizer; E: K fertilizer; F: Seed; G: Diesel; H: CH_4; I: N fertilizer + N_2O.

Agricultural material inputs or sources to the HLRA CF are shown in Figure 5. For rice production, 51% of the CF was derived from CH_4 emissions, followed by the sum of direct N_2O emissions and indirect emissions from N fertilizer use (16.06%), electricity for irrigation (13.98%) and plastic film (8.89%). The remaining five inputs and farming operations amounted to 10.07% of the total CF.

In contrast, the sum of direct N_2O emissions and indirect emissions from N fertilizer use was the largest contributor to the CFs of dry crops (maize, soybeans, and wheat), accounting for 72.9%, 44%, and 49.5% of the CF for maize, soybean, and wheat production, respectively. The second largest contributor to the total CFs for both maize and soybean production was diesel (12.28% and 22%, respectively), and seed was the second largest contributor for wheat production (23.3%), followed by diesel (12.05%). Overall, N fertilizer input and N_2O from N fertilizer use were the dominant sources of GHG emissions in dry crop production, although CH_4 was the dominant source of GHG emissions in rice production. In contrast, pesticides contributed a small amount to each crop's CF, especially for rice production (0.4%).

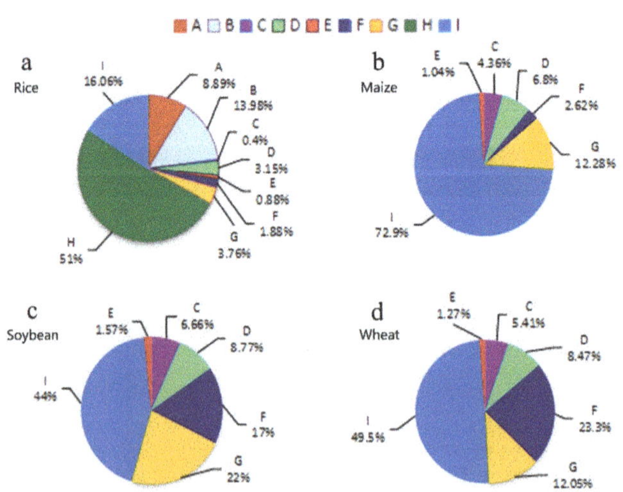

Figure 5. Composition of CFs based on crop structure in the HLRA. (Average of the years 2001–2015). A: Plastic film; B: Electricity; C: Pesticide; D: P fertilizer; E: K fertilizer; F: Seed; G: Diesel; H: CH_4; I: N fertilizer + N_2O.

4.3. Analysis of the CF of Rice Production

As reported above, rice production played an important role in the HLRA and constituted the vast majority of the CF in this region (80%). In recent decades, eight branches (except the SH branch) experienced a weak decoupling between crop production and GHG emissions (Table 6). Here, we take the JSJ branch and the SH branch of the HLRA for comparative analysis (Figures 6 and 7).

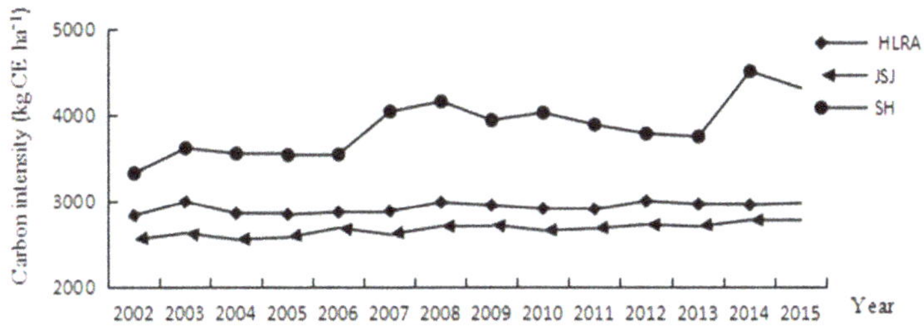

Figure 6. Carbon intensity in area for rice production in the JSJ branch and the SH branch (2001–2015).

The JSJ branch is the largest branch in the HLRA, and its rice planting area and rice yield occupied 41% and 43% of the HLRA total. In contrast, the rice planting area and rice yield in the SH branch each occupied 2% of the HLRA total. There was a distinct difference in trends between CF and rice yield between these two branches (Figure 6). According to Equation (8) for carbon intensity per area, the CF_A of rice production in the JSJ branch fluctuated from 2539 kgCEha^{-1} to 2775 kgCEha^{-1}, which was below the average CF_A in the HLRA (2919 kgCEha^{-1}), whereas the CF_A of rice production in the SH branch fluctuated from 3323 kgCEha^{-1} to 4503 kgCEha^{-1}.

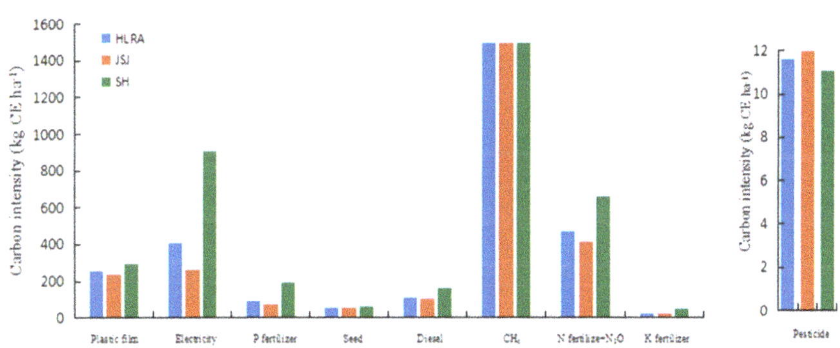

Figure 7. Carbon intensity per area for rice production in the JSJ branch and the SH branch (average of the years 2001–2015).

The SH branch required more electricity for irrigation, more fertilizer input (especially more N fertilizer), and more diesel input per unit area, all of which contributed to a higher CF_A for rice production (Figure 7). It is clear that high material inputs with low utilization efficiency contributed to its degree of expansive coupling. Based on this result, we suggest targeted measures for the SH branch to mitigate GHG emissions from rice production, such as decreasing agricultural material inputs (including fertilizers, electricity for irrigation, diesel, and plastic films), improving the utilization efficiency of agricultural material inputs and increasing agricultural productivity.

5. Conclusions

In this paper, a decoupling index based on carbon footprint and crop yield was used to examine the relationship between crop production and GHG emissions in the HLRA during the years 2001–2015. The results indicated that various decoupling degrees (including strong decoupling, weak decoupling, and recessive decoupling) occurred during more than half of the study phase across the entire HLRA, although each branch showed weak decoupling based on the average value from 2001 to 2015. In addition, rice production constituted 80% of the total CF in the HLRA, and weak decoupling occurred more frequently at the scale of the entire study area and at the branch scale (except for the SH branch, which showed expansive coupling).

Seen from the results of the decoupling analysis, although a high appearance frequency of weak decoupling occurred during 2001–2015 in the HLRA, the status of weak decoupling was not steady, which highlights both pressures and challenges for the HLRA as it develops towards green growth. We also found that high material inputs with low utilization efficiency contributed to a poor relationship between crop production and GHG emissions and that fertilizer was an important contributor to the total CF of crop production. Since it is the major source of GHG emissions from agriculture in the HLRA, we should pay more attention to rice production, in particular for the SH branch.

The current work of decoupling analysis aims to examine the relationship between GHG emissions and crop production, using HLRA as an example. In fact, there is a limitation to the decoupling concept, which lacks a direct contact with the environmental process. Based on the results of decoupling analysis, next we will borrow from the experience of others and use the LMDI decomposition methodology to analyze factors that affect GHG emissions in crop production processes, in view of the activity effect, the structure effect, and the intensity effect. Further integrating more detailed information about GHG emissions from crop production processes could contribute to more targeted suggestions for low-carbon agriculture.

Author Contributions: Q.Y. designed the research. C.X. collected the data and X.Z. analyzed the data. Y.Z. wrote the manuscript. All authors read, revised, and approved the final manuscript.

Funding: The work in this paper was supported by the National Natural Science Funds of China (Grant No. 41571115, 41271555, 41630749, and 41571405) and the Fundamental Research Funds for the Central Universities (2412018ZD012).

Conflicts of Interest: The authors declare no conflict of interest.

References

1. Arrow, K.; Bolin, B.; Costanza, R.; Dasgupta, P.; Folke, C.; Holling, C.S.; Jansson, B.O.; Levin, S.; Maler, K.G.; Perrings, C.; et al. Economic Growth, Carrying Capacity and the Environment. *Science* **1995**, *268*, 520–521. [CrossRef] [PubMed]
2. Holdren, J.P. Science and Technology for Sustainable Well-being. *Science* **2008**, *319*, 424–434. [CrossRef]
3. Magazzino, C. The Relationship among Economic Growth, CO_2 Emissions and Energy Use in the APEC Countries: A Panel VAR Approach. *Environ. Syst. Decis.* **2017**, *37*, 353–366. [CrossRef]
4. United Nations Environment Programme. *Options for Decoupling Economic Growth from Water Use and Water Pollution*; Report of the International Resource Panel Working Group on Sustainable Water Management; UNEP: Nairobi, Kenya, 2015; pp. 1–73.
5. United Nations Environment Programme. *Decoupling 2: Technologies, Opportunities and Policy Options*; A Report of the Working Group on Decoupling to the International Resource Panel; UNEP: Nairobi, Kenya, 2014; pp. 1–158.
6. Grossman, G.M.; Krueger, A.B. Economic Growth and the Environment. *Q. J. Econ.* **1995**, *110*, 353–377. [CrossRef]
7. Organisation for Economic Co-operation and Development. *Indicators to Measure Decoupling of Environmental Pressure from Economic Growth*; OECD: Paris, France, 2002; pp. 1–3.
8. Tapio, P. Towards a Theory of Decoupling: Degrees of Decoupling in the EU and the Case of Road Traffic in Finland between 1970 and 2001. *J. Transp. Policy* **2005**, *12*, 137–151. [CrossRef]

9. Diakoulaki, D.; Mandaraka, M. Decomposition Analysis for Assessing the Progress in Decoupling Industrial Growth from CO_2 Emissions in the EU Manufacturing Sector. *Energy Econ.* **2007**, *29*, 636–664. [CrossRef]
10. Roman, R.; Cansino, J.M.; Botia, C. How far is Colombia from Decoupling? Two-level decomposition analysis of energy consumption changes. *Energy* **2018**, *148*, 687–700. [CrossRef]
11. Schandl, H.; Hatfield-Dodds, S.; Wiedmann, T.; Geschke, A.; Cai, Y.; West, J.; Newth, D.; Baynes, T.; Lenzen, M.; Owen, A. Decoupling Global Environmental Pressure and Economic Growth: Scenarios for Energy Use, Materials Use and Carbon Emissions. *J. Clean. Prod.* **2016**, *132*, 45–56. [CrossRef]
12. Roman, R.; Cansino, J.M.; Rodas, J.A. Analysis of the Main Drivers of CO_2 Emissions Changes in Colombia (1990–2012) and its Political Implications. *Renew. Energy* **2018**, *116*, 402–411. [CrossRef]
13. United Nations Environment Programme. *Decoupling Natural Resource Use and Environmental Impacts from Economic Growth*; A Report of the Working Group on Decoupling to the International Resource Panel; UNEP: Nairobi, Kenya, 2011; pp. 1–174.
14. Intergovernmental Panel on Climate Change. Climate change 2007: The Physical Science Basis. In *Contribution of the Working Group I to the Fourth Assessment Report of the Intergovernmental Panel on Climate Change*; Cambridge University Press: Cambridge, UK; New York, NY, USA, 2007; Available online: http://www.ipcc.ch/pdf/assessment-report/ar4/syr/ar4_syr.pdf (accessed on 6 June 2018).
15. Intergovernmental Panel on Climate Change. Climate Change 2013: The Physical Science Basis. In *Contribution of Working Group I to the Fifth Assessment Report of the Intergovernmental Panel on Climate Change*; Cambridge University Press: Cambridge, UK; New York, NY, USA, 2013; pp. 1–1535.
16. Perry, S.; Klemes, J.; Bulatov, I. Integrating Waste and Renewable Energy to Reduce the Carbon Footprint of Locally Integrated Energy Sectors. *Energy* **2008**, *33*, 1489–1497. [CrossRef]
17. Hillier, J.; Hawes, C.; Squire, G.; Hilton, A.; Wale, S.; Smith, P. The Carbon Footprints of Food Crop Production. *Int. J. Agric. Sustain.* **2009**, *7*, 107–118. [CrossRef]
18. Liu, Y.; Xiao, H.W.; Zikhali, P.; Lv, Y.K. Carbon Emissions in China: A Spatial Econometric Analysis at the Regional Level. *Sustainability* **2014**, *6*, 6005–6023. [CrossRef]
19. Smith, P.; Martino, D.; Cai, Z.C.; Gwary, D.; Janzen, H.; Kumar, P.; McCarl, B.; Ogle, S.; O'Mara, F.; Rice, C.; et al. Greenhouse Gas Mitigation in Agriculture. *Philos. Trans. Biol. Sci.* **2008**, *363*, 789–813. [CrossRef] [PubMed]
20. McDougall, F.; White, P.; Franke, M.; Hindle, P. *Integrated Solid Waste Management: A Life Cycle Inventory*; Blackwell Science: London, UK, 2001.
21. Lal, R. Carbon Emission from Farm Operations. *Environ. Int.* **2004**, *30*, 981–990. [CrossRef] [PubMed]
22. Ponsioen, T.C.; Blonk, T.J. Calculating Land Use Change in Carbon Footprints of Agricultural Products as an Impact of Current Land Use. *J. Clean. Prod.* **2012**, *28*, 120–126. [CrossRef]
23. Clair, S.S.; Hiller, J.; Smith, P. Estimating the pre-harvest greenhouse gas costs of energy crop production. *Biomass Bioenergy* **2008**, *32*, 442–452. [CrossRef]
24. Knudsen, M.T.; Meyer-Aurich, A.; Olesen, J.E.; Chirinda, N.; Hermansen, J.E. Carbon footprints of crops from organic and conventional arable crop rotations—Using a life cycle assessment approach. *J. Clean. Prod.* **2014**, *64*, 609–618. [CrossRef]
25. Yan, M.; Cheng, K.; Luo, T.; Yan, Y.; Pan, G.; Robert, M.R. Carbon footprint of grain crop production in china-based on farm survey data. *J. Clean. Prod.* **2015**, *104*, 130–138. [CrossRef]
26. Huang, X.; Chen, C.; Chen, M.; Song, Z.; Deng, A.; Zhang, J.; Zheng, C.; Zhang, W. Carbon Footprints of Major Staple Grain Crops Production in Three Provinces of Northeast China during 2004–2013. *Chin. J. Appl. Ecol.* **2016**, *27*, 3307–3315.
27. Dubey, A.; Lal, R. Carbon Footprint and Sustainability of Agricultural Production Systems in Punjab, India, and Ohio, USA. *J. Crop Improv.* **2009**, *23*, 332–350. [CrossRef]
28. Cheng, K.; Yan, M.; Nayak, D.; Smith, P.; Pan, G.; Zheng, J. Carbon Footprint of Crop Production in China: An Analysis of National Statistics Data. *Sci. Agric. Sin.* **2015**, *153*, 422–431. [CrossRef]
29. Xue, J.; Pu, C.; Liu, S.; Zhao, X.; Zhang, R.; Chen, F.; Xiao, X.; Zhang, H. Carbon and Nitrogen Footprint of Double Rice Production in Southern China. *Ecol. Indic.* **2016**, *64*, 249–257. [CrossRef]
30. Wang, X.; Zhao, X.; Wang, Y.; Xue, J.; Zhang, H. Assessment of the Carbon Footprint of Rice Production in China. *Resour. Sci.* **2017**, *39*, 713–722.
31. Huang, Y.; Zhang, W.; Zheng, X.; Han, S.; Yu, Y. Estimates of Methane Emission from Chinese Paddy Fields by Linking A Model to GIS Database. *Acta Ecol. Sin.* **2006**, *6*, 980–988. [CrossRef]

32. Gan, Y.; Liang, B.; May, W.; Malhi, S.S.; Niu, J.; Wang, X. Carbon Footprint of Spring Barley in Relation to Preceding Oilseeds and N Fertilization. *Int. J. Life Cycle Assess.* **2012**, *17*, 635–645. [CrossRef]
33. Wang, Y.; Zhao, X.; Li, K.; Wang, X.; Xue, J.; Zhang, H. Dynamics of Carbon Footprint for Wheat Production in the North China Plain. *China Popul. Resour. Environ.* **2015**, *S2*, 258–261.
34. Dyer, J.A.; Kulshreshtha, S.N.; McConkey, B.G.; Desjardins, R.L. An assessment of fossil fuel energy use and CO_2 emissions from farm field operations using a regional level crop and land use database for Canada. *Energy* **2010**, *35*, 2261–2269. [CrossRef]
35. Koga, N.; Tsuruta, H.; Tsuji, H.; Nakano, H. Fuel Consumption-derived CO_2 Emissions under Conventional and Reduced Tillage Cropping Systems in Northern Japan. *Agric. Ecosyst. Environ.* **2003**, *99*, 213–219. [CrossRef]
36. Tian, Y.; Zhang, J.; Li, B. Intensities of Agricultural Carbon Emissions and Their Causes in the Major Grain Producing Areas in China. *Prog. Geogr.* **2012**, *31*, 1546–1551.
37. Cheng, K.; Pan, G.; Smith, P.; Luo, T.; Li, L.; Zheng, J.; Zhang, X.; Han, X.; Yan, M. Carbon Footprint of China's Crop Production–An Estimation Using Agro-statistics Data over 1993–2007. *Agric. Ecosyst. Environ.* **2011**, *142*, 231–237. [CrossRef]
38. Yang, Q.; Liu, J.; Zhang, Y. Decoupling Agricultural Nonpoint Source Pollution from Crop Production: A Case Study of Heilongjiang Land Reclamation Area, China. *Sustainability* **2017**, *9*, 1024. [CrossRef]
39. Zou, J.; Huang, Y.; Zheng, X.; Wang, Y. Quantifying Direct N_2O Emissions in Paddy Fields during Rice Growing Season in Mainland China: Dependence on Water Regime. *Atmos. Environ.* **2007**, *41*, 8030–8042. [CrossRef]
40. Intergovernmental Panel on Climate Change. *2006 IPCC Guidelines for National Greenhouse Gas Inventories*; Institute for Global Environmental Strategy: Kanagawa, Japan, 2006.
41. Yan, X.; Yagi, K.; Akiyama, H.; Akimoto, H. Statistical Analysis of the Major Variables Controlling Methane Emission from Rice Fields. *Glob. Chang. Biol.* **2005**, *11*, 1131–1141. [CrossRef]
42. Magazzino, C. A Panel VAR Approach of the Relationship among Economic Growth, CO_2 Emissions and Energy Use in the ASEAN-6 Countries. *Int. J. Energy Econ. Policy* **2014**, *4*, 546–553.
43. Jeong, K.; Kim, S. LMDI Decomposition Analysis of Greenhouse Gas Emissions in the Korean Manufacturing Sector. *Energy Policy* **2013**, *62*, 1245–1253. [CrossRef]
44. Ren, S.; Yin, H.; Chen, X. Using LMDI to Analyze the Decoupling of Carbon Dioxide Emissions by China's Manufacturing Industry. *Environ. Dev.* **2014**, *9*, 61–75. [CrossRef]
45. Timilsina, G.R.; Shrestha, A. Factors Affecting Transport Sector CO_2 Emissions Growth in Latin American and Caribbean Countries: An LMDI Decomposition Analysis. *Int. J. Energy Res.* **2009**, *33*, 396–414. [CrossRef]
46. Wang, X. Changes in CO_2 Emissions Induced by Agricultural Inputs in China over 1991–2014. *Sustainability* **2016**, *8*, 414. [CrossRef]
47. Snyder, C.S.; Bruulsema, T.W.; Jensen, T.L.; Fixen, P.E. Review of Greenhouse Gas Emissions from Crop Production Systems and Fertilizer Management Effects. *Agric. Ecosyst. Environ.* **2009**, *133*, 247–266. [CrossRef]
48. Gan, Y.; Liang, C.; Wang, X.; McConkey, B. Lowering Carbon Footprint of Durum Wheat by Diversifying Cropping Systems. *Field Crops Res.* **2011**, *122*, 199–206. [CrossRef]
49. Pathak, H.; Jain, N.; Bhatia, A.; Patel, J.; Aggarwal, P.K. Carbon Footprints of Indian Food Items. *Agric. Ecosyst. Environ.* **2010**, *139*, 66–73. [CrossRef]

energies

MDPI

Article

Environmental Impact and Carbon Footprint Assessment of Taiwanese Agricultural Products: A Case Study on Taiwanese Dongshan Tea

Allen H. Hu [1,*], Chia-Hsiang Chen [1], Lance Hongwei Huang [1], Ming-Hsiu Chung [1], Yi-Chen Lan [2] and Zhonghua Chen [3]

1 Institute of Environmental Engineering and Management, National Taipei University of Technology, Taipei 106, Taiwan; jager@pchome.com.tw (C.-H.C.); t103609001@ntut.org.tw (L.H.H.); stevechung@itri.org.tw (M.-H.C.)

2 School of Business, University of Western Sydney, Penrith South DC, NSW 2750, Australia; Y.Lan@westernsydney.edu.au

3 Natural Science, University of Western Sydney, Penrith South DC, NSW 2750, Australia; Z.Chen@westernsydney.edu.au

* Correspondence: allenhu@mail.ntut.edu.tw; Tel.: +886-2-2771-2171 (ext. 4151)

Received: 27 November 2018; Accepted: 26 December 2018; Published: 1 January 2019

Abstract: Climate change is an important global environmental threat. Agriculture aggravates climate change by increasing greenhouse gas (GHG) emissions, and in response, climate change reduces agricultural productivity. Consequently, the modern agricultural development mode has progressively transformed into a kind of sustainable development mode. This study aimed to determine the environmental impact and carbon footprint of Dongshan tea from Yilan County. Environmental impact was assessed with use of SimaPro version 8.0.2 and IMPACT2002+. Results showed that climate change has the largest impact upon it in general, followed by human health, natural resources, and ecosystem quality. Furthermore, with use of the IPCC 2007 100a method for carbon footprint of products (CFP), conventional tea was found to have a CFP of 7.035 kgCO$_2$-e, and its main contributors are the raw material (35.15%) and consumer use (45.58%) phases. From this case study, we found that the hotspots of the life cycle of environmental impact of Taiwanese tea mainly come from fertilizer input during the raw material phase, electricity use during manufacturing, and electricity use during water boiling in the consumer use phase (which contributes the largest impact). We propose the ways for consumers to use of highly efficient boiling water facilities and heating preservation, and the government must market the use of organic fertilizers in the national policy subsidies, and farmers have to prudent use of fertilizers and promote the use of local raw fertilizers, and engagement in direct sales for reducing the environmental impacts and costs of agricultural products and thus advancing sustainable agriculture development.

Keywords: tea; climate change; sustainable agriculture; environmental impact; carbon footprint

1. Introduction

The IPCC Fifth Assessment Report states that greenhouse gas (GHG) emissions from human activity have been the major factor for global warming since the middle of the 20th century. Agriculture and relevant land use transition contributed 17% of the world's anthropogenic GHG emissions in 2010 [1]. FAO [2] predicts a population of nine billion people by 2050, and to have sufficient food supply, agricultural production should increase by 60% by the same year. This assumption has resulted in the over-intensification of agriculture production systems that fail to consider the environmental impact of agricultural activities, causing several adverse effects on environment, such as water pollution, soil degradation and erosion, biodiversity loss, and deforestation [3,4]. Environmental sustainability is a challenge for agriculture, given that the latter is a major contributor of global environmental impacts, especially land degradation, freshwater depletion, nutrient and pesticide pollution, and GHG emissions [5,6]. Climate change can interfere with food availability; for example, temperature rise, precipitation pattern changes, extreme weather events, crop pests, disease outbreaks, and water shortage may result in the reduction of agricultural productivity [7,8]. Therefore, weakening the risks posed by climate change to food security is a major challenge.

Sustainable agriculture is based on emphasizing environmental quality, improving agronomic productivity, and minimizing global climate change. It is a type of agriculture that uses external energy inputs lightly and efficiently and may involve decreases in industrial mineral fertilizers, agrochemical input, and increases in the profit margins of farming systems [9,10]. Sustainable agricultural systems will require a conversion from the dominant industrial agriculture formation to one that conserves water and land, along with plant and animal genetic resources, and that is environmentally non-degrading, economically viable, technically appropriate, and socially acceptable [11,12]. Sustainable agriculture can raise productivity and meet sustainability criteria to satisfy increasing human needs meanwhile contributing to the recovery and sustainability of landscapes, the biosphere, and the earth systems [13].

Life cycle assessment (LCA) is a suitable and powerful means to evaluate environmental impact. It links up with a product, producing process, or activity during its life cycle from raw material extraction or production to final disposal, namely, "cradle to grave." Moreover, LCA is the method that can assess the whole life cycle of a product or service. In the last few years, this methodology has begun to concentrate on agriculture and its affected environmental impacts, such as climate change, eutrophication, acidification, nutrients, fertilizers, and crops [14–16]. However, studies rarely consider the entire agricultural system, which comprises various activities (e.g., cropping, breeding, nutrient leaching) and materials (e.g., fertilizer, feeds), which would provide a systematic analysis and comprehensive strategies [17]. Environmental LCA is a significant method for presenting environmental improvements, given that it quantifies sources of impacts throughout a product's life cycle for various environmental impacts, thereby allowing environmental improvements to be determined and ranked; this method has been confirmed to be useful [18–21]. The concept of circular economy is changing our awareness on waste. Life cycle assessment (LCA) is a method to assess environmental impacts by recycling, recycling and recycling from cradle to cradle to narrow the generation of waste [22].

Taiwan imports more than 90% of its energy and suffers from the effects of climate change, including sea level rise and the resulting energy instability and GHG emission offset, which are now serious problems facing the country [23]. Environmental sustainability is a challenge for agriculture, given that the latter substantially contributes to global environmental impacts, especially land degradation, freshwater scarcity, pesticide pollution, and GHG emissions [24]. Tea is an important domestic economic crop in Taiwan [25]. Therefore, Dongshan tea was selected as the object of this study, and the study region was Yilan County. This work specifically aimed to accomplish the following:

(1) Comprehend the environmental impact from the carbon emission and life cycle of Dongshan tea via LCA;
(2) Determine the carbon emission sources from Dongshan tea from cradle to grave according to the results of LCA;
(3) Assess the study results and propose proper countermeasures for reducing environmental impact and carbon emission.

Tea is primarily produced from the leaves of the plant *Camellia sinensis* and is the oldest and a commonly consumed beverage in the world because of its refreshing effect and mild stimulant properties, as well as the medicinal and general health-promoting properties produced by three major characteristic secondary metabolites (catechins, theanine, and caffeine); tea has had enormous medical, economic, and cultural importance since ancient times [26,27]. Tea production is geographically limited to a few areas around the world. In 2009 to 2013, global tea production grew 20 percent rate to reach 5.06 million tonnes, as shown in Figure 1 [28]; furthermore, world tea production is projected to reach 8.07 million tonnes in 2027 [29]. Recently, the production and consumption of tea have dramatically increased. Over two billion people drink tea in more than 125 countries [30]. Its remarkable health benefits are the main reason for its consumption [31,32]. However, many studies have discovered that tea expansion causes disturbances to ecosystems, threatens biodiversity preservation, and increases carbon dioxide emissions [33]. The interaction between agriculture and the natural environment is strong, and investigations on the contribution of farming systems to environmental degradation have been increasing gradually in regions with intensive agriculture practices. As for the increasing consumer awareness and interest in sustainability issues, the assessment of environmental impact and the usage of resources in distribution systems and food production have become indispensable [34].

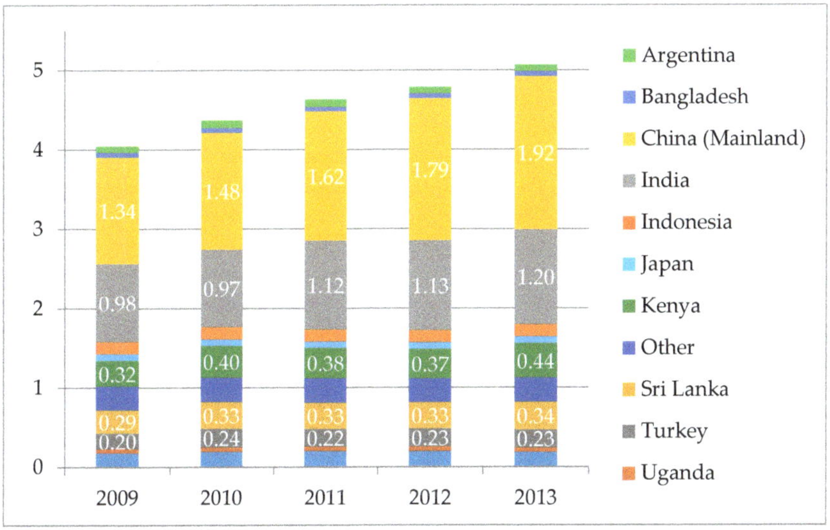

Figure 1. System boundaries and process of Dongshan tea.

The life cycle concept is being increasingly emphasized and regarded as a main idea in ensuring the transition to sustainable production and consumption patterns. LCA has been used broadly for assessing agricultural systems and food processing and manufacturing activities for ensuring sufficient and thorough support in decision-making under business and policy development circumstances [15]. Therefore, an extensive review of literature was conducted by the current work to sort out studies that contained subjects related to LCA for the same agricultural products. Then, their assessment tools were assessed for the main applying direction for the cases that use LCA. Results are shown in Table 1.

Table 1. Comparison of LCA in agricultural products.

Reference	Product/Country	System Boundaries	Method	Results
Farshad et al. [35]	Tea/Iran	Cradle to grave	CML-IA	Most pollutant inputs were machinery and diesel fuel.
Munasinghe et al. [36]	Tea/Sri Lanka	Cradle to grave	LCA	Energy use was the highest in the consumer use phase, whereas CO_2 emission was highest in the packaging phase.
Azapagic et al. [37]	Tea/Kenyan	Cradle to grave	IPCC 100	CO_2 emission was highest in the consumer use phase.

Table 1. *Cont.*

Reference	Product/Country	System Boundaries	Method	Results
Chen et al. [38]	Tea/Taiwan	Cradle to grave	PAS 2050	Fertilizer use in raw material phase and energy use in consumer use phase were hotspots.
Li et al. [39]	Vegetables/China	Farm gate to farm gate	USEtox 2.01	Vegetable multicropping system use would result in reduced environmental impacts.
Shen et al. [40]	Vegetables/China	Farm gate to farm gate	LCA	Venlo-greenhouse environmental impact was the most serious.
Theurl et al. [41]	Vegetables/Austria	Cradle to market	Ecoinvent v2.2	Unheated winter vegetable production was feasible.
Liu et al. [42]	Pear/China	Cradle to market	IPCC 2007	Using manure for biogas production and organic farming can reduce GHG emission.
Ingrao et al. [43]	Peach/Sicilian	Cradle to grave	IMPACT2002+	Largest impacts were due to huge volumes of water and energy used by irrigation.
Longo et al. [44]	Apple/Italy	Cradle to gate	ILCD 2011	Largest energy and environmental impacts were due to fertilizers, pesticides, and diesel.

Researches on LCA for tea have generally focused on cradle to grave in environmental impact and overall sustainability analyses. Farshad et al. [35] worked on an environmental–economic analysis of tea's life cycle in Iran from cradle to grave. LCA results indicated that the major pollutant inputs were machinery and diesel fuel in farms and factories, whereas the three-layer packaging design had the smallest environmental impact. Munasinghe et al. [36] focused on environmental impacts, economic, social, and overall sustainability of the tea sector in Sri Lanka from cradle to grave. Their results showed that energy use was highest during the consumer use phase; CO_2 emission was highest (44–47%) during the packaging phase; labor use was highest during the cultivation phase; and cost was highest in the cultivation and purchasing phases. Chen et al. [38] studied the Organic Tea Product Supply Chain Process Map and Carbon Footprint of Taiwan from cradle to grave. LCA results found that the carbon footprint of tea was 12.53 CO_2eq/kg, which was highest (48.87%) during the raw material phase, followed by the consumer use phase (31.8%). Fertilizer use during the raw material phase and energy use during the consumer use phase were the identified hotspots.

In the field of LCA research on vegetables, Li et al. [39] studied a highly diverse vegetable multi-cropping system in Fengqiu County (China) from farm gate to farm gate. Results showed that a vegetable multi-cropping system would cause even fewer environmental impacts compared with a single-cropping system. Shen et al. [40] studied the three facility modes of vegetable production in China from farm gate to farm gate. LCA results indicated that the serious impact on fresh water depletion and human, fresh water, and terrestrial toxicity is solar greenhouse. Furthermore, the venlo-greenhouse environmental impacts are 101 and 740 times more serious than the solar greenhouse and pollution-free approaches, respectively. Theurl et al. [41] studied unheated, soil-grown winter vegetables in Austria via LCA from "cradle to gate". LCA results found that unheated winter vegetable production was more feasible than existing systems in Austria and Italy.

In the field of LCA research on fruits, Liu et al. [42] studied fossil energy use and greenhouse emission in Chinese pear production from cradle to gate. LCA results showed that GHG emissions in the pear production chain can be reduced by the use of manure in biogas production, transition from conventional farming to organic farming, and reducing of mechanical cultivation. For reducing environmental impacts, LCA could be applied as a means to conduct selections of agricultural inputs. The work of Ingrao et al. [43] highlighted environmental hotspots in Sicilian peach production systems from cradle to grave. LCA results found that irrigation methods had the most serious impact due to the use of large volumes of water and energy. With improvements in irrigation methods, the production process and usage of agricultural machinery can reduce GHG emissions. Longo et al. [44] studied organic and conventional apple supply chains in the north of Italy from cradle to gate. LCA results showed that a considerable share of the overall energy and environmental impacts in farming was because of the diesel consumption of agricultural machines and the use of fertilizers and pesticides.

Research on the sustainability of tea production and consumption shall produce important information about means of improving the global tea industry and about output key experiences for a broad range of other agri-industries and contribute greatly to make development more sustainable [36]. According to Table 1, environmental LCA is a useful means for reporting environmental improvements, given that it quantifies causes of impacts across the product's life cycle for a range of environmental impacts and provides relevant facts and information that can guide decisions on practice change. By contrast, substantial research methods of LCA are available, but most of them are complex and time consuming. Therefore, the simplification and standardization of LCA methods will assist in the development of sustainable agri-industries.

2. Materials and Methods

2.1. Study Scope and Goal

Defining system boundaries on the basis of related goals is the first step in LCA (Figure 2). Dongshan Township (Yilan County) in the east of Taiwan was the main field in this study. Tea is the major crop in the area, which measures approximately 0.41 hectare and tea production volume is approximately 150 kg. Teas have been produced here since 1987, and tea saplings originate from Nantou. Even without applications for organic certification, mountain spring water is being used for irrigation, and soybean meal is being applied as base fertilizer, aided with a few chemical fertilizers and pesticides. The cultivation method is relatively eco-friendly [45].

Field investigation and literature collection were conducted for assessing the cradle-to-grave data, including materials obtained, manufacturing, distribution and transportation, consumer use, and disposal and recycling phases. A PAS 2050-based LCA approach was used to evaluate the environmental impact and carbon emission from each phase through SimaPro 8.0.2 (PRé Consultants B.V., Amersfoort, the Netherlands), and the concept of LCA was employed for determining the environmental impact and carbon emission of the product and for proposing countermeasures for carbon emission reduction.

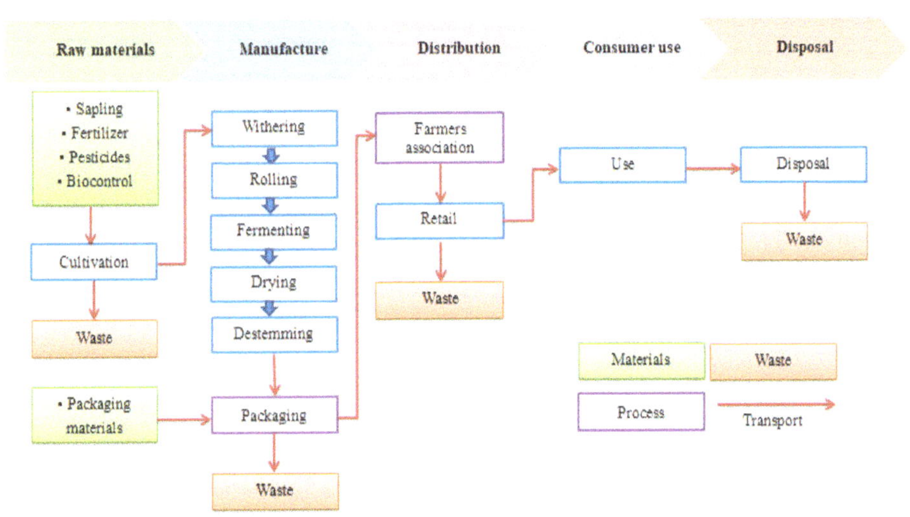

Figure 2. System boundaries and process of Dongshan tea.

2.2. Life Cycle Inventory (LCI) Analysis and Limited

LCI analysis is the second step in LCA; it aims to obtain an accurate study of product carbon footprint. Although PAS 2050, TS-Q 0010, and Product Life Cycle Accounting and Reporting Standard and ISO14067 all consider that treatments of specific emissions and removals are given of land use change, renewable power resources, and carbon storage, as well as delayed emissions, these approaches are different and incomplete [46]. Inventory data for the study were obtained from face-to-face interviews, databases, and other studies; however, the inventory of land use change, delayed emissions, renewable power resources, and carbon storage was excluded from this study due to limited manpower and material resources [46], and CO_2 absorption by growing plants was not considered [47].

Data collected from tea sapling acquisition, transport, and disposal phases were from the factories. Data for cultivation (Compound fertilizer (15-15-15 = 15% nitrogen, 15% phosphorus, and 15% potassium), Soybean meal (6.8-1.5-2.3), pesticide (herbicide), gasoline for transport, PE barrel, and PVC tube) and manufacturing (withering, rolling, fermenting, drying, sorting, and packaging) were collected from factories and other in-country studies [48]. At the consumer use phase, relevant emissions, such as boiling water used to brew tea, were assessed. In this study, 1 kg tea was assigned as the functional unit; 10 g tea and 0.5 L water were assigned for the consumer use phase; and 0.06 kWh from using an electric kettle with grid electricity for boiling a pot of water and 0.35 L wastewater was assumed to be energy consumption for assessment. The LCI result for a 150 kg production in 0.41 ha land use is shown in Table 2.

Table 2. Inventory of data for Dongshan tea LCI.

Processing	Material/Energy	Quantity	Unit
Raw materials	Sapling transport	23.25	TKM
	Machine (diesel)	1.5	L
	Machine (gasoline)	55.6	L
	Compound fertilizer (15-15-15)	240	kg
	Soybean meal (6.8-1.5-2.3)	60	kg
	Pesticide (herbicide)	0.5	L
	PE (tube and barrel)	13.99	kg
	PVC tube	1.99	kg
Manufacture	Machine (electricity)	172.42	kWh
	Kerosene	5	L
	Liquefied petroleum gas	51.81	L
	Polythene bag	1.5	kg
Distribution	Finished product transport	8.25	TKM
Consumer use	Boiling of water	900	kWh
	Tap water	7.5	m^3
	Wastewater	5.63	m^3
Disposal	Retort pouch (landfill)	50	kg
	Papers (incineration plant)	60	kg

2.3. Carbon Footprint Analysis

For quantifying the GHG impact of a product, PAS 2050, TS-Q 0010, Product Life Cycle Accounting and Reporting Standard, and ISO14067 provide principles and requirements. Although methodologies and procedures of these standards are similar, some differences still exist; quantization evolving from GHG activity data multiplied by GHG emission or removal factors is recommended and in common use [46]. In this study, this equation was used to quantify the carbon footprint for Dongshan tea from cradle to grave.

Emission factors were mainly obtained from the database of Taiwan Environmental Protection Administration, and the rest were based on the SimaPro 8.0.2 database and other public data. The GWP value was used on the basis of IPCC 2007 100a, and the carbon dioxide emissions generated from electricity consumption (0.532 kg CO_2e/KWh) were determined by Taiwan's National Greenhouse Gas Registry website [49].

2.4. Life Cycle Impact Assessment (LCIA)

Environmental impacts were assessed using IMPACT2002+, with an assessment pattern involving five steps, namely, characterization, damage assessment, normalization, weighting, and single score computation [50]. This assessment method indicates a workable implementation of a combined midpoint and damage approach, integrating all kinds of life cycle inventory results from 13 midpoint categories to four damage categories. These midpoint categories can make the interpretation easier and more useful for optimizing the damage categories [51].

The 13 midpoint categories of IMPACT 2002+ are carcinogens, respiratory, noncarcinogens, inorganics, ozone layer depletion, ionizing radiation, respiratory organics, terrestrial ecotoxicity, aquatic ecotoxicity, terrestrial acidification/nutrification, land occupation, nonrenewable energy, global warming potential, and mineral extraction. The four damage categories are climate change, human health, ecosystem quality, and resources. Jolliet et al. [51] introduced the set of normalization factors, which were applied to change each category value into a new damage unit to conquer the problems. The relations between impact categories and damaged categories are shown in Table 3.

Table 3. IMPACT 2002+ damage unit values.

Damage Categories Value/Damage Unit	Midpoint Category	Value	Damage Unit
Human health 0.0077 DALY/pers/yr	Carcinogens	1.45×10^{-6}	DALY/kg C_2H_3Cl
	Noncarcinogens	1.45×10^{-6}	DALY/kg C_2H_3Cl
	Respiratory inorganics	7.00×10^{-4}	DALY/kg $PM_{2.5}$
	Ozone layer depletion	1.05×10^{-3}	DALY/kg CFC-11
	Ionizing radiation	2.10×10^{-10}	DALY/Bq C-14
	Respiratory organics	2.13×10^{-6}	DALY/kg C_2H_4
Ecosystem quality 4650 PDF × m^2 × yr/pers/yr	Aquatic ecotoxicity	8.86×10^{-5}	PDF × m^2 × yr/kg·TEG water
	Terrestrial ecotoxicity	8.86×10^{-5}	PDF × m^2 × yr/kg·TEG soil
	Terrestrial acidification/nutrification	1.04	PDF × m^2 × yr/kg SO_2
	Land occupation	1.09	PDF × m^2 × yr/m^2org.arable
Climate change 9950 Kg CO_2/pers/yr	Global warming potential	1	kg CO_2/kg CO_2
Resources 152,000 MJ primary/pers/yr	Nonrenewable energy	5.10×10^{-2}	MJ primary/MJ primary
	Mineral extraction	45.6	MJ primary/MJ surplus

3. Results and Discussion

3.1. Carbon Footprint Analysis Results

In the case study of Dongshang tea, the carbon footprint is approximately 7.035 CO_2eq/kg, and the ratio is roughly in accordance with the study of cradle to grave from local experts [38]. Details are specified in Figure 3. Our study found that consumer use, accounting for 45.58% of the total, is the major source of carbon emissions in the tea product life cycle. The raw material phase is the second main source of carbon emissions, accounting for 35.15% of the total, followed by the manufacturing phase, which accounts for 18.67% of the total. Distribution and disposal phases have a low percentage of emissions.

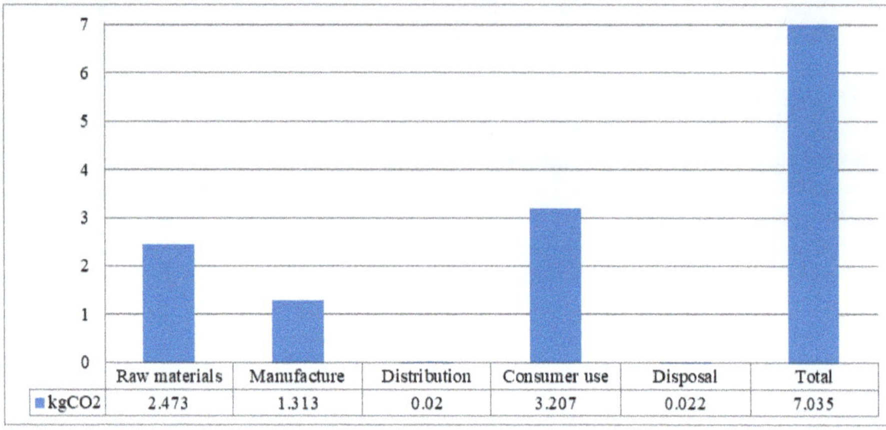

Figure 3. Overall carbon emission ($kgCO_2eq/kg$) contribution.

In this study, the raw material phase, which involves machinery use, fertilizer application, sapling transport, and pesticide use, contributes 2.473 CO_2eq/kg per kg of tea, and nitrogen fertilizer inputs into planting and growing was identified as a hotspot. Since the 1850s, a large volume of anthropogenic nitrogen fertilizer has been applied to agricultural land to facilitate crop production. Improper nitrogen fertilizer management causes various ecological and environmental problems, and compared with the production of phosphate and potash fertilizers, that of nitrogen fertilizers demands more energy requirement. Therefore, carbon footprint can be reduced via the efficient use of nitrogen fertilizers [52,53].

The manufacturing phase contributes 1.313 CO_2eq/kg per kg of tea, and the use of LPG roller fixation machine and electricity consumption for hot air drying were identified as hotspots. The four main types of tea are black, green, white, and Oolong. Although they all originate from *C. sinensis*, they are produced with varying fermenting degrees. Black tea is fully fermented; Oolong tea is semifermented; white tea is low fermented, and green tea is nonfermented [54].

At the consumer use phase, tap water, boiling of water, and wastewater contribute 3.207 CO_2eq/kg from 10 g tea and 0.5 L water. The main source of carbon emissions in the tea product life cycle is electricity consumption from boiling a pot of water at 0.06 KWh using an electric kettle with grid electricity. Results in this study confirmed the findings of Munasinghe et al. [36], Azapagic et al. [37], and Doublet and Jungbluth [55]. Therefore, carbon footprint can be reduced through the minimization of the frequency of boiling water (such as heat preservation) or use of highly efficient boiling water facilities. The Taiwan Bureau of Energy, Ministry of Economic Affairs initiated the voluntary Energy Label program in order to urge manufacturers to invest in research and development of energy-efficient products and promote the deployment of energy efficiency technologies. A consumer environment that values highly energy-efficient products could be created, given that consumers could easily recognize such products through the "Energy Label."

3.2. *LCIA Results*

The software SimaPro was used in this study to calculate the environmental impact of Dongshan tea, and the calculation was performed via the IMPACT 2002+ assessment method. Figure 4 shows the normalization environmental impact category of Dongshan tea. The figure indicates that the biggest environmental impact categories are human health, climate change, resources in the raw material phase, and climate change in the consumer use and manufacturing phases.

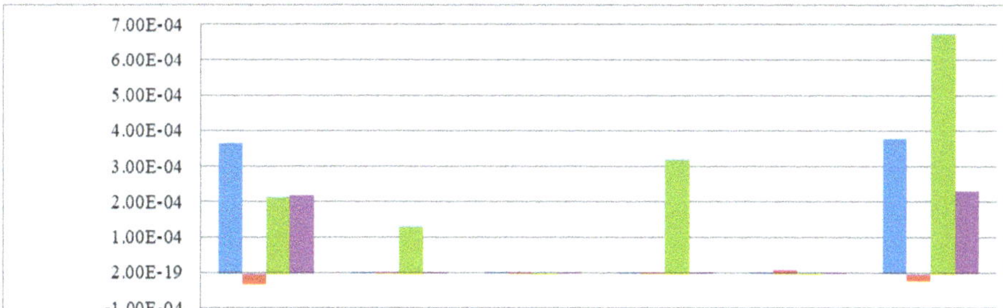

	Raw materials	Manufacture	Distribution	Consumer use	Disposal	Total
Human health	3.68E-04	3.19E-06	3.60E-06	2.58E-07	3.65E-06	3.79E-04
Ecosystem quality	-2.85E-05	6.69E-08	5.50E-07	3.69E-08	8.71E-06	-1.91E-05
Climate change	2.18E-04	1.33E-04	1.99E-06	3.24E-04	2.08E-06	6.79E-04
Resources	2.22E-04	5.30E-06	1.98E-06	1.11E-06	1.86E-06	2.32E-04

Figure 4. Normalization environmental impact of Dongshan tea.

This study found that 96% of the impact of resource categories comes from the raw material phase, the main midpoint being nonrenewable energy. The impact of climate change categories is quite similar with carbon footprint analysis; 46%, 35%, and 19% come from the consumer use, raw material, and manufacturing phases, respectively. The distribution and disposal phases have a lower percentage of environmental impacts due to the concept of local use and consumption, which minimizes food miles and waste materials through the use of large packing sizes for tea. Ecosystem quality yielded a negative value in the raw material phase, given that mountain spring water and soybean meal are used as base fertilizer and only minimal amounts of chemical fertilizers and pesticides are applied; the cultivation method in this tea farm is relatively eco-friendly.

This study found that 97% of the impact of human health categories comes from the raw material phase, and the largest midpoint is respiratory inorganics, which contributes 76%. Paramesh et al. [56] also reported that the life cycle assessment pointed out on-farm emissions are the hotspot for respiratory inorganics, whereas fertilizers are a potential hotspot. With increased chemical fertilizer inputs and adoption of new technologies, crop yields have increased steadily, and food security has improved, although it results in soil deterioration, GHG emissions, and water contamination [57,58]. Fertilizers are held responsible as the main factor for that category; however, compared to climate change and fossil fuel depletion, these emissions are relatively low. With an optimized fertilization strategy (including use compound fertilizer, and shorten transportation distances), the environmental burden can be reduced [47]. Although the influence of organic fertilizers on crop yield is gradual and changeable in a short period of time [59], the application of organic fertilizers, instead of chemical fertilizers, is economically practical, contributes considerably to environmental sustainability, and increases agricultural production [60]. In addition to spreading awareness about the importance of environmental sustainability, national policy subsidies must market the use of organic fertilizers [61,62]. For encouraging the development of organic agriculture in the country, organic agriculture and eco-friendly farming promotion have been included in 10 key policies of the new agriculture policy by the Taiwan Council of Agriculture. Since 2017, subsidies for organic and eco-friendly farming have been in place; a user receives NTD 3 per kg by using the recommended fertilizer brand, and for every 10 t organic fertilizer applied on a hectare of land, a user receives NTD 30,000.

4. Conclusions and Perspectives

This study investigated the environmental impact of Taiwanese Dongshan tea, the carbon footprint assessment of growth in Yilan, and the consumption of tea in Taipei, covering the raw material, manufacturing, distribution and transportation, consumer use, and disposal and recycling phases. LCA results showed that energy used in the consumer use phase is the main hotspot in the case of

Dongshan tea, and the biggest environmental impact in the human health category comes from the use of fertilizers in the raw material phase. Therefore, a national policy of subsidizing use of organic fertilizers and optimized fertilization strategy can reduce the environmental impact.

The largest energy consumption and life cycle environmental impact during tea processing is contributed by black tea (versus Oolong and green) [63]. Dongshan tea is a kind of black tea, where the emissions of LCA is the remaining 18.67% in the manufacture phase. In this case, solar power can be utilized directly or indirectly during leaf drying and withering, thus reducing energy use.

Tea is the most widely consumed nonalcoholic beverage in the world apart from water [64]; therefore, improving sustainability in the tea industry will facilitate sustainable production and consumption. Hence, we propose some measures for reducing carbon footprint and environmental impact, including the use of highly efficient boiling water facilities and heating preservation. The implementation of these measures will minimize the frequency of boiling water in the consumer use phase and decrease the use of nitrogen fertilizers in the raw material phase.

An entire change of the economic model is impracticable; thus, a progressive conversion of consumer behavior is feasible in achieving an environmentally sustainable society [65]. In summary, LCA is an internationally recognized approach for the environmental assessment of products and processes [66]. The methods, results, and conclusions in this study can be used as a reference by future researchers. In addition, this study provides a complete impact analysis and identifies relevant hotspots. Results provide essential data for policymakers, tea producers, and consumers, and the suggested measures for the reduction of environmental impact can contribute toward a low-carbon and sustainable agricultural development and consumption.

Author Contributions: C.-H.C. and A.H.H. analyzed the data and compiled the article. L.H.H. and M.-H.C. supported the data. Y.-C.L. and Z.C. provided technical comments on the whole work.

Funding: This research was funded by Youngsun Cultue & Education Foundation under the project of Carbon Footprint Inventory of Agricultural products in Yilan County.

Acknowledgments: The study was supported by Youngsun Cultue & Education Foundation under the project of Carbon Footprint Inventory of Agricultural products in Yilan County. In addition, it's many thanks to the reviewers for their valuable recommendation to improve this research.

Conflicts of Interest: The authors declare no conflict of interest.

References

1. Edenhofer, O.; Pichs-Madruga, R.; Sokona, Y.; Farahani, E.; Kadner, S.; Kadner, K.; Seyboth, A.; Adler, I.; Baum, S.; Myhre, G.; et al. *Climate Change 2014: Mitigation of Climate Change*; Working Group III Contribution to the IPCC Fifth Assessment Report; Cambridge University Press: Cambridge, UK, 2015.
2. Food and Agriculture Organization of the United Nations (FAO). *Regional Strategy for Sustainable Hybrid Rice Development in Asia*; Food and Agriculture Organization of the United Nations Regional Office for Asia and the Pacific: Bangkok, Thailand, 2014.
3. Irani, Z.; Sharif, A.M. Sustainable food security futures: Perspectives on food waste and information across the food supply chain. *J. Enterp. Inf. Manag.* **2016**, *29*, 171–178. [CrossRef]
4. Runhaar, H. Tools for integrating environmental objectives into policy and practice: What works where? *Environ. Impact Assess. Rev.* **2016**, *59*, 1–9. [CrossRef]
5. Godfray, H.C.J.; Beddington, J.R.; Crute, I.R.; Haddad, L.; Lawrence, D.; Muir, J.F.; Pretty, J.; Robinson, S.; Thomas, S.M.; Toulmin, C. Food security: The challenge of feeding 9 billion people. *Science* **2010**, *327*, 812–818. [CrossRef] [PubMed]
6. Meyfroidt, P. Trade-offs between environment and livelihoods: Bridging the global land use and food security discussions. *Glob. Food Secur.* **2018**, *16*, 9–16. [CrossRef]
7. Lobell, D.B.; Schlenker, W.; Costa-Roberts, J. Climate trends and global crop production since 1980. *Science* **2011**, *333*, 616–620. [CrossRef] [PubMed]

8. Kangalawe, R.Y.M.; Mungongo, C.G.; Mwakaje, A.G.; Kalumanga, E.; Yanda, P.Z. Climate change and variability impacts on agricultural production and livelihood systems in Western Tanzania. *Clim. Dev.* **2017**, *9*, 202–216. [CrossRef]
9. Sá, J.C.M.; Lal, R.; Cerri, C.C.; Lorenz, K.; Hungria, M.; Carvalho, P.C.F. Low-carbon agriculture in South America to mitigate global climate change and advance food security. *Environ Int.* **2017**, *98*, 102–112. [CrossRef]
10. Moraes, A.; Carvalho, P.C.F.; Anghinoni, I.; Lustosa, S.B.C.; Andrade, S.E.V.G.; Kunrath, T.R. Integrated crop–livestock systems in the Brazilian subtropics. *Eur. J. Agron.* **2014**, *57*, 4–9. [CrossRef]
11. Prost, L.; Berthet, E.T.; Cerf, M.; Jeuffroy, M.H.; Labatut, J.; Meynard, J.M. Innovative design for agriculture in the move towards sustainability: Scientific challenges. *Res. Eng. Des.* **2017**, *28*, 119–129. [CrossRef]
12. Pigford, A.A.E.; Hickey, G.M.; Klerkx, L. Beyond agricultural innovation systems? Exploring an agricultural innovation ecosystems approach for niche design and development in sustainability transitions. *Agric. Syst.* **2018**, *164*, 116–121. [CrossRef]
13. Rockström, J.; Williams, J.; Daily, G.; Noble, A.; Matthews, N.; Gordon, L.; Wetterstrand, H.; De Clerck, F.; Shah, M.; Steduto, P.; et al. Sustainable intensification of agriculture for human prosperity and global sustainability. *Ambio* **2017**, *46*, 4–17. [CrossRef] [PubMed]
14. Pergola, M.; Persiani, A.; Pastore, V.; Palese, A.M.; Arous, A.; Celano, G. A comprehensive Life Cycle Assessment (LCA) of three apricot orchard systems located in Metapontino area (Southern Italy). *J. Clean. Prod.* **2017**, *142*, 4059–4071. [CrossRef]
15. Notarnicola, B.; Salomone, R.; Petti, L.; Renzulli, P.A.; Roma, R.; Cerutti, A.K. *Life Cycle Assessment in the Agri-Food Sector: Case Studies, Methodological Issues and Best Practices*; Springer International Publishing: Cham, Switzerland, 2015; pp. 123–184, ISBN 978-3-319-11940-3.
16. Bartzas, G.; Zaharaki, D.; Komnitsas, K. Life cycle assessment of open field and greenhouse cultivation of lettuce and barley. *Inf. Process. Agric.* **2015**, *2*, 191–207. [CrossRef]
17. Wu, H.; Wang, S.; Gao, L.; Zhang, L.; Yuan, Z.; Fan, T.; Wei, K.; Huang, L. Nutrient-derived environmental impacts in Chinese agriculture during 1978–2015. *J. Environ. Manag.* **2018**, *217*, 762–774. [CrossRef] [PubMed]
18. Hellweg, S.; Mila Canals, L. Emerging approaches, challenges and opportunities in life cycle assessment. *Science* **2014**, *344*, 1109–1113. [CrossRef] [PubMed]
19. Ali, S.A.; Tedone, L.; De Mastro, G. Optimization of the environmental performance of rainfed durum wheat by adjusting the management practices. *J. Clean. Prod.* **2015**, *87*, 105–118. [CrossRef]
20. Kulak, M.; Nemecek, T.; Frossard, E.; Gaillard, G. Eco-efficiency improvement by using integrative design and life cycle assessment. The case study of alternative bread supply chains in France. *J. Clean. Prod.* **2016**, *112*, 2452–2461. [CrossRef]
21. Renouf, M.A.; Renaud-Gentie, C.; Perrin, A.; Kanyarushoki, C.; Jourjon, F. Effectiveness criteria for customised agricultural life cycle assessment tools. *J. Clean. Prod.* **2018**, *179*, 246–254. [CrossRef]
22. Oldfield, T.L.; White, E.; Holden, N.M. The implications of stakeholder perspective for LCA of wasted food and green waste. *J. Clean. Prod.* **2018**, *170*, 1554–1564. [CrossRef]
23. Kung, C.C.; McCarl, B.A.; Chen, C.C.; Chen, L.J. Environmental Impact and Bioenergy Potential: Evaluation of Agricultural Commodity and Animal Waste Based Biochar Application on Taiwanese Set-aside Land. *Energy Procedia* **2014**, *61*, 679–682. [CrossRef]
24. Tilman, D.; Balzer, C.; Hill, J.; Befort, B.L. Global food demand and the sustainable intensification of agriculture. *Proc. Natl. Acad. Sci. USA* **2011**, *108*, 20260–20264. [CrossRef] [PubMed]
25. Lee, S.C.; Wang, C.H.; Yen, C.E.; Chang, C. DNA barcode and identification of the varieties and provenances of Taiwan's domestic and imported made teas using ribosomal internal transcribed spacer 2 sequences. *J. Food Drug Anal.* **2017**, *25*, 260–274. [CrossRef]
26. Stadler, R.H.; Hughes, G.; Guillaume-Gentil, O. Safety of food and beverages: Coffee, tea and herbals, cocoa and derived products. *Food Mater. Technol. Risks* **2014**, *3*, 371–383.
27. Xia, E.H.; Zhang, H.B.; Sheng, J.; Li, K.; Zhang, Q.J.; Kim, C.; Zhang, Y.; Liu, Y.; Zhu, T.; Li, W.; et al. The Tea Tree Genome Provides Insights into Tea Flavor and Independent Evolution of Caffeine Biosynthesis. *Mol. Plant* **2017**, *10*, 866–877. [CrossRef] [PubMed]
28. FAO. *World Tea Production and Trade Current and Future Development, Trade and Markets Division*; Food and Agriculture Organization: Rome, Italy, 2015.

29. FAO. *Committee on the Commodity Problems Intergovernmental Group on Tea, Current Market Situation and Medium-Term Outlook*; Food and Agriculture Organization: Beijing, China, 2015.
30. Mei, Y. China tea production and marketing report in 2014 and situation forecast in 2015. *Tea World* **2015**, *6*, 50–59.
31. Pinto, M.S. Tea: A new perspective on health benefits. *Food Res. Int.* **2013**, *53*, 558–567. [CrossRef]
32. Shen, C.L.; Chyu, M.C. Tea flavonoids for bone health: From animals to humans. *J. Investig. Med.* **2016**, *64*, 1151–1157. [CrossRef] [PubMed]
33. Zhang, Q.; Gao, W.; Su, S.; Weng, M.; Cai, Z. Biophysical and socioeconomic determinants of tea expansion: Apportioning their relative importance for sustainable land use policy. *Land Use Policy* **2017**, *68*, 438–447. [CrossRef]
34. Rivera, X.C.S.; Bacenetti, J.; Fusi, A.; Niero, M. The influence of fertiliser and pesticide emissions model on life cycle assessment of agricultural products: The case of Danish and Italian barley. *Sci. Total Environ.* **2017**, *592*, 745–757. [CrossRef]
35. Farshad, S.F.; Hamed, K.P.; Mahmoud, G.N.R.; Chen, G. Cradle to grave environmental-economic analysis of tea life cycle in Iran. *J. Clean. Prod.* **2018**, *196*, 953–960. [CrossRef]
36. Munasinghe, M.; Deraniyagala, Y.; Dassanayake, N.; Karunarathna, H. Economic, social and environmental impacts and overall sustainability of the tea sector in Sri Lanka. *Sustain. Prod. Consum.* **2017**, *12*, 155–169. [CrossRef]
37. Azapagic, A.; Bore, J.; Cheserek, B.; Kamunya, S.; Elbehri, A. The global warming potential of production and consumption of Kenyan tea. *J. Clean. Prod.* **2016**, *112*, 4031–4040. [CrossRef]
38. Chen, L.C.; Yang, T.F. Carbon Footprint Study of Organic Tea Product. *J. TungNan Univ.* **2011**, *37*, 203–214.
39. Li, L.; Wu, W.; Giller, P.; O'Halloran, J.; Liang, L.; Peng, P.; Zhao, G. Life cycle assessment of a highly diverse vegetable multi-cropping system in Fengqiu county, China. *Sustainability* **2018**, *10*, 983. [CrossRef]
40. Shen, J.; Gao, L.; Chen, Q.; Zhang, Z.; Wang, Z. Life cycle assessment (LCA) on three facility modes of vegetable production. *J. Jiangsu Univ.* **2013**, *34*, 650–657.
41. Theurl, M.C.; Hörtenhuber, S.J.; Lindenthal, T.; Palme, W. Unheated soil-grown winter vegetables in Austria: Greenhouse gas emissions and socio-economic factors of diffusion potential. *J. Clean. Prod.* **2017**, *151*, 134–144. [CrossRef]
42. Liu, Y.; Langer, V.; Høgh-Jensen, H.; Egelyng, H. Life Cycle Assessment of fossil energy use and greenhouse gas emissions in Chinese pear production. *J. Clean. Prod.* **2010**, *18*, 1423–1430. [CrossRef]
43. Ingrao, C.; Matarazzo, A.; Tricase, C.; Clasadonte, M.T.; Huisingh, D. Life cycle assessment for highlighting environmental hotspots in Sicilian peach production systems. *J. Clean. Prod.* **2015**, *92*, 109–120. [CrossRef]
44. Longo, S.; Mistretta, M.; Guarino, F.; Cellura, M. Life Cycle Assessment of organic and conventional apple supply chains in the North of Italy. *J. Clean. Prod.* **2017**, *140*, 654–663. [CrossRef]
45. Zhang, L.; Li, X.; Yu, J.; Yao, X. Toward Cleaner Production: What drives farmers to adopt eco-friendly agricultural production? *J. Clean. Prod.* **2018**, *184*, 550–558. [CrossRef]
46. Gao, T.; Liu, Q.; Wang, J. A comparative study of carbon footprint and assessment standards. *Int. J. Low Carbon Technol.* **2014**, *9*, 237–243. [CrossRef]
47. Hasler, K.; Broring, S.; Omta, S.W.F.; Olfs, H.W. Life cycle assessment (LCA) of different fertilizer producer types. *Eur. J. Agron.* **2015**, *69*, 41–51. [CrossRef]
48. Hsieh, H.W.; Tsai, I.T.; Hsieh, C.L. Survey and Analysis of Fuel Consumption of Popular Gasoline-Type Agricultural Machinery in Taiwan. *J. Agric. Mach.* **2007**, *16*, 1–14.
49. Taiwan Environmental Protection Administration (Taiwan EPA). *Environmental Resource Database in Taiwan*; Environmental Protection Administration (EPA): Taiwan, 2012. Available online: https://erdb.epa.gov.tw/DataRepository/Other/Carbon_Footprint_Emission_Factor.aspx (accessed on 18 December 2018).
50. Prado, V.; Wender, B.A.; Seager, T.P. Interpretation of comparative LCAs: External normalization and a method of mutual differences. *Int. J. Life Cycle Assess.* **2017**, *22*, 2018–2029. [CrossRef]
51. Jolliet, O.; Margni, M.; Charles, R.; Humbert, S.; Payet, J.; Rebitzer, G.; Rosenbaum, R. IMPACT 2002+: A new life cycle impact assessment methodology. *Int. J. Life Cycle Assess.* **2003**, *8*, 324–330. [CrossRef]
52. Maraseni, T.N.; Cockfield, G.; Maroulis, J.; Chen, G. An assessment of greenhouse gas emissions from the Australian vegetables industry. *J. Environ. Sci. Health Part B* **2010**, *45*, 578–588. [CrossRef] [PubMed]

53. Cao, P.; Lu, C.; Yu, Z. Historical nitrogen fertilizer use in agricultural ecosystems of the contiguous United States during 1850-2015: Application rate, timing, and fertilizer types. *Earth Syst. Sci. Data* **2018**, *10*, 969–984. [CrossRef]
54. Sanlier, N.; Goksen, B.B.; Altug, M. Tea consumption and disease correlation. *Trends Food Sci. Technol.* **2018**, *78*, 98–106. [CrossRef]
55. Doublet, G.; Jungbluth, N. *Life Cycle Assessment of Drinking Darjeeling Tea*; ESU-Services Ltd.: Schaffhausen, Switzerland, 2010.
56. Paramesh, V.; Arunachalam, V.; Nikkhah, A.; Das, B.; Ghnimi, S. Optimization of energy consumption and environmental impacts of arecanut production through coupled data envelopment analysis and life cycle assessment. *J. Clean. Prod.* **2018**, *203*, 674–684. [CrossRef]
57. Duan, Y.; Xu, M.; Gao, S.; Liu, H.; Huang, S.; Wang, B. Long-term incorporation of manure with chemical fertilizers reduced total nitrogen loss in rain-fed cropping systems. *Sci. Rep.* **2016**, *6*, 33611. [CrossRef]
58. Norman, U.; Dazzo, F.B. Making rice production more environmentally-friendly. *Environments* **2016**, *3*, 12. [CrossRef]
59. Hu, H.; Yang, Y. Research on farmers' chemical fertilizers use based on resources substitution. *J. Agron. Technol. Econ.* **2015**, *3*, 84–91.
60. Ning, C.C.; Gao, P.D.; Wang, B.Q.; Lin, W.P.; Jiang, N.H.; Cai, K.Z. Impacts of chemical fertilizer reduction and organic amendments supplementation on soil nutrient, enzyme activity and heavy metal content. *J. Integr. Agric.* **2017**, *16*, 1819–1831. [CrossRef]
61. Jaza Folefack, A. The determinants for the adoption of compost from household waste for crop production by farmers living nearby Yaounde, Cameroon: Descriptive and logit model approaches of analysis. *Int. J. Brain Cognit. Sci.* **2015**, *9*, 308–328. [CrossRef]
62. Lim, S.L.; Lee, L.H.; Wu, T.Y. Sustainability of using composting and vermicomposting technologies for organic solid waste biotransformation: Recent overview, greenhouse gases emissions and economic analysis. *J. Clean. Prod.* **2016**, *111*, 262–278. [CrossRef]
63. Khanali, M.; Mobli, H.; Hosseinzadeh-Bandbafha, H. Modeling of yield and environmental impact categories in tea processing units based on artificial neural networks. *Environ. Sci. Pollut. Res.* **2017**, *24*, 1–17. [CrossRef]
64. Pelvan, E.; Özilgen, M. Assessment of energy and exergy efficiencies and renewability of black tea, instant tea and ice tea production and waste valorization processes. *Sustain. Prod. Consum.* **2017**, *12*, 59–77. [CrossRef]
65. Matuštík, J.; Kočí, V. Environmental impact of personal consumption from life cycle perspective—A Czech Republic case study. *Sci Total Environ.* **2019**, *646*, 177–186. [CrossRef]
66. Dassisti, M.; Intini, F.; Chimienti, M.; Starace, G. Thermography-enhanced LCA (Life Cycle Assessment) for manufacturing sustainability assessment. The case study of an HDPE (High Density Polyethylene) net company in Italy. *Energy* **2016**, *108*, 7–18. [CrossRef]

energies

Article

A Study on the Strategy for Departure Aircraft Pushback Control from the Perspective of Reducing Carbon Emissions

Xinhua Zhu [1,2], Nan Li [3], Yu Sun [3], Hongfei Zhang [4], Kai Wang [5,*] and Sang-Bing Tsai [5,6,*]

1 College of Economics and Management, Civil Aviation University of China, Tianjin 300300, China; xhzhu@cauc.edu.cn
2 College of Civil Aviation, Nanjing University of Aeronautics and Astronautics, Nanjing 211106, China
3 College of Air Traffic Management, Civil Aviation University of China, Tianjin 300300, China, nanli@cauc.edu.cn (N.L.); sunyu8787@gmail.com (Y.S.)
4 Operation and Control Center, Shandong Airlines, Jinan 250014, China; zhanghf5@sda.cn
5 College of Business Administration, Capital University of Economics and Business, Beijing 100070, China
6 Zhongshan Institute, University of Electronic Science and Technology of China, Zhongshan 528400, China
* Correspondence: wangkai@cueb.edu.cn (K.W.); sangbing@hotmail.com (S.-B.T.)

Received: 2 August 2018; Accepted: 13 September 2018; Published: 17 September 2018

Abstract: In order to reduce the taxiing time of departing aircraft and reduce the fuel consumption and exhaust emissions of the aircraft, Shanghai Hongqiao Airport was taken as an example to study the control strategy for aircraft departure. In this paper, the influence of the number of departure aircraft on the runway utilization rate, the takeoff rate, and the departure rate of flight departures under the conditions of airport runway capacity constraints are studied. The influence of factors, such as the number of departure aircraft, the gate position of the aircraft, and the configuration of airport arrival and departure runways, on the aircraft taxiing time for departure is analyzed. Based on a multivariate linear regression equation, a time prediction model of aircraft departure taxiing time is established. The fuel consumption and pollutant emissions of aircraft are calculated. The experimental results show that, without reducing the utilization rate of the runway and the departure rate of flights, implementing a reasonable pushback number for control of departing aircraft during busy hours can reduce the departure taxiing time of aircraft by nearly 32%, effectively reducing the fuel consumption and pollutant emissions during taxiing on the airport surface.

Keywords: aircraft; taxi time; takeoff rate; pushback control; green transportation; carbon emissions; reducing carbon emissions

1. Introduction

With the rapid development of air transportation, the number of flights at major airports in China has been increasing, making airport surface runways congested. This, in particular, causes the aircraft on airport surfaces to take a long time to taxi, and an excessive number of flights to wait in line at the entrance to the runway. Due to premature pushback of aircraft and waiting on crowded taxiways, an additional 10–20 kg of fuel consumption is added for each additional minute of taxi time [1,2], resulting in an increase in aircraft exhaust emissions affecting the air quality around the airport.

A series of studies have been conducted by scholars at domestic and foreign terminals to control the number of aircraft departing from airport surfaces and reduce the congestion time in aircraft taxiing. In 2007, Balakrishnan H and Jung Y [3] studied the airport surface operation of Dallas–Fort Worth airport by establishing an integer programming model. This study shows that using the method of delaying the pushback time of departing aircraft can reduce the number of airport surface taxiing aircraft and reduce congestion, thereby reducing the average taxiing time of departing aircraft, and using airport surface aircraft taxiway optimization methods can significantly reduce the waiting time for aircraft crossing the runway, thereby reducing the average taxiing time of arriving aircraft. In 2009, Simaiakis I et al. [4] analyzed the key factors affecting the taxi time of departing aircraft. Taking the Boston International Airport (BOS) as an example, a forecasting analysis of the departure taxiing time of aircraft was made, and a queuing theory model based on the departure process of aircraft was proposed. Using this queue-pushback strategy for departing aircraft can reduce the departure taxiing time and, thus, reduce aircraft pollutant emissions. In 2010, Jung YC et al. proposed the Spot Release Planner (SRP) and Runway Scheduler (RS) [5]. The SRP aims to reduce an aircraft's taxiing time by keeping the runway productivity at the maximum level by sorting the order of departure of the departing aircraft on the apron and controlling the pushback time of each aircraft to control the time that the aircraft enters the maneuvering area. The RS is designed to sort and time-allocate departing aircraft and arriving flights across the takeoff runway to achieve maximum runway utilization. Jung YC et al. combined the proposed two strategies to optimize the operation of busy airport surfaces. In 2010, Lee H [6] proposed two ways to optimize the operation of airport surfaces: delayed pushback and path optimization. Delayed pushback refers to the control of off-board aircraft that is applied during an airport congested period to control the airport's congestion. The path optimization optimizes the taxi path of all aircraft at the airport based on the delayed pushback. With the application of airport surface monitoring equipment, it makes it possible to analyze the airport surface trajectory of the aircraft in detail using the airport surface monitoring data. In 2011, I. Simaiakis et al. [7] applied Airport Surface Detection Equipment Model-X (ASDE-X) data from the monitoring equipment at Boston International Airport. They considered the impact of the airport runway configuration, the different type series in the fleet, meteorological conditions, and other factors on the airport runway capacity. A study on aircraft pushback rate control at the airport was conducted. In 2013, S Ravizza [8] and others calculated the required taxiing distance, the total steering angle, the type of departing and arriving aircraft, the number of aircraft in operation on the airport surface, the usage configuration of departure and arrival runways, and the position of the gate. The establishment of a multiple linear regression model helped to provide a more accurate prediction of the aircraft into and out of the required taxi time. In 2015, Tang Y [9] elaborated on the concept of the Advanced Surface Movement Guidance and Control System (A-SMGCS) proposed by International Civil Aviation Organization (ICAO) in his doctoral dissertation and conducted a comprehensive study on an aircraft's initial taxi route planning, real-time optimization of aircraft taxiing routes, and A-SMGCS three-dimensional (3D) simulation. In addition, Xiangling Z et al. [10] studied the issue of virtual pushback queues for departing flights at the gate position and the issue of decision-making for collaborative pushback of departing flights. In 2016, Nan L and Hongzhe L [11] analyzed the surveillance data of Hongqiao Airport, used support vector machines to classify and determine the trajectory of taxi aircraft, and applied data mining technologies to the prediction of airport surface aircraft taxi time, the determination of airport surface taxi hotspots, and conflict zone determination.

Under the same runway configuration conditions, as the number of aircraft pushback into the apron and taxi systems increases, more flights are added to the takeoff queue, resulting in a gradual increase in runway utilization and departures from flights. However, due to the effect of aircraft wake spacing, when a certain number of taxiing departing aircraft is reached, the runway capacity becomes the limiting factor and the number of taxiing departing aircraft will continue to increase. The runway utilization rate and takeoff and departure rate of flights will only tend to change smoothly. Therefore, in practice, the tower controllers are more concerned with a reasonable number of departing aircraft operations in a given runway usage configuration. This paper uses the airport surface monitoring data from Shanghai Hongqiao Airport to study the influence of the number of different departing aircraft within the apron and taxi systems on the takeoff and departure rate of the flights, the departure taxiing time, and the runway utilization rate under runway capacity constraints. A departure time prediction model for departing aircraft is established. A reasonable control strategy is implemented for departing aircraft within the busy airport departure period without reducing the operating efficiency of the runway, thereby reducing the aircraft departure taxiing time and reducing aircraft fuel consumption and pollutant emissions.

2. Airport Surface Operation Data Analysis and Definition

2.1. Airport Surface Monitoring Data Analysis

While the airport surface monitoring system assists controllers in performing aerodrome control services more safely and efficiently, the equipment also records real-time aircraft trajectory data. The aircraft movement trajectory data recorded by the airport surface monitoring system includes a time stamp and the aircraft's position, altitude, speed, and so forth, and by combining these with the airport topology data, we can identify the operational status of the aircraft, such as its pushback, taxiing, takeoff, and landing; calculate parameters, such as taxi distance and taxi time; identify the taxi path; and provide a data foundation for studying aircraft airport surface operation and optimization [12,13].

Through the data analysis of the Shanghai Hongqiao Airport's March 2015 airport surface monitoring system, we have sorted out the full arrival and departure trajectories for flights when using the 18 L/18 R configuration on the departure and arrival runway (18 L runway for approach and 18 R runway for departure calculated as two complete tracks during the stop-and-depart process).

2.2. Airport Surface Operation Data Definition

Aircraft departure taxiing refers to the entire process of the pushback of the aircraft from the gate position, the taxiing to the departure runway, and the wait for takeoff. In order to analyze the operation of airport surfaces, this paper gives the following definitions of the quantitative indicators for measuring airport surface operations:

1. Departure taxiing time: the total time of the aircraft's pushback from the parking position to the time of taxiing to the departure runway, including the aircraft's pushback, apron and taxiway taxiing, and the holding time the entrance of the runway (unit: minute).
2. Number of airport surface aircraft: the total number of departing and arriving aircraft taxiing (including taxi wait) or undergoing pushback in the apron and taxiway systems (unit: flight).
3. Number of departure aircraft: the number of departing aircraft that are taxiing (including the taxi wait) or undergoing pushback in the apron and taxiway systems (unit: flight).
4. Runway utilization rate: the ratio of the length of time the runway is accumulatively occupied over a period of time to the length of the time period of the calculation.
5. Takeoff and departure rate of flights: the number of departing aircraft per unit of time (unit: flights/minute).

When calculating the runway utilization rate over a period of time, first of all, it is necessary to find out the total time taken for the runway to be occupied during this period. In general, the time spent on the following operations of the aircraft is accumulated into the occupied time of the runway.

6. Takeoff running: the duration from the point where a departing aircraft accelerates for takeoff on the runway until the aircraft's landing gear tires are off the ground;
7. Departure waiting: the departing aircraft waits for takeoff clearance on the runway;
8. Final Approach: the duration from an arriving aircraft's being at its final approach phase of 2.5 nm (nautical miles) from the runway's end to the landing of the aircraft on the runway;
9. Landing: the duration of an arriving aircraft's starting to land to when it is off the runway;
10. Cross-taxi: the aircraft crosses over the runway.

For an airport with only one runway, when calculating the runway utilization rate of the airport, the cumulative time of the five operations (6)–(10) of the aircraft needs to be taken into account as the occupied time of the runway.

For airports with multiple runways, if aircraft arrivals and departures run on different runways, the time taken on the departure runway will only take into account the aircraft operations (6), (7), and (10), and only the aircraft operations (8)–(10) need to be considered when the approach runway is occupied.

3. Runway Capacity Analysis

This section applies the airport surveillance data to analyze the actual airport operational data. Hongqiao Airport is a narrow-distance, dual-runway airport; 18 L is mainly used for arrival, 18 R is mainly used for departure, and its taxiway system network is shown in Figure 1.

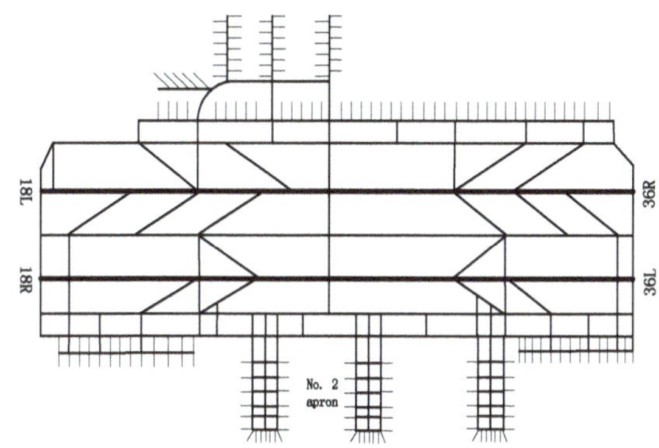

Figure 1. Network diagram of the Shanghai Hongqiao Airport taxiway system.

3.1. Runway Utilization Rate Analysis

This paper analyzes the main runway configuration of Hongqiao Airport, namely the 18 L approach and the 18 R departure configuration. Figure 2 shows the relationship between the utilization of the departure 18 R runway and the number of departing aircraft on the airport surface. The calculation of the runway utilization rate is based on a statistical period of every 15 min, because the number of departing aircraft on the airport surface changes slightly every 15 min and this can avoid statistical errors caused by excessive changes in the number of departing aircraft. Additionally, a 15 min statistical period is not too short and the average runway utilization rate can be calculated during this period so as to avoid statistical errors due to excessive statistical fluctuations caused by the statistical time being too short.

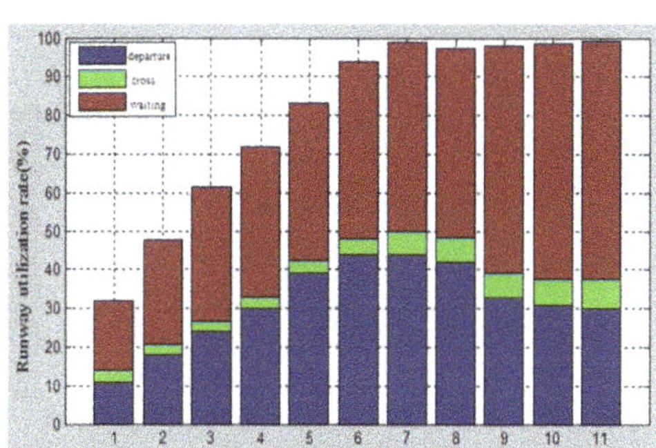

Figure 2. The relationship between runway utilization and the number of departing aircraft.

The horizontal axis in Figure 2 shows the number of departure aircraft on the airport surface and the vertical axis shows the runway utilization rate of 18 R. It can be seen that when the number of departure aircraft on the airport surface is $N \leq 6$, the number of aircraft departing the airport surface at this time is small, and the runway utilization rate of the departure runway 18 R increases with the number of airport surface departure aircraft. When the number of airport surface departure aircraft is $N = 7$, the runway utilization rate is already close to 1, indicating that under the operating scale of the number of departing aircraft, the 18 R departure runway has been basically used efficiently. When the number of departure aircraft in operation on the airport surface is $N \geq 8$, the runway utilization rate fluctuates at a value close to 1, indicating that the departure runway 18 R is continuing to be used efficiently.

3.2. *Analysis of Flight Takeoff Rate*

Figure 3 shows the takeoff rate of flights from departure runway 18 R as a function of the number of departure aircraft on the airport surface. The takeoff rate statistics for the runway are also based on a 15 min statistical period, and then the takeoff rate average of each departing aircraft quantity is calculated.

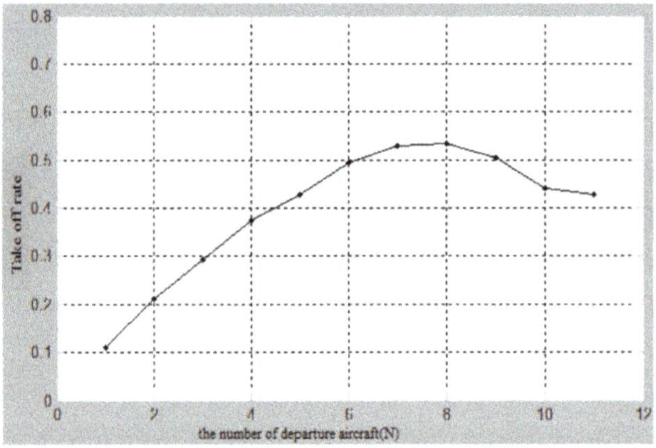

Figure 3. The relationship between the takeoff rate and the number of departure aircraft.

When the number of departing aircraft is $N \leq 6$, the number of departure aircraft on the airport surface at this time is small, and the number of arriving and departing flights per unit of time increases with the increase in the number of departure aircraft on the airport surface. When the number of departure aircraft is $N = 7$, the takeoff rate of flights has reached its maximum. Correspondingly, from Figure 2, it can also be seen that the runway utilization rate is close to 1, indicating that the arrival and departure of the flights began to be restricted by the runway capacity; the arrival and takeoff rates of flights reach 0.52 flights/min (31 flights/h). When the number of departure aircraft on the airport surface is $N \geq 9$, the takeoff rate of flights calculated by the statistics shows a slight downward trend. However, this does not simply indicate that the takeoff rate will decrease when the number of departure aircraft on the airport surface is large in correspondence with Figure 2. When the performance is $N \geq 9$ compared to $N = 7$–8, the runway utilization is also close to 1, indicating that the runway is still fully utilized. However, the percentage of waiting time for the aircraft at the runway's end increases when $N \geq 9$ in Figure 2 compared with the case when $N = 7$–8, indicating that the decrease in the takeoff rate is due to the increase in the holding time for takeoff at the runway's end. Through further analysis of the arrival and departure flight data, it was found that because the double runways of Hongqiao Airport are narrow parallel runways, the arriving and departing aircraft of the two runways cannot take off and land at the same time because of the aircraft wake spacing.

This means that the two runways cannot operate independently. Therefore, the continuous arrival of multiple aircraft will affect the aircraft taking off, resulting in a decrease in the takeoff rate of the runway 18 R and a cumulative increase in the number of departure aircraft on the airport surface, such as the number of departure aircraft on the airport surface reaching 12. This situation is reflected in the performance of $N \geq 9$ in Figures 2 and 3. Although the takeoff rate of flights has decreased, the percentage of time that an aircraft is waiting to take off at the runway entrance has increased in the runway utilization histogram.

Figures 2 and 3 can be summarized as follows: in the fixed runway configuration, the number of departure aircraft that undergo pushback into the apron and taxiway systems increases as more flights are added to the takeoff queue. The runway utilization rate and takeoff rate have gradually increased. However, aircraft takeoff will be limited by the time interval. When the number of departure aircraft on the airport surface reaches a critical value or if the number of departure aircraft on the airport surface continues to increase, the runway capacity will become a limiting factor. The runway utilization rate and the takeoff rate will only tend to have stable fluctuations and will no longer increase significantly.

4. Aircraft Departure Time Prediction

4.1. Factors Affecting Departure Taxiing Time

Analysis of the departure taxiing process shows that the aircraft departure taxiing time is related to the airport current runway configuration, the gate of the aircraft apron, and the congestion status of the departure taxiway through the apron and the taxiway systems. Under certain runway usage configurations, the influence of the gate of the apron on the taxi time of the departing aircraft can be expressed by a taxi distance parameter. The influence of the apron and taxiway systems' congestion conditions on the taxi time of departing aircraft can be expressed by the parameter of the number of aircraft on the airport surface.

Through the data analysis of the Shanghai Hongqiao Airport March 2015 airport surface monitoring system, we select the flights in UTC time 04:00–06:00 (Local Time 12:00–18:00), which is the busy time. The full sample contains 1469 departure flights and 1399 arrival flights. The data in Table 1 summarizes the statistics for the departure flight sample.

Table 1. Summary of Statistics for the departure flight Sample.

	Mean	Median	Standard Deviation	Max.	Min.
the Departure Taxiing Time (Minutes)	16.32	14.19	7.05	30.94	2.73
the Number of the Departure Aircraft (Flights)	5.90	6	2.55	11	1
the Taxiing Distance (Meters)	2232.37	2278	458.06	3055	1005

4.2. Impact of the Number of Aircraft on the Airport Surface

Figure 4 is a scatter plot of the departure taxiing time of aircraft and the number of aircraft on the airport surface when each aircraft underwent pushback on 9 March 2015, UTC time 4:00–6:00. From the scatter plot, it can be clearly seen that the departure taxiing time of aircraft gradually increases with the increase in the number of aircraft on the airport surface.

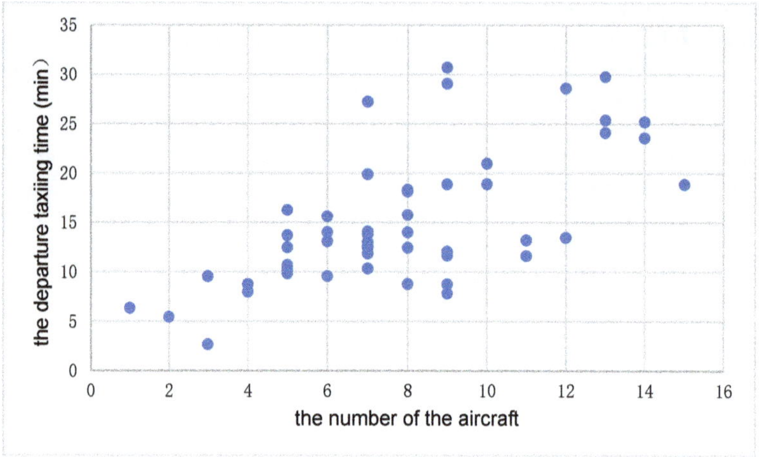

Figure 4. The scatter diagram of the relationship between the departure taxiing time and the number of aircraft.

Considering that there are large differences in the taxi path between the arriving and departing aircraft, the number of arriving aircraft has little influence on the taxiing time of the departing aircraft. Therefore, all the arriving aircraft in the statistical data of Figure 4 are excluded and the scatter plot of the departure taxiing time of aircraft and the number of departing aircraft on the aircraft surface is obtained in Figure 5.

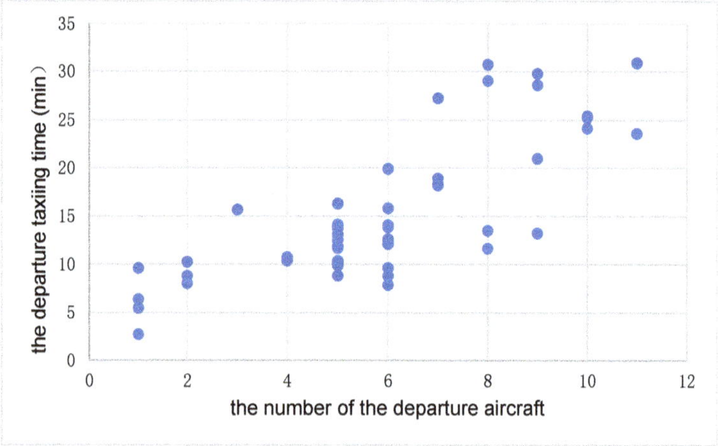

Figure 5. The scatter diagram of the relationship between the departure taxiing time and the number of departing aircraft.

The linear correlation coefficient between the aircraft departure taxiing time and the number of aircraft on the airport surface is $R_1 = 0.62$; the linear correlation coefficient between the aircraft departure taxiing time and the number of departure aircraft on the airport surface is $R_2 = 0.79$. By comparing R_1 and R_2, it is shown that the aircraft departure taxiing time has a stronger linear relationship with the number of departure aircraft on the airport surface at the time of aircraft pushback. Therefore, the number of departure aircraft on the airport surface can be used as a predictor variable of the aircraft departure taxiing time.

4.3. Effect of Aircraft Departure Taxiing Distance

The effect of the airport runway configuration and an aircraft's gate position on the aircraft's departure taxiing time is reflected in the distance required for the aircraft to taxi from the apron to the takeoff runway entrance. Figure 6 is the scatter plot of the aircraft departure taxiing time and the required taxi distance for departure on 9 March 2015, UTC time 4:00–6:00. From the scatter plot, it can be seen that the departure taxiing time increases gradually with an increase in the taxiing distance.

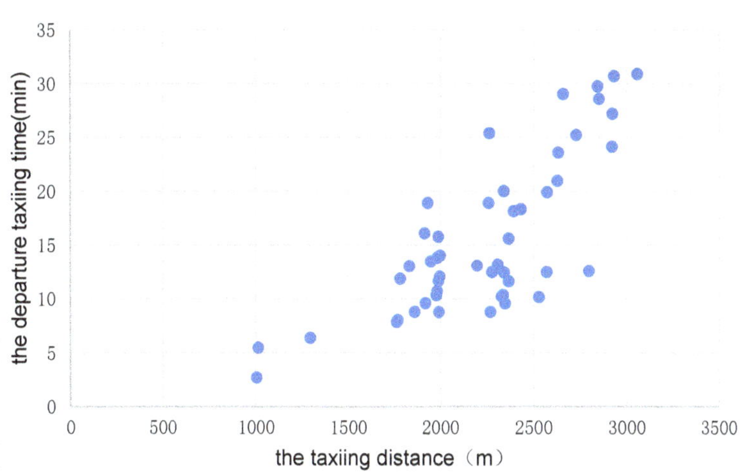

Figure 6. The scatter diagram of the relationship between the flight taxi time and the taxi distance.

In apron and taxiway systems, the aircraft generally taxis at a low and uniform speed. Therefore, under an ideal no-taxiing-collision condition, when the aircraft is taxiing at a constant speed, the taxiing time is positively related to the taxiing distance. The linear correlation coefficient between the aircraft departure taxiing time and departure taxiing distance is $R_3 = 0.77$. This indicates that the aircraft departure taxiing time and the departure taxiing distance have a strong linear relationship. Therefore, the aircraft departure taxiing distance can be used as a predictor of aircraft departure taxiing time.

4.4. Departure Taxiing Time Prediction Model

The aircraft departure taxiing time T can be divided into two parts:

$$T = t_{tw} + t_c, \tag{1}$$

where t_{tw} represents the time taken by the aircraft to taxi from the apron to the departure runway without conflict, and the magnitude of the value is related to the taxiing distance d; and t_c indicates the amount of time spent escaping and waiting for each aircraft during the taxiing process due to mutual influence. The magnitude of the value reflects the degree of airport congestion and is related to the number (N) of aircraft departures on the airport surface.

From the analysis in the previous section, the departure taxiing time of aircraft is linearly related to the number of aircraft departing the airport surface and the departure taxiing distance. Table 2 shows the correlation data. According to the correlation coefficient r of the independent variable, when r is close to 1, there is a strong linear relationship between the two independent variables. It represents only a judgment on collinearity between two independent variables. Therefore, multiple linear regression models could be used to predict the aircraft departure taxiing time.

Table 2. Correlation Data.

	the Number of Departure Aircraft	the Taxiing Distance
the Number of Departure Aircraft	1	0.66551
the Taxiing Distance	0.66551	1

The multiple linear regression equation can be expressed as

$$y = m(x_1, x_2, \ldots, x_p) + \varepsilon, \tag{2}$$

Since the linear regression assumes that $m(x_1, x_2, \ldots, x_p)$ is a linear function of the random variables $(x_1, x_2, \ldots, x_p)$, in this paper, the aircraft departure taxiing time T is linearly related to the number of departure aircraft on the airport surface N and the departure taxiing distance d. Therefore, the multivariate linear regression equation expression of the aircraft departure taxiing time prediction model can be expressed as

$$T = \beta_0 + \beta_1 N + \beta_2 d + \varepsilon, \tag{3}$$

In this formula, β_0, β_1, and β_2 are the linear regression coefficients to be solved. For convenience of description, Equation (3) is represented by the matrix expression below (Equation (4)):

$$Y = X\beta + \varepsilon, \tag{4}$$

To ensure correct statistics, it is usually necessary to make multiple observations on the independent variable and the dependent variable corresponding to the independent variable. Assume that the observation statistics are performed n times, where Y and ε are n-dimensional column vectors, β is a (p + 1)-dimensional column vector, and the independent variable X is an n × (p + 1)-dimensional matrix whose first column is all 1. Additionally, take p = 2 corresponding to Equation (3).

In order to obtain the best-estimated vector parameter β, we make the sample $X\beta$ estimation as close as possible to the observed value Y, making the error term ε as small as possible. Using least squares estimation, we can see that when $\beta = \left(X^T X\right)^{-1} X^T Y$, the square of ε-mode

$$\begin{aligned} \|\varepsilon\|^2 &= (Y - X\beta)^T (Y - X\beta) \\ &= \sum_{i=1}^{n} \left(y_i - \beta_0 - \sum_{j=1}^{p} x_{ij}\beta_j \right)^2 \end{aligned} \tag{5}$$

reaches the minimum, so the best linear unbiased estimate is

$$\hat{\beta} = \left(X^T X\right)^{-1} X^T Y, \tag{6}$$

The adjusted coefficient of determination R^2_{Adj} can be used to measure how well the model fits the data. The expression is as follows:

$$R^2_{\text{Adj}} = 1 - \frac{\sum_{i=1}^{n} (\hat{y}_i - \overline{y})^2 / (n - p - 1)}{\sum_{i=1}^{n} (y_i - \overline{y})^2 / (n - 1)}, \tag{7}$$

In this formula, n is the number of observations, that is, the number of departure aircraft counted; y_i is each observation value of the dependent variable, that is, the ith aircraft departure taxiing time; $\overline{y}$ is the average of the dependent variable observations, that is, the average departure taxiing time of the aircraft calculated; and $\hat{y}_i$ is the estimated value of the dependent variable for each observation, that is, the multiple linear regression model prediction of the departure time of the *i*th aircraft. R^2_{Adj} values between 0 and 1, with values closer to 1 indicating a better fit [14].

To sum up, we assume that the number of airport surface departure aircraft is N and the required departure taxiing distance is d. Equations (3) and (6) can then be used to obtain the fitting prediction formula for the aircraft departure taxiing time T as

$$T = -8 + 1.35N + 6.9d, \tag{8}$$

Table 3 shows the parameters of the multiple linear regression. By Equation (7), the goodness of fit is described by an Adjusted R^2 = 0.835, which shows that the aircraft departure taxiing time prediction model is reasonable.

Table 3. The tables of the parameters for regression.

Variable	Coefficient	*t*-Statistic	Sig
C	−8.078409	−2.81201	**
N	1.353452	4.748534	***
d	6.958	4.259463	***

Sig. indicates if the *p*-value is 0.05 (*), 0.01 (**), or 0.001 (***).

5. Pushback Strategy for Departure on the Airport Surface

5.1. Implementation of Control Strategies

Taking the departing flight of the No. 2 apron of Hongqiao Airport shown in Figure 1 as an example, data analysis is conducted to compare the flight departure taxiing time, the runway utilization rate, and the takeoff rate under different departure aircraft numbers on the airport surface.

When the airport runway configuration is 18 L/18 R, the average departure taxiing distance of the No. 2 apron departing flight is 2.25 km, and we substitute the taxiing distance into Equation (8). Thus, the correspondence between the departure taxiing time of the flight and the number of departure aircraft on the airport surface can be obtained. Then, the corresponding relationship between the runway utilization rate, the flight takeoff and departure rate, and the number of departure aircraft on the airport surface can be calculated according to the statistics of Figures 2 and 3. Thus, Table 4 can be obtained.

Table 4. The operating parameters for different numbers of departure aircraft.

Airport Surface Departure Aircraft Number (N)	Taxi Time (min)	Takeoff and Departure Rate (flight/min)	Runway Utilization (%)
1	8.87	0.11	32.0
2	10.2	0.21	47.5
3	11.6	0.29	61.5
4	12.9	0.37	72.1
5	14.3	0.43	83.3
6	15.6	0.49	94.0
7	16.9	0.53	98.9
8	18.3	0.53	97.3
9	19.7	0.50	97.0
10	21.0	0.44	98.5
11	22.3	0.42	100

According to Table 4, when the number of aircraft departures on the airport surface is $N \leq 6$, the number of departure aircraft is small and the flight departure taxiing time, runway utilization rate, and takeoff and departure rate of flights increase with an increase in the number of aircraft departures on the airport surface. When the number of aircraft departures on the airport surface is N = 7, the average departure taxiing time of No. 2 apron flights is 16.9 min, the runway utilization rate begins to be close to 1, the takeoff and departure rate is 0.53 flight/min, and the flight departure process begins to be controlled by the runway capacity limits. When the number of airport surface departure aircraft is $N \geq 8$, the runway utilization rate and the flight takeoff and departure rate no longer increase, while the average taxi time of departure flights on the No. 2 apron no longer increases. Therefore, when the number of airport surface departure aircraft is $N \leq 7$, the tower aircraft controller can act in accordance with the first-come first-service (FCFS) principle based on the pushback request clearance issued by the flight. When the number of airport surface departure aircraft is $N \geq 8$, the controller can first control the aircraft pushback at the gate position, and then set up a virtual pushback sequence for these departure flights. When $N \leq 7$, the departure flights will be queued according to the virtual pushback sequence. The implementation of the departure control strategy for the aircraft did not reduce the runway utilization rate and flight takeoff and departure rate, but it slowed the airport surface congestion so that the departure taxiing time can be effectively reduced without increasing the total delay in departure flights.

5.2. SIMMOD Simulation

The SIMMOD software was used to simulate the use of the departure control strategy. SIMMOD is a discrete-time simulation software released by the U.S. Federal Aviation Administration. SIMMOD provides dynamic decisions based on user-defined rules, and each process of a flight is controlled based on user rules. Its performance indicators are: the flight time of the aircraft, the capacity per unit of time, delays, etc. [15–19].

The SIMMOD simulation model relies mainly on a detailed description of the airport and airspace network, and the traffic flow moves on the nodes and connections of the network. The operating path of the aircraft can be specified either by the user or automatically by the Dijkstra Algorithm [20–23].

First, the Computer Aided Design (CAD) map of Hongqiao Airport is imported into SIMMOD to establish the airport topology map and the waypoints are inputted to establish the arrival and departure procedures and routes. Then, according to the flight plan of on 9 March 2015, UTC time 4:00–6:00 (local time 12:00–14:00), the arrival and departure flight information is established. The departure control strategy simulation is implemented.

The simulation results show that the average wait time of the 52 departure flights was 2.24 min, the total taxi time was decreased by 119 min, and the maximum waiting time for the gate was 14 min. As shown in Figure 7, the blue line represents the case where the departure control is not used, and the red line represents the case where the departure control is used. It can be clearly seen from the figure that by the use of the departure control strategy, the total departure taxi time is reduced.

The taxi time predicted in Section 5.1 by the Section 4 multiple linear regression prediction method is basically consistent with the taxi time obtained by the Section 5.2 SIMMOD simulation

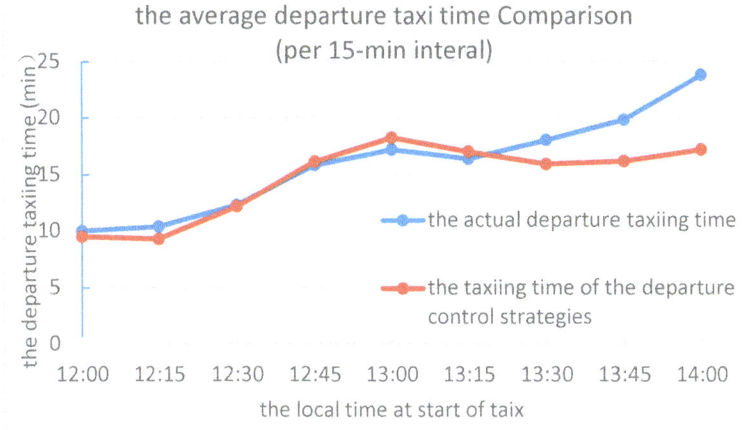

Figure 7. The scatter diagram of the relationship between the flight taxi time and the taxi distance.

5.3. Analysis of Fuel Saving and Emission Reduction Data

During the busy hours of airport operation, the number of departure aircraft on the airport surface was controlled at N = 7. Compared with N = 9–11, the runway utilization rate and the takeoff and departure rate did not change substantially; however, the average departure taxiing time of each aircraft on the No. 2 apron decreased by 2.8–5.4 min, accounting for 16.7–31.9% of the aircraft departure taxiing time at this time.

Taking the CFM56-5B4/P engine of the Airbus A320 as an example, the total amount of fuel oil and pollutant gas emissions consumed for each departing aircraft's taxiing on the No. 2 apron is calculated when the aircraft implements different departure control strategies [24,25] as shown in Table 5.

Table 5. The total fuel consumption and total pollutant emissions of departing aircraft under different departure control strategies.

Airport Surface Departure Aircraft Number (N)	Fuel Consumption (kg)	Pollutants Total Emissions (kg)
6	194.6	6.28
7	210.9	6.81
8	228.4	7.37
9	245.9	7.94
10	262.1	8.46
11	278.3	8.98

As can be seen from Table 2, when the number of airport surface departure aircraft is controlled to be N = 7, the fuel consumption per A320 departure flight is reduced by 35 kg to 67 kg compared to N = 9–11. Gas emissions decreased by 1.13–2.17 kg.

For other common aircraft types on the apron, when the airport implements a departure control strategy with N = 7 during busy hours, the reduction in fuel consumption and total pollutant emissions per flight is shown in Figure 8.

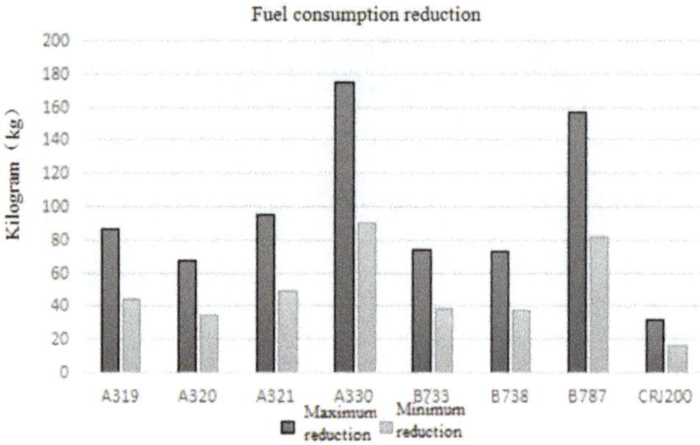

Figure 8. The fuel consumption reduction graph.

From Figures 8 and 9, it can be seen that the fuel consumption and pollutant emissions of each type are greatly reduced. Therefore, the departure control strategy was adopted during the busy hours of airport operation, which effectively reduced the fuel consumption during the taxiing stage of the aircraft and reduced their pollutant emissions.

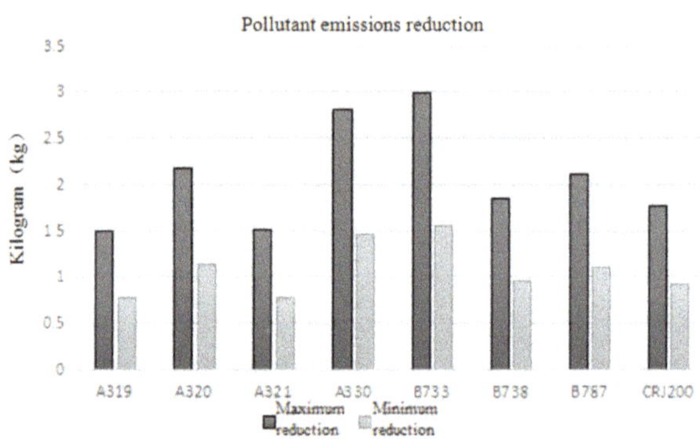

Figure 9. The reduction in the pollutant emissions.

6. Conclusions

In this paper, the Shanghai Hongqiao Airport is taken as an example to study the control strategy for departure aircraft pushback on the airport surface. The influence of the different numbers of departure aircraft within the apron and taxiway systems on the runway utilization rate and the takeoff rate was studied under airport runway capacity constraints. Additionally, the influence of factors, such as the number of departure aircraft in the apron and taxiway systems, the position of the apron, and the configuration of airport arrival and departure runways, on the departure taxiing time of aircraft was analyzed. Multiple linear regression equations were used to establish an aircraft taxi departure time prediction model and the reductions in fuel consumption and pollutant emissions were calculated. The results show that reasonable control of the pushback of departing aircraft during the airport's busy hours can reduce the aircraft departure taxiing time without reducing the runway utilization rate and takeoff and departure rate of the aircraft, thereby reducing aircraft fuel consumption and pollutant emissions during the taxiing phase.

Author Contributions: Writing: X.Z. and N.L.; Providing the case study and the idea: X.Z., N.L., Y.S., and H.Z.; Revising and editing: S.-B.T. and K.W.

Funding: This research was funded by the National Natural Science Foundation (No. U1533112) and the National Social Science Foundation (No.13CGL005).

Acknowledgments: The authors would like to thank the National Natural Science Foundation (No. U1533112) and the National Social Science Foundation (No.13CGL005) for the support.

Conflicts of Interest: The authors declare no conflict of interest.

References

1. María, V.M.; Hall, C.A. Modelling Performance and Emissions from Aircraft for the Aviation Integrated Modelling project. *J. Aircr.* **2010**, *47*, 812–819.
2. Jung, Y. Fuel Consumption and Emissions from Airport Taxi Operations. Available online: https://core.ac.uk/display/10559073 (accessed on 17 August 2018).
3. Balakrishnan, H.; Jung, Y. A Framework for Coordinated Surface Operations Planning at Dallas-Fort Worth International Airport. Available online: https://arc.aiaa.org/doi/abs/10.2514/6.2007-6553 (accessed on 13 September 2108).
4. Simaiakis, I.; Balakrishnan, H. Queuing Models of Airport Departure Processes for Emissions Reduction. Available online: https://arc.aiaa.org/doi/10.2514/6.2009-5650 (accessed on 13 September).
5. Jung, Y.C.; Hoang, T.; Montoya, J. A Concept and Implementation of Optimized Operations of Airport Surface Traffic. Available online: https://arc.aiaa.org/doi/abs/10.2514/6.2010-9213 (accessed on 13 September).
6. Lee, H.; Simaiakis, I.; Balakrishnan, H. A comparison of aircraft trajectory-based and aggregate queue-based control of airport taxi processes. In Proceedings of the 29th Digital Avionics Systems Conference, Salt Lake City, UT, USA, 3–7 October 2010; pp. 1–15.
7. Simaiakis, I.; Khadilkar, H.; Balakrishnan, H.; Hansman, R.G. Demonstration of reduced airport congestion through pushback rate control. *Transp. Res. Part A* **2014**, *66*, 251–267. [CrossRef]
8. Ravizza, S.; Atkin, J.A.D.; Maathuis, M.H.; Burke, E.K. A combined statistical approach and ground movement model for improving taxi time estimations at airports. *J. Oper. Res. Soc.* **2013**, *64*, 1347–1360. [CrossRef]
9. Tang, Y. Aircraft Taxi Routes Planning and 3D Simulation for A-SMGCS. Ph.D. Thesis, Nanjing University of Aeronautics and Astronautics, Nanjing, China, December 2014.
10. Zhao, X.; Tang, J.; Lu, F.; Han, B. Strategy Analysis for Delayed Fights Pushback and Sensitivity Analysis of the Length of Virtual Queue. *J. Sichuan Univ.* **2016**, *3*, 115–123.
11. Li, N.; Lv, H. Method to Find out Path of Aircraft on Surface Based on Surface Movement Radar Data. *Aeronaut. Comput. Tech.* **2016**, *46*, 6–9.
12. Lv, H. The Analysis of Aircraft Taxiing and Prediction Study Based on Surveillance Data. Master's Thesis, Civil Aviation University of China, Beijing, China, April 2016. Available online: http://kns.cnki.net/KCMS/detail/detail.aspx?dbcode=CMFD&dbname=CMFD201701&filename=1016776031.nh&uid=WEEvREcwSlJHSldRa1FhdkJkVWI0UTA0a3RxcEdTZnNJVW1ZVVFkM3gvVT0=$9A4hF_YAuvQ5obgVAqNKPCYcEjKensW4IQMovwHtwkF4VYPoHbKxJw!!&v=MDAwNDFyQ1VSTEtmYnVkb0Z5bmhWYjdMVkYyNkdMUy9HTkhQcnBFYlBJUjhlWDFMdXhZUzdEaDFUM3FUcldNMUY= (accessed on 17 August 2018).
13. Eurocontrol Standard Document for Surveillance Data Exchange Part 9: Category 062 SDPS Track Messages. Available online: https://www.google.com.tw/url?sa=t&rct=j&q=&esrc=s&source=web&cd=1&ved=2ahUKEwiHtb3gtLndAhWLyrwKHdT3AxoQFjAAegQIARAC&url=https%3A%2F%2Fwww.eurocontrol.int%2Fsites%2Fdefault%2Ffiles%2Fcontent%2Fdocuments%2Fnm%2Fasterix%2Farchives%2Fasterix-cat062-system-track-data-part9-v1.11-042010.pdf&usg=AOvVaw389n3cb5i1jElDR9PsFJ3s (accessed on 14 September 2018).
14. Anderson, R.C.; Reeb, D.M. Founding-family Ownership and Firm Performance: Evidence from the S&P 500. *J. Financ.* **2003**, *58*, 1301–1328.
15. Tsai, S.-B. Using the DEMATEL Model to Explore the Job Satisfaction of Research and Development Professionals in China's Photovoltaic Cell Industry. *Renew. Sustain. Energy Rev.* **2018**, *81*, 62–68. [CrossRef]
16. Lee, Y.-C.; Hsiao, Y.-C.; Peng, C.-F.; Tsai, S.-B.; Wu, C.-H.; Chen, Q. Using Mahalanobis-Taguchi System, Logistic Regression and Neural Network Method to Evaluate Purchasing Audit Quality. Available online: http://journals.sagepub.com/doi/10.1177/0954405414539934 (accessed on 13 September 2108).

17. Liu, B.; Li, T.; Tsai, S.B. Low carbon strategy analysis of competing supply chains with different power structures. *Sustainability* **2017**, *9*, 835.
18. Qu, Q.; Tsai, S.-B.; Tang, M.; Xu, C.; Dong, W. Marine ecological environment management based on ecological compensation mechanisms. *Sustainability* **2016**, *8*, 1267. [CrossRef]
19. Lee, Y.-C.; Wang, Y.-C.; Chien, C.-H.; Wu, C.-H.; Lu, S.-C.; Tsai, S.-B.; Dong, W. Applying revised gap analysis model in measuring hotel service quality. *SpringerPlus* **2016**, *5*, 1191. [CrossRef] [PubMed]
20. Wang, J.; Yang, J.; Chen, Q.; Tsai, S.-B. Collaborative Production Structure of Knowledge Sharing Behavior in Internet Communities. *Mob. Inf. Syst.* **2016**, *2016*, 8269474. [CrossRef]
21. Tsai, S.-B.; Lee, Y.-C.; Guo, J.J. Using modified grey forecasting models to forecast the growth trends of green materials. *Proc. Inst. Mech. Eng. Part B J. Eng. Manuf.* **2014**, *228*, 931–940. [CrossRef]
22. Ge, B.; Jiang, D.; Gao, Y.; Tsai, S.-B. The influence of legitimacy on a proactive green orientation and green performance: A study based on transitional economy scenarios in china. *Sustainability* **2016**, *8*, 1344. [CrossRef]
23. Lee, S.-C.; Su, J.-M.; Tsai, S.-B.; Lu, T.-L.; Dong, W. A comprehensive survey of government auditors' self-efficacy and professional Development for improving audit quality. *SpringerPlus* **2016**, *5*, 1263. [CrossRef] [PubMed]
24. ICAO Aircraft Engine Emissions Databank. Available online: https://www.easa.europa.eu/easa-and-you/environment/icao-aircraft-engine-emissions-databank (accessed on 17 August 2018).
25. Li, N.; Zhang, H.F. Calculating aircraft pollutant emissions during taxiing at the airport. *Acta Sci. Circumstantiae* **2017**, *37*, 1872–1876.

energies

MDPI

Article

Evaluation of the CO_2 Emissions Reduction Potential of Li-ion Batteries in Ship Power Systems

César O. Peralta P. [1,*,†], **Giovani T. T. Vieira** [1,†], **Simon Meunier** [2,†], **Rodrigo J. Vale** [1,†], **Mauricio B. C. Salles** [1,†] and **Bruno S. Carmo** [3,†]

1 Laboratory of Advanced Electric Grids (LGrid), Escola Politécnica, University of São Paulo, São Paulo 05508-010, Brazil; cesar.op@usp.br (C.O.P.P); giovanigtvieira@usp.br (G.T.T.V.); rodrigojvale@usp.br (R.J.V.); mausalles@usp.br (M.B.C.S.)

2 GeePs | Group of Electrical Engineering—Paris, CNRS, CentraleSupélec, University of Paris-Sud, University of Paris-Saclay, Sorbonne University, 91192 Gif-sur-Yvette, France; simon.meunier@centralesupelec.fr

3 Department of Mechanical Engineering, Escola Politécnica, University of São Paulo, São Paulo 05508-030, Brazil; bruno.carmo@usp.br

* Correspondence: cesar.op@usp.br; Tel.: +57-301-266-2909

† Current address: [1] Av. Prof. Luciano Gualberto, travessa 3 nº 158, Butantã, São Paulo-SP, CEP 05508-010, Brazil. [2] 11 rue Joliot Curie, 91192 Gif-sur-Yvette, France. [3] Av. Prof. Mello Moraes, 2231, Butantã, São Paulo–SP, CEP 05508-030, Brasil.

Received: 11 December 2018; Accepted: 3 January 2019; Published: 24 January 2019

Abstract: Greenhouse gas emissions are one of the most critical worldwide concerns, and multiple efforts are being proposed to reduce these emissions. Shipping represents around 2% of global CO_2 emissions. Since ship power systems have a high dependence on fossil fuels, hybrid systems using diesel generators and batteries are becoming an interesting solution to reduce CO_2 emissions. In this article, we analyze the potential implementation of Li-ion batteries in a platform supply vessel system through simulations using HOMER software (Hybrid Optimization Model for Multiple Energy Resources). We evaluate the impact of battery characteristics such as round trip efficiency, rated power, and energy capacity. We also evaluate the potential CO_2 emissions reduction that could be achieved with two of the most common types of Li-ion batteries (lithium titanate, lithium iron phosphate). Furthermore, we consider that the Li-ion batteries are installed in a 20 ft container. Results indicate that the lithium iron phosphate battery has a better performance, even though the difference between both technologies is lower than 1% of total emissions. We also analyze the potential emissions reduction for different parts of a mission to an offshore platform for different configurations of the ship power system. The most significant potential CO_2 emissions reduction among the analyzed cases is 8.7% of the total emissions, and it is achieved by the configuration including the main and auxiliary diesel engines as well as batteries. Finally, we present managerial implications of these results for both companies operating ships and ship building companies.

Keywords: CO_2 emissions; HOMER software; hybrid ship power systems; Li-ion battery; shipping

1. Introduction

In recent years, concerns about greenhouse gas emissions have risen and discussions took place to decide each country's strategy to reduce these emissions. Kyoto Protocol and Paris Agreement are the results of these discussions.

The Kyoto protocol was adopted in 1997 at the Kyoto conference. This protocol was the first to introduce emissions limits and an agenda to prevent further global warming [1]. Moreover, the Kyoto protocol brought in some mechanisms such as the International Emission Trading (IET), the Joint

Implementation Mechanism, and the Clean Development Mechanism (CDM) [2]. IET allowed industrialized countries, included in Annex B of the protocol, to buy and sell their assigned emissions.

The Joint Implementation Mechanism allowed governments to develop projects that reduced emissions in other countries to earn emission reduction units (ERUs). In addition, the Clean Development Mechanism focuses on the relations between countries of Annex B and developing countries. Thanks to this mechanism, developing countries could sell Certified Emission Reduction (CER) units to an industrialized nation. Certified Emission Reduction units could be earned by obtaining the certification that a given project, promoted by a developing country, reduced emissions. Unfortunately, the final deal did not set an objective for emissions reduction in the international shipping sector.

The Paris agreement entered into force on the 4th of November 2016; its main ambition is to keep the temperature increase below 2 °C [3]. Countries have presented their national situation and their target to reduce global greenhouse gases in the intended nationally determined contributions [4]. Moreover, the International Panel on Climate Change (IPCC) also wrote a report about the impacts that would be caused by an increase of 1.5 °C of the global temperature. Every potential solution is required to attain this reduction level.

These international agreements helped to bring out the discussion about emissions in the transports sector and more particularly shipping emissions, which are the focus of this work. As can be seen in Figure 1, shipping accounts for 2% of global CO_2 emissions, shipping emissions are equivalent to the ones of Canada, and shipping emits 20% more than aviation. A recent study by the European Parliament shows that shipping emissions should be reduced of 13% by 2030 and 63% by 2050 from 2005 level, in order to stay below 2 °C. However, maritime emissions have increased by 3% per year between 1990 and 2010 [5], this is higher than the increase of global Greenhouse Gas (GHG) emissions, which is of 1.1%.

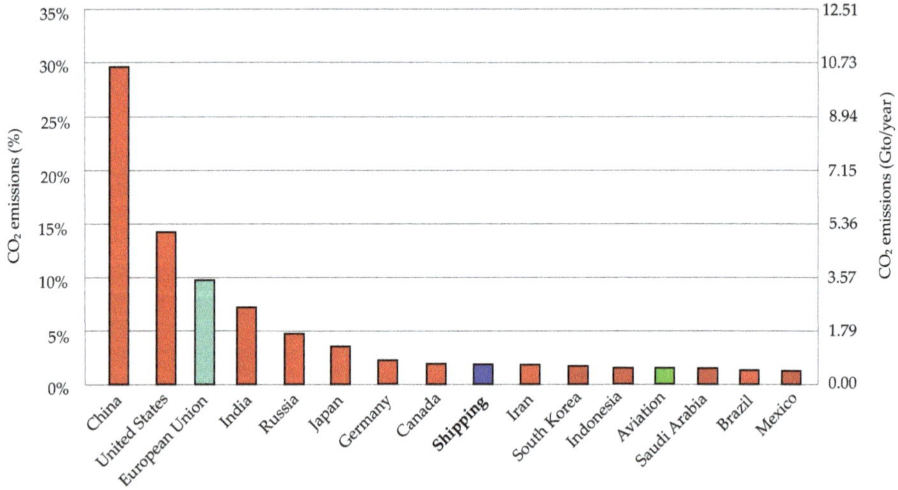

Figure 1. Global CO_2 emissions ranking according to the Emission Database for Global Atmospheric Research (EDGAR) report [6].

Figure 2 shows the rise of the world seaborne trade. We observe that the crude oil loaded trade has remained stable during the 37 years analyzed and that the trade involving petroleum products and gas has slightly increased. We also observe that the dry cargo ship loaded trade has increased very significantly.

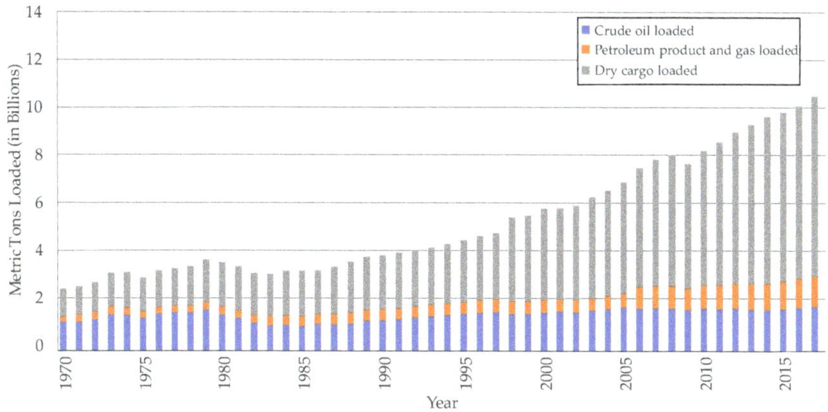

Figure 2. Annual world seaborne trade from 1970 to 2017 according to the United Nations Conference on Trade and Development [7].

The fact that ships are responsible for 90% of global trade [8] combined with the rapid growth in maritime trade (Figure 2) highlights the urgency of lowering CO_2 emissions from this sector.

The following articles have studied the potential CO_2 emissions reduction in the shipping sector, notably by investigating energy dispatch strategies in ships.

In [9], Miyasaki et al. proposed a model to calculate fuel savings and the emissions reduction potential, considering various constraints. Even though this article presents good results, such as a fuel consumption reduction of around 45%, it neither clearly describes the missions that are considered nor evaluates CO_2 emissions for a full routine of the ship. Compared to [9], our article presents a mission routine with a load curve that clearly shows the power variations that the generators need to support during the whole mission. Moreover, we pursue the analysis for each part of the mission highlighting the parts that present the highest CO_2 emissions reduction potential.

In [10], Miyasaki et al. presented a model of a hybrid power system and validated it experimentally. The authors evaluated the effect of the battery efficiency, considering a variation of the efficiency from 80% to 100%. However, since there is no commercially available battery for ships that achieves 100% efficiency and battery efficiency can drop below 80%, we consider a variation of the round trip efficiency from 70 to 96%. Moreover, our article also studies the impact of different C-rates and discharge times on emissions reduction.

In [11], Kanellos presented an algorithm to optimize the energy dispatch in an all-electric ship considering some constraints such as power balance, generator loading, GHG emissions, and ramp rates. The article included a cost to start the generators, which is not considered in our article. On the other hand, [11] neither studied the potential of hybrid power systems nor investigated the influence of the minimum load level of generators.

Section 2 presents the methodology used in this work. In section 3, sensitivity analyses are performed showing the effects on CO_2 emissions reduction of changes in power system characteristics such as round trip efficiency, battery power, battery capacity, and minimum generators load level. Section 4 presents a comparison between two Li-ion battery technologies (i.e., lithium-titanate, lithium iron phosphate). In Section 5, the potential reduction in CO_2 emissions that can be achieved for each part of a mission is evaluated. We notably study the potential impact of using auxiliary generators during low load parts of the mission, since they could allow to disconnect bigger generators which are working at lower efficiency points.

The originalities of this article include the consideration of different technologies of batteries, a sensitivity analysis over parameters that were not considered previously, the study of new power system architectures including batteries and auxiliary generators, and the computation of CO_2 emissions reduction during different parts of a mission to an offshore platform.

These aspects are worth considering since there are multiple battery technologies with different round trip efficiencies and several ship power system architectures. In addition, requirements for each part of the mission are different, which implies that the use of batteries and auxiliary generators could also differ for each part.

2. Methodology

This section presents the general methodology for the simulations, the software used and the range of analyzed parameters. The different architectures used in the simulation cases are shown in Figure 3.

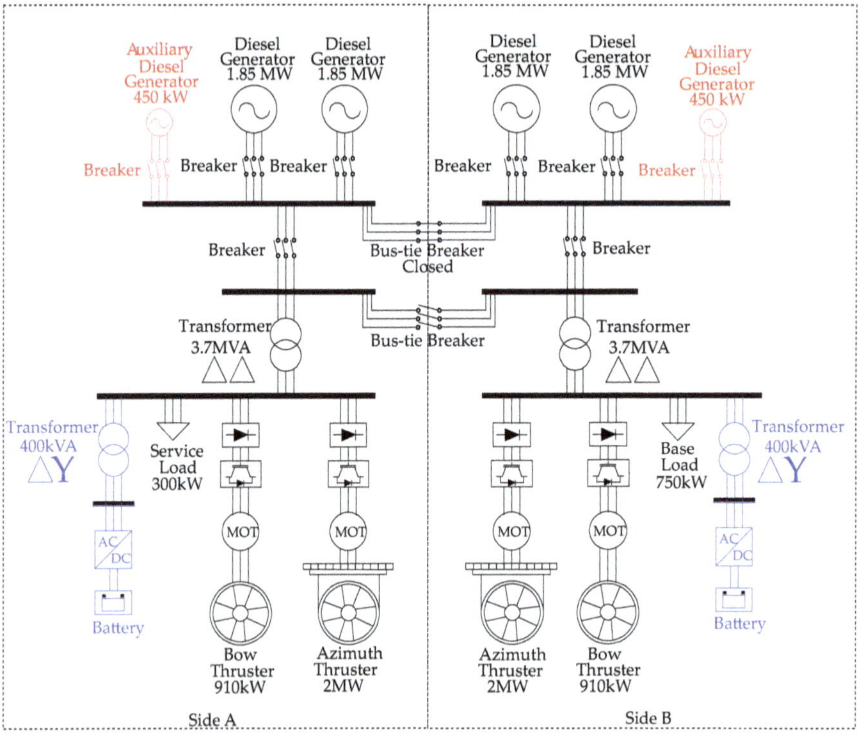

Figure 3. Power system diagram of the Platform Supply Vessel (PSV) considered. The base case analyzed is in black. The components connected to the base case to evaluate the potential of auxiliary generators are in red. The ones related to batteries are in blue. The last analysis pursued (auxiliary generators + batteries) comprises all the elements of this figure.

2.1. Simulation Cases

The performance of a generic Platform Supply Vessel (PSV) is analyzed by simulating multiple architectures of a hybrid system including diesel generators, energy storage systems, and auxiliary generators. Figure 3 shows the electrical diagram of the hybrid system [12]. The main group of generators is composed of four 1850 kW diesel generators and is connected to a bus-bar providing energy to different types of loads such as thrusters (bow and azimuth thrusters), base load, service load, and to batteries.

The optimal dispatch of the available energy to supply the demand is performed using HOMER (Hybrid Optimization Model for Multiple Energy Resources) software. HOMER uses a proprietary derivative-free optimization method to find the system with the lowest cost considering combinations of available energy sources and the load profile.

Following general procedures to analyze isolated systems such as island electrical systems [13,14] and general recommendations to implement Li-ion batteries in ships [15], we developed a general methodology to evaluate CO_2 emissions.

A base case is defined using the load demand profile shown in Figure 4. The minimum load level for three 1850 kW diesel generators is set at 50%, for the other diesel generator the minimum load level is set at 10%. Minimum load means that the generator will be switched on only if the power it would produce is higher than 50% or 10% of its rated power. No batteries or auxiliary generators are used in the base case. In this article, we use "Diesel" to refer to the main generators and "Diesel Aux" to refer to the auxiliary ones. We simulate four identical 1850 kW diesel generators, two additional auxiliary generators (450 kW) and multiple Li-ion batteries with different round trip efficiency, rated power, rated capacity, and C-rate. CO_2 emissions are evaluated for all cases. The time step of the simulations is 30 min.

Sensitivity analyses compare CO_2 emissions reduction potential that can be achieved by four main parameters:

- Round trip efficiency. The total round trip efficiency of batteries and charge converter is combined under one parameter which varies from 70% to 96%.
- Minimum load level of diesel generators. We simulate minimum load levels of 30%, 40%, and 50%.
- Battery C-rate. The energy capacity of the battery system is fixed at 1000 kWh to simulate multiple cases with different C-rates from 0.2 to 2.
- Battery energy capacity. The energy storage system is fixed at 1000 kW of rated power. Multiple cases are simulated for different energy capacities, therefore influencing the time that the storage would last if we discharge it at the rated power. This discharge time varies from 12 min to 120 min.

We investigate the implementation of lithium iron phosphate and lithium titanate batteries in a typical 20 ft container. Some considerations are taken into account to estimate the effective volume available for batteries inside the container.

In Section 5, we present the results of the individual simulations for each part of the mission (loading in port, laden voyage, DP operation, partial load voyage, standby). It allows the evaluation of the CO_2 emissions reduction potential of each configuration proposed in this article for each part of the mission.

The following configurations including auxiliary generators and batteries are evaluated:

- Diesel + Battery. There are four 1850 kW diesel generators and a 1000 kW–1000 kWh battery system.
- Diesel + Diesel Aux. There are four 1850 kW diesel generators and two 450 kW auxiliary generators at minimum load level 50%, which are included only during loading in port and standby operations.
- Diesel + Battery + Diesel Aux. There are four 1850 kW diesel generators, a 1000 kW–1000 kWh battery system, and two 450 kW generators during loading in port and standby.

The full mission and individual mission parts are simulated for each configuration. For configurations including batteries, the state of charge at the end of one part of the mission is used as initial state of charge for the next part of the mission.

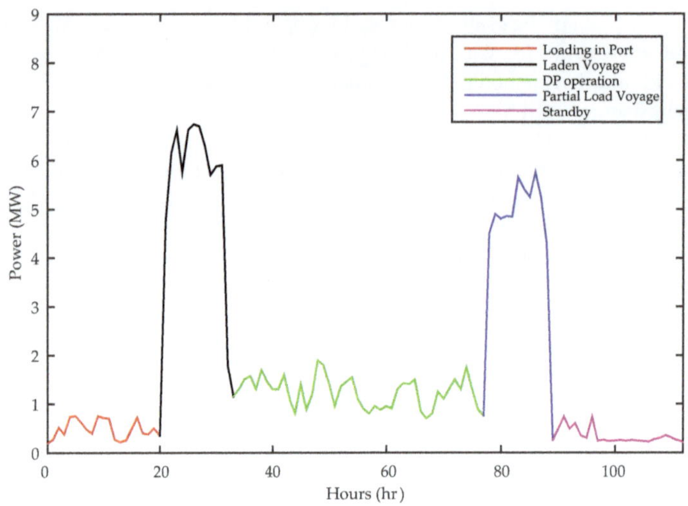

Figure 4. Power demand during the different parts of the mission.

2.2. Mission Profile Considered

The power demand during the different parts of the mission is presented in Figure 4. This load profile is considered for the different simulations performed in this article. The mission starts in the port, where the ship is fully loaded with goods that should be transported to the offshore platform. Then it begins a voyage to the platform and, since the vessel is fully loaded, this part of the mission is called "laden voyage". When the ship stops at the platform, if the climate conditions are appropriate, the ship starts the Dynamically Positioned (DP) operation. During this operation, the DP control system receives all the information about wind and tidal current speed and calculates the power that each thruster should dispatch to keep the ship steady.

The thrusters of this PSV are the bow thrusters, that only move the ship forward and backward, and the azimuth thrusters, that can drive the vessel in any direction. The DP control system sends the exact direction towards which the azimuth thrusters should move the ship to counterbalance the forces that tend to make the ship drift, keeping the ship steady.

After the DP operation the ship starts a voyage back to the port. During this voyage, the ship is not fully loaded and it is therefore called partial load voyage. As can be seen in Figure 4, the partial load voyage requires less power than the laden voyage. Finally, the ship arrives close to the port area but receives a message to wait because the port does not have a space available to receive the vessel. During this time, the vessel is in standby operation.

As shown in Figure 4, even though the ship is keeping a position in DP operation and standby, the restriction of movement during DP operation is more power demanding. Indeed, the tidal and wind forces are stronger during the DP operation than during the standby operation, as the ship is further away from the shore. Moreover, the ship does not have to remain completely steady during the standby operation.

2.3. Batteries Charge-Discharge Cycles

An additional consideration related to batteries is their total number of full charge-discharge cycles. Since batteries present erratic states of charge (SoC) through the whole mission, it is difficult to directly determine the total number of cycles for a battery. We decided to estimate the number of cycles by calculating equivalent battery cycles per mission. For this step we take the total energy E_T cycled through the battery and divide it by the actual available battery energy. The result, shown in

Equation (1), is an equivalent of total full charge-discharge cycles during the mission and is one of the variables evaluated for multiple cases in this article.

$$Number\ of\ cycles = \frac{E_T}{E_{90\%} - E_{20\%}} \tag{1}$$

where:

E_T: Total cycled energy through the battery (kWh)
$E_{90\%}$: Battery capacity at 90% SoC (kWh)
$E_{20\%}$: Battery capacity at 20% SoC (kWh)

Battery states of charge 90% and 20% are selected to increase battery life, following the recommendations provided by Xu et al. [16]. Indeed, using batteries at very high state of charge increases their capacity loss and deeply discharging batteries decreases their lifetime.

2.4. Optimization Performed with HOMER

HOMER [17] is a software for modeling and optimizing small grids connected or not connected to the main grid. HOMER software is a commercial program developed by the National Renewable Energy Laboratory (NREL) and now managed by HOMER Energy LLC. HOMER uses available information about:

1. Energy resources (e.g., fuel-powered generators, solar photovoltaic, wind turbine, biomass power, fuel cell);
2. Energy storage (e.g., batteries, flywheel, hydrogen);
3. Loads (i.e., load demand profile)

Figure 5 shows the 1850 kW generators efficiency curve used in HOMER simulations. HOMER allows the evaluation of multiple system configurations, which combine elements of these three types, calculating energy balance for each time step and then calculating the energy flow for each component ensuring that the demand is always met.

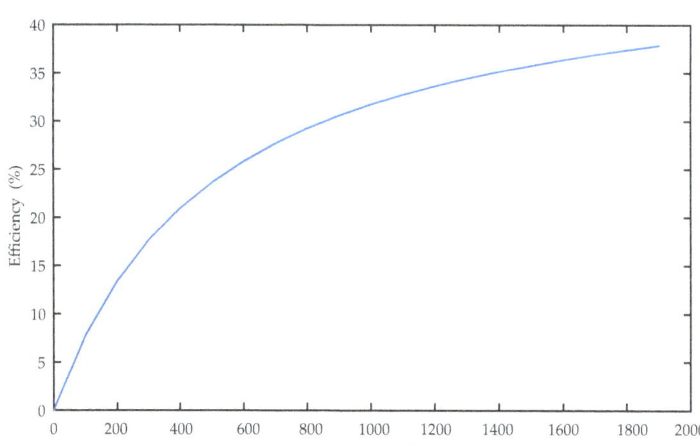

Figure 5. Efficiency curve for the 1850 kW diesel generators used in simulations.

To determine which element is dispatched at each time step, HOMER compares the energy cost of each element considered in the configuration, dispatching elements with the lowest energy cost first. For example, for fuel generators [18], the required data include the capital cost ($) the replacement cost ($), the O&M cost ($/h/year), and the fuel cost ($/l). With this information, HOMER calculates fixed and marginal costs. Fixed energy cost is the cost of running the generator for one hour. Marginal energy cost is the cost of every kilowatt-hour produced. A similar procedure to generators is followed for batteries, calculating fixed and marginal costs (from capital cost, replacement cost, O&M cost, cost of stored energy when it comes from a fuel generator). However, in this article battery fixed costs and O&M costs are not considered.

Thanks to all these information, HOMER performs an optimization to find the cheapest configuration amongst all simulated cases. HOMER also provides the CO_2 emissions associated to the different simulated cases. In this article, we always select the cheapest configuration and give the associated CO_2 emissions.

3. Sensitivity Analysis

3.1. Round Trip Efficiency

This section describes a sensitivity analysis considering a variation from 70% to 96% in the battery round trip efficiency. The minimum load of diesel generators for simulations is 50%, since it is recommended to operate generators above this level [19]. The sensitivity analysis aims at finding out a round trip efficiency value which presents better results in comparison to the base case of vessel operation without battery. Figure 6 shows CO_2 emissions comparing the with and without battery cases.

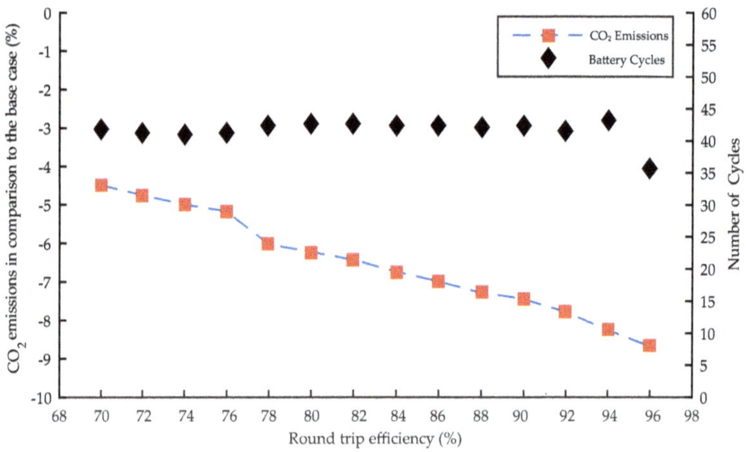

Figure 6. Changes in CO_2 emissions and number of cycles caused by a variation in the round trip efficiency. Battery power and capacity fixed at 1000 kW–1000 kWh. In red: CO_2 emissions reduction in the case of four diesel generators with batteries in comparison to the base case (four diesel generators and no batteries). In black: Number of battery cycles.

CO_2 emissions reduction vary from 4.5% to 8.6%. The number of battery cycles remains almost constant for the whole range.

The total energy generated varies with the round trip efficiency, as can be seen in Figure 7. Even though the total energy generated is higher in the case of the hybrid system than in the case with diesel only, batteries enable the operation of generators at higher efficiency points. This explains why the introduction of batteries allows to reduce fuel consumption and CO_2 emissions.

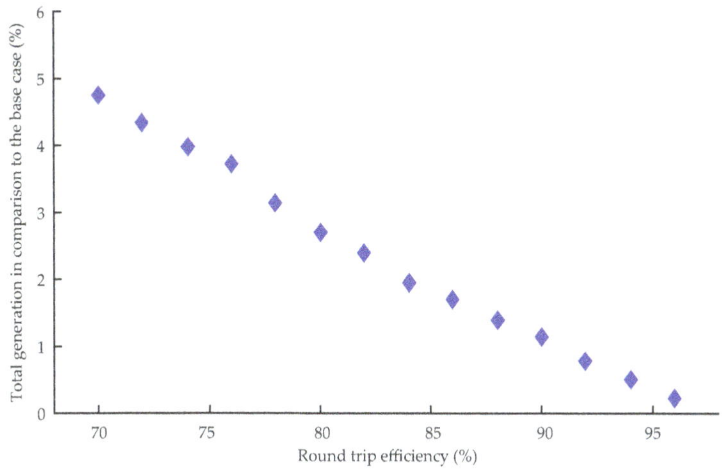

Figure 7. Total generation variation with the round trip efficiency for the case of four diesel generators with batteries in comparison to the base case.

3.2. *Minimum Load Level of Diesel Generators*

Diesel generators have predefined minimum load levels below which they cannot operate. To evaluate the effect of minimum load level on CO_2 emissions reduction, full mission simulations for three cases are performed. These simulations consider minimum load operation of 30%, 40%, and 50% for the 1850 kW diesel generators. Moreover, the batteries parameters are set as following: power 1000 kW, capacity 1000 kWh, round trip efficiency 92%. A value of 92% is considered for the efficiency because it is the average between the round trip efficiencies of lithium titanate and lithium iron phosphate batteries (see Table 2), two very common battery technologies. We compare the results with the base case.

According to Tufte [19], marine engineers sometimes operate with diesel generator load lower than 50%, even though low-load operations (i.e., under 50%) for more than 8 hours are not recommended by some engine manufacturers. There is evidence that operating under that limit can cause a negative impact related with the incomplete combustion of fuel in the inner part of the engine components due to low cylinder pressure [19]. If the low-load operation lasts for a long time, it is recommended to operate for at least 30 min at 50% after the low-load operation. Results on CO_2 emissions and the total energy produced for the three minimum loads are shown in Figure 8.

The evolution of CO_2 emissions with the load level is explained by the fact that, for lower minimum levels, generators operate more frequently at lower power resulting in overall higher fuel consumption and CO_2 emissions. Indeed, generators efficiency decreases as operating power decreases (see Figure 5). In any case, results do not show a high impact of the generators minimum load level on CO_2 emissions and generated energy. Given this result and the problems that may be caused by running at a low load level, lowering the minimum load level does not appear to be an interesting strategy.

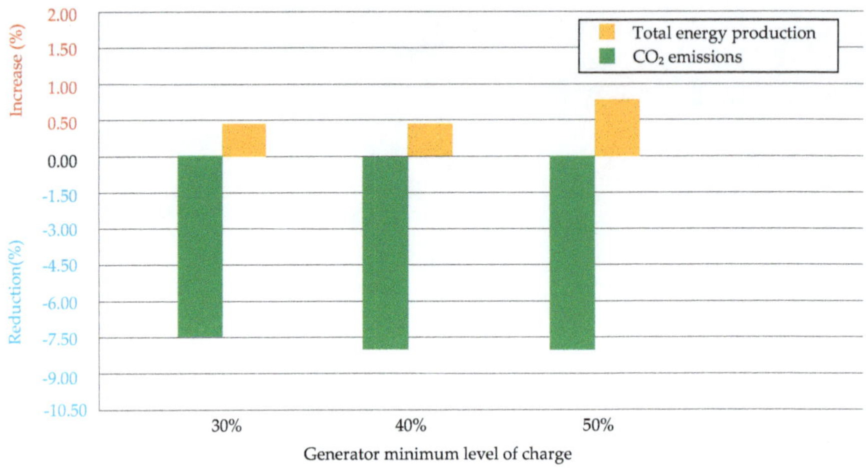

Figure 8. CO_2 emissions reduction and total energy generated increase for the case with four diesel generators and batteries in comparison to the case with four diesel generators without battery. The round trip efficiency is kept at 92% for the three minimum load levels simulated. The generators minimum load levels simulated are 30%, 40%, and 50%, the equivalent number of battery cycles during the operation are 43.8, 41.9, and 41.6, respectively.

3.3. Battery C-rate and Energy Capacity

Another sensitivity analysis is made in order to account for the impact of battery power and energy capacity. First, the energy capacity of the battery is fixed at 1000 kWh to simulate multiple cases with different C-rates from 0.2 to 2. Results of this evaluation are shown in Figure 9.

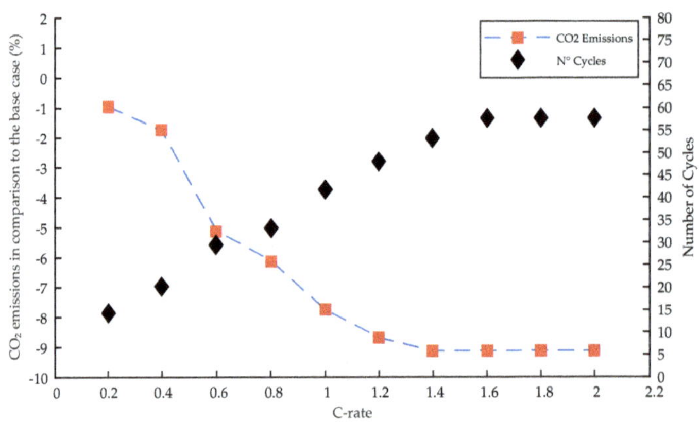

Figure 9. CO_2 emissions reduction in the case of four diesel generators with batteries in comparison to the base case (four diesel generators and no battery). The energy capacity of the battery system is fixed at 1000 kWh and the round trip efficiency at 92%. Multiple cases with C-rates from 0.2 to 2 are simulated.

A 0.2 C-rate is equivalent to a rated power of 200 kW for the current battery capacity of 1000 kWh. Similarly, a 2.0 C-rate is equivalent to a rated power of 2000 kW. For C-rates under 0.6, CO_2 emissions reduction is lower than 2%. However, emissions reduction is higher than 5% for C-rates higher or equal to 0.6.

Figure 10 shows the CO_2 emissions variation with rated battery energy capacity. Indeed, as the rated power is fixed at 1000 kW, a rated energy capacity of 200 kWh is equivalent to 12 min of discharge time and 2000 kWh to 120 min.

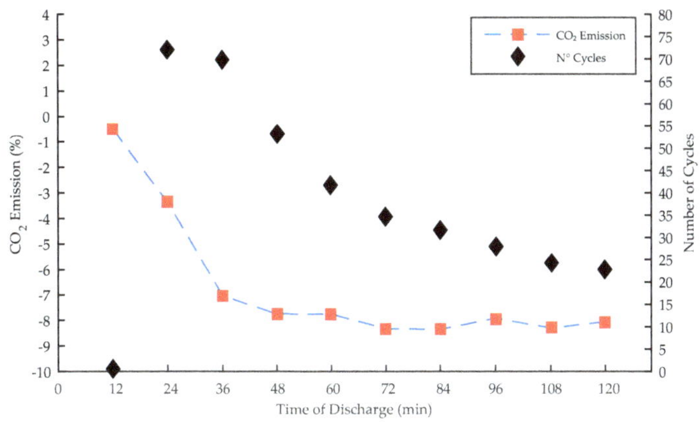

Figure 10. CO_2 emissions reduction in the case of four diesel generators with batteries in comparison to the base case (four diesel generators and no battery). Battery rated power is fixed at 1000 kW. Multiple cases are simulated for different energy capacities, varying the time that the storage would last if we discharge it at the rated power. Every 12 min of discharge time is equivalent to 200 kWh of battery capacity.

As can be seen in Figure 10, the reduction of CO_2 emissions considerably increases from 12 to 36 min, but after that the rate of increase is reduced. The gain obtained by the increase of the discharge time after 36 min is triggered by the reduction of the number of cycles. The reduction of the number of cycles has a significant impact on the increase of the battery life. Results show that a battery with 12 min capacity of discharging at rated power is almost not used, indicating that for the current load demand a 200 kWh–1000 kW battery system is considered not worth to dispatch by the optimization method. At the same time, since energy is limited by feasible generators and load demand, it is expected that there is a saturation level over which increasing battery energy capacity and rated power would not affect CO_2 emissions.

4. Lithium Iron Phosphate and Lithium Titanate Batteries Comparison

With the increasing installation and use of energy storage systems in PSV and other ship power systems, space and standardization of batteries have become relevant to minimize their time for installation and maintenance. Overall, the most suitable installation method of batteries that has also gained popularity in the last years is their installation in a 20 ft container. Li-ion batteries are among the preferred technologies used in power systems around the world. Figure 11 shows examples of rated power and operating duration of Li-ion battery systems connected to the grid in operation worldwide [20].

Zubi, Dufo-López, Carvalho and Pasaoglu [21] highlight that the production structure of lithium-ion batteries can be divided into three tiers. Tier 1 includes the battery cells and the battery pack usually used in Battery Management Systems (BMS). Tier 2 comprises the cell components such as: Cathode, anode, separator, and electrolyte. Tier 3 covers materials such as lithium, aluminum, graphite, and cobalt. Valence Technology, Sony, ATL, Panasonic, A123 System, GS Yuasa, Lishen, Hitachi Vehicle Energy, Samsung, Kokam, SK Innovation, BYD Company, Tesla, Johnson Controls, EnerDel, and LG Chem are the major manufacturers of the Li-ion battery industry [21].

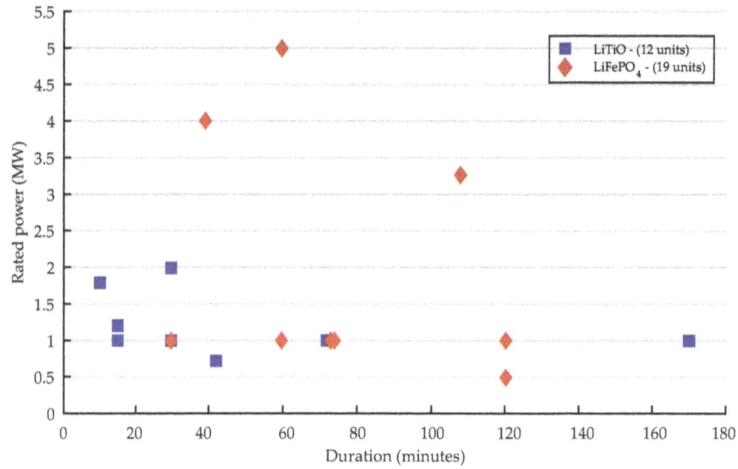

Figure 11. Li-ion battery technologies from 500 kW to 5 MW of power and 10 to 180 min of operating duration that are in operation worldwide [20].

For our analysis, a standard 20 ft container is used with the following dimensions: L × W × H 6.058 × 2.438 × 2.591 m. Li-ion batteries used are lithium titanate (LiTiO) and lithium iron phosphate (LiFePO4). The main characteristics of these batteries are given in Table 1.

Table 1. Parameters of Li-ion batteries.

Li-ion Battery	Volumetric Energy (kWh/m^3)	Cycle Life [22,23]	Round Trip Efficiency (%) [23]	Recharge Rates [24]
LiFePO4	211–620 [23,25]	1000–2000	92	2C-1C
LiTiO	120–620 [22,23]	3000–7000	96	3C-1C

The effective container volume used is estimated to determine the characteristics of these two Li-ion batteries containers. Considering distribution [26–28] and arrangement of battery racks in a container [29–31], a typical battery container layout is shown in Figure 12.

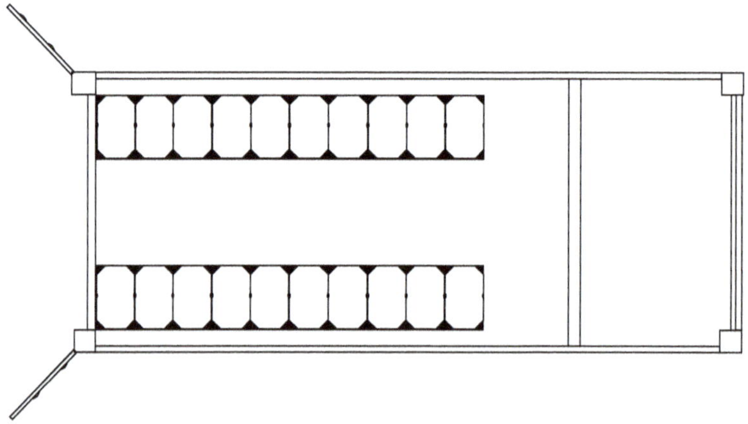

Figure 12. Plane view of a typical arrangement of battery racks in a 20 ft container.

Two volume factors are used to calculate the rated power and energy of Li-ion batteries inside the 20 ft container. One factor represents the space occupied by cells in the battery rack, and the other factor considers the space occupied by the battery racks in the container. Since battery cells inside racks need additional space for proper electrical and thermal installation [32], the effectively utilized rack volume by battery cells is of 57%. The estimated effective volume occupied by battery racks varies from 30% for layouts similar to Figure 12 with two battery racks up to 52% for containers with three battery racks [29–31]. A layout with two battery racks is used with a volume factor of 30%. Considering these two factors and the parameters of lithium iron phosphate and lithium titanate batteries, the rated parameters of the energy storage system in a 20 ft container, used in simulations, are shown in Table 2.

Table 2. Rated parameters of Li-ion batteries in a 20 ft container.

Li-ion Battery	Rated Power (kW)	Energy Capacity (kWh)	Total Round Trip Efficiency (%)
LiFePO4	1285	1285	90%
LiTiO	948	948	94%

Additionally, efficiency losses due to the converter are considered. It is estimated that for different topologies, converter losses vary from 0.4% up to 1.56% [33]. A conservative value of 2.0% for efficiency losses is therefore used in simulations. This explains why the round trip efficiencies presented in Table 2 are 2.0% lower than the ones of Table 1. Similar to the analysis in previous sections, an entire mission is considered and a minimum load of 50% for the diesel generators is chosen.

CO_2 emissions and total energy produced by the diesel generators are shown in Figure 13.

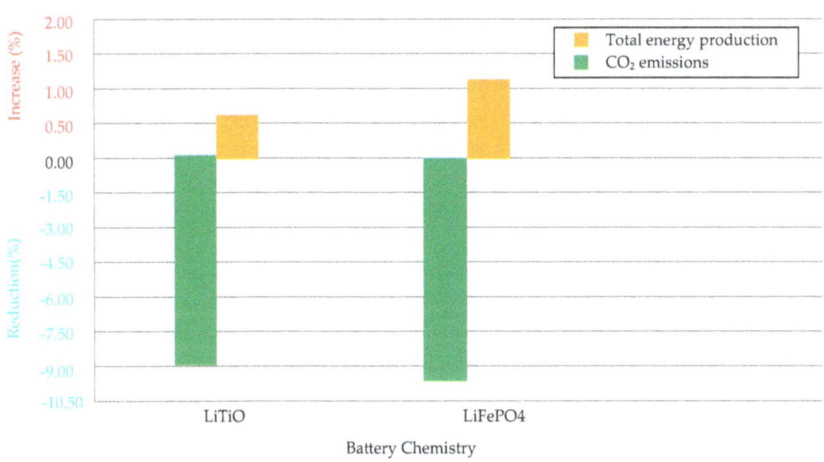

Figure 13. Variation of CO_2 emissions and total energy production using Lithium Titanate (LiTiO) and Lithium Iron Phosphate (LiFePO4) batteries in a 20 ft container.

LiFePO4 battery presents better results compared to LiTiO battery. For the considered ship load demand, although lithium titanate has the highest round trip efficiency, the lithium iron phosphate battery with higher capacity and power presents lower CO_2 emissions. All things considered, the difference between these two batteries is relatively small; other elements should be taken into consideration when selecting a suitable battery for a ship system. For instance, maintenance and replacement cost may also play essential roles in the choice of a battery technology and they are out of the scope of this article.

5. Evaluation of CO_2 Emissions Reduction per Part of the Mission

In this section, an analysis is performed to evaluate the CO_2 emissions reduction that can be achieved in each part of the mission pursued by the platform supply vessel.

Figure 4 shows the variation in power demand of each part of the mission of the PSV. This mission load demand is designed according to [34].

Four cases are simulated considering each part of the PSV routine, separately. The first case considers the elements shown in black in Figure 3. These elements include 4 × 1850 kW diesel generators, one service load, one base load, 2 × 910 kW bow thrusters, and 2 × 2 MW azimuth thrusters. This configuration is based on a real platform supply vessel [12]. It is used to measure the difference in CO_2 emissions triggered by adding components highlighted in blue and red in Figure 3.

The real configuration, in black in Figure 3, is not optimal for the different missions that a standard PSV pursues. Considering that during a 112 hours mission, the ship pursues low power demand 35% of the time (loading in port and standby) and that the rated power of the PSV's main diesel engines is 1850 kW, one of the generators is forced to operate with the minimum load set at 10%. This operation at low load is not advisable [19]. Battery connection allows all diesel generators to operate with minimum load of 50% since the excess of power generation can be used to charge the batteries. Later, the stored energy can be used to power the ship, which allows to disconnect diesel generator for some time.

The second case considers the inclusion of batteries, shown in blue in Figure 3. The battery system considered in this analysis has 1000 kW of rated power and 1000 kWh of energy. During the missions that require a low level of power from generators, this battery system allows generators to operate at higher efficiency points. It also permits to disconnect generators at times when the battery can run the operation alone. Moreover, batteries can offer support of reliability for the ship power system during DP operation.

The third case evaluates the connection of two small auxiliary diesel generators of 450 kW each, in red in Figure 3, that will operate only during loading in port and standby. This case does not include batteries since the focus is to analyze the potential of the small auxiliary diesel generators to reduce CO_2 emissions.

The last case appraises the reduction capacity that two auxiliary diesel generators of 450 kW each can offer when combined with a 1000 kW/1000 kWh battery system. It comprises all the elements shown in Figure 3.

Battery and Auxiliary Generators Configurations

Results for the three additional cases per part of the mission are shown in Figure 14.

As can be seen in Figure 14, batteries have a higher potential to reduce CO_2 emissions when used in loading in port and standby. When we compare the difference of slope in the loading in port and standby operations, we can see that the case that considers the connection of auxiliary diesel generators and batteries to the real PSV power system has the lowest slope, on the other hand the real PSV that only considers the four 1850 kW diesel generators has the highest slope. Implementation of batteries and auxiliary generators decreases CO_2 emissions growth over time. However, batteries have a lower impact on CO_2 emissions during laden voyage, partial load voyage and DP operation. During loading in port and standby, the use of the batteries allows the disconnection of generators during a period of time. When these generators are connected they operate at a higher power than the power required by the load in order to charge the batteries. Figure 14 also indicates that the operations that require more power, laden voyage and partial load voyage, are also the missions that generate the largest share of the CO_2 emissions, considering the slope of both lines.

A results summary for the three cases is shown in Figure 15.

Loading in port and standby have the highest reduction in CO_2 emissions compared to other parts of the mission. However, despite the fact that the use of batteries and auxiliary generators achieves a reduction of 34%, these two parts of the mission have overall low energy demand resulting in a low impact on the total reduction for the whole mission (lower than 9%).

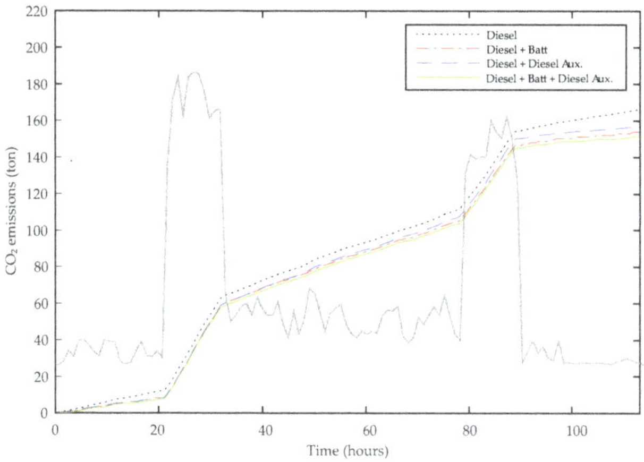

Figure 14. Accumulated CO_2 emissions for the four cases simulated in contrast to the demand curve. The most inclined segments represent the mission parts that have the highest CO_2 emissions, while the difference between the curves at the beginning and at the end of the mission part shows the periods that have the highest reduction of CO_2 emissions. The round trip efficiency of the battery is kept at 92%. In the case that includes only the diesel generators, the minimum load level of the generators is 10% for one diesel generator and 50% for the other three diesel generators when the ship is in loading in port and standby and 50% for all diesel generators when the ship is in laden voyage, DP operation, and partial load voyage. In the case that includes diesel generators and batteries, the minimum load level is 50% for all generators. The case considering diesel generators and auxiliary diesel generators has a minimum load level fixed at 50% for all generators. In the case that considers the diesel generators, batteries, and the auxiliary diesel generators, the minimum load level is also 50%.

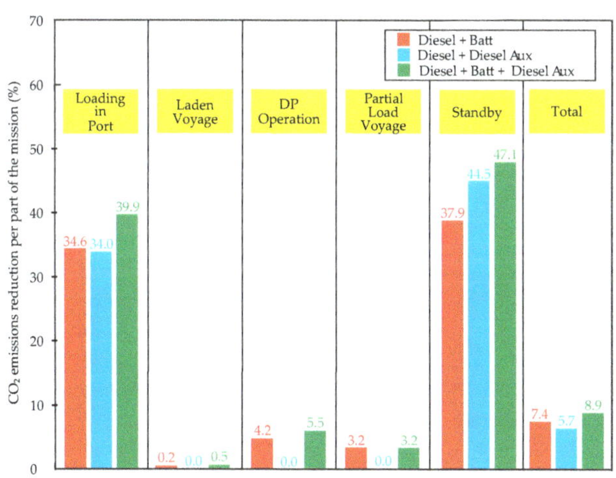

Figure 15. Reduction of CO_2 emissions in each mission part of the ship routine. Loading in port and standby, which are the operations with the lowest power requirements, also present the highest reduction when the batteries and/or auxiliary generators are used. When only diesel generators are used (base case), these mission parts are run at low efficiency points in the specific fuel oil consumption curve of the 1850 kW diesel generators. The impact of the reduction in these mission on the total reduction remains small because the sum of the CO_2 emitted during loading in port and standby represents less than 15% of the total CO_2 emitted during the mission. The low reduction during laden voyage and partial load voyage pulls the total emissions results down since these operations represents more than 56% of the emissions in all of the cases simulated.

Figure 15 also shows that the system using diesel + auxiliary generators has a lower reduction of the CO_2 emissions when compared to the case that comprises diesel + batteries. When auxiliary generators and batteries are combined, they present the highest CO_2 emissions reduction. However, the reduction achieved by this combination is much lower when compared to the sum of the reductions achieved independently by the auxiliary generators and by the batteries. The CO_2 emissions reduction achieved are: 5.7% for diesel + auxiliary generators; 7.4% for diesel + battery system; and 8.9% for a diesel + battery + diesel aux.

For the PSV mission studied in this article, during laden voyage and partial load voyage the four diesel generators operate close to their full rated power; hence there is no available energy to charge batteries and generators already operate at high-efficiency points. This explains why the presence of batteries does not affect CO_2 emissions during laden voyage and partial load voyage (see Figure 15).

6. Conclusions, Managerial Implications and Discussion

We investigated several CO_2 emissions mitigation strategies in a platform supply vessel. We simulated several architectures of the power system including batteries and auxiliary generators. In addition, we performed sensitivity analysis considering several battery parameters: battery power ranging from 200 kW to 2000 kW, battery capacity ranging from 200 kWh to 2000 kWh, and round trip efficiency (70% to 96%). We also considered the use of two Li-ion technologies: LiTiO and LiFePO4. Additional analyses for the different parts of the mission (e.g., loading-in-port, DP operation, etc) were performed to evaluate CO_2 emissions with batteries and auxiliary generators on every segment separately.

Batteries can reduce CO_2 emissions by enabling a more efficient use of diesel generators. Indeed, diesel generators can charge batteries at times of low demand (where generators had to operate at lower power low-efficiency points) allowing generators to operate at higher power high-efficiency points. Although the total energy generated by diesel generators is higher using batteries compared to a system with no batteries, energy storage elements allow generators to operate at higher efficiency points, reducing fuel consumption and therefore CO_2 emissions. Later, this stored energy is released allowing the disconnection of generators.

Sensitivity analysis show the impact on CO_2 emissions of different characteristics such as battery round trip efficiency, minimum generators load, rated battery power, and rated battery capacity. Our simulations indicate that round trip efficiency has a direct impact on CO_2 emissions. The analyses on generators minimum load level do not show a high impact on generated energy and reduction of CO_2 emissions, although at lower load levels it is more likely that generators operate at lower efficiency points.

Evaluating impact of maximum battery C-rate shows that for 0.2 and 0.4 C-rate there is a small reduction of CO_2 emissions (lower than 2%). However, for C-rates higher than 0.6, CO_2 emissions reduction is higher than 5%. For C-rates higher than 1.4, CO_2 emissions reduction begins to stabilize around 9%. Low rated power batteries have a lower number of total cycles during the mission since they take more time to complete a full charge discharge cycle. The total number of cycles per mission varies from 15 to 55 approximately.

Energy capacity variation shows increments of CO_2 reduction at higher available discharge times. However for discharge times higher than 60 min, CO_2 emissions variation remains lower than 1%. At this energy capacity level, the rated power of the battery restricts the amount of energy that can be dispatched by the energy storage system during each cycle. The total number of cycles per mission varies from 22 to 75 approximately, not taking into account the 200 kWh battery that does not have a significant number of equivalent cycles.

Two Li-ion technologies were studied and simulated. Lithium iron phosphate and lithium titanate batteries were used as if they were installed in a 20 ft container. Although both batteries present reduction of CO_2 emissions compared to the base case, lithium iron phosphate battery (1289 kW-kWh) has higher CO_2 emissions reduction compared to its lithium titanate counterpart (948 kW-kWh).

Overall, Li-ion batteries present benefits for all of the studied cases and represent a viable solution to reduce CO_2 emissions in ship power systems.

During parts of the mission of low demand such as loading-in-port and standby, configurations using batteries and auxiliary generators present a reduction of CO_2 emissions ranging from 34.6% to 47%. However, since these two part of the mission have lower energy consumption compared to other parts of the mission, the total CO_2 emissions reduction varies from 5.7% to 8.9%. The use of auxiliary generators increases the reduction of CO_2 emissions, allowing the disconnection of the main generators during times of low load. However, when auxiliary generators and batteries are combined, their impacts on CO_2 emissions do not add up linearly.

The results provided in this study have significant managerial implications for both companies operating ships and ship building companies. This article shows that the use of batteries and/or auxiliary generators reduces fuel consumption, equipment renewal and CO_2 emissions. Moreover, auxiliary generators and batteries can be easily integrated into current ship power systems, notably since batteries can be installed in 20 ft containers.

The reduction of fuel consumption and equipment renewal allows to lower ship operational cost. In addition, low levels of CO_2 emissions is now a requirement in many ports. Therefore hybrid power systems, including batteries and auxiliary generators, may allow companies operating ships to increase their revenue. It may also give them the opportunity to enter more ports and extend their operations. As a result, the demand for ships powered by hybrid systems may increase. Ship building companies may therefore be interested in retrofitting existing ships by integrating batteries and/or auxiliary generators in their power system or in building new ships with hybrid power systems. To this end, this article highlights the influence of key design parameters of hybrid power systems on fuel consumption.

Future work could consider different load mission profiles, diesel generators efficiencies, and operating times for batteries, allowing shorter charge-discharge time while operating generators during longer times to avoid frequent connections and disconnections. All these elements align toward our mutual goal to reduce global CO_2 emissions.

Author Contributions: Conceptualization, C.O.P.P., G.T.T.V., M.B.C.S., B.S.C., S.M.; methodology, C.O.P.P., M.B.C.S.; software, C.O.P.P.; validation, C.O.P.P., S.M., M.B.C.S.; formal analysis, C.O.P.P. and G.T.T.V.; investigation, C.O.P.P., G.T.T.V. and R.J.V.; resources, C.O.P.P. and G.T.T.V.; data curation, C.O.P.P.; writing—original draft preparation, C.O.P.P. and G.T.T.V.; writing—review and editing, C.O.P.P. and S.M.; visualization, G.T.T.V.; supervision, M.B.C.S.; project administration, B.S.C. and M.B.C.S.; funding acquisition, B.S.C.

Funding: This research was funded by FAPESP and Shell through the Research Centre for Gas Innovation, FAPESP Grant Processes 2014/50279-4 and 2014/05261, TOSHIBA Scholarship Program—grant 2014/Dr-01 (TOSHIBA-EPUSP), the French National research Agency (ANR) as part of the « Investissement d'Avenir » program, through the "IDI 2015" project funded by the IDEX Paris-Saclay, ANR-11-IDEX0003-02 and the Coordenação de Aperfeiçoamento de Pessoal de Nível Superior—Brazil (CAPES)—Finance Code 001.

Conflicts of Interest: The authors declare no conflict of interest. The founding sponsors had no role in the design of the study; in the collection, analysis, or interpretation of data; in the writing of the manuscript, and in the decision to publish the results.

Abbreviations

The following abbreviations are used in this manuscript:

CDM	Clean Development Mechanism
CER	Certified Emission Reduction
IET	International Emission Trading

GHG	Greenhouse Gas
HOMER	Hybrid Optimization Model for Multiple Energy Resources
ERU	Emission Reduction Unit
IPCC	International Panel on Climate Change
DP	Dynamically Positioned
PSV	Platform Supply Vessel
SoC	State of Charge
LiTiO	Lithium Titanate
LiFePO4	Lithium Iron Phosphate

References

1. Springer, U. The market for tradable GHG permits under the Kyoto Protocol: A survey of model studies. *Energy Econ.* **2003**, *25*, 527–551. [CrossRef]
2. Böhringer, C. The Kyoto protocol: A review and perspectives. *Oxf. Rev. Econ. Policy* **2003**, *19*, 451–466. [CrossRef]
3. Rogelj, J.; Den Elzen, M.; Höhne, N. Paris Agreement climate proposals need a boost to keep warming well below 2 C. *Nature* **2016**, *534*, 631. [CrossRef] [PubMed]
4. Hulme, M. 1.5 C and climate research after the Paris Agreement. *Nat. Clim. Chang.* **2016**, *6*, 222. [CrossRef]
5. Cames, M.; Graichen, J.; Siemons, A. *Emission Reduction Targets for International Aviation and Shipping*; Technical Report; European Parliament: Brussels, Belgium, 2015.
6. Janssens-Maenhout, G.; Crippa, M.; Guizzardi, D. *Fossil CO_2 & GHG Emissions of All World Countries*; Technical Report; European Commission: Brussels, Belgium, 2017.
7. United Nations Conference on Trade and Development. World Seaborne Trade by Types of Cargo and by Group of Economies, Annual. Available online: http://unctadstat.unctad.org/wds/TableViewer/tableView.aspx?ReportId=32363 (accessed on 26 October 2018).
8. Green, J.F. Why do We Need New Rules on Shipping Emissions? Well, 90 Percent Of Global Trade Depends on Ships. Available online: https://www.washingtonpost.com/news/monkey-cage/wp/2018/04/17/why-do-we-need-new-rules-on-shipping-emissions-well-90-of-global-trade-depends-on-ships/?noredirect=on&utm_term=.2b5726dc9e76 (accessed on 5 October 2018).
9. Miyazaki, M.R.; Sørensen, A.J.; Vartdal, B.J. Reduction of Fuel Consumption on Hybrid Marine Power Plants by Strategic Loading With Energy Storage Devices. *IEEE Power Energy Technol. Syst. J.* **2016**, *3*, 207–217. [CrossRef]
10. Miyazaki, M.R.; Sørensen, A.J.; Lefebvre, N.; Yum, K.K.; Pedersen, E. Hybrid modeling of strategic loading of a marine hybrid power plant with experimental validation. *IEEE Access* **2016**, *4*, 8793–8804. [CrossRef]
11. Kanellos, F.D. Optimal Power Management With GHG Emissions Limitation in All-Electric Ship Power Systems Comprising Energy Storage Systems. *IEEE Trans. Power Syst.* **2014**, *29*, 330–339. [CrossRef]
12. Fujian Shipbuilding. 87M Platform Supply Vessel. Available online: http://www.fujianshipbuilding.com/87m-platform-supply-vessel (accessed on 26 September 2018).
13. Peralta, C.; Salles, M. Advanced energy storage systems to increase the penetration of renewable energies in Fernando de Noronha Island. In Proceedings of the 2017 6th International Conference on Clean Electrical Power (ICCEP), Santa Margherita Ligure, Italy, 27–29 June 2017; pp. 392–396.
14. Kroposki, B.; Burman, K.; Keller, J.; Kandt, A.; Glassmire, J.; Lilienthal, P. Integrating high levels of renewables into the lanai electric grid. *Contract* **2012**, *303*, 275–3000.
15. Lloyd, D.N.V.G. *Handbook for Maritime and Offshore Battery Systems*; Technical Report; DNV-GL: Oslo, Norway, 2016.
16. Xu, B.; Oudalov, A.; Ulbig, A.; Andersson, G.; Kirschen, D.S. Modeling of lithium-ion battery degradation for cell life assessment. *IEEE Trans. Smart Grid* **2018**, *9*, 1131–1140. [CrossRef]
17. HOMER Energy. HOMER Pro User Manual; 2016. Available online: https://www.homerenergy.com/pdf/HOMERHelpManual.pdf (accessed on 15 November 2018).
18. Lambert, T.; Gilman, P.; Lilienthal, P. Micropower system modeling with HOMER. *Integr. Alternative Sources Energy* **2005**, 379–418. [CrossRef]

19. Tufte, E.D. Low Load Operation of Modern Four-Stroke Diesel Engines in Generator Configuration. Master's Thesis, Norwegian University of Science and Technology, Trondheim, Norway, 2014.
20. Department of Energy of the United States of America. Global Energy Storage Database. Available online: https://www.energystorageexchange.org/ (accessed on 27 October 2018).
21. Zubi, G.; Dufo-López, R.; Carvalho, M.; Pasaoglu, G. The lithium-ion battery: State of the art and future perspectives. *Renew. Sustain. Energy Rev.* **2018**, *89*, 292–308. [CrossRef]
22. Linden, D.; Reddy, T.B. *Handbook of Batteries*; McGraw-Hill Education: New York, NY, USA, 2002.
23. International Renewable Energy Agency IRENA. Electricity Storage and Renewables: Costs and Markets to 2030. Technical Report; 2017. Available online: http://www.irena.org/-/media/Files/IRENA/Agency/Publication/2017/Oct/IRENA_Electricity_Storage_Costs_2017.pdf (accessed on 5 October 2018).
24. DNV GL. Battery Energy Storage Study for the 2017 IRP. Technical Report; 2016. Available online: http://www.pacificorp.com/content/dam/pacificorp/doc/Energy_Sources/Integrated_Resource_Plan/2017_IRP/10018304_R-01-D_PacifiCorp_Battery_Energy_Storage_Study.pdf (accessed on 20 September 2018).
25. EEMB CO., LTD. Lithium Iron Phosphate Battery Specification. Technical Report; 2014. Available online: https://datasheetspdf.com/pdf-file/1281808/EEMB/LP903395F/1 (accessed on 14 September 2018).
26. Corvus Energy. Containerized Energy Storage System. Available online: https://corvusenergy.com/containerized-energy-storage-system (accessed on 14 November 2018).
27. Delta Electronics Inc. Delta Lithium-ion Battery Energy Storage Container. Available online: http://www.deltaww.com/filecenter/Products/Download/18/1805/0803%20DM05-Container-201807.pdf (accessed on 14 November 2018).
28. Sonoda, M. Development of Containerized Energy Storage System with Lithium-ion batteries. In *Mitsubishi Heavy Industries Technical Review*; 2013; pp. 36–41. Available online: https://www.mhi.co.jp/technology/review/pdf/e503/e503036.pdf (accessed on 18 November 2018).
29. Tesvolt. Tesvolt TPS 200-864kWh Lithium Battery Storage. Available online: https://zerohomebills.com/product/tesvolt-tps-200-864kwh-lithium-battery-storage-all-in-one-20ft-container (accessed on 15 November 2018).
30. Corvus Energy. Containerized: Energy Storage System. Available online: https://corvusenergy.com/wp-content/uploads/2016/04/Corvus-Energy_Containerized-Solution_Nov2015.pdf (accessed on 17 November 2018).
31. Dudek. Technical Memorandum. Available online: https://www.sandiegocounty.gov/content/dam/sdc/pds/ceqa/Soitec-Documents/Final-EIR-Files/00_AIS_Combined_OPT_January%202015_Part3.pdf (accessed on 13 November 2018).
32. Avo Reinap, IEA/LU. Battery Pack Design. Available online: http://www.ht.energy.lth.se/fileadmin/ht/Kurser/MVKF25/MVKF25-vt17_BatPackDes.pdf (accessed on 17 November 2018).
33. Trintis, I. Grid Converters for Stationary Battery Energy Storage Systems. Ph.D. Thesis, Aalborg University, Aalborg, Denmark, 2011.
34. Morales Vásquez, C.A. A methodology to select the electric propulsion system for Platform Supply Vessels (PSV). Master's Thesis, Universidade de São Paulo, São Paulo, Brazil, 2014.

energies

MDPI

Article

A Relational Analysis Model of the Causal Factors Influencing CO_2 in Thailand's Industrial Sector under a Sustainability Policy Adapting the VARIMAX-ECM Model

Pruethsan Sutthichaimethee * and Kuskana Kubaha

Division of Energy Management Technology, School of Energy, Environment and Materials, King Mongkut's University of Technology Thonburi, 126 Pracha Uthit Rd., Bang Mod, Thung Khru, Bangkok 10140, Thailand; kuskana.kub@kmutt.ac.th

* Correspondence: pruethsan.sut@gmail.com; Tel.: +66-639-645-195

Received: 2 June 2018; Accepted: 27 June 2018; Published: 1 July 2018

Abstract: Sustainable development is part and parcel of development policy for Thailand, in order to promote growth along with economic growth, social advancement, and environmental security. Thailand has, therefore, established a national target to reduce CO_2 emissions below 20.8%, or not exceeding 115 Mt CO_2 Equivalent (Eq.) by 2029 within industries so as to achieve the country's sustainable development target. Hence, it is necessary to have a certain measure to promote effective policies; in this case, a forecast of future CO_2 emissions in both the short and long run is used to optimize the forecasted result and to formulate correct and effective policies. The main purpose of this study is to develop a forecasting model, the so-called VARIMAX-ECM model, to forecast CO_2 emissions in Thailand, by deploying an analysis of the co-integration and error correction model. The VARIMAX-ECM model is adapted from the vector autoregressive model, incorporating influential variables in both short- and long-term relationships so as to produce the best model for better prediction performance. With this model, we attempt to fill the gaps of other existing models. In the model, only causal and influential factors are selected to establish the model. In addition, the factors must only be stationary at the first difference, while unnecessary variables will be discarded. This VARIMAX-ECM model fills the existing gap by deploying an analysis of a co-integration and error correction model in order to determine the efficiency of the model, and that creates an efficiency and effectiveness in prediction. This study finds that both short- and long-term causal factors affecting CO_2 emissions include per capita GDP, urbanization rate, industrial structure, and net exports. These variables can be employed to formulate the VARIMAX-ECM model through a performance test based on the mean absolute percentage error (MAPE) value. This illustrates that the VARIMAX-ECM model is one of the best models suitable for the future forecasting of CO_2 emissions. With the VARIMAX-ECM model employed to forecast CO_2 emissions for the period of 2018 to 2029, the results show that CO_2 emissions continue to increase steadily by 14.68%, or 289.58 Mt CO_2 Eq. by 2029, which is not in line with Thailand's reduction policy. The MAPE is valued at 1.1% compared to the other old models. This finding indicates that the future sustainable development policy must devote attention to the real causal factors and ignore unnecessary factors that have no relationships to, or influences on, the policy. Thus, we can determine the right direction for better and effective development.

Keywords: causal factors; CO_2 emissions forecasting; VARIMAX-ECM model; sustainable development; economic growth; population growth

1. Introduction

Thailand is currently in the midst of accelerating economic growth in order to develop the country. Along the way, it has found that the current GDP (gross domestic product) has increased as a result of the promotion and expansion of various areas, such as the support of export activities, a continual increase of private consumption, a rise in government spending, an acceleration of foreign investment, and the promotion of industrialization and urbanization. Throughout these enforcements, the environment is being affected as the amount of CO_2 emissions from the country's energy consumption rose by 1.3% in 2016. CO_2 emissions have been seen to increase in almost all economic sectors, including industrial, transportation, and other economic sectors. Concerning CO_2 emissions per unit of electricity production (kWh) and per GDP in the sectors, they continue to increase beyond the global average. Among the different economic sectors, CO_2 emissions in the industrial sectors are at the highest rate, equivalent to 27%, while their growth rate is at 4.3%. During the year of 2017, compared to 2016, the petroleum sector contributed the highest CO_2 emissions [1].

Thailand produces total CO_2 emissions of 69.9 Mt CO_2 Eq. under the industrial sectors, with a growth rate (2017/2016) of 4.3% due to economic growth. Moreover, CO_2 is emitted by the energy sector at 88.7% with a 10.3% growth rate (2017/2016). This reflects that this sector produces the highest amount of greenhouse gas. Generally, it releases up to 90% of carbon dioxide and 75% of other greenhouse gases out of the total greenhouse gases [1,2].

Sustainable development is the national roadmap that Thailand aims to follow. It aims to boost the economy, along with social improvement, while the environment is simultaneously enhanced. The above roadmap has to be given full attention and carefully implemented. This is because economic and social growth are likely to negatively affect the environment. Nevertheless, the vital action of creating efficiency in planning and sustainability in implementation is to analyze the relationship of various variables which can influence, and have an impact on, policy-making. Thus, the analysis outcome can provide future predictions so as to facilitate in both short- and long-term policy-making and action planning.

Energy consumption evolves around producing more and more CO_2 that is emitted into the air, causing natural damage and climate change. Thus, forecasting future energy consumption is becoming an important task, as it represents another way to determine what actions need to be taken in order to minimize CO_2 emissions and achieve the national reduction goal. By reviewing various studies across the region, it is evident that CO_2 emissions are associated with various forces, and energy consumption is an integral part of the emission level. Therefore, a forecasting strategy would be instrumental for the energy consumption industry.

Many studies have attempted to generate different approaches and applications to support energy consumption, production, and optimization. For example, the studies of Ren et al. [3], Xu et al. [4], Jeong and Kim [5], González et al. [6,7], Xu et al. [8], Wang et al. [9], Tian et al. [10], and Lin and Long [11] focused on the attributes or characteristics of energy consumption by using an analysis of logarithmic mean Divisia index (LMDI) factor decomposition. Among them, Wang et al. [9] also proposed a new method of LMDI, and this method was structured based on five perspectives of effect: labor, economic structure, investment, energy mix, and energy intensity. This study was conducted in China's energy consumption sector, and its result showed that the energy intensity does help to decrease energy consumption. As the energy intensity plays an important role in energy consumption, Baležentis et al. [12] started exploring the energy intensity trends in the Lithuanian economy under different economic sectors from 1995 to 2009, and their study reported that energy efficiency increased when the economy exhibited a downward trend. Therefore, certain measures should be issued as policies in order to enhance the energy intensity in Lithuania, as suggested by the study. González et al. [6] explored the underlying factors causing changes in aggregate energy consumption by using LMDI, and their study showed that the enhancement in energy efficiency was not sufficient to lower the economic pressure of European activity with regard to aggregate energy consumption. In recent years, many countries have put forth efforts to increase production, which requires a higher energy consumption, so as to boost their economic growth. However, a study by Mulugeta et al. [13] showed that energy consumption is an important driving force towards growth in the economy, which they investigated by forming an economic growth hypothesis. For a particular country, such as Saudi Arabia, Alkhathlan and Javid [14] investigated the relationship between economic growth, energy consumption, and CO_2 emissions, and they found that the rise of CO_2 emissions was influenced by the increment of income per capita. On the other hand, Khan et al. [15] analyzed the relationship of studied variables for the period of 1975 to 2011, and they witnessed that energy consumption had a significant impact on the CO_2 emissions in Pakistan in particular.

Other studies, such as that of Arouri et al. [16], studied the relationship between the real GDP, CO_2 emissions, and energy consumption in 12 selected Middle East and North African countries (MENA) using a bootstrap panel method. They found clear evidence that CO_2 emissions are significantly affected by energy consumption. Additionally, Acaravci and Ozturk [17] initiated a study of the causality between various factors, including energy use, economic growth, and CO_2 emissions, with a sample size of 19 European countries. By using a technique of autoregressive distributed lag (ARDL) and the error-correction Granger causality test, they were able to find only the long-run relationship between those factors in certain countries, such as Iceland, Switzerland, Denmark, Portugal, Germany, Greece, and Italy. In addition, Menyah and Wolde-Rufael [18] conducted a similar study on the causality between energy consumption, pollutant emissions, and economic growth in South Africa with the same approach of ARDL. As of the result, a long-run relationship between the variables was revealed. Ohlan [19] performed an analysis of the impact of energy consumption, population density, trade openness, and economic growth on the emissions of CO_2 in India for the period of 1970–2013. For this analysis, the researcher employed the ARDL approach, and its result showed that those three studied factors had a great positive influence on CO_2 emissions in both the short and long term. With the same method of analysis, the ARDL method, Sulaiman and Abdul-Rahim [20] conducted an investigation of a three-way linkage relationship between economic growth, CO_2 emissions, and energy consumption in Malaysia during the period of 1975–2015. The examination's result revealed that the rise of both factors; energy consumption and economic growth, do contribute to the rise of CO_2 emissions.

In order to determine other evidence of association with CO_2 emissions, Akpan and Akpan [21] found in their study conducted in Nigeria that economic growth improves when carbon emissions are rising, and this rise of CO_2 emissions is positively associated with electricity consumption. In the same area of study with the application of the Toda and Yamamoto causality test, Sulaiman [22] claimed that CO_2 emissions do support economic growth, while energy consumption contributes to the increase of CO_2 emissions. However, Manu and Sulaiman [23] adapted the simple ordinary least squares (OLS) approach to examine the relationship between economic growth, energy consumption, and CO_2 emissions in Malaysia. This study covered the period of 1965–2015, and found that CO_2 emissions are reduced when the income is raised. In the meantime, it increases when the trade openness increases.

In addition to those factor relation studies, it is necessary to mention the grey system and autoregressive integrated moving average by Lotfalipour, Falahi, and Bastam [24]. They optimized the above model to predict CO_2 emissions in Iran. Their findings showed that the models could produce a more accurate result than any other method, and estimated up to 925.68 million tons of carbon dioxide emissions by 2020, equivalent to 66% growth compared to 2010. Liang [25] discussed China's multi-region energy consumption and CO_2 emissions under an input-output model. Additionally, his findings were portrayed through a scenario analysis for 2010 and 2020. For a shorter-term forecasting coverage, Li [26] evaluated the CO_2 emissions reduction under different scenarios for the years of 2016 and 2020 in Beijing. He applied a back propagation (BP) neural network optimized by the improved particle swarm optimization algorithm. However, his investigation showed that the model was not effective enough to provide high precision. Meanwhile, Zhao, Huang, and Yan [27] forecasted CO_2 emissions in China from 2017 to 2020 with the deployment of some selected models: the single LSSVM model, the LSSVM model enhanced by the particle swarm optimization algorithm

(PSO-LSSVM), and the back propagation (BP) neural network model. The above prediction verified that structural factors will have a significant impact on CO_2 emissions by 2020. Potentially, this allows China to keep its promise to reduce greenhouse gas emissions by 2030. Consequently, Dai, Niu, and Han [28] proposed to adapt the MSFLA-LSSVM model in CO_2 emissions prediction in China from 2018 to 2025. They concluded that China's CO_2 emissions would exhibit a slow growth trend for the next few years. With this in mind, China's CO_2 emissions could be effectively controlled in the future, which could start to reduce the greenhouse effect. In another approach, Lin et al. [29] incorporated the grey forecasting model to estimate CO_2 emissions from 2010 to 2012 in Taiwan. According to the forecasting results, they found that the CO_2 emissions of Taiwan would decline for the next three years.

The Government of Thailand aims to establish a future reduction goal for CO_2 emissions, whereby Thailand should reduce emissions below 20.8% or not exceed 115 Mt CO_2 Eq. by 2029. However, over the years, CO_2 emissions produced from energy consumption have been continuously increasing. Industrial sectors, in particular, have the highest increase of up to 27%, while the growth rate is increasing continuously every year. Also, it is observed that the petroleum sector is the major contributor and is emitting the most CO_2. This is seen to contradict Thai government policy and planning, and the CO_2 emissions reductions are not improving [2]. Hence, the author sees this as an issue that needs to be tackled, and this study has, therefore, been carried out. The study focuses on the policy framework, which reflects the fact that Thailand still lacks a forecasting model which can produce good results and make effective predictions in both the short- and long-term. As for the existing forecasting models used in Thailand's policy formulation, they are models without proper processing and with ineffective research. In addition, most of the models are too common, such as multiples regression, the ARMA model, and many more. As a result, the previous predictions have become spurious and erroneous. In the same model forecasts, the causal factors that actually affect the CO_2 emissions have not been analyzed or taken into account.

Based on a review of previous studies, many studies share similarities in metrology, research methodologies, and various analytical outcomes. In this study, unlike any other studies, a new research focus is introduced, which constitutes an investigation of the relationship of causal factors of various variables. The analytical outcome is later driven into further forecasting for both short- and long-term use. In fact, this research is designed to support sustainable development policy-making, create analysis guidelines, as well as to open new areas for those interested in exploring and expanding sustainable development in the future; be it Thailand or any other country. This research provides guidance in the process of establishing the country's sustainable development policy as it allows the determination of effective management and working processes. The research's guideline flow is as follows.

(1) Analyze the causal variables that can influence the change of CO_2 emissions with the Augment Dickey Fuller theory [30] only at the same level. This analysis is within the framework of sustainable development, using data from 1990 to 2017. Moreover, only crucial and influential variables are used in the forecasting model.

(2) Place the stationary causal variables at the same level in the analysis of long-term relationship based on the Johansen Juselius concept [31].

(3) Create a forecasting model by adapting the advance statistics of the so-called vector autoregressive model, with full consideration of the relationship of all causal variables, both in terms of the error correction model and co-integration, consisting of significant causal variables towards the change of CO_2 emissions. Additionally, a forecasting pattern for both the short- and long-term must be taken into account so as to produce the best and most effective model with the least errors. The average relative errors between the simulation and actual data are measured through an output comparison of relevant models, namely the ARMA model, ARIMA model, and GM-ARIMA model.

(4) Forecast CO_2 emissions from the VARIMAX-ECM model for the period of 2018 to 2029, totaling 12 years, with certain selected causal factors. Discard unnecessary variables.

The flowchart of the VARIMAX-ECM model is shown in Figure 1.

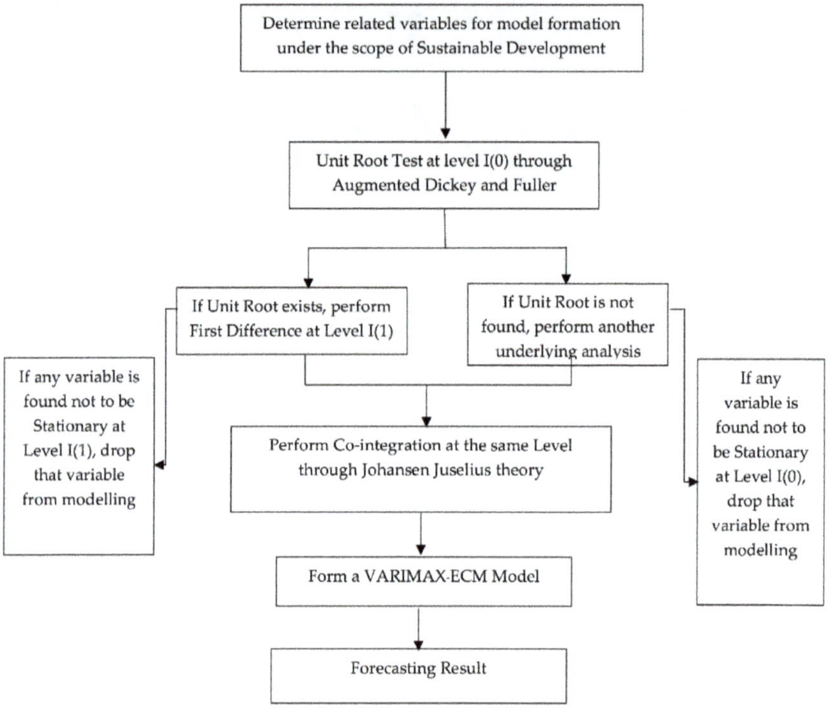

Figure 1. The flowchart of the VARIMAX-ECM model.

The main structures of this article flow as follows: the second section introduces the forecasting model of VARIMAX-ECM. The third section carries out the empirical analysis to prove the practicality and validity of the proposed model for CO_2 emissions forecasting, and to predict the CO_2 emissions in Thailand's industrial sector from 2018 to 2029. The fourth section summarizes the discussion.

2. The Forecasting Model

2.1. Unit Root Test

We analyze the data for the stationary process by testing the unit root according to the Augment Dickey Fuller concept [30].

Stationary Process

The stationary, or stationary stochastic, process [32,33] is the series of time data with the mean or expected value, variance, constant overtime, and covariance. The expected value and constant variance in the context of εt lacks the property of being white noise, meaning that it has the autocorrelation property where the correlations are high or the order of the autoregressive process is higher. Hence, a test like the Augmented Dickey Fuller test (ADF) is required. The lagged variables are added into the equation in the higher level to eliminate the autocorrelation, heteroskedasticity, and multicollinearity, as shown below:

$$\Delta Y_t = \delta_1 Y_t + \sum_{i=2}^{p} \beta_i \Delta Y_{t-i+1} + \varepsilon_t \quad (1)$$

$$\Delta Y_t = \alpha_1 + \delta Y_{t-1} + \sum_{i=2}^{p} \beta_i \Delta Y_{t-i+1} + \varepsilon_t \tag{2}$$

$$\Delta Y_t = \alpha_1 + \alpha_2 T + \delta Y_{t-1} + \sum_{i=2}^{p} \beta_i \Delta Y_{t-i+1} + \varepsilon_t \tag{3}$$

From the equations above, the value of p seems to be the lagged values of first difference to the variable, which is verified by testing the unit root with the Augmented Dickey Fuller method. With the above equation, three problems are considered and taken into account. In particular, the autocorrelation in ε_t is set to have the property of white noise, and the error term has the mean of 0 and is constant under the following hypotheses:

Hypotheses 1 (H_0). *$\delta = 0$, non-stationary;*

Hypotheses 2 (H_1). *$\delta < 0$, stationary.*

If tau-statistics of the efficiency δ are in the form of the absolute term, there must be more critical values appearing in the ADF table. This denies the major hypothesis, meaning that the time series of the variables are stationary. Thus, it can be said that ΔY_t integrated numbered is represented by $\Delta Y_t \sim I(d)$.

2.2. VARIMAX-ECM Model

The VARIMAX-ECM model is a new model adapted from the vector autoregressive model, incorporating influential variables in both short-term and long-term relationships so as to produce the best prediction model with the maximum performance and least error.

2.2.1. VARIMAX-ECM and Co-Integrating Vector

In this section, we consider the segment of the deterministic component in a time series of the VAR model [34]. In order to simplify the concept for a better understanding, we consider the VAR model as follows:

$$X_t = A_1 X_{t-1} + \mu_0 + \mu_1 t + u_t \tag{4}$$

where μ_0 is the vector of the parameter representing a constant value in the VAR(p) model, μ_1 is the vector of the parameter indicating a defined trend in the VAR(p) model, and vectors μ_0 and μ_1 are shown below:

$$\mu_0 = \begin{bmatrix} \mu_{01} \\ \mu_{02} \\ \vdots \\ \mu_{0n} \end{bmatrix}_{nx1} ; \mu_1 = \begin{bmatrix} \mu_{11} \\ \mu_{12} \\ \vdots \\ \mu_{1n} \end{bmatrix}_{nx1}$$

When vectors μ_0 and μ_1 are not zero, Equation (4) reflects that at least one time series in the VAR (1) model must be a deterministic component, in which it can either be a constant or a defined trend, or both forms. The above VAR(p) model can be converted into the VARIMAX-ECM model as shown below:

$$\Delta X_t = \alpha\beta' X_{t-1} + \mu_0 + \mu_1 t + u_t \tag{5}$$

From the above equation, it can be observed that vectors μ_0 and μ_1 exist in both the VAR and VARIMAX–ECM models, and that both ΔX_t and $\beta' X_{t-1}$ have to be stationary in the deterministic area.

However, to observe a deviation out of the long-term co-integration of j (j = 1, 2, ... , r) denoted as vector $\beta' X_{t-1}$, the mean of the above deviation must be zero. In order to obtain such a result, the deterministic component must be eliminated from the deviation out of long-term balance $(\beta' X_{t-1})$ by separating vectors μ_0 and μ_1 in the VARIMAX-ECM model, as illustrated in Equation (6), and by combining them into $\beta' X_{t-1}$ as explained below.

Vector μ_0 and vector μ_1 can be separated into the sum of the two vectors by using the following equation:

$$\alpha(\beta'\alpha)^{-1}\beta' + \beta_\perp(\alpha'_\perp\beta_\perp)^{-1}\alpha'_\perp = I \quad (6)$$

where $\beta_\perp$ and $\alpha_\perp$ are the orthogonal matrices with β and α, respectively. Here, it is seen that $\beta'\beta_\perp = 0$ and $\alpha'\alpha_\perp = 0$.

When we multiply vector μ_0 with Equation (6), the result is obtained from:

$$\alpha(\beta'\alpha)^{-1}\beta'\mu_0 + \beta_\perp(\alpha'_\perp\beta_\perp)^{-1}\alpha'_\perp\mu_0 = \mu_0 \quad (7)$$

If given:

$$\beta_0 = (\beta'\alpha)^{-1}\beta'\mu_0 \quad (8)$$

$$\gamma_0 = \beta_\perp(\alpha'_\perp\beta_\perp)^{-1}\alpha'_\perp\mu_0 \quad (9)$$

We substitute Equations (8) and (9) into Equation (7), we obtain:

$$\mu_0 = \alpha\beta_0 + \gamma_0 \quad (10)$$

At the same time, if we use vector μ_1 to multiply with Equation (6), obtaining:

$$\mu_1 = \alpha\beta_1 + \gamma_1 \quad (11)$$

where $\beta_1 = (\beta'\alpha)^{-1}\beta'\mu_1$ and $\gamma_1 = \beta_\perp(\alpha'_\perp\beta_\perp)^{-1}\alpha_1\mu_1$.

If we substitute Equations (10) and (11) into Equation (5), we obtain:

$$\Delta X_t = \alpha\beta' X_{t-1} + \alpha\beta_0 + \alpha\beta_1 t + \gamma_0 + \gamma_1 t + \mu_t \quad (12)$$

Equation (12) can be restructured as follows:

$$\Delta X_t = \alpha(\beta' X_{t-1} + \beta_0 + \beta_1 t) + \gamma_0 + \gamma_1 t + \mu_t \quad (13)$$

where ΔX_t is the $n \times 1$ vector, X_t is the $n \times 1$ vector, α is the $n \times r$ matrix, and β is the $n \times r$ matrix. β_0 is the $r \times 1$ matrix, β_1 is the $r \times 1$ matrix, γ_0 is the $n \times 1$ matrix, γ_1 is the $n \times 1$ matrix, and μ_t is the $n \times 1$ matrix. n is the number of time series in vector X_t.

Equation (13) shows that if vector X_t determines $(\mu_0 + \mu_1 t)$, there is a possibility that the VARIMAX-ECM model determines $(\gamma_0 + \gamma_1 t)$ and the long-term co-integration determines $(\beta_0 + \beta_1 t)$. In addition, Equation (13) can be rewritten as:

$$\Delta X_t = \alpha\begin{bmatrix} \beta' & \beta_0 & \beta_1 \end{bmatrix}_{r\times(n+2)} \begin{bmatrix} X_{t-1} \\ 1 \\ t \end{bmatrix}_{(n+2)\times 1} + \gamma_0 + \gamma_1 t + \mu_t \quad (14)$$

Let $\widetilde{\beta}' = \begin{bmatrix} \beta' & \beta_0 & \beta_1 \end{bmatrix}, \widetilde{X}_{t-1} = \begin{bmatrix} X_{t-1} & 1 & t \end{bmatrix}'$, and we can then structure another equation as:

$$\Delta X_t = \alpha\widetilde{\beta}'\widetilde{X}_{t-1} + \gamma_0 + \gamma_1 t + u_t \quad (15)$$

The above equation contains the following characteristics:

$$E(\Delta X_t) = \gamma_0 + \gamma_1 t;\ E\left(\widetilde{\beta}'\widetilde{X}_{t-1}\right) = 0 \quad (16)$$

Moreover, Equation (15) explains the connection between the deterministic area of the VARIMAX-ECM model and the long-term co-integrating vector, which can be classified into five situations.

Situation 1: If $\gamma_0 = \gamma_1 = \beta_0$ or $(\mu_0 = \mu_1 = 0)$ for both the VARIMAX-ECM model and the long-term co-integrating vector (deviation out of long-term balance), they are not deterministic or can be expressed as $E(\Delta X_t) = 0$ and $E(\beta'\widetilde{X}_{t-1}) = 0$. Therefore, the VARIMAX-ECM model in this position is:

$$\Delta X_t = \alpha\beta' X_{t-1} + u_t \tag{17}$$

The case of $\mu_0 = \mu_1 = 0$ indicates a time series in vector X_t and is not deterministic (it is not constant nor a defined trend) in the equation.

Situation 2: If $\gamma_0 = 0, \gamma_1 = \beta_1 = 0$ (or $\mu_1 = 0$) but $\beta_0 \neq 0$, then the vector of the long-term co-integration reflects a constant value ($\beta_0 \neq 0$) or can be written as $E(\beta' X_{t-1}) = \beta_0$. Meanwhile, the VARIMAX-ECM model is not deterministic at all, or can be written as $E(\Delta X_t) = 0$. In order to remove the constant value out of the long-term co-integration, the VARIMAX-ECM model must be in the form of:

$$(\Delta X_t) = \alpha\widetilde{\beta}'\widetilde{X}_{t-1} + u_t \tag{18}$$

where $\widetilde{\beta}' = \begin{bmatrix} \beta' & \beta_0 \end{bmatrix}$ and $\widetilde{X}_{t-1} = \begin{bmatrix} X_{t-1} & 1 \end{bmatrix}'$. Thus, we can retrieve $E(\widetilde{\beta}'\widetilde{X}_{t-1}) = 0$.

The case of $\mu_1 = 0$ and $\gamma_0 = 0$, but $\beta_0 \neq 0$, indicates at least one time series in vector X_t and is constant (but it is not a defined trend) in the equation.

Situation 3: If $\gamma_1 = \beta_1 = 0$ (or $\mu_1 = 0$) but $\gamma_0 \neq 0$ and $\beta_0 \neq 0$, the vector of the long-term co-integration is not a defined trend but is constant ($\beta_0 \neq 0$), or can be written as $E(\beta' X_{t-1}) = \beta_0$. If the VARIMAX-ECM model is found to be constant, $\gamma_0 \neq 0$, or can be written as $E(\Delta X_t) = \gamma_0$, and the above fixed value in the long-term co-integrating vector can be removed by using the long-term co-integration $\widetilde{\beta}'\widetilde{X}_{t-1}$ in the VARIMAX-ECM model, as illustrated below:

$$\Delta X_t = \alpha\widetilde{\beta}'\widetilde{X}_{t-1} + \gamma_0 + u_t \tag{19}$$

where $\widetilde{\beta}' = \begin{bmatrix} \beta' & \beta_0 \end{bmatrix}$ and $\widetilde{X}_{t-1} = \begin{bmatrix} X_{t-1} & 1 \end{bmatrix}'$.

The case of $\mu_1 = 0$, but $\gamma_0 \neq 0$ and $\beta_0 \neq 0$ indicates that at least one time series is a defined trend.

Situation 4: If $\gamma_1 = 0$, but $\gamma_0 \neq 0$, $\beta_0 \neq 0$ (or $\mu_0 \neq 0$), and $\beta_0 \neq 0$, the long-term co-integration $\beta' X_{t-1}$ cannot eliminate the constant value and defined trend, and it can be rewritten as $E(\beta' X_{t-1}) = \beta_0 + \beta_1 t$. This can be described in such a way that the long-term co-integration is a stationary trend, while the VARIMAX-ECM model is found to have a fixed value of $\gamma_0 \neq 0$ or can be written as $E(\Delta X_t) = \gamma_0$. The fixed value and defined trend that exist in the long-term co-integrating vector could be removed by using a $\widetilde{\beta}'\widetilde{X}_{t-1}$ long-term co-integration in the VARIMAX-ECM model as follows:

$$\Delta X_t = \alpha\widetilde{\beta}'\widetilde{X}_{t-1} + \gamma_0 + u_t \tag{20}$$

where $\widetilde{\beta}' = \begin{bmatrix} \beta' & \beta_0 & \beta_1 \end{bmatrix}$ and $\widetilde{X}_{t-1} = \begin{bmatrix} X_{t-1} \\ 1 \\ t \end{bmatrix}$, which can also be written as $\begin{bmatrix} X_{t-1} & 1 & t \end{bmatrix}'$.

The case of $\mu_1 \neq 0$ and $\gamma_1 = 0$ but $\beta_1 \neq 0$ demonstrates that at least one time series in vector X_t has to be constant and a defined linear trend, but it is not a quadratic trend.

Situation 5: If $\gamma_1 \neq 0$, $\gamma_0 \neq 0$, $\beta_0 \neq 0$, and $\beta_0 \neq 0$, this shows that the long-term co-integration has to be a stationary trend $(\beta_0 + \beta_1 t)$, while the VARIMAX-ECM model has to be constant and defined trend $(\gamma_0 + \gamma_1 t)$ which can be written as below:

$$\Delta X_t = \alpha\widetilde{\beta}'\widetilde{X}_{t-1} + \gamma_0 + \gamma_1 t + u_t \tag{21}$$

where $\widetilde{\beta}' = \begin{bmatrix} \beta' & \beta_0 & \beta_1 \end{bmatrix}$ and $\widetilde{X}_{t-1} = \begin{bmatrix} X_{t-1} & 1 & t \end{bmatrix}'$. The case of $\mu_0 \neq 0$ and $\mu_1 \neq 0$ occurs when at least one time series in vector X_t has to be defined by a quadratic trend $(\mu_0 + \mu_1 t + \mu_2 t^2)$.

2.2.2. An Estimation of the Co-Integrating Vector with the Use of Various Equations [33]

Consider the VARIMAX-ECM model as follows [35]:

$$\Delta X_t = \alpha\widetilde{\beta}'\widetilde{X}_{t-1} + \Gamma_1\Delta X_{t-1} + \Gamma_2\Delta X_{t-2} + \ldots + \Gamma_{p-1}\Delta X_{t-(p-1)} + \varphi D_t + u_t \quad (22)$$

where $\widetilde{\beta}' = \begin{bmatrix} \beta' & \beta_0 & \beta_1 \end{bmatrix}$ is the $r \times (n+2)$ matrix, β is the $n \times r$ matrix, β_0 and β_1 are the $r \times 1$ vector, $\widetilde{X}_{t-1} = \begin{bmatrix} X_{t-1} & 1 & t \end{bmatrix}'$ is the $(n+2) \times 1$ vector, α is the $n \times r$ matrix, and rank (α) = rank $(\widetilde{\beta}) = r$. Additionally, D_t is the matrix indicating a deterministic component.

The estimation of the parameter of the long-term co-integrating vector $\widetilde{\beta}$ can be achieved with the application of maximum likelihood by assuming vector $u_t \approx$ Normal $(0, \sum)$ 0 is zero, and $\sum$ is the variant matrix of u_t. Johansen (1995) proved that the estimation of vector $\widetilde{\beta}_{n\times r}$ with this method would result in an eigenvector in accordance with the eigenvalue from the minimum to maximum value. This is achieved using the equation below:

$$\left|\lambda S_{11} - S_{10}S_{00}^{-1}S_{01}\right| = 0 \quad (23)$$

$S_{ij} = \frac{1}{T}R_{it}R'_{jt}$, $i = 0, 1$, and $j = 0, 1$; where T is the number of data used in the VARIMAX-ECM model. R_{0t} is the $n \times T$ matrix of the residual retrieved from a regression equation with a variable of ΔX_t, and the independent variable is $\Delta X_{t-1}, \Delta X_{t-2}, \ldots, \Delta X_{t-p+1}, D_t$. R_{1t} is the $(n+2) \times T$ matrix of the residual retrieved from a regression equation with a variable of $\widetilde{X}_{t-1}$, and the independent variable is $\Delta X_{t-1}, \Delta X_{t-2}, \ldots, \Delta X_{t-p+1}, D_t$.

If $\hat{\lambda}_i(i = 1,2,\ldots,n)^{11}$ is the eigenvalue computed from Equation (24) where $1 > \hat{\lambda}_1 > \hat{\lambda}_2 > \ldots > \hat{\lambda}_n \geq 0$, let the eigenvector consistant with the eigenvalue $\hat{\lambda}_1, \hat{\lambda}_2, \ldots, \hat{\lambda}_n$ be written as $\hat{V} = \begin{bmatrix} \hat{V}_1 & \hat{V}_2 & \ldots & \hat{V}_n \end{bmatrix}_{(n+2)\times(n+2)}$. Therefore, we can obtain the estimator of the co-integrating vector as follows:

$$\hat{V} = \begin{bmatrix} \hat{V}_1 & \hat{V}_2 & \ldots & \hat{V}_r \end{bmatrix}_{(n+2)\times r} \quad (24)$$

Commonly, there are two popular patterns of forming primary and secondary assumptions pertaining to the number of the long-term co-integration.

Pattern 1: H_0 is the maximal number of vectors indicating the long-term co-integration equivalent to r. H_1 is the number of vectors indicating the long-term co-integration greater than r.

In the above, $r = 0, 1, 2, \ldots, n-1$, and the statistical value to testify the above assumption is trace statistic λ_{trace}, which can be computed using the equation below:

$$\lambda_{trace}(r) = -T\sum_{i=r+1}^{n}(1 - \hat{\lambda}_i) \quad (25)$$

Pattern 2: H_0 is the maximal number of vectors indicating the long-term co-integration equivalent to r. H_1 is the number of vectors indicating the long-term co-integration equivalent to $r+1$.

In the above, $r = 0, 1, 2, \ldots, n-1$, and the statistical value to testify the above assumption is maximum eigenvalue λ_{trace}, which can be computed using the equation below:

$$\lambda_{max}(r, r+1) = -T(1 - \hat{\lambda}_{r+1}) \quad (26)$$

$$\hat{A}_i = \begin{cases} I + \widehat{\Pi} + \widehat{\Gamma} & ,i = 1 \\ \widehat{\Gamma}_i - \widehat{\Gamma}_{i-1} & ,2 \leq i \leq -1 \\ -\widehat{\Gamma}_{p-1} & ,i = p \end{cases} \tag{27}$$

After that, we use the VARIMAX-ECM forecasting model of the time series in vector X_t by using the same concept, which is the forecasting of the minimum mean square error. Hence, the forecast of $1, 2, \ldots, h$ pre-timing of the time series in the vector X_t can be illustrated as:

$$\hat{X}_{T+1} = \hat{A}_1 X_T + \hat{A}_2 X_{T-1} + \hat{A}_p X_{T-p+1} \tag{28}$$

$$\hat{X}_{T+2} = \hat{A}_1 X_{T+1} + \hat{A}_2 X_{T-1} + \hat{A}_p X_{T-p+2} \tag{29}$$

$$\hat{X}_{T+h} = \hat{A}_1 X_{T+h-1} + \hat{A}_2 X_{T+h-2} + \ldots + \hat{A}_p X_{T-p+h} \tag{30}$$

where $\hat{X}_{T+j} = \hat{A}_1 X_{T+j}$ if $j < 0$.

2.2.3. Measurement of the Forecasting Performance

In order to evaluate the forecasting effect of each model, we employ the mean absolute percentage error (MAPE) to compare the forecasting accuracy of each model. The calculated equations are shown as follows:

$$\text{MAPE} = \frac{1}{n}\sum_{i=1}^{n}\left|\frac{\hat{y}_i - y_i}{y_i}\right| \tag{31}$$

3. Empirical Analysis

3.1. Screening of Influencing Factors for Model Input

In this paper, we tested the causal factors in the context of Thailand's sustainable development policy. Here, we deploy the time series data of the period of 1990–2017. The tested factors consist of seven variables, namely CO_2 emission ($\ln(CO_2)$), population growth (ln(Population)), per capita GDP (ln(GDP)), urbanization rate (ln(UR)), industrial structure (ln(IS)), total coal consumption (ln(CC)), and total exports and imports ($\ln(X - E)$). The test was conducted based on the Augment Dickey Fuller theory at Level I (0) and Fist Difference I (1), as illustrated in Table 1.

Table 1. Unit root test at Level I (0) and First Difference I (1).

ADF Test at Level I (0)		ADF Test at First Difference I (1)		MacKinnon Critical Value		
				1%	5%	10%
$\ln(CO_2)$	−3.41	$\Delta \ln(CO_2)$	−4.90	−4.12	−3.27	−3.05
ln(Population)	−2.05	Δ ln(Population)	−3.02	−4.12	−3.27	−3.05
ln(GDP)	−3.81	Δ ln(GDP)	−5.69	−4.12	−3.27	−3.05
ln(UR)	−3.25	Δ ln(UR)	−4.71	−4.12	−3.27	−3.05
ln(IS)	−3.72	Δ ln(IS)	−4.65	−4.12	−3.27	−3.05
ln(CC)	−2.45	Δ ln(CC)	−3.01	−4.12	−3.27	−3.05
ln(X − E)	−3.64	Δ ln(X − E)	−4.64	−4.12	−3.27	−3.05

Note: $\ln(CO_2)$ is the natural logarithm of CO_2 emissions; ln(Population) is the natural logarithm of population growth; ln(GDP) is the natural logarithm of per capita GDP; ln(UR) is the natural logarithm of urbanization rate; ln(IS) is the natural logarithm of industrial structure; ln(CC) is the natural logarithm of total coal consumption; ln(X − E) is the natural logarithm of total exports and imports, and Δ is the first difference.

Table 1 shows that all variables under the unit root test are non-stationary at Level I (0), and this explains the non-significance at 5% and 1%. Therefore, the First Difference I (1) is required to carry on. The finding here indicates that when the variables are tested through the unit root test at Level I (1) with a significance level of 5% and 1%, or stationary identification, the variables appear to be CO_2 emissions, per capita GDP, urbanization rate, industrial structure, and total exports and imports. Thus, these variables are carried forward for a co-integration analysis. The other two variables, population growth and total coal consumption, are non-stationary at Level I (1). Therefore, the researcher tests the two variables in pairs with other variables. The outcome shows that the variables do not represent any correlation to the changes in CO_2 emissions at significance levels of 5% and 1%. Accordingly, the two variables are dropped out of the model. Meanwhile, those stationary variables at the First Difference are brought forth to investigate the long-term relationship (co-integration) as demonstrated in Table 2.

3.2. Analysis of Co-Integration

Table 2 shows that all variables have a long-term relationship (co-integration), because the results of the trace test are 210.25 and 70.55, which are higher than the critical values at significance levels of 1% and 5%. The maximum eigenvalue test results are 130.55 and 75.46, which are higher than the critical values at the same significance levels. Consequently, those variables are used to form a forecasting model by adapting the ARIMAX-ECM model and applying short- and long-term relationships into the model. For a better understanding, the model is presented in the form of a regression line, so as to show the influence of variables as seen in Table 3.

Table 2. Co-integration test by Johansen Juselius.

Variables	Hypothesized No of CE(S)	Trace Statistic Test	MacKinnon Critical Value		Max-Eigen Statistic Test	MacKinnon Critical Value		Status
			1%	5%		1%	5%	
$\Delta\ln(CO_2)$, $\Delta\ln(GDP)$, $\Delta\ln(UR)$, $\Delta\ln(IS)$, $\Delta\ln(X-E)$	None **	210.25	19.75	15.41	130.55	15.68	14.07	I (1)
	At Most 1 **	70.55	5.75	3.16	75.46	5.75	3.16	I (1)

** denotes significance $\alpha = 0.01$.

3.3. Formation of Analysis Modeling with the VARIMAX-ECM Model

Table 3 illustrates the parameters of the VARIMAX-ECM Model at a statistically significant level of 1% and 5%. The findings show that per capita GDP (ln(GDP)) causes a change in CO_2 emissions ($\ln(CO_2)$), which covers both short- and long-terms at a statistically significant level of 1%. At the same significant level and effect coverage, the urbanization rate (ln(UR)), total coal consumption (ln(CC)) and total exports and imports ($\ln(X-E)$) are also found to cause changes in CO_2 emissions ($\ln(CO_2)$). Hence, this study suggests that the above causal factors have an influence over changes in CO_2 emissions with the parameter size shown in the table.

Table 3. The result of the VARIMAX-ECM model.

Dependent Variables	Direction of Causality					
	Short Term					Long Term
	$\sum\Delta\ln(CO_2)$	$\sum\Delta\ln(GDP)$	$\sum\Delta\ln(UR)$	$\sum\Delta\ln(IS)$	$\sum\Delta\ln(X-E)$	ECM_{t-1}
$\Delta\ln(CO_2)$		6.43 **	4.76 **	3.42 **	5.13 **	−2.15 **
$\Delta\ln(GDP)$	4.31 *		3.05 **	5.77 **	6.65 **	−2.05 **
$\Delta\ln(UR)$	3.76 *	3.44 *		6.59 **	4.61 **	−1.97 **
$\Delta\ln(IS)$	4.71 **	2.78 *	3.49 **		7.11 **	−1.51 **
$\Delta\ln(X-E)$	2.45 *	2.98 **	6.78 **	4.62 **		−2.77 **

In the above, ** denotes significance $\alpha = 0.01$, * denotes significance $\alpha = 0.05$, R-squared is 0.92, adjusted R-squared is 0.91, the Durbin-Watson statistic is 2.02, the F-statistic is 275.05 (probability is 0.00), the ARCH test is 30.45 (probability is 0.1), the LM test is 1.55 (probability is 0.10), and the response test ($\chi^2 >$ critical) represents the significance.

However, this study also reveals that the changes in per capita GDP (ln(GDP)), urbanization rate (ln(UR)), total coal consumption, and total exports and imports ($\ln(X-E)$) are caused by the factors shown in Table 4 at a statistically significant level of 1% and 5%, respectively, for both short- and long-terms with the parameter size stated in the table.

In addition to this, the author has compared some selected forecasting models in terms of their effectiveness with MAPE as indicated in Table 4. The comparison takes the VARIMAX-ECM model compared with other models, including the ARMA model, ARIMA model, and GM-ARIMA model, as follows.

Table 4. The performance monitoring of the forecasting model.

Forecasting Model	MAPE (%)
ARMA Model	7.44
ARIMA Model	5.75
GM-ARIMA Model	2.25
VARIMAX-ECM Model	1.01

Table 4 shows that the VARIMAX-ECM model has the lowest MAPE value at 1.01%. Accordingly, the GM-ARIMA model, ARIMA model, and ARMA model have MAPE values of 2.25%, 5.75%, and 7.44%, respectively. Based on the findings of the study, it has shown that the VARIMAX-ECM model used by the author is the most effective one. This can be observed from the value of the mean absolute percentage error (MAPE) was found to be lowest compared to the old model. Moreover, the study has also found that the VARIMAX-ECM model is suitable for long-term forecasting unlike other previously conducted studies, which were mostly old models proved only to be feasible for short-term forecasting. In addition to this, the VARIMAX-ECM model is a forecasting model which captures, with detail and prudency in the analysis process, by selecting the only stationary causal variables at the same level, as well as securing the same level of co-integration in order to create the best model. If any of the variables does not meet the set conditions, it would not be taken into account. Therefore, the VARIMAX-ECM model becomes the right forecasting model suitable for long-term policy-making and management planning in order to achieve a sustainable development in the future. Therefore, the VARIMAX-ECM model is used to forecast CO_2 emissions in the following step.

3.4. CO_2 Emissions Forecasting Based on the VARIMAX-ECM Model

Figure 2 shows that CO_2 emissions from 2018 to 2029 in Thailand are continuously increasing with changes up to 14.68% or 289.58 Mt CO_2 Eq. The result shown is beyond the target and deviates from the reduction policy; CO_2 emissions are expected to reduce by 20.8%, or be less than 115 Mt CO_2 Eq. in the industrial sectors so as to attain sustainable development.

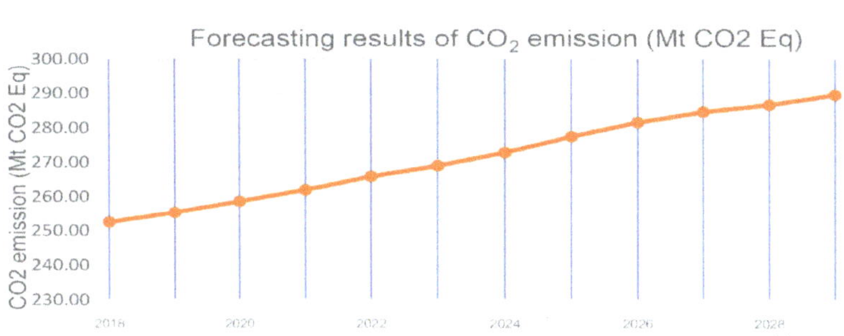

Figure 2. The forecasting results of CO_2 emissions from 2018 to 2029 in Thailand.

4. Conclusions and Discussion

This study disclosed new knowledge and guidelines for future research. The forecasting model must emphasize the causal factors that can influence CO_2 emissions in both the short- and long-term. In addition, the to-be-used variables must be stationary at the same level. It is important to drop or ignore unnecessary variables, which have no direct influence on the dependent variables, so as to produce the best performing model with the most effective prediction outcomes. At the same time, this will facilitate the formulation of effective sustainable development policies. The newly-introduced model in this study attempts to fill the gaps or weaknesses of most existing forecasting models. Additionally, it provides more accurate output with fewer errors, which is instrumental for both the academic world and the country in enhancing policy-making for future sustainable development.

From this study's findings, both the short- and long-term causal factors affecting CO_2 emissions are per capita GDP, urbanization rate, industrial structure, and total exports and imports. These variables can be employed to formulate the VARIMAX-ECM model through a performance testing based on MAPE values. Here, the test's results indicate this model's higher quality and efficiency compared to other existing models, such as the ARMA, ARIMA, and GM-ARIMA models. This illustrates that the VARIMAX-ECM model is one of the best models suitable for the future forecasting of CO_2 emissions. Deploying the data of 2018 to 2029, we found that CO_2 emissions continue to increase by 14.68%, which is not in line with Thailand's reduction policy, in which Thailand aims to reduce CO_2 emissions to be lower than 20.8% by 2029.

This study produced new findings and, thus, differentiates itself from other existing studies, including those studies in the above literature review. Specifically, this study generated a forecasting model with the ability to provide a long-term forecast over more than 10 years (2018–2029) and perform effectively. In addition, this study is one of the first reports to introduce the VARIMAX-ECM model. This model is basically adapted from the existing concept and theory. Based on previous studies, the VARIMAX-ECM model is the best model appropriate for long-term forecasting. Unlike many existing and relevant studies, this study makes long-term forecasting possible. This can be observed from the review of relevant studies with the capability of only short-term prediction. For instance, Dai, Niu, and Han [28] put forth the GM (grey model) and least squares support vector machine (LSSVM), along with the optimization of the modified shuffled frog-leaping algorithm (MSFLA) (MSFLA-LSSVM), to forecast CO_2 emissions in China. Their study was conducted only for the period of 2018 to 2025, which is less than 10 years of evaluation. Lin et al. [29] used the grey model to estimate CO_2 emissions in Taiwan for only three years, from 2010 to 2012. Additionally, Zhao, Huang, and Yan [27] proposed a CO_2 forecasting model called SSA-LSSVM, which was structured based on the Salp Swarm Algorithm (SSA) and least squares support vector machine (LSSVM) model to forecast CO_2 emissions in China from 2017 to 2020, covering only four years. For five years of prediction coverage, Li [26] used a BP neural network with the improved particle swarm optimization algorithm to examine CO_2 emissions reduction in Beijing under different scenarios for 2016 and 2020. Meanwhile, Liang [25] obtained a longer forecast from 2010 until 2020 with the application of the input-output model on China's multi-region energy consumption and CO_2 emissions. With the same coverage of prediction, Lotfalipour, Falahi, and Bastam [24] employed the grey and ARIMA models in their study to forecast CO_2 emissions in Iran for the period of 2010 to 2020.

With those studies taken into consideration, it can be observed that the efficiency of the VARIMAX-ECM model is superior, that it is suitable for long-term, yet accurate, forecasting, and that it produces fewer errors (absence of heteroskedasticity, multicollinearity, and autocorrelation). These findings are in parallel with those of Manu and Sulaiman [23]. Additionally, this study differs from other studies in term of the causal factors, as it focuses and selects only the true influencing factors for CO_2 emissions.

Hence, unnecessary factors, such as population growth and total coal consumption, are eliminated from the study in order to reduce potential errors. The reason behind this elimination is because the variables are non-stationary factors at the level and first difference, and incompetent for the co-integration. If the said variables are included in this research the model will be false and it may incur errors denoted by issue alignment to heteroskedasticity, multicollinearity, and autocorrelation at the same time. If the above issues become problematic, it will affect, and have a negative influence over, the forecasting process. However, from the previous policy-making of Thailand (in 1970–2017), the mentioned factors were used in the model and, as a result of that inclusion, there was an absolute failure because the application failed in the forecasting and future planning. Thus, the government should emphasize the issue and prioritize on those causal factors with a direct influence on CO_2 emission to be used in the forecasting model. This is to create the best forecasting model capable for both short- and long-term predictions, though the factors share the same characteristics under the sustainable development policies of many other countries and Thailand, as claimed by Dai, Niu, and Han [28], and Chindo and Abdul-Rahim [20]. This study opens another arena to explore, which can be further developed for future study. At the same time, the findings of this study can be deployed in formulating long-term development strategies so as to boost both the economy and environment in the most efficient and effective way possible.

However, the limitation of this research is that the author is not able to apply the energy price in the model. This is due to the government's continuous control of energy prices and the use of energy funds. Therefore, it has become impossible to perceive the true changes in energy prices, which may affect energy consumption. In addition to this, past policies have not deployed the energy price factor as a causal factor in its policy formulation. Nonetheless, if the government allows the energy price to change according to the current global trend and market movements, it would enable us to know the impact of changes in energy prices on CO_2 emissions forecasting.

As for future research, it is suggested to consider more causal influential variables that are relevant to the national policies of particular countries, so as to align sustainable development policies with the national management and direction of the country. This research indicates that both variables, population growth and total coal consumption, should not be included by Thailand in its VARIMAX-ECM model, as evidenced by the relevant studies. Through the study of the policy framework of Thailand, the author instead recommends that other variables need to be taken into account so as to have an appropriate and most effective forecasting model. Some of these variables are like domestic and foreign private investment, energy consumption structure, energy intensity, carbon emissions intensity, and many more. In fact, encouraging the use of low carbon technologies, like energy utilization efficiency, abatement equipment, and renewal energy, would greatly help in CO_2 emissions reduction with an energy consumption amount maintained and, therefore, simultaneously obtaining sustainable economic growth.

Author Contributions: In this research activity, P.S. and K.K. were involved in the data collection and preprocessing phase, model constructing, empirical research, results analysis and discussion, and manuscript preparation. All authors have approved the submitted manuscript.

Funding: This research received no external funding.

Acknowledgments: This work was performed with the approval of King Mongkut's University of Technology Thonburi and Office of the National Economic and Social Development Board.

Conflicts of Interest: The authors declare no conflict of interest.

References

1. Achawangkul, Y. Thailand's Alternative Energy Development Plan. Available online: http://www.unescap.org/sites/default/files/MoE%20_%20AE%20policies.pdf (accessed on 27 June 2018).
2. Office of the National Economic and Social Development Board (NESDB). Available online: http://www.nesdb.go.th/nesdb_en/more_news.php?cid=154&filename=index (accessed on 27 June 2018).
3. Ren, S.; Yin, H.; Chen, X. Using LMDI to analyze the decoupling of carbon dioxide emissions by China's manufacturing industry. *Environ. Dev.* **2014**, *9*, 61–75. [CrossRef]
4. Xu, J.; Fleiter, T.; Eichhammer, W.; Fan, Y. Energy consumption and CO_2 emissions in China's cement industry: A perspective from LMDI decomposition analysis. *Energy Policy* **2012**, *50*, 821–832. [CrossRef]

5. Jeong, K.; Kim, S. LMDI decomposition analysis of greenhouse gas emissions in the Korea manufacturing sector. *Energy Policy* **2013**, *62*, 1245–1253. [CrossRef]
6. González, P.F.; Landajo, M.; Presno, M.J. Multilevel LMDI decomposition of changes in aggregate energy consumption. A cross-country analysis in the EU 27. *Energy Policy* **2014**, *68*, 576–584. [CrossRef]
7. González, P.F.; Landajo, M.; Presno, M.J. Tracking European Union CO_2 emissions through LMDI decomposition. The activity revaluation approach. *Energy* **2014**, *73*, 741–750. [CrossRef]
8. Xu, S.; He, Z.; Long, R. Factors that influence carbon emissions due to energy consumption in China: Decomposition analysis using LMDI. *Appl. Energy* **2014**, *127*, 182–193. [CrossRef]
9. Wang, W.; Liu, X.; Zhang, M.; Song, X. Using a new generalized LMDI method to analyze China's energy consumption. *Energy* **2014**, *67*, 617–622. [CrossRef]
10. Tian, Y.; Zhu, Q.; Geng, Y. An analysis of energy-related greenhouse gas emissions in the Chinese iron and steel industry. *Energy Policy* **2013**, *56*, 352–361. [CrossRef]
11. Lin, B.; Long, H. How to promote energy conservation in China's chemical industry. *Energy Policy* **2014**, *73*, 93–102. [CrossRef]
12. Baležentis, A.; Baležentis, T.; Streimikiene, D. The energy intensity in Lithuania during 1995–2009: A LMDI approach. *Energy Policy* **2011**, *39*, 7322–7334. [CrossRef]
13. Mulugeta, S.K.; Nondo, C.; Schaeffer, P.V.; Gebremedhin, T.G. Does level of income matter in the energy consumption and GDP Nexus: Evidence from Sub-Saharan African countries. *Energy Econ.* **2012**, *34*, 739–746.
14. Alkhathlan, K.; Javid, M. Energy consumption, carbon emissions and economic growth in Saudi Arabia: An aggregate and disaggregate analysis. *Energy Policy* **2013**, *62*, 1525–1532. [CrossRef]
15. Khan, M.A.; Khan, M.Z.; Zaman, K.; Khan, M.M.; Zahoor, H. Causal links between greenhouse gas emissions, economic growth and energy consumption in Pakistan. A fatal disorder of society. *Renew. Sustain. Energy Rev.* **2013**, *25*, 166–176. [CrossRef]
16. Arouri, M.E.H.; Youssef, A.B.; M'henni, H.; Rault, C. Energy consumption, economic growth and CO_2 emissions in Middle East and North African countries. *Energy Policy* **2012**, *45*, 342–349. [CrossRef]
17. Acaravci, A.; Ozturk, I. On the relationship between energy consumption, CO_2 emissions and economic growth in Europe. *Energy* **2010**, *35*, 5412–5420. [CrossRef]
18. Menyah, K.; Wolde-Rufael, Y. Energy consumption, pollutant emissions and economic growth in South Africa. *Energy Econ.* **2010**, *32*, 1374–1382. [CrossRef]
19. Ohlan, R. The impact of population density, energy consumption, economic growth and trade openness on CO_2 emissions in India. *Nat. Hazard.* **2015**, *79*, 1409–1428. [CrossRef]
20. Sulaiman, C.; Abdul-Rahim, A.S. The relationship between CO_2 emission, energy consumption and economic growth in Malaysia: A three-way linkage approach. *Environ. Sci. Pollut. Res.* **2017**, *24*, 25204–25220. [CrossRef] [PubMed]
21. Akpan, G.E.; Akpan, U.F. Electricity consumption, carbon emissions and economic growth in Nigeria. *Int. J. Energy Econ. Policy* **2012**, *2*, 292–306.
22. Sulaiman, C. The causality between energy consumption, CO_2 emissions and economic growth in Nigeria: An application of Toda and Yamamoto Procedure. *Adv. Nat. Appl. Sci.* **2014**, *8*, 75–81.
23. Manu, S.B.; Sulaiman, C. Environmental Kuznets curve and the relationship between energy consumption, Economic growth and CO_2 emissions in Malaysia. *J. Econ. Sustain. Dev.* **2017**, *8*, 142–148.
24. Lotfalipour, M.R.; Falahi, M.A.; Bastam, M. Prediction of CO_2 Emissions in Iran using Grey and ARIMA Models. *Int. J. Energy Econ. Policy* **2013**, *3*, 229–237.
25. Liang, Q.M.; Fan, Y.; Wei, Y.M. Multi-regional input–output model for regional energy requirements and CO_2 emissions in China. *Energy Policy* **2007**, *35*, 1685–1700. [CrossRef]
26. Li, J.; Shi, J.; Li, J. Exploring Reduction Potential of Carbon Intensity Based on Back Propagation Neural Network and Scenario Analysis: A Case of Beijing, China. *Energies* **2016**, *9*, 615. [CrossRef]
27. Zhao, H.; Huang, G.; Yan, N. Forecasting Energy-Related CO_2 Emissions Employing a Novel SSA-LSSVM Model: Considering Structural Factors in China. *Energies* **2018**, *11*, 781. [CrossRef]
28. Dai, S.; Niu, D.; Han, Y. Forecasting of Energy-Related CO_2 Emission in China Based on GM (1,1) and Least Squared Support Leaping Vector Machine Optimized by Modified Shuffled Frog Leaping Algorithem for Sustainability. *Sustainability* **2018**, *10*, 957. [CrossRef]
29. Lin, C.S.; Liou, F.M.; Huang, C.P. Grey forecasting model for CO_2 emissions: A Taiwan study. *Adv. Mater. Res.* **2011**, *88*, 3816–3820. [CrossRef]

30. Dickey, D.A.; Fuller, W.A. Likelihood ratio statistics for autoregressive time series with a unit root. *Econometrica* **1981**, *49*, 1057–1072. [CrossRef]
31. Johansen, S.; Juselius, K. Maximum likelihood estimation and inference on cointegration with applications to the demand for money. *Oxford Bull. Econ. Stat.* **1990**, *52*, 169–210. [CrossRef]
32. Enders, W. *Applied Econometrics Time Series*; Wiley Series in Probability and Statistics; University of Alabama: Tuscaloosa, AL, USA, 2010.
33. MacKinnon, J. Critical Values for Cointegration Test. In *Long-Run Economic Relationships*; Engle, R., Granger, C., Eds.; Oxford University Press: Oxford, UK, 1991.
34. Johansen, S. *Likelihood-Based Inference in Cointegrated Vector Autoregressive Models*; Oxford University Press: New York, NY, USA, 1995.
35. Harvey, A.C. *Forecasting, Structural Time Series Models and the Kalman Filter*; Cambridge University Press: Cambridge, UK, 1989.

energies

MDPI

Article

Influencing Factors and Scenario Forecasts of Carbon Emissions of the Chinese Power Industry: Based on a Generalized Divisia Index Model and Monte Carlo Simulation

Lin Zhu [1], Lichun He [2], Peipei Shang [3], Yingchun Zhang [4] and Xiaojun Ma [1,*]

1 School of Statistics, Dongbei University of Finance and Economics, Dalian 116025, China; zhulinzl@outlook.com

2 School of Public Administration, Dongbei University of Finance and Economics, Dalian 116025, China; lichunhelch@outlook.com

3 Editorial Department, Dongbei University of Finance and Economics, Dalian 116025, China; peipeishangpps@outlook.com

4 School of Economics, Qingdao University, Qingdao 266071, China; zhangyc@qdu.edu.cn

* Correspondence: maxiaojun@dufe.edu.cn; Tel.: +86-189-009-54292

Received: 18 August 2018; Accepted: 8 September 2018; Published: 11 September 2018

Abstract: The power industry is the industry with the most direct uses of fossil fuels in China and is one of China's main carbon industries. A comprehensive and accurate analysis of the impacts of carbon emissions by the power industry can reveal the potential for carbon emissions reductions in the power industry to achieve China's emissions reduction targets. The main contribution of this paper is the use of a Generalized Divisia Index Model for the first time to factorize the change of carbon emissions in China's power industry from 2000 to 2015, and gives full consideration to the influence of the economy, population, and energy consumption on the carbon emissions. At the same time, the Monte Carlo method is first used to predict the carbon emissions of the power industry from 2017 to 2030 under three different scenarios. The results show that the output scale is the most important factor leading to an increase in carbon emissions in China's power industry from 2000 to 2015, followed by the energy consumption scale and population size. Energy intensity levels have always promoted carbon emissions reduction in the power industry, where energy intensity and carbon intensity effects of energy consumption have great potential to mitigate carbon levels. By setting the main factors affecting carbon emissions in the future three scenarios, this paper predicts the carbon emissions of China's power industry from 2017 to 2030. Under the baseline scenario, the maximum probability range of the potential annual growth rate of carbon emissions by the power industry in China from 2017 to 2030 is 1.9–2.2%. Under the low carbon scenario and technological breakthrough scenario, carbon emissions in China's power industry continue to decline from 2017 to 2030. The maximum probability range of the potential annual drop rate are measured at 1.6–2.1% and 1.9–2.4%, respectively. The results of this study show that China's power industry still has great potential to reduce carbon emissions. In the future, the development of carbon emissions reduction in the power industry should focus on the innovation and development of energy saving and emissions reduction technology on the premise of further optimizing the energy structure and adhering to the low-carbon road.

Keywords: power industry; carbon emissions; Generalized Divisia Index; scenario forecast; Monte Carlo method

1. Introduction

The increase of greenhouse gas emissions is the main reason for the sharp rise in global temperatures. Compared with the 19th century, the global temperature increased by 0.4–0.6 °C at the end of the 20th century. The United Nations Committee of Experts on Climate Change proposes that the threshold for global climate change is 2 °C. If climate warming is not controlled, the global average temperature will rise by 4–6 °C in the 21st century, which is 6–10 times that of the 20th century [1]. Global warming will bring a series of hazards, such as rising sea levels, melting glaciers, extreme weather, and so on, thus leading to a series of economic and political conflicts. By the summer of 2017, the area of Arctic sea ice had shrunk significantly [2], leaving only 4.1 million km^2, which is 40% less than the average level in September 2000 (about 6.7 million km^2) [3]. The rate of sea level rise in the recent 50 years is about 1.0–2.5 mm/year. With global warming, the sea level will continue to rise in the next ≈50–100 years, and it will increase by about 12–50 cm by 2050 [4]. Climate warming has become the focus problem of the world's attention, and it is also a major challenge for humanity. Countries around the world have basically reached an agreement on climate change. On 4 November 2016, the Paris Agreement for global warming came into force. The goal is to control the global average temperature rise to within 2 °C by 2100, compared with the pre-industrial period, and strive to control the temperature rise within 1.5 °C [5]. Greenhouse gas emissions are an important cause of global warming. The greenhouse gases produced by humans are mainly carbon dioxide and other gases produced by the burning of fossil fuels. China is the world's largest carbon emitter. In 2015, China's carbon emissions reached 10.4 billion tons, accounting for 29% of global carbon emissions and exceeding the values of the United States and the 28 countries of the European Union [6]. As proposed in the "Global Carbon Budget 2016" [7] report, in 2016, China's carbon emissions reached 10.136 billion tons, accounting for 28% of the global total. According to the report, global carbon emissions were expected to increase by 2% in 2017 mainly because China's carbon emissions were expected to increase by 3.5% [7]. As the world's largest carbon emitter, China is actively shouldering its responsibility to reduce its emissions and has set a series of emission reduction targets. According to the "The Twelfth Five-Year Plan" period, China planned to pursue energy development efforts that involve accelerating the transformation of modes of energy development and controlling total energy consumption levels. "The Thirteenth Five-Year Plan" proposes a decline of energy intensity levels of 15% [8]. In addition, the Chinese government has also proposed to substantially reduce its use of fossil fuels and to increase its clean energy consumption levels by 2030. The country's total energy consumption amounts to less than 6 billion tons of standard coal, and its unit gross domestic product (GDP) energy consumption level is in line with the global average. Carbon emissions generated per unit of GDP have been reduced by 60–65% from 2005, and the country's carbon emissions are set to peak by 2030. By 2050, the proportion of fossil fuel used will drop to below 50% and ideal energy-consumption levels will be achieved [9].

The production of electricity mainly relies on the use of thermal power, while the production of thermal power is mainly reliant on the use of coal. Nearly half of the coal that China develops each year is used to generate electricity, meaning that half of the carbon dioxide emitted by burning coal comes from the power industry. Furthermore, 70% of China's energy comes from coal. According to the proportion of existing thermal power, at least one-quarter of the country's carbon emissions comes from the power industry [10]. As the basis of China's economic development, the power industry also directly relies on the sector that uses the most fossil fuels, and the proportion of its energy consumption in China's energy consumption has been rising. In 2015, the carbon emissions by the power industry accounted for about 32% of the total carbon emissions in the country, making it the largest source of greenhouse gas emissions in China. In 2015, fossil energy accounted for 74.2% of the energy consumption structure in the power industry, and low-carbon energy accounted for less than 30%. Therefore, the adjustment of energy structure of the power industry has great carbon emission reduction potential. In 2015, the government proposed an ultra-low emissions action plan for coal-fired power plants: by 2020, coal-fired power generation will reduce carbon emissions by 180 million tons, and the total emissions of major pollutants in the power industry will be reduced by about 60% [11]. In December 2017, China started to operate a nationwide carbon market explicitly targeting the power industry while taking the lead in launching a nationwide carbon emissions trading system and then gradually expanded its industrial scope, denoting the importance of energy conservation and emissions reduction in the power industry. Therefore, effective control of the carbon emissions by the power industry is central to achieving low-carbon development in China. In addition, the carbon emissions reduction targets proposed by the Chinese government will greatly accelerate the pace of energy structure optimization in China's power industry. Developing low-carbon power generation will become an important way to achieve carbon emissions reduction targets.

As an important carrier of a low-carbon economy, the development and use of low-carbon energy will fundamentally reduce human consumption of fossil energy and reduce greenhouse gas emissions. At present, low-carbon energy in the world mainly includes solar energy, wind energy, biomass energy, geothermal energy, nuclear energy, water energy, ocean energy, etc. With the progress of science and technology, the application of low-carbon energy gradually became widespread, and low-carbon energy generation technology has become a necessary choice to deal with the current energy crisis and environmental problems. China's current low-carbon energy generation technologies include wind power generation, solar photovoltaics power generation, biomass power generation, hydropower generation, and nuclear power generation. Among them, hydropower generation is the most widely used. China is rich in small hydropower resources, with an exploitable capacity of 87 million kW [12], ranking first in the world. Hydropower generation has a great development space. Nuclear power and wind power generation are also the main low-carbon energy technologies in China. For nuclear power, China has been actively developing nuclear power technology, and the advantages of nuclear power development in the third generation have been basically formed. For wind power generation, there are abundant wind resources in north China, northeast grassland, the northwest Gobi region, and southeast coastal areas. Wind power is stable in shallow coastal areas of eastern China, with the potential to build land and offshore wind farms. Solar photovoltaics is the fourth largest low carbon energy generation technology in China. China has a vast territory, and 70% of the land's annual average solar radiation accounts for more than 20% of the total time [13]. In addition, the western region is sparsely populated, which provides a very convenient condition for the collection and utilization of solar energy. A summary of China's major low carbon energy generation technologies is shown in Table 1.

According to Table 1, the proportion of China's low carbon energy generation is still lower than that of fossil energy, and in low-carbon energy, compared with traditional fossil energy power generation, the carbon emissions of solar photovoltaic power generation with the largest carbon emissions per unit of power generation is only 25.64% of coal power generation (975.3 g-CO_2/kWh). Thus, low-carbon energy has great potential for reducing carbon emissions.

Table 1. A summary of China's major low carbon energy generation technologies.

Low Carbon Energy Generation Technology	Advantage	Disadvantage	Application (Data in 2017) [14]	Carbon Emissions ($g\text{-}CO_2/kWh$) [15,16]
Wind power generation	High environmental benefits; Renewable; Short construction period; Flexible application.	Large area occupied; Poor controllability; High cost.	Installed capacity is 164 million kW. The power generation capacity is 305.7 billion kWh, accounting for 4.8% of the total power generation.	≈9.7–123.7
Hydropower generation	High power generation efficiency; Low power generation cost; Base group starts fast; Easy to adjust.	Long construction period; Small installed capacity; High investment in infrastructure.	Installed capacity is 341 million kW. The power generation capacity is 1119.45 billion kWh, accounting for 18.6% of the total power generation.	≈3.7–237
Solar photovoltaics power generation	No resource distribution area restriction; High energy quality; Simple energy conversion process.	Low energy density; Large area occupied; Vulnerable to climate.	Installed capacity is 130 million kW. The power generation capacity is 118.2 billion kWh, accounting for 1.8% of the total power generation.	≈53.4–250
Biomass power generation	Low-sulfur; High efficiency of investment and utilization; Strong production adjustment.	High investment and power generation costs; Unstable fuel supply.	Installed capacity is 14.8 million kW. The power generation capacity is 79.5 billion kWh, accounting for 1.2% of the total power generation.	≈35–178
Nuclear power	Huge energy; No air pollution; Easy to transport and store.	Easy to produce radioactive waste; Serious heat pollution; High investment cost.	Installed capacity is 35.8 million kW. The power generation capacity is 247.469 billion kWh, accounting for 3.94% of the total power generation.	≈10–130

To control the carbon emissions of the power industry more effectively, this paper decomposes the carbon emissions in the power industry, identifies the influencing factors of the carbon emissions in the power industry, and makes scenario analyses based on the future development trend of each influencing factor. At the same time, this paper fully considers the possibility of future technology development, and sets up the current scenario, low-carbon scenario, and technical breakthrough scenario, aiming to select the appropriate path of carbon emissions reduction for the power industry. Based on the existing research, this paper makes the following innovations in the decomposition of carbon emissions and scenario prediction in the power industry:

1. In applying the factor decomposition method, this study is the first to apply the Generalized Divisia Index Model (GDIM) to factor decomposition of carbon emissions by the power industry. The GDIM model can not only make up the factor dependence of the existing exponential decomposition model, but also considers the potential factors in the decomposition process and investigates the effect of multiple absolute factors and relative indices on carbon emissions. This paper uses the GDIM model to decompose the change of carbon emissions in China's power industry into absolute and relative value indices, which are not realized by other exponential decomposition methods. As the decomposition results reveal correlations among factors, no double-counting applies, which can make up for shortcomings of existing research methods while allowing a more complete and accurate analysis of the actual impacts of various factors on carbon emissions by the power industry.
2. In the selection of factor decomposition variables, this paper gives full consideration to the influence of the economy, population, energy consumption factors, and relative factors produced by absolute factors on the change of carbon emissions in the power industry. The paper examines the three absolute factors of the economy, population size, and energy consumption in measuring carbon emissions by the power industry and identifies relevant relative factors derived from three absolute factors, which make up the insufficiency of the analysis of carbon emissions from the electric power production and industrial angle. In addition, absolute factors and relative factors are taken into consideration, rendering our analysis of carbon emissions by the power industry more complete.
3. In the prediction of carbon emissions in the power industry, the first use of Monte Carlo simulation technology, and the uncertainty is considered in the scenario analysis. The Monte Carlo method is a dynamic simulation method that randomly applies and combines model variables on a probabilistic basis. The advantage of this method is that it can estimate the future change trend according to the assumption value of each use factor based on the relevant research. Because the uncertainty is considered, the pre-judgment results that are generated are more scientific and reasonable than those of other simulation methods. This paper combines China's development planning with a Monte Carlo simulation to forecast the situation of China's power industry, which can not only embody the inertial logic in a static scenario analysis, but also considers the uncertainty of the related influencing factors in the future evolution, such that the basic logic and scientific methods can be combined organically.

2. Literature Review

At present, academia has conducted extensive and in-depth research on carbon emissions. Quesada and Molina et al. [17] studied fifteen kinds of industrial carbon emissions in Spain and analyzed the gap between industrial carbon emissions and carbon emissions allocation rights. Quesada et al. [18] analyzed emissions and subsidies of the industries in different regions for the period between 2005 and 2009. Statistical analysis shows that there are significant differences between emissions and distribution in several regions. Mikayilov [19] investigated the relationship between the economic growth and CO_2 emissions in Azerbaijan. Bollen [20] used the IMAGE 2 model to calculate regional carbon emissions and costs, and examined the emission reductions needed to allocate industrialized areas in a cost-effective manner. Carbon emission factors research is an important component of carbon emissions research. Remuzgo [21] analyzed the determinants of the global inequality distribution of carbon dioxide emissions across the regions considered by the International Energy Agency during the period 1990–2010. Zhu et al. [22] analyzed factor decomposition of carbon emissions generated in China from 1980 to 2007 and concluded that the output effect makes the greatest positive contribution to carbon emissions, and that the energy intensity effect makes the strongest negative contribution to carbon emissions. Xia et al. [23] considered the energy-related carbon emissions in Zhejiang Province from 2000 to 2014, and used Logarithmic Mean Divisia Index (LMDI) to factorize energy carbon emissions of Zhejiang Province. Kopidou [24] divided the study period into two sub-periods, 2000–2008 and 2008–2011, and decomposition of the carbon emissions of the industrial sectors of Greece, Italy, Spain, and Portugal to investigate how changes in the production and consumption of industrial products during the period 2000–2011 affected the industrial carbon emissions of the four southern European countries. Mahony [25] used the LMDI method to decompose the factors of Ireland's carbon emissions changes from 1990 to 2010 and studied the driving forces of carbon emissions. Wang et al. [26] employed the LMDI model to divide China's carbon emissions for 1995–2007 into 11 drivers and to study them. Wang and He [27] conducted a factor decomposition of China's carbon emissions for 1990 to 2007 using the LMDI factorization method and concluded that the economic growth and energy intensity effects, respectively, were the main factors that contributed to an increase and decrease in carbon emissions. Population and structural effects are not significantly affected. Du et al. [28], based on provincial panel data, studied the influence factors and potential emission reductions of China's carbon emissions from 1995 to 2009. Dong [29] used the LMDI model to categorize changes in China's carbon emissions increment and used the co-integration method to establish a co-integration relationship between each variable and carbon emissions. Based on the equation developed, carbon emissions were estimated via a Monte Carlo simulation.

Research on the carbon emissions by the power industry represents another important area of research. H. Ali et al. [30] used an Energy PLAN optimization model to assess the impact of Singapore's various strategies in the electricity generation sector on CO_2 emissions from Singapore's electricity generation sector through to 2020. A comparison of a business-as-usual (BAU) scenario in 2020 (BAU 2020), 2020 target emissions reduction trajectory, and three emissions reduction alternative policy scenarios (denoted APS-I, APS-II, and APS-III) was carried out. Sun et al. [31] applied the Stochastic Impacts by Regression on Population, Affluence and Technology (STIRPAT) model to analyze the influencing factors of carbon emissions in the power industry. Wu and Peng [32] constructed the Long-Range Energy Alternatives Planning (LEAP) model to simulate China's national power needs in six scenarios and estimated carbon emissions and carbon intensity by 2030. Cai et al. [33] used a long-term alternative energy planning system to identify three scenarios for the future development of carbon emissions generated by China's power industry. The potential for carbon emissions reduction in the power industry was analyzed through a comparison of different scenarios, and the costs of key measures were quantified. Ari and Koksal [34] studied the carbon emissions from Turkey's electricity production and developed four scenarios based on different fuel mixtures. The results from these scenarios show that if the proportion of fossil energy inputs is reduced, carbon emissions from electricity production will be significantly reduced. Hou and Tan [35] used a logarithmic average weight method to divide the carbon emissions by the power industry into income, power production intensity, electricity production structure, population, and power generation coal consumption effects. For a typical period, carbon emissions factors were individually analyzed. Steenhof and Weber [36] developed a decomposition model for the Canadian electricity industry to assess the effects of various factors, particularly climate and energy policies, on emissions from 1990–2008, which affect greenhouse gas emissions in the sector. Based on the mechanism of carbon emissions in the power system, Chen et al. [37] analyzed carbon emissions results and the influencing factors of carbon emissions. The authors also established means of carbon emission structure identification and evaluation, and a way to estimate contributions of low levels of carbonization based on carbon emission structures was proposed and verified. Huo et al. [38] used LMDI decomposition methods to determine the carbon emissions by the power industry from 1991 to 2010 and analyzed the impacts of changes in plant electricity consumption, power generation structures, power generation coal consumption, and line loss on changes in carbon emissions generated by the power industry. A decrease in coal consumption levels is the main positive effect of carbon emission reduction. A short-term change in carbon emission intensity levels is mainly affected by the power generation structure used. A summary of research methods on carbon emissions generated by the power sector is given in Table 2.

While prior studies have made substantial contributions to the understanding of carbon emissions by the power industry in China, there are still some deficiencies:

1. Regarding research methods, at present, the exponential decomposition method used in power industry factor decomposition is based on a Kaya identity. Kaya identity is determined by multiplying factors such that they are interdependent in form, and thus, decomposition results are affected by the factors selected. When the same target variable is decomposed, the selection of different factors will lead to contradictory conclusions. In addition, the exponential decomposition method used in the prior studies can only be used to analyze the impacts of a change in the absolute number of factors, and it cannot take other related absolute and implicit factors into account, rendering the analysis unilateral.
2. On the selection of research factors, the factor selected for the factor decomposition of the power industry presents a certain degree of one-sidedness. In the literature, two forms of factor selection are mainly used when factoring the power industry: (1) The first approach involves considering the impact of a single factor or policy on carbon emissions by the power industry, such that while research elements or policies can be studied through an accurate and in-depth analysis, the potential of carbon emissions by the power industry cannot be explored. (2) From an industrial point of view, they only consider the impact of production-side factors on carbon emissions by the power industry while neglecting the impact of output scales and population sizes, thus producing incomplete research results.
3. Regarding carbon emissions projections, most scenario analyses of various factors apply a fixed rate of change. In fact, the future development of various factors is largely uncertain and the potential rate of change should fall within a range of values. Using a fixed rate of change for forecasting can generate large deviations in forecasted results from future developments.

In this paper, the factor decomposition and scenario analysis are fully combined to avoid the singleness of only the factor decomposition or the integrated scenario analysis. By setting the changing range of the main influencing factors in different situations in the future, a Monte Carlo simulation is used to obtain the average annual conversion rate of carbon emissions from 2017 to 2030 in China's power industry in different scenarios. The shortcomings of the existing research are alleviated.

The article's structure is as follows: The third part constructs a model for the decomposition of carbon emissions in the power industry and its future scenario prediction. The fourth section decomposes factors that influence carbon emissions by the power industry and predicts changes in carbon emissions that will occur in the power industry from 2017–2030 via the Monte Carlo method. The fifth section contrasts the research results of this article with those of existing research, expounds on the significance of this article, and puts forward shortcomings of this article and avenues for improvement. The sixth section presents conclusions and policy recommendations.

Table 2. Prior research on the carbon emissions by the power industry.

Author	Research Methods	Research Approach	Research Conclusions	Research Limitations
Shrestha et al. [39]	LMDI	This paper analyzes the carbon emissions of the power sector in 15 countries in Asia and the Pacific using the LMDI model from the perspectives of changes in economic output, electricity intensity of the economy, fuel intensity of power generation, and power generation structure.	The increasing electricity intensity of the economy was the main factor in Bangladesh, Indonesia, and Philippines. Structural changes in power generation were found to be the main contributor to changes in the CO_2 emissions in the case of Sri Lanka and New Zealand. Economic growth was the dominant factor behind the increase in CO_2 emissions in the other countries.	This paper uses the LMDI model to define each contributing factor as interdependent, and the results are found to be dependent on the choice of factors. Thus, different forms of factor decomposition may lead to the opposite conclusion, and potential impacts of the absolute quantity factor selected cannot be determined.
Malla et al. [40]	LMDI	This paper uses the LMDI of complete decomposition to examine the role of three factors (electricity production, electricity generation structure and energy intensity of electricity generation) affecting the evolution of CO_2 emissions from electricity generation in seven Asia-Pacific and North American countries.	The production effect is the main factor of CO_2 emissions increase during 1990–2005, and the energy intensity effect is the main factor to reduce the CO_2 emissions. In 2005–2030, the structure effect of power generation plays a role in reducing carbon emissions. The impact of the factors varies by country.	This paper mainly analyzes carbon emissions by the power industry starting with the production of electric power enterprises, but it does not consider relevant factors such as economic growth or population size with one-sidedness. The LMDI model presents limitations.
Zhang et al. [41]	Meta-frontier non-radial directional distance function	After considering the non-uniformity, non-radial relaxation, and unwanted output of the power generation group, this paper estimates the changes in the energy use and CO_2 emissions of the Korean power industry under different power generation assumptions.	The coal-fired power plant's total factor energy efficiency and carbon dioxide emissions performance is higher than the fuel power plant, under the technical assumptions, the technical difference of coal-fired power plants is smaller than the fuel power plant.	N. Zhang et al. only consider the single factor's impact on the carbon emissions in the power industry. Although it can well quantify the carbon emissions reduction potential brought by technology, it cannot grasp the carbon emissions of the power industry as a whole situation.
Hu et al. [42]	LMDI	This paper uses the LMDI exponential decomposition method to examine the effects of China's differential pricing policy on carbon emissions by the power industry. Compare the effectiveness of different pricing policies in different industries.	Electricity consumption in China's energy-intensive industries is still growing. Output effects make the largest contributions to carbon reduction outcomes while intensity effects represent the most adverse inverse factor. Policies have different impacts on different industries.	J. Hu et al. take the difference pricing policy as a starting point and only consider the potential of carbon emissions by the power industry based on the policy of differential electricity prices. The authors do not explore the potential of carbon emissions by the power industry.
Özer et al. [43]	LEAP model	This paper comprehensively considers the various policies of Turkey and analyzes the carbon emission reduction in Turkey's power sector based on the scenario analysis of the Long-Range Energy Alternatives Planning system (LEAP) model.	Carbon dioxide (CO_2) emissions will rise significantly under the baseline scenario. In the mitigation scenario, electricity-related CO_2 emissions grew by 5.8% annually between 2006 and 2030, while electricity output grew at an average of 6.6% per annum in this period.	Scenario analysis does not consider the degree of impact of various factors affecting the carbon emissions of the power sector, which may lead to an increase in the bias of future scenario estimates.
Liu et al. [44]	Multivariable linear regression (MLR) model	Taking Chongqing as an example, this paper applies a multivariate linear analysis of factors (e.g., industrial structure optimization and power supply structure optimization) to design three scenarios to examine the potential for carbon emissions reduction in the power industry.	Optimization of industrial structures and promotion of emissions reduction technologies have considerable effects on carbon emissions reduction outcomes. Hydropower resources will limit the power structure optimization of carbon emissions reduction potential.	This paper uses multiple linear analysis to predict carbon emissions that will be generated by the power industry in the future. As the study does not consider various factors or uncertainties related to future changes, the results of the analysis are limited.

3. Methods

3.1. Electricity Industry Carbon Emissions Calculations

This paper refers to the "Intergovernmental Panel on Climate Change (IPCC) 2006" [45] published carbon emissions calculation method to estimate carbon emissions generated by China's power industry from the perspective of fossil fuel consumption. The following formula is used (1):

$$C = \sum_{i=1}^{8} E_i \times CV_i \times CCF_i \times COF_i \times (44/12) \tag{1}$$

In Equation (1), C represents carbon emissions by the power industry in 10^4 tons; i denotes the type of energy use. According to energy distinctions outlined by the IPCC, this paper examines eight energy sources (raw coal, coke, crude oil, gasoline, kerosene, diesel oil, fuel oil, and natural gas). E_i is consumption of the ith energy source, and the unit is 10^4 tons of Standard Coal Equivalent (SCE). CV_i is the low calorific value of the ith energy source in kJ/kg or kJ/m^3, CCF_i is the carbon content of the ith energy source in units of kg/10^6 kJ, COF_i is the carbon oxidation rate of the ith energy source, and 44/12 is the molecular weight ratio of carbon dioxide to carbon.

In calculating carbon emissions generated by China's power industry, the energy consumption data used were drawn from the "China Energy Statistical Yearbook"(2001–2016) [46]; lower heating value, carbon content levels, and carbon oxidation rates for various energy sources were sourced from the "IPCC Guidelines for National Greenhouse Gas Inventories" [45].

3.2. Model Construction for the Decomposition of Carbon Emission Factors by the Power Industry

In this paper, GDIM is used to decompose carbon emissions by the power industry. The model establishes a multi-dimensional factorization model by transforming a Kaya identity that can reveal causes of carbon emission changes. We create Equations (2)–(4) based on the GDIM model

$$C = G \times (C/G) = E \times (C/E) = P \times (C/P) \tag{2}$$

$$E/G = (C/G)/(C/E) \tag{3}$$

$$(G/P) = (C/P)/(C/G) \tag{4}$$

In Equations (2)–(4), C denotes carbon emissions, G denotes added value, E denotes total energy consumption, P denotes total population, G/P denotes GDP per capita, C/P denotes per capita carbon emissions, C/G denotes output carbon intensity, C/E denotes energy consumption carbon intensity, and E/G denotes energy intensity.

Equations (2)–(4) are transformed to obtain Equations (5)–(9):

$$C = G \times (C/G) \tag{5}$$

$$G \times (C/G) - E \times (C/E) = 0 \tag{6}$$

$$G \times (C/G) - P \times (C/P) = 0 \tag{7}$$

$$G - P \times (G/P) = 0 \tag{8}$$

$$E - G \times (E/G) = 0 \tag{9}$$

Function $C(X)$ denotes the contribution of factor X to changes in carbon emissions. According to Equations (5)–(9), we establish a Jacobian matrix Φ_X composed of various factors, as given in Equation (10):

$$\Phi_X = \begin{pmatrix} C/G & G & -C/E & -E & 0 & 0 & 0 & 0 \\ C/G & G & 0 & 0 & -C/P & -P & 0 & 0 \\ 1 & 0 & 0 & 0 & -G/P & 0 & -P & 0 \\ -E/G & 0 & 1 & 0 & 0 & 0 & 0 & -G \end{pmatrix}^T \tag{10}$$

According to the basic principle of the GDIM model, the variation in carbon emissions ΔC can be expressed as the sum of the contribution of each factor according to Equation (11):

$$\Delta C\langle X|\Phi\rangle = \int_L \nabla C^T (\mathrm{I} - \Phi_X \Phi_X^+) dX \tag{11}$$

In Equation (11), L is the time span, $\nabla C = \begin{pmatrix} C/G & G & 0 & 0 & 0 & 0 & 0 & 0 \end{pmatrix}^T$, I is its unit matrix, and "+" is the generalized matrix. When the columns of the matrix Φ_X are linearly independent, $\Phi_X^+ = \left(\Phi_X^T \Phi_X\right)^{-1} \Phi_X^T$.

This paper divides the influencing factors of carbon emissions into three absolute amount factors and into five relative amount factors. The absolute factors include factors that reflect effects of changes in output scale on carbon emissions ΔG, factors that reflect effects of changes in energy consumption scales on carbon emissions ΔE, and factors that reflect effects of changes in population size on carbon emissions ΔP. The relative quantity factor includes factor $\Delta C/G$, which reflects the low-carbon degree of the power industry, and namely a change in the carbon productivity of carbon emissions to reflect the energy use and substitution degree of the power industry. The influence of the energy structure changes on carbon emissions $\Delta C/E$ reflects the effect of changes in carbon emissions per capita on carbon emissions $\Delta C/P$. The effect of changes in per capita GDP on carbon emissions is denoted as $\Delta G/P$, and the extent to which energy production is dependent on energy use in the power sector and the impacts of its use efficiency on carbon emissions are denoted as $\Delta E/G$.

This paper presents a factor analysis of carbon emissions by the power industry for 2000–2015 as a sample interval. The added value of the power industry and the total population were derived from the "China Statistical Yearbook" (2001–2016) [47], and the total energy consumption levels for the power industry were derived from the "China Energy Statistical Yearbook" (2001–2016) [46]. To eliminate effects of price factors on the variables and to ensure the comparability of data, the paper deducts the added value of the power industry for 2000 to a constant price.

3.3. Scenario Forecast Model Construction and Scenario Setting for Carbon Emissions by the Power Industry

The following factor analysis shows that in the evolution of carbon emissions by the power industry, the most important growth-inducing factor is the output scale. Carbon intensity of energy consumption and energy intensity can effectively reduce carbon emissions by the power industry and in the future, the power industry should focus on these two aspects when formulating carbon emission reduction policies. To further our scenario analysis, we construct Equation (12):

$$C = G * \frac{E}{G} * \frac{C}{E} \tag{12}$$

Suppose that the rate of the change in carbon emissions (C), output scale (G), energy intensity level (E/G), and energy consumption carbon intensity level (C/E) are, respectively, written as c, a, d, and e in Equation (13):

$$\begin{aligned} C_{t+1} &= G_{t+1} \times \left(\tfrac{E}{G}\right)_{t+1} \times \left(\tfrac{C}{E}\right)_{t+1} \\ &= G_t \times (1+a) \times \left(\tfrac{E}{G}\right)_t \times (1+d) \times \left(\tfrac{C}{E}\right)_t \times (1+e) \end{aligned} \tag{13}$$

Therefore, the rate of change in carbon emissions (C) is given by Equation (14):

$$c = (1 + a) \times (1 + d) \times (1 + e) - 1 \quad (14)$$

The evolution of carbon emissions in the power industry is closely related to the evolution trend of the output scale, energy intensity, and energy consumption carbon intensity. To predict future carbon emissions by the power industry, this paper sets the three following future scenarios according to the changing trends and development potentials of various factors for before 2015 and of various emission-reduction policies: The baseline scenario, low-carbon scenario, and technical breakthrough scenario.

(1) Baseline scenario: In this scenario, the power industry develops from the inertia of past development trends. Current economic development, policies, and measures related to carbon emission reduction and technological levels will not change. Economic changes are generally more dependent on inertia. The farther along the cycle is, the lesser its impact on the future will be. The closer a cycle is, the stronger its influence on the future will be [48]. Therefore, taking the annual average rate of the change in potential factors for the recent time frame as a reference, we took the annual average rate of change for each potential factor for 2010–2015 as the intermediate value of the factor for the baseline scenario (i.e., the most likely value). According to Lin and Liu [49], set the changes of various factors, the change range of the potential change rate of the energy intensity index is 0.3–0.5%, and 0.3% was used in this article. From the potential rate of change in the energy consumption carbon intensity change rate of 0.2%, combined with actual trends for the power industry, we applied a unified 0.15%. The rate of change in the output scale refers to the rate of change in economic growth by 1%. The average annual rate of change in each factor for the baseline scenario is shown in Table 3.

Table 3. Annual average rate of change in each factor for the baseline scenario. Unit: %.

Factor	2017–2030		
	Min	Med	Max
G	7.69	8.69	9.69
E/G	−4.62	−4.32	−4.02
C/E	−2.02	−1.87	−1.72

(2) Low carbon scenario: Under this scenario, the government intensifies its efforts to optimize the energy consumption structure, to improve energy efficiency levels, and begin to enter the green development road. The World Bank expects the average GDP growth rate of 2017–2020 to be 6.27% [50], which was used as the median of the average annual variability of the output scale in this period under low carbon scenarios. A new economic report on China forecasts that China's GDP will grow at an average yearly rate of 3.28% from 2020 to 2030 [51]. Morgan Stanley predicts that China's real GDP will grow by 4.6% each year from 2021 to 2025 [52]. Mr. Teng notes that low carbon targets will have a negative impact on GDP of 1% [53]. Therefore, average annual change rates for the output scales of 2021–2025 and 2026–2030 were set at 3.6% and 2.95%, respectively. The "Strategy for Energy Production and Consumption Revolution" [9] (hereinafter referred to as the "Strategy") proposes that by 2020, energy consumption levels per unit of GDP will drop by 15% from 2015, and thus calculate the potential annual average change rate of energy intensity for 2017–2020 as −3.2%. According to the "China and New Economic Report," the annual average rate of change in energy intensity levels for 2020 to 2030 under an accelerated emission-reduction scenario is expected to be 2.69% [51]. With China's increasing levels of investment in renewable energy development under a low-carbon scenario, the potential annual average change rates in energy intensity for 2021–2025 and 2026–2030 were estimated at −2.51% and −2.69%, respectively. Reference the "Strategic Plan of Action for Energy Development" [54], assuming that the proportions of coal, oil, and natural gas consumption in 2020 will, respectively, reach 62%, 13%, and 10%, the potential annual average rate of change in energy consumption carbon intensity for 2015–2020 is expected to be −2.6%. For the calculation method used, refer to Lin and Liu [49]. The "Strategy" proposes that by 2020–2030, the share of fossil fuels of total

energy consumption be reduced to 80%, while that for natural gas will decline to 15% [9]. If uniform changes will occur in proportions of fossil energy and natural gas, coal and oil levels will remain the same with proportions of coal, oil, natural gas consumption in 2025 and 2030 estimated at 58%, 12%, and 12%; and 54%, 11%, and 15%, respectively. Potential annual rates of change of −2.72% and −2.89% were, respectively, anticipated for 2021–2025 and 2026–2030. The method of selecting the maximum and minimum values of each factor was in reference to the baseline scenario. The result is shown in Table 4.

Table 4. Annual rate of change in each factor for the low-carbon scenario. Unit: %.

Factor	2017–2020			2021–2025			2026–2030		
	Min	Med	Max	Min	Med	Max	Min	Med	Max
G	5.27	6.27	7.27	2.6	3.6	4.6	1.95	2.95	3.95
E/G	−3.5	−3.2	−2.9	−2.8	−2.5	−2.2	−2.99	−2.89	−2.39
C/E	−2.75	−2.6	−2.45	−2.87	−2.72	−2.57	−3.04	−2.89	−2.74

(3) Technical breakthrough scenario: Under this scenario, assuming that the average annual growth rate of output for all stages is in line with the low-carbon scenario, the power sector has intensified its research and development and investments in energy-saving and emission-reduction technologies to further promote carbon emissions reduction in the power industry. With the development of technologies, energy consumption carbon intensity levels will decline. In considering the time lag of technological breakthroughs, we set an annual average rate of change in energy consumption for 2017–2020 that is in line with the low-carbon scenario. From 2021, the optimization effect of technological breakthroughs will gradually emerge. Therefore, for energy consumption carbon intensity in 2021–2025 and 2026–2030, the average annual change rate of relatively low carbon scenario was decreased by 0.2%. In 2017, the goal set out in the government's "Strategy" is to reduce energy intensity by 15% in 2020, compared with 2015 [9], assuming that energy intensity can continue to fall by as much as 15 percent in the event of a breakthrough in energy efficiency technologies. The potential annual change rate in energy intensity for 2017–2020, 2021–2025, and 2026–2030 was calculated to be −3.2%. The method of selecting the maximum and minimum values of each factor was in reference to the baseline scenario. Results are shown in Table 5.

Table 5. Average annual rate of change in each factor under the technological breakthrough scenario. Unit: %.

Factor	2017–2020			2021–2025			2026–2030		
	Min	Med	Max	Min	Med	Max	Min	Med	Max
G	5.27	6.27	7.27	2.6	3.6	4.6	1.95	2.95	3.95
E/G	−3.5	−3.2	−2.9	−3.5	−3.2	−2.9	−3.5	−3.2	−2.9
C/E	−2.75	−2.6	−2.45	−3.07	−2.92	−2.77	−3.24	−3.09	−2.94

3.4. Summary of the Research Methods

A summary of the research methods used for the decomposition and scenario prediction of carbon emissions for China's power industry is summarized in Figure 1.

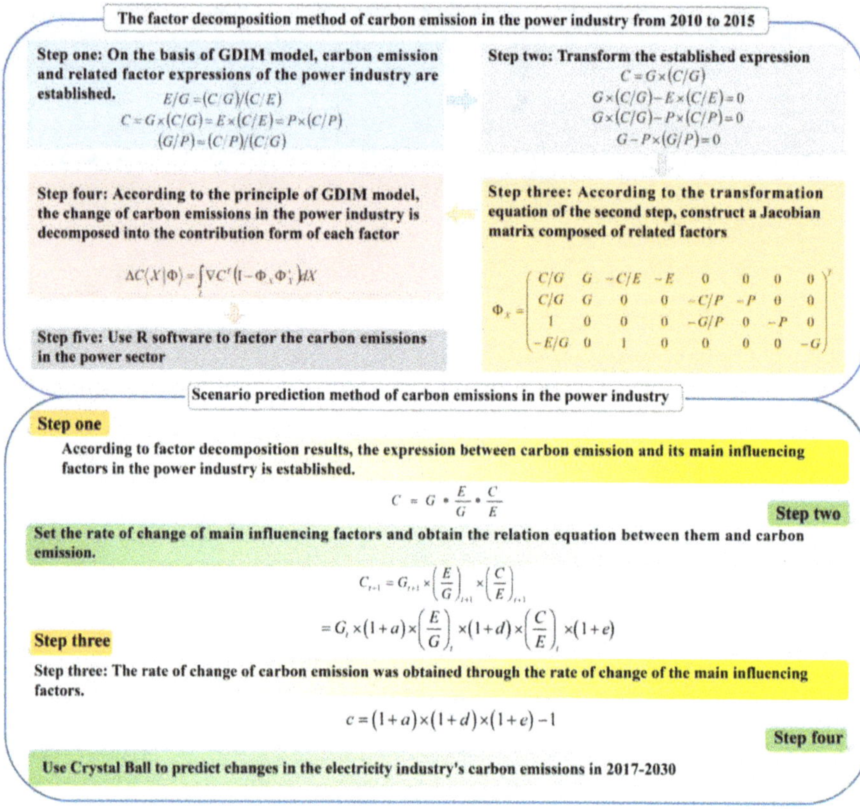

Figure 1. The summary of research methods and models.

4. Result Analysis

4.1. Result Analysis of the Factor Decomposition of Carbon Emissions by the Power Industry

In this paper, the period of 2000–2015 was divided into three stages: 2000–2005, 2005–2010, and 2010–2015. According to Equations (7) and (11), using R software (Version 3.4.1, MathSoft), the GDIM decomposition of the drivers of carbon emissions from the power sector for each time segment is shown in Figure 2. Figure 2 shows that from 2000 to 2015, the output scale, energy consumption scale, population size, and per capita carbon emissions are major growth factors of carbon emissions in the power industry, while per capita GDP and energy intensity have a restraining effect on the growth of carbon emissions by the power industry. Output carbon intensity and energy consumption carbon intensity have a two-way change in the role of carbon emissions in the power industry.

(1) From 2000 to 2015, output scale, energy consumption scale, population size, and per capita carbon emissions always increased the carbon emissions in the power industry. Of factors that contribute to this increase, the growth effects of the output scale first weakened and then enhanced in three time periods, and was the most significant contributor to increases in carbon emissions observed in 2000–2005 and 2010–2015, causing carbon emissions to reach 360 million tons and 483 million tons, respectively. The growth effect of the output scale for 2005–2010 declined relative to those of the other two time periods, but the contribution of the output scale to the increase in carbon emissions was still significant relative to the other factors. Enhancing effects of the energy consumption scale contrasted

with those of the output scale, showing slight fluctuations that first increased and then decreased the scale of energy consumption. This generates a maximum volume of 251 million tons of carbon emissions for 2005–2010. Levels were then reduced to 197 million tons in 2010–2015 because for its "The Twelfth Five-Year Plan" period, the Chinese government proposed a national energy development approach to accelerating energy development and to rationally controlling energy consumption. At the same time, the Electricity Industry Federation proposed its "The Twelfth Five-Year Plan" for power generation, which led to a decrease in the scale of energy consumption and of corresponding carbon emissions by the power industry. The growth effect of population size on carbon emissions by the power industry was minor compared to those of the above two factors, but it has been on the rise. Effects of population size on carbon emissions gradually grew more pronounced, rising from 14 million tons for 2000–2005 to 23 million tons for 2010–2015, and was mainly due to changes in the size of Chinese households and due to urbanization. According to census results for 2000 and 2010, family size decreased from 3.44 persons/household to 3.1 persons/household [55], which has led to an increase in the use of household appliances and thus in carbon emissions use in the electricity sector. In the meantime, rates of urbanization in China have increased at an average annual rate of 1.35% from 36.22% in 2000 to 56.1% in 2015 [47]. Compared with rural areas, the perfect level of urban power facilities makes urban residents' electricity consumption more convenient. The increasing effect of carbon emissions per capita on carbon emissions showed a continuing downward trend, from 352 million tons in 2000–2005 to 89 million tons in 2010–2015; this was because the Chinese government set low-carbon targets in its "The Twelfth Five-Year Plan" for 2011. The low-carbon concept was deeply rooted in people's minds, leading to a significant decline in the contribution of per capita carbon emissions to the carbon emissions from the power industry.

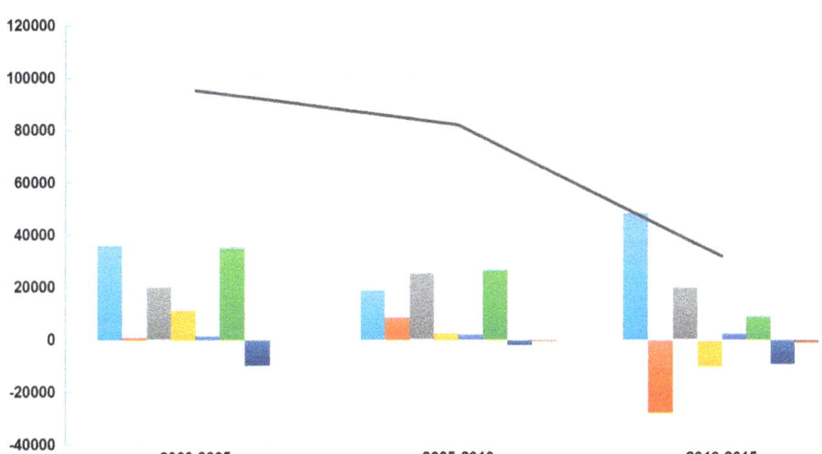

	2000-2005	2005-2010	2010-2015
G	36025.25	18955.5	48338.65
C/G	972.6772	8616.858	-27798.9
E	19851.92	25119.57	19654.49
C/E	11404.79	2698.82	-9878.204
P	1391.785	1981.017	2311.95
C/P	35211.64	26653.95	8941.399
G/P	-9734.24	-1970.535	-9152.203
E/G	-142.48	-386.0805	-974.8945
C	94981.34	81669.1	31442.29

Figure 2. Stages of the decomposition of carbon emissions in the electricity industry. Unit: 10^4 tons.

(2) From 2000 to 2015, energy intensity and per capita GDP have always played a catalytic role in reducing carbon emissions by the power industry. Among them, the declining effect of energy

intensity on carbon emissions, although relatively weak, showed a trend of continuous increase from 1.4 million tons for 2000–2005 to 10 million tons for 2010–2015. This phenomenon could be attributed to the fact that for the first time, in China's "The Eleventh Five-Year Plan," the energy intensity constraint indicator had been clearly set forth, and targets of a 20% and 16% reduction in energy intensity were proposed in the "The Eleventh Five-Year Plan" and "The Twelfth Five-Year Plan" periods, respectively [56]. In addition, "The Thirteenth Five-Year Plan for Energy Development" [8] proposed that energy intensity levels in "The Thirteenth Five-Year Plan" period should drop by more than 15%. Therefore, it is expected that a reduction in energy intensity will continue to promote carbon emissions reduction in the power sector in the coming period. Per capita GDP has always had a catalytic effect on carbon emissions. This appears to be an unreasonable phenomenon that must be clarified. Relative indicators of GDP per capita explicitly include two quantitative indicators (GDP and population), which have a significant impact on carbon dioxide emissions. Changes in these indicators affect their carbonization and are also energy related. As per capita GDP correlates with several indicators, changes in these indicators are affected by calculations of the GDIM model such that changes in per capita GDP are assigned to all these indicators. Only part of the change in these indicators is attributable to per capita GDP, and it is calculated as the effect of changes in carbon dioxide emissions while others are included in the impacts of additional indicators. The results show that the nature of this interconnection had a negative impact on carbon dioxide emissions even when GDP per capita was increasing. In addition, from an environmental point of view, the negative impact of per capita GDP on carbon emissions is also desirable, indicating that the dynamic of population welfare lags behind the gross domestic product [57].

(3) From 2000 to 2015 the carbon intensity of output and the carbon intensity of energy consumption had a two-way effect on carbon emissions by the power industry. Output carbon intensity in 2000–2005 and 2005–2010 contributed to increased carbon emissions in the power industry, and the resulting carbon emissions were 10 million tons and 80 million tons, respectively. The carbon intensity of output had an inhibiting effect on carbon emissions by the power industry for 2010 to 2015, reaching 278 million tons, thus rendering it the most significant contributor to curbing carbon emissions in 2010–2015. This is attributable to new progress made in the reform and development of the power sector from 2010–2015, to structural readjustments, and to remarkable improvements made in technological innovation [58]. As a result, carbon productivity was effectively enhanced (i.e., output carbon intensity was significantly reduced) and carbon emissions dropped significantly. The carbon-strengthening effect of energy consumption carbon intensity continued to decline in 2000–2010. From 2010 to 2015, energy consumption carbon intensity played a catalytic role in reducing carbon emissions by the power industry and reached 100 million tons. This shows that in the "The Twelfth Five-Year Plan" period, energy consumption structures underwent limited carbon transformation. By the end of 2015, non-fossil energy power generation installed capacities accounted for 35%, representing an increase of 8.1% from 2010 levels, and thermal power installed capacity levels decreased by roughly 9% from 2010 levels [59]. This shows that the declining carbon intensity of energy consumption will make an important contribution to carbon emission reduction in the power industry.

To more clearly reflect the dynamic impacts of various factors on carbon emissions for 2000 to 2015, this article took 2000 years as the base period, and obtaining accumulate the contribution value of each factor in the power industry each year, as shown in Figure 3. Figure 3 shows that carbon emissions by the power industry from 2000 to 2013 continuously increased, reaching a maximum level of 2.533 billion tons in 2013–2015, with a downward trend in 2013–2015, and showing a cumulative increase in carbon emissions of 2.081 billion tons in 2015. After 2013, the power industry's decline in carbon emissions has a realistic background. In 2013, the Shenzhen carbon emission trading market, a national carbon trading pilot project, was launched. Subsequently, carbon emissions exchanges, such as those in Beijing, Shanghai, and Tianjin, were successively established, playing an important role in promoting a reduction in carbon emissions. Since the establishment of the four pilot exchanges, enterprises involved in the transactions have mainly focused on the electric power, steel, and chemical industries, and more enterprises have started to participate [60]. In December 2017, the government led the power industry through a breakthrough in carbon market construction and launched a nationwide carbon trading system for the power industry, highlighting the importance of reducing carbon emissions by the power industry.

Figure 3. Cumulative contributions of various facets of the power industry to changes in carbon emissions. Unit: 10^4 tons.

As can be observed from Figure 3, carbon emissions generated by the output scale in the power industry increased from 57 million tons in 2001 to 997 million tons in 2015, representing an average annual growth rate of 22.7%. The cumulative increase in carbon emissions generated by per capita carbon emissions in 2001–2013 continued to increase from 2013 to 2015 and gradually decreased from 2000 to 2015 to a cumulative increase of 650 million tons of carbon emissions, representing an average annual growth rate of 28.8%. The cumulative increase in carbon emissions resulting from the scale of energy consumption continued to grow, and relatively was stable in 2013–2015. From 2000 to 2015, the average annual growth rate reached 26.9% with a cumulative increase of 664 million tons of carbon emissions. The cumulative increase in population size and energy consumption carbon intensity from 2000 to 2015 (with less carbon emissions) reached 65 and 31 million tons, respectively. Emissions resulting from energy intensity and GDP per capita were reduced. Among them, the cumulative increase in carbon emissions resulting from energy intensity decreased rapidly at an annual average rate of 31.4% from 2000 to 2015, representing a cumulative reduction of carbon emissions of 29 million tons. Carbon emissions generated by per capita GDP in aggregate declined by 92 million tons from 2000 to 2015. The cumulative increase in carbon emissions resulting from the output intensity of carbon continued to increase from 2001 to 2009, reaching a maximum value of 182 million tons in 2009, and the cumulative increase in 2012 turned negative while a cumulative reduction of 209 million tons of carbon emissions occurred in 2000–2015.

The above results show that China has put forward relevant proposals for energy development and that a series of policies and measures aimed at achieving carbon emissions reduction has achieved some success. However, relative to projections, there is still much room for improvement in the realm of power industry carbon reduction. At this stage, China is still a developing country. The overall goal of the development plans is thus to form a developed society. China plans to eliminate absolute poverty through its "The Thirteenth Five-Year Plan" and to achieve economic growth necessary to support a developed society. Thus, the strategy developed to reduce carbon emissions by slowing the rate of economic growth is not suited to China's current development plans. In the future, China's power industry should reduce carbon emissions by optimizing energy consumption structures, by increasing levels of clean energy use, by enhancing technologies to improve energy efficiency levels, by increasing investment in energy-saving emissions reduction equipment, and by investing other areas that support the realization of future goals.

In summary, the output scale was the most important factor to promote the carbon emissions in China's power industry, followed by population size and energy consumption scale; energy intensity had a positive effect on carbon emissions reduction in the power industry; and output carbon intensity and energy consumption carbon intensity had a two-way effect on carbon emissions reduction in the power industry.

4.2. Dynamic Predictions of the Evolution of Future Carbon Emissions by the Power Industry

Because of the uncertainty of the future development trend and the average annual change rate of the factors affecting the carbon emissions, it can be regarded as the risk variable at the same time, and the carbon emissions are forecasted using a Monte Carlo simulation [29,61,62]. At present, academia has done a lot of meaningful and referential work on the Monte Carlo method [63–66]. The advantage of this method in predicting carbon emissions lies in the assumption of possible values, rather than fixed amplitude changes, based on the relevant literature and research on the future trend of influencing factors, and the probability distributions of different evolution paths of carbon emission are given to identify the most probable evolution paths [49,67].

Based on the above listed annual average rates of change for each factor under different scenarios, we used Monte Carlo simulations to obtain the potential annual average rate of change in carbon emissions in combination with Equation (14) to predict carbon emissions by the power industry for 2017–2030. Given the need to assign a value range to model variables in Monte Carlo simulations, the annual average rate of change for each factor was taken as a variable. The median values of each variable in different time periods under the baseline scenario, low carbon scenario, and technology breakthrough scenario were taken as the minimum, median, and maximal values of the variable in the corresponding scenario in 2017–2030. Corresponding results are shown in Figure 4.

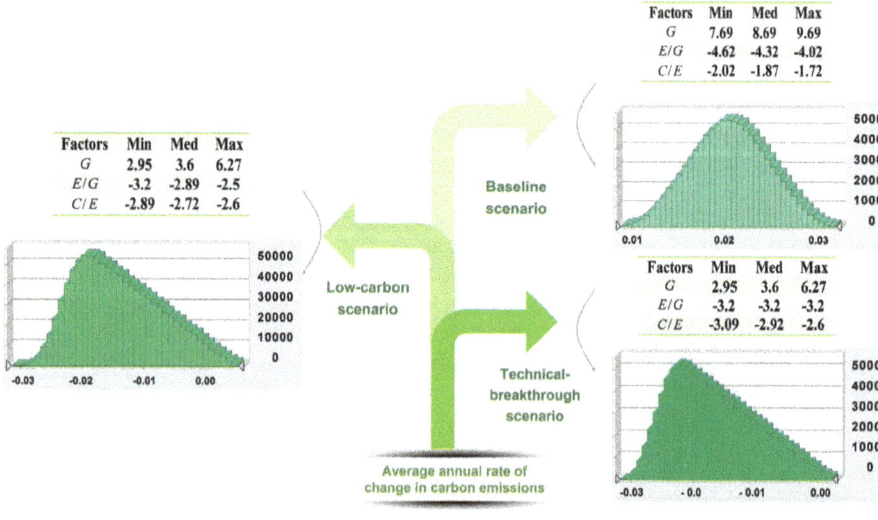

Figure 4. Average annual rate of change in carbon emissions by the power industry for different scenarios of 2017 to 2030.

The distribution of triangles is most suited for the random selection of variables when the most probable result of known variables and the interval of values are known but when the shape of the probability distribution is unknown [68]. The most likely result of each variable in this article is the middle value. We use the triangle distribution as minimum, median, and maximum values to establish the relationship between the probability distribution, and then the simulation variable value is arbitrarily selected from a random range, which is more realistic. We used the Crystal Ball tool to simulate the average annual rate of change in carbon emissions by the power industry for three scenarios through 1 million simulations, the results of which are shown in Figure 4.

Figure 4 shows all the scenarios based on the assumption that the potential average annual growth rate in carbon emissions generated by the power sector from 2017 to 2030 is likely to be high. The figure shows the following:

1. Under the baseline scenario, the average yearly growth rate of the maximum probability of carbon emissions in China's power industry for 2017 to 2030 is expected to range between 1.9% and 2.2%. The forecasted growth rate of carbon emissions under the baseline scenario shows that in the current development of economic and carbon emissions, the power industry is expected to have a rebound in carbon emissions, with an annual growth probability of 1.9–2.2% in 2017–2030. When calculated according to the minimum rate of increase, carbon emissions generated by the power sector in 2030 are expected to reach 4079.14 million tons under the baseline scenario.
2. Under the low carbon scenario, the average annual growth rate of carbon emissions in China's power industry for 2017 to 2030 is likely to range between −2.1% and −1.6%. Compared to the baseline scenario, carbon emissions will continue to decline in the low-carbon scenario. As the low-carbon scenario was based on a series of related goals set forth by the country's current carbon emissions, this shows that the government's green development plan for the future of the country is expected to have a catalytic effect in reducing carbon emissions generated by the power sector.
3. With regard to technological breakthroughs, the average annual growth rate of the maximum probability in carbon emissions in China's power industry for 2017 to 2030 is expected to range between −2.4% and −1.9%. Compared to those of the baseline and low-carbon scenarios, reductions in carbon emissions achieved under the technology breakthrough scenario are expected to be the greatest. At this time, carbon emissions reduction achieved in the power sector under this scenario are expected to be the best among the three scenarios.

In addition, emissions levels can decline faster or slower under each scenario, though with a low probability. However, low probability events can still occur. From the changes in carbon emissions by the power industry found under the three scenarios, future declines in carbon emissions are predicted to be ordered as follows: baseline scenario < low-carbon scenario < technological breakthrough scenario. If in the low-carbon scenario, the power industry was to actively implement the concept of low-carbon development and adhere to the green development path, carbon emissions are predicted to continue to decline. However, over the long term there may not be a significant breakthrough in carbon emissions reduction relative to China's proposed carbon reduction targets. With regard to technological breakthroughs, we predict a breakthrough in technologies of low-carbon energy and energy conversion efficiency, in power generation equipment, and in other realms, and we anticipate significant reductions of carbon emissions accordingly. In summary, according to these predictions, the long-term development of carbon emissions by the power industry should adhere to low-carbon green development while at the same time focusing on low-carbon technological research and development and equipment upgrading, which are expected to make a big contribution to the future development of carbon emissions reduction.

Summarize: Under the baseline scenario, low-carbon scenario, and technological breakthrough scenario, the average annual growth rate of the maximum probability of carbon emissions in China's power industry from 2017 to 2030 are expected to be: [1.9%, 2.2%], [−2.1%, −1.6%], and [−2.4%, −1.9%].

5. Discussion and Analysis

5.1. The Results of This Paper Are Compared to Those of Previous Studies

Wang et al. [69] and Hou et al. [55] studied the carbon emissions of the power industry in the same way, but the method used was different from this article. Therefore, this article selects the two studies and compares these results with the results of this study. The common factors that Wang and this paper chose are economic scale, population size, and energy structures; the common factors that Hou and this paper chose are population size and energy structures. From the common factors shaping carbon emissions by the power industry, the results are shown in Table 6.

Table 6. Contributions of each common factor shaping carbon emissions by the power industry. Unit: %.

Common Factor	Wang et al.	Hou et al.	This Paper
Contribution of economic scale	109.51		102.7
Contribution of population size	7.92	8.1	3.03
Contribution of energy structures	0.51	3.9	7.8

As is shown in Table 6, although the influence factors selected by the representative literature overlap with this article, the contributions of the measured variables to carbon emissions by the power industry vary. At an economic scale, contribution shown in the compared literature are slightly greater than those shown in this paper and mainly because the LMDI model used in the representative literature for factorization can present slightly higher economic impacts on carbon emissions. For population size, contributions presented in the compared literature are greater than those shown in this article. At present, China's population growth rate is declining, and awareness of issues of energy conservation and environmental protection among residents is on the rise. Contributions of population size to carbon emissions by the power industry should thus be relatively low. For energy structures, the results of the contribution is 7.8% in this paper while the compared studies present values of 0.51% and 3.9%. Although the optimization of energy structures will significantly decrease carbon emissions by the power industry, several technical problems related to the development and application of clean energy must still be resolved, and while energy structure optimization has achieved success in recent years, challenges remain. Therefore, the past optimization of energy structures has made strong contributions to the reduction of carbon emissions by the power industry.

5.2. The Significance of Research Results Presented in This Paper

This paper presents main factors that shape carbon emissions generated by the power industry. Future changes in carbon emissions generated from this industry are predicted to help governments and related enterprises take measures to reduce carbon emissions while seizing opportunities and fully tapping into potential for carbon emissions reduction based on the influencing factors identified. At the same time, future plans can be made based on the predicted changes in future carbon emissions, thus contributing to the overall energy conservation and emissions reduction efforts. In this paper, the research ideas and methods of carbon emissions in the power industry have some reference value to the research of carbon emissions in other industries.

5.3. Study Limitations and Avenues for Future Research

This paper does not consider impacts of electricity consumption in the study of carbon emissions by the power industry. The research content related to the power industry is very extensive. For example, J.H. Huh et al. discussed the hybrid advanced metering infrastructure design for a micro grid based on the game theory model [70], which is a low-reliability problem of the power line communication (PLC) caused by the signal interference and attenuation, and the solution to the method was described in the results of the test bench experiment [71] and smart grid test bed that used OPNET software (14.5, OPNET Technologies, Inc., Washington, DC, USA) and power line communication [72]. The study of carbon emissions in the power industry is also very extensive. This paper involves the factor decomposition of carbon emissions in the power industry and sets relevant scenarios for prediction. There are other aspects of carbon emissions in the power industry that can be further studied. Future research can be carried out in the following aspects:

1. Future research can estimate the responsibility for carbon emissions in the provincial power industry, and synthetically consider the responsibilities of producers and consumers. It is beneficial for all regions to undertake the responsibility of carbon emissions reduction fairly and achieve coordinated emissions reduction in all regions.
2. Track and analyze the carbon intensity of demand-side power such that users can understand their carbon emissions responsibility and increase the enthusiasm of users for clean energy consumption.
3. For the generator sets with different carbon emission factors, study the unit's power generation, market electricity price, and carbon emissions, and establishing a carbon emission right allocation mechanism with strong emissions reduction incentives.
4. Compare the international carbon emissions trading system with China's carbon emissions trading system, and analyze the shortcomings and advantages of China's carbon emissions trading system compared with other countries. Furthermore, draw lessons from countries and improving China's carbon emissions trading system combined with China's reality.
5. Analyze the impact of the implementation of carbon tax policy on carbon emissions reduction in the power industry, demonstrate the feasibility of incorporating the carbon tax into the future carbon emissions reduction policy, and provide reference for the government to introduce a carbon tax policy.

6. Main Conclusions and Recommendations

6.1. Main Conclusions

As a major carbon emissions-generating industry in China, the electric power industry is a major target of China's energy conservation and emissions reduction plan. Therefore, it is important to find a suitable path toward carbon emissions reduction in the power industry. First, this paper applies the GDIM model to the electric power industry from 2000–2015 to examine the factors that shape carbon emissions levels. To set the future annual average rate of change of major influencing factors according to government policy objectives, we used the Monte Carlo method to simulate the potential average annual change rate in carbon emissions from 2017 to 2030 for the power industry. The main conclusions of this study are as follows.

1. Under the reality of China's rapid economic development and thermal power generation as the main force, the output scale was the most important factor leading to the increase of carbon emissions in the power industry, and the cumulative increase of carbon emissions in 2000–2015 reached 996.89 million tons. Energy consumption scales and population size were also important factors for the increase of carbon emissions in the power industry, and the cumulative increase of carbon emissions reached 664.340 million tons and 65.204 million tons, respectively. Because fossil energy consumption still accounts for a large proportion in the energy structure of China's power industry, energy intensity and carbon intensity effects of energy consumption have great carbon emission reduction potential. The energy intensity had reduced the carbon emissions by 29.4241 million tons in 2000–2015, and carbon intensity of energy consumption achieved a breakthrough in carbon emissions from positive to negative effects. In the future, the two are expected to become the main factors to promote carbon emission reduction in the power sector. At present, the power industry in China should focus on the optimization of energy structure to further reduce carbon emissions in the power industry.

2. In different scenarios, changes in carbon emissions of the power industry are different from 2017 to 2030. In the baseline scenario, the maximum probability of a potential annual growth rate

of carbon emissions is expected to be between 1.9% and 2.2%. Under the low-carbon scenario and technological breakthrough scenario, the maximum probability of the potential annual drop rate of carbon emissions are expected to be between 1.6–2.1% and 1.9–2.4%, respectively. Thus, in the current state of development, China's power industry carbon emissions will likely continue to increase. If there is a breakthrough in the low carbon policy and energy-saving and emission-reduction technologies, the carbon emissions of the power industry are expected to reduce. Under the technological breakthrough scenario, the power industry is predicted to experience the strongest reduction in carbon emissions and thus, technological breakthroughs are considered to be the main path for future carbon emission reduction development in the power industry.

The summary of the main conclusions in this paper can be seen in Table 7.

6.2. Recommendations

1. Promote green economics and reduce total energy consumption. Accelerate the transformation of economic development modes; focus on the quality of economic growth; and gradually adopt new economic models that limit energy consumption, pollution and emissions generation while upgrading industrial structures to change the structure of electricity consumption. In addition, change the traditional method of coal combustion, realize secondary processing of coal, and reduce end-use energy consumption, especially in energy intensive sectors, such as steel manufacturing.
2. Improve public awareness of energy conservation issues and optimize energy consumption structures. Encourage energy conservation policies, formulate relevant energy efficiency standards and norms, enhance public awareness of energy saving and emissions reduction, and reduce electricity consumption. At the same time, limit thermal power production using coal and improve the ratio of natural gas to fossil energy use. To protect the environment, develop solar energy, hydropower, wind energy, and low-carbon energy generation options by encouraging low-carbon energy power generation and reduce carbon emissions by the power industry.
3. Encourage the use of low-carbon equipment and boost technological innovation research and development. At present, the international investment in innovation and research and development continues to increase. In 2017, the European Union issued the third phase of "Horizon 2020" [73], which comprehensively proposed the goals of climate action and sustainable development, and invested 2.205 billion euros to support the research and innovation of energy technology. As the world's largest carbon emitter, China should upgrade energy-saving and emissions-reduction equipment in the power industry, strengthen international exchanges and cooperation in environmental technologies, and increase investment in innovative technologies for emissions reduction on the basis of adhering to a series of low-carbon development measures. At the same time, the government must encourage enterprises to adopt incentives and invest more in continuous research and development in new energy savings and emissions-reducing technologies as well as production equipment. Low-carbon technologies must be developed to realize further reductions in electric carbon emissions by the power industry.
4. Explore the market potential of biomass power generation. Based on Molina's research [74–76], and combined with the current situation of China's agriculture, biomass power generation should be vigorously developed. The utilization of agricultural and forestry wastes by biomass power generation not only eliminates methane emissions from biomass stacking or landfill fermentation, but also promotes the development of carbon emissions reduction in the power industry, which fully embodies the concept of a circular economy. The government should adopt an active incentive mechanism to enhance the enthusiasm and market activity of biomass power generation industry to a greater extent.

Table 7. Main conclusions of this paper.

Factor	The Contribution of Various Decomposition Factors to the Changes of Carbon Emissions in China's Power Industry (10^4 Tons)				Effect	Degree of Influence	
	Stage Contributions			Cumulative Contributions			
	2000–2005	2005–2010	2010–2015				
Output scale	36,025.25	18,955.5	48,338.65	99,689.402	Positive	Biggest	Factor decomposition
Energy consumption scale	19,851.92	25,119.57	19,654.49	66,434.07	Positive	Bigger	
Population size	1391.785	1981.017	2311.95	6524.081	Positive	Small	
Per capita carbon emissions	35,211.64	26,653.95	8941.399	65,413.974	Positive	Bigger	
Energy intensity	−142.48	−386.081	−974.895	−2942.41	Negative	Smaller	
Carbon intensity of output	972.6772	8616.858	−27,798.9	−20,934.45	Double-sided	Bigger	
Per capita GDP	−9734.24	−1970.54	−9152.2	−9254.794	Negative	Small	
Carbon intensity of energy consumption	11,404.79	2698.82	−9878.2	3171.344	Double-sided	Bigger	
Simulation results of the average annual growth rate of carbon emissions in China's power industry under different scenarios from 2017 to 2030							
	Characteristics of the scenario				**The most probable value**		
Baseline scenario	According to the current development trend				1.9% to 2.2%		Scenario forecasts
Low-carbon scenario	Follow the path of green development				−2.1% to −1.6%		
Technical breakthrough scenario	Strengthen technological R&D investment				−2.4% to −1.9%		

Author Contributions: L.Z. was mainly responsible for the writing of the full text. X.M. conceived and designed the study. L.H., P.S., and Y.Z. built the models of the paper.

Acknowledgments: This paper is supported by Humanities and Social Sciences Research Project of Ministry of Education (18YJC910013), National Natural Science Foundation (71573034), Liaoning Economic and Social Development Research Project (2018lslktzd-010), Liaoning Social Science Fund (L17CTJ001, L17BJY042), China Postdoctoral Science Foundation special fund (2017T100180), China Postdoctoral Science Foundation (2016M601318), and Research Project of Dongbei University of Finance and Economics (DUFE2017Q16), National Social Science Foundation Key Project (17ATJ002), National Statistical Research Projects (2016LY42).

Conflicts of Interest: The authors declare no conflict of interest.

References

1. Global Warming. Available online: https://www.pishu.cn/psbk/psgd/zyhj/383725.shtml (accessed on 30 August 2018).
2. Ge, Q.S.; Wang, F.; Wang, S.W.; Cheng, B.B. Determination and uncertainty of seven questions about global warming. *China Popul. Resour. Environ.* **2014**, *24*, 1–6.
3. Blunden, J.; Arndt, D.S.; Baringer, M.O. State of the Climate in 2010. *Bull. Am. Meteorol. Soc.* **2011**, *92*, S1–S236. [CrossRef]
4. Fu, Y. The causes and effects of global warming. *J. Cap. Norm. Univ.* **2007**, *06*, 11–21.
5. Liu, Y.; Zheng, Y.Y.; Ge, Q.S.; Wang, F. China's greenhouse gas emission changes under the background of low carbon development and its contribution to global emission reduction. *Resour. Sci.* **2017**, *39*, 2399–2407.
6. Global Carbon Project Releases "Global Carbon Budget 2016", Global Carbon Emissions Increase Continued to Decline for the Third Year. Available online: http://cnc-fe.cast.org.cn/ (accessed on 15 December 2017).
7. "Global Carbon Budget Report": Strong Global Carbon Emissions Rise 2017 | China Meteorological Administration. Available online: http://m.cma.gov.cn/2011xwzx/2011xqhbh/2011xdtxx/201711/t20171122_456610.html (accessed on 15 December 2017).
8. Energy Bureau Released "The Eleventh Five-Year Plan" and "The Twelfth Five-Year Plan". Available online: http://www.gov.cn/xinwen/2017-01/05/content_5156795.htm#1 (accessed on 20 November 2017).
9. Du, X.W. Interpretation and reflection on China's "Strategy of Energy Production and Consumption Revolution (2016–2030)". *China Econ. Trade Herit.* **2017**, *15*, 44–45.
10. Zhang, W. Sampling and Separation Performance of Coal Fly Ash in Power Plant. Master's Thesis, Beijing Jiaotong University, Beijing, China, 2010.
11. Yu, Y.H.; Lv, D.W. China's power industry emissions reduction policy—An empirical analysis based on power demand function. *Technol. Econ.* **2017**, *36*, 110–118.
12. Cheng, H.Z. China's small hydropower and rural electrification. *Small Hydro Power* **2001**, *17*, 1–6.
13. Han, D.; Zhao, Y. Talking about new energy power generation and its development prospects. *Tech. Wind* **2018**, *30*, 123.
14. National Energy Administration. Available online: http://www.nea.gov.cn/ (accessed on 30 August 2018).
15. Bhat, I.K.; Prakash, R. LCA of renewable energy for electricity generation systems—A review. *Renew. Sustain. Energy Rev.* **2009**, *13*, 1067–1073.
16. Lenzen, M. Life cycle energy and greenhouse gas emissions of nuclear energy: A review. *Energy Convers. Manag.* **2008**, *49*, 2178–2199. [CrossRef]
17. Quesada-Rubio, J.M.; Villar-Rubio, E.; Madrid-Salvador, V.; Molina-Moreno, V. The gap between CO_2 emissions and allocation rights in the Spanish industry. *Environ. Eng. Manag. J.* **2010**, *9*, 1161–1164. [CrossRef]
18. Quesada-Rubio, J.M.; Villar-Rubio, E.; Mondéjar-Jiménez, J.; Molina-Moreno, V. Carbon dioxide emissions vs. allocation rights: Spanish case analysis. *Int. J. Environ. Res.* **2011**, *5*, 469–474.
19. Mikayilov, J.I.; Galeotti, M.; Hasanov, F.J. The impact of economic growth on CO_2 emissions in Azerbaijan. *J. Clean. Prod.* **2018**, *197*, 1558–1572. [CrossRef]
20. Bollen, J.C.; Toet, A.; Vries, H.D. Evaluating cost-effective strategies for meeting regional CO_2 targets. *Glob. Environ. Chang.* **1996**, *6*, 359–373. [CrossRef]
21. Remuzgo, L.; Sarabia, J.M. International inequality in CO_2 emissions: A new factorial decomposition based on Kaya factors. *Environ. Sci. Policy* **2015**, *54*, 15–24. [CrossRef]
22. Zhu, Q.; Peng, X.Z.; Lu, Z.M.; Wu, K.Y. Factor decomposition and empirical analysis of China's carbon emissions from energy consumption. *Resour. Sci.* **2009**, *31*, 2072–2079.

23. Xia, C.; Li, Y.; Ye, Y.; Shi, Z.; Liu, J. Decomposed driving factors of carbon emissions and scenario analyses of low-carbon transformation in 2020 and 2030 for Zhejiang province. *Energies* **2017**, *10*, 1747. [CrossRef]
24. Kopidou, D.; Diakoulaki, D. Decomposing industrial CO_2 emissions of southern European Countries into production- and consumption-based driving factors. *J. Clean. Prod.* **2017**, *112*, 4159–4172. [CrossRef]
25. Mahony, T.O. Decomposition of Ireland's carbon emissions from 1990 to 2010: An extended Kaya identity. *Energy Policy* **2013**, *59*, 573–581. [CrossRef]
26. Wang, F.; Wu, L.H.; Yang, C. Study on the driving factors of the growth of carbon emissions in China's economic development. *J. Econ. Res.* **2010**, *45*, 123–136.
27. Wang, J.S.; He, C.F. Consumption of energy, economic growth and CO_2 emissions in China: A decomposition analysis based on LMDI. *Resour. Environ. Yangtze Basin* **2010**, *19*, 18–23.
28. Du, L.; Wei, C.; Cai, S. Economic development and carbon dioxide emissions in China: Provincial panel data analysis. *China Econ. Rev.* **2012**, *23*, 371–384. [CrossRef]
29. Dong, F.; Yang, Q.L.; Long, R.Y.; Cheng, S.B. Chinese carbon decomposition and dynamic simulation. *China Popul. Resour. Environ.* **2015**, *25*, 1–8.
30. Ali, H.; Sanjaya, S.; Suryadi, B.; Weller, S.R. Analysing CO_2 emissions from Singapore's electricity generation sector: Strategies for 2020 and beyond. *Energy* **2017**, *124*, 553–564. [CrossRef]
31. Sun, W.; Meng, M.; He, Y.; Chang, H. CO_2 emissions from china's power industry: Scenarios and policies for 13th five-year plan. *Energies* **2016**, *9*, 825. [CrossRef]
32. Wu, Q.; Peng, C. Scenario analysis of carbon emissions of china's electric power industry up to 2030. *Energies* **2016**, *9*, 988. [CrossRef]
33. Cai, W.; Wang, C.; Wang, K.; Zhang, Y.; Chen, J. Scenario analysis on CO_2 emissions reduction potential in China's electricity sector. *Energy Policy* **2007**, *35*, 6445–6456. [CrossRef]
34. Ari, I.; Koksal, M.A. Carbon dioxide emission from the Turkish electricity sector and its mitigation options. *Energy Policy* **2011**, *39*, 6120–6135. [CrossRef]
35. Hou, J.C.; Tan, Z.F. Decomposition of influencing factors on CO_2 emissions from power production. *Electr. Power* **2011**, *44*, 39–42.
36. Steenhof, P.A.; Weber, C.J. An assessment of factors impacting Canada's electricity sector's Ghg emissions. *Energy Policy* **2011**, *39*, 4089–4096. [CrossRef]
37. Chen, X.K.; Zhou, T.R.; Li, X.; Kang, C.Q.; Chen, Q.X. Analysis on decomposition of carbon emission structure and contribution to low carbon in power system. *Autom. Electr. Power Syst.* **2012**, *36*, 18–25.
38. Huo, M.L.; Han, X.Y.; Shan, B.G. An empirical analysis of the influencing factors of carbon emissions intensity in China's power industry. *Electr. Power* **2013**, *46*, 122–126.
39. Shrestha, R.M.; Anandarajah, G.; Liyanage, M.H. Factors affecting CO_2 emission from the power sector of selected countries in Asia and the Pacific. *Energy Policy* **2009**, *37*, 2375–2384. [CrossRef]
40. Malla, S. CO_2 emissions from electricity generation in seven Asia-pacific and North American Countries: A decomposition analysis. *Energy Policy* **2009**, *37*, 1–9. [CrossRef]
41. Zhang, N.; Zhou, P.; Choi, Y. Energy efficiency, CO_2 emission performance and technology gaps in fossil fuel electricity generation in Korea: A Meta-Frontier Non-Radial Directional Distance Function analysis. *Energy Policy* **2013**, *56*, 653–662. [CrossRef]
42. Hu, J.; Kahrl, F.; Yan, Q.; Wang, X. The impact of China's differential electricity pricing policy on power sector CO_2 emissions. *Energy Policy* **2012**, *45*, 412–419. [CrossRef]
43. Özer, B.; Görgün, E.; İncecik, S. The scenario analysis on CO_2 emission mitigation potential in the Turkish electricity sector: 2006–2030. *Energy* **2013**, *49*, 395–403. [CrossRef]
44. Liu, Z.; Zhu, K.W.; Yan, J.M.; Shi, Y.R. Design and analysis of carbon emissions reduction scenarios in the power industry. *Power Syst. Technol.* **2012**, *36*, 1–7.
45. Intergovernmental Panel on Climate Change (IPCC). 2006 IPCC Guidelines for National Greenhouse Gas Inventories. Available online: https://wenku.baidu.com/view/4eb6dad4195f312b3169a5a9.html (accessed on 8 November 2017).
46. National Bureau of Statistics of People's Republic of China. China Energy Statistical Yearbook. 2016. Available online: http://www.stats.gov.cn/tjsj/tjcbw/201706/t20170621_1505833.html (accessed on 10 November 2017).
47. National Bureau of Statistics of People's Republic of China. China Statistical Yearbook. 2011–2016. Available online: http://www.stats.gov.cn/tjsj/ndsj/ (accessed on 10 November 2017).

48. Zheng, S.M. Political cycle, five years planning and environmental pollution—Taking industrial sulfur dioxide emission as an example. *Cass J. Political Sci.* **2016**, *2*, 80–94.
49. Lin, B.Q.; Liu, X.Y. Carbon emissions in China's urbanization: Influencing factors and emission reduction strategy. *Econ. Res. J.* **2010**, *45*, 66–78.
50. IMF World Economic Outlook (WEO) update, January 2017: A Shifting Global Economic Landscape. Available online: http://www.imf.org/external/pubs/ft/weo/2017/update/01/ (accessed on 23 December 2017).
51. Niu, J.M.; Dai, R.F.; Dong, W.J.; Ban, J.H.; Hu, C.Y. China's future CO_2 emissions scenarios under the US new deal background climate prediction. *Adv. Clim. Chang. Res.* **2018**, *01*, 95–105.
52. Morgan Stanley: Avoid the Banking Crisis, We Are Optimistic about China! Available online: http://oversea.huanqiu.com/article/ (accessed on 25 December 2017).
53. China's Carbon Pricing Is Expected to Reach 10–25 Yuan/Ton in 2020–2030. Available online: http://www.tanpaifang.com/tanqiquan/2014/1115/40146.html (accessed on 26 December 2017).
54. General Office of the State Council of the People's Republic of China issued "Action Plan for Energy Development Strategy (2014–2020)". *Constr. Sci. Tech.* **2014**, *22*, 6.
55. Hou, J.C.; Shi, D. Study on the driving factors of carbon emissions changes in China's power industry. *China Ind. Econ.* **2014**, *06*, 44–56.
56. People's Republic of China National Economic and Social Development Twelfth Five-Year Plan. *People's Daily* **2011**, *1*, 1–56.
57. Vaninsky, A. Factorial decomposition of CO_2 emissions: A Generalized Divisia Index Approach. *Energy Econ.* **2014**, *45*, 389–400. [CrossRef]
58. China Electricity Council. Condense together the past to create a new brilliant power industry—A Summary of China electricity council service industry development since 2010. *China Power Enterp. Manag.* **2016**, 12–15.
59. China Electricity Council. *China's Power Industry Annual Development Report 2016*; China Electronic Production Equipment Industry Association: Beijing, China, 2016; pp. 18–23.
60. Zeng, M.; Yao, J.; Zhao, X.L. Effects of carbon emissions trading on the power industry. *China Power Enterp. Manag.* **2014**, *5*, 23–26.
61. Lin, B.; Wang, X. Promoting energy conservation in China's Iron & Steel Sector. *Energy* **2014**, *73*, 465–474.
62. Wu, Y.Z.; Shen, J.; Zhang, X. The Impact of urbanization on carbon emissions in developing countries: A Chinese study based on the U-kaya method. *J. Clean. Prod.* **2016**, *135*, 589–603. [CrossRef]
63. Salakhutdinov, R.; Mnih, A. Bayesian probabilistic matrix factorization using Markov chain Monte Carlo. In Proceedings of the 25th International Conference on Machine Learning, Helsinki, Finland, 5–9 July 2008; pp. 880–887.
64. Frieze, A.; Kannan, R.; Vempala, S. Fast Monte-Carlo algorithms for finding low-rank approximations. *J. ACM* **2004**, *51*, 1025–1041. [CrossRef]
65. Veach, E. Optimally combining sampling techniques for Monte Carlo rendering. In Proceedings of the 22nd Annual Conference on Computer Graphics and Interactive Techniques, Los Angeles, CA, USA, 6–11 August 1995; pp. 419–428.
66. Procopiuc, C.M.; Jones, M.; Agarwal, P.K.; Murali, T.M. A Monte Carlo algorithm for fast projective clustering. In Proceedings of the 2002 ACM SIGMOD International Conference on Management of Data, Madison, WI, USA, 2–6 June 2002; pp. 418–427.
67. Shao, S.; Zhang, X.; Zhao, X.R. Empirical decomposition and peak path of carbon emissions in China's manufacturing industry—Generalized Divisia Index decomposition and dynamic scenario analysis. *China Ind. Econ.* **2017**, *34*, 44–63.
68. Ramírez, A.; Keizer, C.D.; Sluijs, J.P.V.D.; Olivier, J.; Brandes, L. Monte Carlo analysis of uncertainties in the netherlands greenhouse gas emission inventory for 1990–2004. *Atmos. Environ.* **2008**, *42*, 8263–8272. [CrossRef]
69. Wang, C.K.; Cui, W.J. Construction, intensity, scale and electricity carbon emissions: A study based on LMDI decomposition method. *Sci. Technol. Manag.* **2015**, *35*, 220–241.
70. Huh, J.H.; Seo, K. Hybrid advanced metering infrastructure design for micro grid using the game theory model. *Int. J. Softw. Eng. Appl.* **2015**, *9*, 257–268. [CrossRef]
71. Huh, J.H.; Otgonchimeg, S.; Seo, K. Advanced metering infrastructure design and test bed experiment using intelligent agents: Focusing on the plc network base technology for smart grid system. *J. Supercomput.* **2016**, *72*, 1862–1877. [CrossRef]

72. Huh, J.H. Smart Grid Test Bed Using OPNET and Power Line Communication. Available online: https://ieeexplore.ieee.org/document/7801748/ (accessed on 5 September 2018).
73. Horizon 2020. Available online: http://ec.europa.eu/programmes/horizon2020/en/area/energy (accessed on 5 September 2018).
74. Molina-Moreno, V.; Núñez-Cacho Utrilla, P.; Cortés-García, F.J.; Peña-García, A. The Use of Led Technology and Biomass to Power Public Lighting in a Local Context: The Case of Baeza (Spain). *Energies* **2018**, *11*, 1783. [CrossRef]
75. Nuñez-Cacho, P.; Górecki, J.; Molina-Moreno, V.; Corpas-Iglesias, F.A. What Gets Measured, Gets Done: Development of a Circular Economy Measurement Scale for Building Industry. *Sustainability* **2018**, *10*, 2340. [CrossRef]
76. Villar Rubio, E.; Quesada Rubio, J.M.; Molina Moreno, V. Convergence Analysis of Environmental Fiscal Pressure Acrosss EU-15 Countries. *Energy Environ.* **2015**, *26*, 789–802. [CrossRef]

energies

Article

Can China Achieve the 2020 and 2030 Carbon Intensity Targets through Energy Structure Adjustment?

Ying Wang [1], Peipei Shang [2], Lichun He [3], Yingchun Zhang [4] and Dandan Liu [1,*]

1 School of Statistics, Dongbei University of Finance and Economics, Dalian 116025, China; wangyingxyz0710@outlook.com
2 Editorial Department, Dongbei University of Finance and Economics, Dalian 116025, China; peipeishangpps@outlook.com
3 School of Public Administration, Dongbei University of Finance and Economics, Dalian 116025, China; lichunhelch@outlook.com
4 School of Economics, Qingdao University, Qingdao 266071, China; zhangyc@qdu.edu.cn
* Correspondence: liudandan@dufe.edu.cn; Tel.: +86-0411-84710484

Received: 27 September 2018; Accepted: 9 October 2018; Published: 11 October 2018

Abstract: To mitigate global warming, the Chinese government has successively set carbon intensity targets for 2020 and 2030. Energy restructuring is critical for achieving these targets. In this paper, a combined forecasting model is utilized to predict primary energy consumption in China. Subsequently, the Markov model and non-linear programming model are used to forecast China's energy structure in 2020 and 2030 in three scenarios. Carbon intensities were forecasted by combining primary energy consumption, energy structure and economic forecasting. Finally, this paper analyzes the contribution potential of energy structure optimization in each scenario. Our main research conclusions are that in 2020, the optimal energy structure will enable China to achieve its carbon intensity target under the conditions of the unconstrained scenario, policy-constrained scenario and minimum external costs of carbon emissions scenario. Under the three scenarios, the carbon intensity will decrease by 42.39%, 43.74%, and 42.67%, respectively, relative to 2005 levels. However, in 2030, energy structure optimization cannot fully achieve China's carbon intensity target under any of the three scenarios. It is necessary to undertake other types of energy-saving emission reduction measures. Thus, our paper concludes with some policy suggestions to further mitigate China's carbon intensities.

Keywords: carbon intensity target; energy structure; gray model (GM (1, 1)); generalized regression neural network (GRNN); Markov forecasting model; non-linear programming

1. Introduction

As the greenhouse effect continues to increase on a global scale, the warming climate has become a universal challenge facing modern human society [1]. In recent years, as China's economy has continued to develop, its energy consumption and carbon emissions have also risen. In 2007, China's total carbon emissions surpassed those of the United States, making China the world's largest carbon emitter [2]. At present, China's carbon emissions account for approximately one-quarter of the total global carbon emissions, and the country's participation in climate change mitigation actions is essential [3]. In 2009, the Chinese government made a commitment at the Copenhagen Global Climate Conference: by 2020, carbon dioxide emissions per unit of gross domestic product (GDP) in China will decrease by 40–45% compared to 2005 levels [4]. In 2015, China submitted a UN self-determination document on climate change. By 2030, the country intends to reduce carbon dioxide emissions per unit of GDP by 60–65% compared to 2005 [5]. These carbon intensity targets are not only voluntary actions for China to combat climate change but also a commitment to the international community. China's energy structure is lagging behind that of developed countries, and coal consumption has continued at a high level for many years. The slow development of renewable energy sources has led to high total carbon emissions, high carbon intensity and low energy efficiency in China. Simultaneously, the unreasonable structure of energy consumption has also put considerable pressure on China's ecological environment. As the largest developing country in the world, China remains in a stage of industrialization and rising urbanization with immense energy consumption. One of the great challenges China faces is how to coordinate economic growth with energy conservation and emission reduction. Optimization of the energy structure not only aids in reducing carbon emissions and carbon intensity, but it also addresses the current situation of China's energy demand. During the process of economic growth, the global community should prevent further deterioration of the ecological environment and promote sustainable economic development.

Forecasting energy consumption and carbon emissions will aid in setting reasonable energy saving and emission reduction policies. Recently, many experts have conducted research on China's carbon emissions. These studies can be classified into two main categories. The first is to factorize carbon emissions and to search for carbon emission factors to predict carbon emissions. The widely used methods include the logarithmic mean divisia index (LMDI) decomposition model [6,7], the divisia index decomposition model [8,9], the input–output analysis model [10], the Kaya model [11,12], stochastic impacts by regression on population, the affluence and technology (STIRPAT) model [13], and so on [14,15]. However, the prediction models do not usually have high accuracy due to the complexity of the selected factors and difficulty in predicting the influencing factors. The second category is based on timing trends, directly establishing mathematical models to predict carbon emissions. The most frequently used methods are the auto-regressive integrated moving average (ARIMA) model [16], gray prediction model [17], and the artificial neural network model [18]. Such models often have high requirements for data quality. In addition, some researchers have used other models to study carbon emissions. Gambhir et al. [19] used a combined model to forecast China's carbon emissions from 2005 to 2050. Choi et al. [20] used a data envelopment analysis (DEA) model to predict the carbon emission reduction potential and energy efficiency in China. When Du et al. [21] evaluated potential carbon emission reductions in China using a non-parametric metafrontier model, the results showed that China's annual carbon emission reduction potential during the 11th five-Year period reached up to 168.7 million tons of carbon dioxide.

Based on the forecasted carbon emissions, several researchers have conducted studies on whether the carbon intensity targets for China in 2020 and 2030 can be achieved [22–26]. Stern et al. [27] evaluated the difficulty of achieving the carbon intensity targets in China and India by decomposing the factors that influence carbon intensity, but the authors did not consider the economic factors in their model. Yi et al. [28] and Xiao et al. [29] used scenario analysis to conclude that the target for carbon intensity in China in 2020 will most likely be realized, while Yuan et al. [30] determined that if China's clean energy accounted for 17% of the total energy in 2020, the carbon intensity target could be achieved by 2020. Starting with a low-carbon policy, Wang et al. [31] conducted an inter-provincial emission reduction path analysis of China's carbon intensity in 2020. According to the principle of fairness and common but differentiated responsibility, Yi et al. [32] selected three indicators—per capita GDP, accumulated carbon emissions from fossil fuel and energy consumption per unit of industrial added value—to establish a provincial carbon intensity distribution model to achieve the 2020 carbon intensity target. Research by Xu et al. [33] showed that under China's existing policies, the carbon intensity targets for both 2020 and 2030 can be achieved, but the overall goals of 840 million tons of carbon dioxide emissions by 2020 and 710 million tons by 2030 cannot be met. Through Monte Carlo simulation and scenario analysis, Zhang et al. [34] observed that China can achieve the carbon intensity targets for 2020 and 2030 on the basis of the existing policies. However, it is not clear whether China can achieve its peak carbon emission goal by 2030. Most of the above studies focus mainly on the relationship between economic development and carbon emissions, and the generation of regional allocations of the carbon intensity targets. There are few studies on the energy consumption structure. In addition, existing research lacks a forecast for China's carbon emissions by 2030, and omits whether the carbon intensity targets can be achieved by 2030. The abovementioned papers are listed in Appendix A; Table A1. This paper also summarizes the above research methods and their advantages and disadvantages in Table 1. Based on these studies, we present research topics and methods.

According to our discussion, there are many ways to predict energy consumption and carbon emissions, but each method has some shortcomings. To overcome these shortcomings, this paper first uses the combined forecasting model to forecast the total primary energy consumption. Then, scenario analysis is utilized to predict the energy consumption structure. Finally, based on the predictions for energy consumption and energy structure, combined with the carbon emission factors, the total carbon emissions and carbon intensities under different scenarios are obtained, and the potential contribution of energy structure optimization to achieve the carbon emission intensity target is calculated.

Compared with the existing research, the innovations in this paper are reflected in the following three main aspects:

(1) First, this paper predicts the primary energy consumption based on a combined forecast model. A primary energy consumption forecast is the basis for a prediction of the energy structure. In this paper, to determine the characteristics of a time series of primary energy consumption that are affected by numerous factors, the gray prediction model and the generalized regression neural network (GRNN) model are combined to predict energy consumption. The gray prediction model predicts future energy consumption based on historical changes, and the exogenous variables considered by this model have less impact. To compensate for defects in the gray prediction model, the GRNN model is introduced. The influencing factors of primary energy consumption are selected as the input layer variables for the GRNN model, and the prediction results are achieved by predicting the input variables. Then, gray relational analysis is used to empower the gray prediction model and GRNN model, and finally, the combined forecasting result is obtained. Compared to the distinct forecasting model, the combined model synthesizes more factors that affect the dependent variable, the forecasting accuracy is higher, and the forecasting result is more closely aligned with reality.

(2) Second, this paper considers energy structure optimization in three scenarios: a natural evolution scenario, a policy planning scenario, and a cost perspective scenario. Firstly, according to the characteristics of China's energy consumption structure, the Markov model is used to predict the natural evolution of the energy consumption structure, and the forecast result is set as an unconstrained scenario. In addition, combined with the energy development plan formulated by the state, the energy structure should be adjusted accordingly to set the situation as a policy-constrained scenario. Finally, from the cost perspective, the minimum external cost of carbon emissions is used as the decision-making target, non-linear programming is performed, and the forecast result for the energy structure is obtained as the minimum cost scenario. Applying different scenarios is conducive to a more comprehensive understanding of future changes in China's energy structure.

(3) Third, this paper combines China's carbon intensity targets for 2020 and 2030 for analysis. The existing research focuses mainly on the target of a 40–45% reduction of carbon intensity by 2020 and less on the goal of a 60–65% reduction by 2030. This paper combines the carbon intensity targets for 2020 and 2030, and then analyzes the potential for optimizing the energy structure to contribute to achieving the carbon intensity targets in order to explore the possibility of reaching the targets in 2020 and 2030; finally, the paper presents several reference suggestions.

Table 1. Summary of each model's advantages and shortcomings.

Model	Purpose	Advantages	Shortcomings
ARIMA model	Forecast energy consumption and greenhouse gas (GHG) emissions for a pig iron manufacturing organization in India.	ARIMA is quite suitable for short-term forecasting.	The model requires a long data series.
Non-linear gray model	Forecast carbon emissions.	(1). Evaluates the non-linear effects of economic growth on carbon emissions in the model. (2). This model can analyze the effect of multiple influencing factor variables on the behavior of the system.	After the model is improved, the newly added power exponents are unknown. There are difficulties in estimating these parameters.
Artificial neural networks model	Forecast carbon emissions.	ANNs are adaptable for optimization and adaptive methods and are computationally efficient.	For some complex systems, artificial neural network prediction results may have considerable errors.
Hybrid model	Use the technology-rich integrated assessment (MESSAGE) model, and a combination of bottom-up models to forecast carbon emissions.	(1). The carbon emission index selected by this model is more concentrated. (2). The model can better describe the overall demand-side uncertainties owing to structural and methodological differences between top-down and bottom-up approaches.	The model does not consider feedback with respect to price-induced changes in energy supply from a detailed energy demand outlook.
Nonparametric metafrontier approach	Forecast carbon emissions.	The model takes into consideration the technology gap among China's regions.	This model does not consider any statistical noise.

Based on the above discussions, this paper first uses the GM (1, 1) model and GRNN model to predict China's primary energy consumption separately, and then a gray relational analysis is used to empower the GM (1, 1) model and GRNN model to obtain the forecasting results of the combined model. Secondly, the evolution of energy structure is divided into "Unconstrained scenario", "Policy-constrained scenario", and "Minimum external costs of carbon emissions scenario" to study the future changes in China's energy structure. Finally, according to the predicted results of energy consumption and structure, China's carbon emissions, and carbon intensity results for 2020 and 2030 are calculated for further analysis. The research process of this paper is shown in Figure 1.

The remainder of this paper is organized as follows. Section 2 discusses the model theory. Section 3 analyzes the forecast results for primary energy consumption. Section 4 analyzes the optimization results of the energy structure in different situations. Section 5 explores the potential contribution of optimizing the energy structure to achieving the carbon intensity targets under different scenarios. Section 6 presents the main conclusions and policy suggestions.

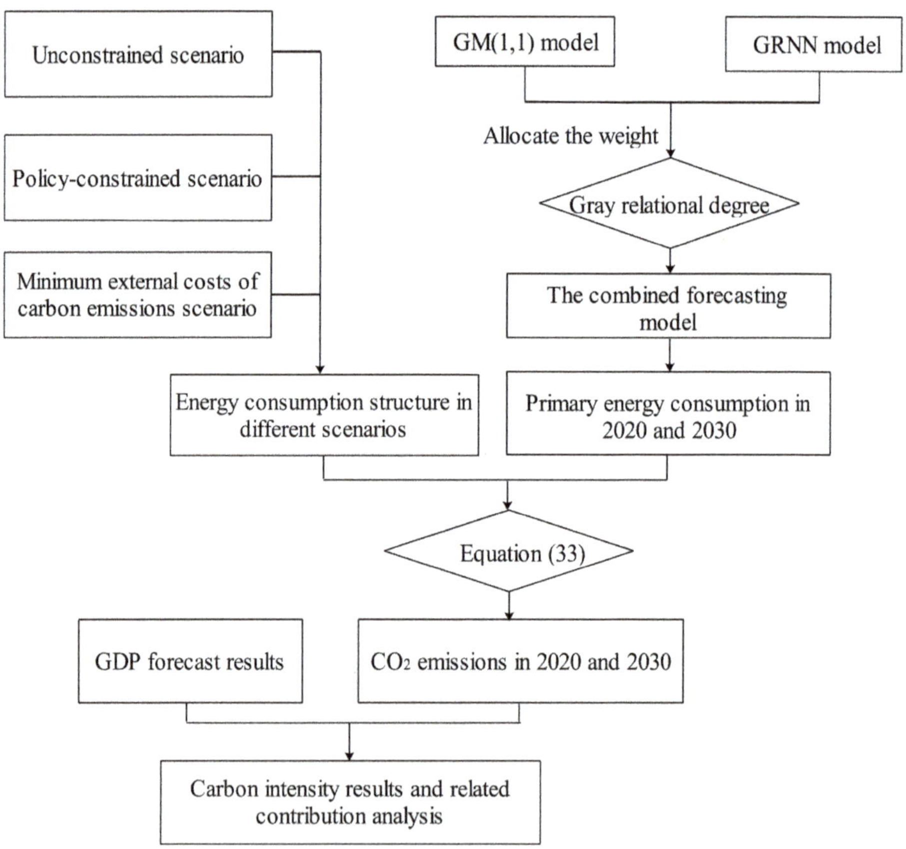

Figure 1. The general flowchart conducted in this paper.

2. Materials and Methods

2.1. Forecast Models for Energy Consumption in China

2.1.1. Model I: Gray Prediction Model

Since the advent of gray system theory, remarkable achievements have been accomplished in predictions regarding military systems, social systems, ecosystems, and commercial systems [35,36]. According to the gray system theory, useful information is extracted from the gray comprehensive sequence in the annual energy consumption time series to predict the future demand for energy consumption. Based on data availability, this paper selects primary energy consumption data [37] from 1953 to 2016 to predict the primary energy consumption in China for 2017 to 2030.

Step 1: Pre-process the primary energy consumption data. We assume that the sequence of primary energy consumption $Y^{(0)}$ from 1953 to 2016 is as follows, Equation (1):

$$Y^{(0)} = [Y^{(0)}(1), Y^{(0)}(2), \ldots Y^{(0)}(64)] \tag{1}$$

To weaken the randomness in the original sequence, prior to the establishment of the gray prediction model, the data for primary energy consumption sequences from 1953 to 2016 is processed, and the cumulative generation, Equation (2), is used to generate a cumulative generated column $Y^{(1)}$:

$$Y^{(1)}(k) = \sum_{i=1}^{k} Y^{(0)}(i) = Y^{(1)}(k-1) + Y^{(0)}(k) \tag{2}$$

$$Y^{(1)} = [Y^{(1)}(1), Y^{(1)}(2), \ldots Y^{(1)}(64)] \tag{3}$$

Step 2: Establish the gray differential, Equation (4):

$$\hat{Y}^{(1)}(k+1) = [Y^{(0)}(1) - \frac{b}{a}]e^{-ak} + \frac{b}{a} \tag{4}$$

Step 3: Determine the values of parameters a and b, according to Equation (5):

$$\begin{cases} \begin{bmatrix} a \\ b \end{bmatrix} = [B^T B]^{-1} B^T X_N \\ B = \begin{bmatrix} -\frac{1}{2}(Y^{(1)}(1) + Y^{(1)}(2)) & 1 \\ -\frac{1}{2}(Y^{(1)}(2) + Y^{(1)}(3)) & 1 \\ \cdots & \cdots \\ -\frac{1}{2}(Y^{(1)}(n-1) + Y^{(1)}(n)) & 1 \end{bmatrix} \\ X_N = [Y^{(0)}(2), Y^{(0)}(3), \ldots, Y^{(0)}(n)]^T \end{cases} \tag{5}$$

After obtaining the cumulative generated column $Y^{(1)}$, according to the available matrices B and X_N knowing that $B^T B$ is a symmetric matrix, the original time series $Y^{(0)}$ and cumulatively generated column $Y^{(1)}$ are entered into Equation (5) using the least squares method to obtain the values of parameter a and parameter b.

According to the above method, the primary energy consumption data from 1953 to 2016 are entered to obtain the gray differential Equation (6):

$$\hat{Y}^{(1)}(k+1) = 161324.24e^{0.0612k} - 155913.24 \tag{6}$$

Step 4: Predict the primary energy consumption for 2017–2030 according to Equation (7):

$$\hat{Y}^{(0)}(k+1) = \hat{Y}^{(1)}(k+1) - \hat{Y}^{(1)}(k) \tag{7}$$

Step 5: Test the residuals of the forecast data and examine the accuracy of the forecast data. The model fitting value $\hat{Y}^{(0)}$ for primary energy consumption in China from 1953 to 2016 is obtained based on the predictive value reduction formula. From the original sequence $Y(0)$ and predicted sequence $\hat{Y}^{(0)}$, the prediction data are tested for residuals and accuracy. The absolute error (AE), mean absolute error (MAE) and mean absolute percentage error (MAPE) of the prediction result are obtained from Equations (8), (9) and (10), respectively:

$$\delta(t) = Y^{(0)}(t) - \hat{Y}^{(0)}(t) \tag{8}$$

$$MAE = \frac{1}{n}\sum_{i=1}^{n} \delta_i(t) \tag{9}$$

$$MAPE = \frac{1}{n}\sum_{i=1}^{n} |\delta_i(t)|/Y_i^0(t) \tag{10}$$

The standard deviation of the predicted value S_1 and standard deviation of sample S_2 are calculated in Equations (11) and (12), respectively. The posterior difference ratio C can be calculated as stated in Equation (13). According to the principle of the posterior difference test, as the posterior difference ratio decreases, the predicting effect improves:

$$S_1 = \sqrt{\frac{\sum_{i=1}^{n} [\delta(t) - \bar{\delta}]^2}{n}} \tag{11}$$

$$S_2 = \sqrt{\frac{\sum_{i=1}^{n} [Y^{(0)}(t) - \bar{Y}]^2}{n}} \tag{12}$$

$$C = S_1/S_2 \tag{13}$$

The correctness of using this model can be judged by observing the average relative error and posterior difference ratio.

2.1.2. Model II: Generalized Regression Neural Network

It is well-known that primary energy consumption is affected by many factors and that the system is complicated. A study of historical data shows that the sample is relatively small and presents a non-linear trend of development. GRNN is a general nonparametric regression model, which is a branch of the Radial Basis Function (RBF) neural network [38], and has a strong nonlinear mapping ability and flexible network structure, as well as a high degree of fault tolerance and robustness [39,40]. Therefore, in order to obtain a higher prediction accuracy, we apply GRNN to conduct the primary energy consumption forecasting.

Supposing the joint probability density functions of random variables x and y is $f(x,y)$, and the observed value of variable x is X, then the regression of y to x, that is, the conditional mean, is shown in Equation (14):

$$\hat{Y} = E[y|X] = \frac{\int_{-\infty}^{+\infty} y f(X,y)dy}{\int_{-\infty}^{+\infty} f(X,y)dy} \tag{14}$$

The unknown probability density function $f(x,y)$ can be estimated from the sample observations x and y, and the non-parametric estimation Equation (15) is as follows:

$$f(X,Y)=\frac{1}{(2\pi)^{(m+1)/2}\sigma^{m+1}n}\times\sum_{i=1}^{n}\exp[-\frac{(X-X_i)^T(X-X_i)}{2\sigma^2}]\exp[-\frac{(Y-Y_i)^2}{2\sigma^2}] \tag{15}$$

where X_i and Y_i are the sample observations of variables x_i and y_i, σ is the kernel width, n is the number of samples, and m is the dimension of variable x.

Substituting $\hat{f}(X,Y)$ with $f(x,y)$, exchange integrals and summation order, the estimated value is obtained from Equation (16):

$$\hat{Y}(X)=\frac{\sum_{i=1}^{n}\exp[-\frac{(X-X_i)^T(X-X_i)}{2\sigma^2}]\int_{-\infty}^{+\infty}y\exp[-\frac{(y-Y_i)^2}{2\sigma^2}]dy}{\sum_{i=1}^{n}\exp[-\frac{(X-X_i)^T(X-X_i)}{2\sigma^2}]\int_{-\infty}^{+\infty}\exp[-\frac{(y-Y_i)^2}{2\sigma^2}]dy} \tag{16}$$

where the estimated value $\hat{Y}(X)$ is the weighted average of all sample observations Y_i, and the weighting factor for each observation Y_i is the index of the Euclidean distance squared between the corresponding samples X_i and X.

The GRNN network structure consists of four layers: the input layer, pattern layer, summation layer, and output layer. The corresponding network input is $X=[x_1,x_2,\ldots,x_m]^T$, and the output is $Y=[y_1,y_2,\ldots,y_k]^T$.

(1) Input layer

The number of neurons in the input layer is equivalent to the dimensions of the input vector in the learning sample. Each neuron is a simple distribution unit that directly passes input variables to the pattern layer. For the analysis of the influencing factors of primary energy consumption, the energy price, population, GDP, household consumption level, industrial energy consumption, and industrial added value are selected as the input variables of the network; that is, the number of neurons in the input layer is six. Table 2 provides the detailed economic implications of these variables.

Table 2. The variables of the input layer.

Input	Economic Implications
Energy price	This variable is the most crucial factor for determining energy demand. Due to the lack of an energy price index, this paper uses the coal price to represent this index because coal is China's foremost consumer energy source, accounting for 60–70% of energy consumption, and the price of coal is more market-oriented.
Population	Energy is the fundamental material on which human beings depend for survival. Both energy production activities and energy consumption activities are intended to meet human needs. Studies such as those of Liu et al. [41] and Guo et al. [42] showed that population is a significant factor affecting energy demand.
GDP	This variable reflects a country's income level, which is a fundamental factor in determining energy demand. Lin et al. [43] and He et al. [44] showed that there is a significant positive correlation between energy demand and GDP.
Household consumption level	Household energy consumption includes two categories: direct and indirect consumption; that is, energy will be directly consumed during residential life and indirectly consumed by producing various goods and services. Residents' rising consumption level will increase not only the direct energy consumption but also the indirect energy consumption.
Energy consumption in the industrial sector	China's industrialization is in the mid-to-late stage. Studies have demonstrated that primary energy consumption in the industrial sector accounts for more than 70% of the total primary energy consumption in China [45].
Industrial added value	In the process of industrialization, the output values of the manufacturing industry and secondary industry have been continuously increasing, and the changes reflect the adjustment of the industrial structure, which is a basic factor affecting the energy demand.

The data for all the input layer variables proposed above come from the China Statistical Yearbook [46] and the China Economic Net Statistics Database [37]. Since the energy consumption data from the industrial sector date back to 1980 at the earliest, the historical data for the previous period is missing; thus, the above six variables range from 1980 to 2016.

(2) Pattern layer

The number of neurons in the pattern layer is equal to the number of learning samples. Here, the number of neurons in the pattern layer is six, and each neuron in the input layer corresponds to a different sample. The transfer function of the neurons $x_i\,(i = 1,\ldots,6)$ is show in Equation (17):

$$p_i = \exp[-\frac{(X - X_i)^T(X - X_i)}{2\sigma^2}]\,(i = 1,\ldots,6) \tag{17}$$

where the output of a neuron x_i is an exponential form of the square of the Euclidean distance between the input variable X and its corresponding sample X_i.

(3) Summation layer

The summation layer is divided into two types of neuron summing: one is $\sum\limits_{i=1}^{n}\exp[-\frac{(X-X_i)^T(X-X_i)}{2\sigma^2}]$; this expression is arithmetic summation of the output of all the pattern layer neurons, which has a connection weight of 1 with each pattern layer neuron, and the transfer function is $s_D = \sum\limits_{i=1}^{n} p_i$.

Another type is $\sum\limits_{i=1}^{n} Y_i\exp[-\frac{(X-X_i)^T(X-X_i)}{2\sigma^2}]$; this weighted expression sums the output of all the pattern layer neurons and the connection weight between the i-th neuron in the pattern layer. The j-th molecule summation neuron in the sum layer is y_{ij}, and the transfer function is $s_{Nj} = \sum\limits_{i=1}^{n} y_{ij}P_i$, $j = 1,2,\ldots,k$.

(4) Output layer

The number of neurons in the output layer is equal to the number of dimensions of the output vector k in the learning sample; each neuron divides the output of the summation layer to obtain the output of the j-th neuron y_j through Equation (18):

$$y_j = \frac{S_{Nj}}{S_D}\;\; j = 1,2,\ldots,k \tag{18}$$

2.1.3. Combined Forecasting Model of China's Energy Consumption Based on the Gray Relation Degree

Since the introduction of the combined forecasting model in 1969, it has been a popular topic in the field of forecasting both domestically and internationally [47]. The traditional single model prediction process has many shortcomings, such as the single structure of the forecast, limited information sources, incomplete factors, and the sensitive model setting. However, the combined forecasting model can comprehensively utilize the information provided by each individual model, collect the advantages of the individual models, enrich the model structure, and ultimately use the weighted average method to obtain the result of the combined forecasting model to improve the fitting accuracy and forecasting ability; therefore, the forecasting method is more effective.

Based on the predictions of primary energy consumption in China for 2017 to 2030 through the gray prediction model and the GRNN model, respectively, this paper uses the gray relational method to allocate the weights of the two models to construct the combined forecasting model.

Step 1: Preprocess the data. To eliminate the order of the magnitude difference between each set of dimension data and avoid the error of the order of magnitude being too large, according to

Equation (19), the real value sequence, the gray forecast value sequence and GRNN predictive value sequence are processed and converted into values between [0, 1]:

$$X_{\mathrm{i}}'(t) = \frac{X_i(t) - \min X_i(t)}{\max X_i(t) - \min X_i(t)} \tag{19}$$

Step 2: Calculate the gray relational degree between the reference sequence and the real value sequence.

Setting the actual value of primary energy consumption as $\{x_t, t = 1, 2, \ldots, N\}$, there is the gray prediction model and GRNN forecasting, which are two single forecasting models for predicting primary energy consumption, where x_{1t} is the forecast value corresponding to the gray prediction model at time t and x_{2t} is the forecast value corresponding to the GRNN forecast model at time t.

$$\xi_{0i} = \frac{1}{N}\sum_{i=1}^{N} \frac{\min\limits_{1\le i\le 2}\min\limits_{1\le t\le N}|e_{it}| + \rho \max\limits_{1\le i\le 2}\max\limits_{1\le t\le N}|e_{it}|}{|e_{it}| + \rho \min\limits_{1\le i\le 2}\min\limits_{1\le t\le N}|e_{it}|} \tag{20}$$

$$e_{it} = x_t - x_{it} \tag{21}$$

In Equation (20), ξ_{0i} represents the gray correlation degree between the predicted value sequence $\{x_{it}, t = 1, 2, \ldots, N\}$ of the i-th single-item prediction method and the real value sequence $\{x_t, t = 1, 2, \ldots, N\}$, as the gray correlation degree of the i-th single forecasting method. Here, e_{it} represents the prediction error of the i-th prediction model at time t. $\rho \in (0, 1)$ is the resolution coefficient, which usually takes the value $\rho = 0.5$. According to Equation (20), the gray relational degree between the gray prediction sequence and real value sequence ξ_{01} and the gray relational degree between the GRNN prediction sequence and real value sequence ξ_{02} can be calculated separately.

$$l_i = \xi_{0i} \Big/ \sum_{i=1}^{2} \xi_{0i} \tag{22}$$

Therefore, we can calculate the weight of the gray prediction model, and the weight of the GRNN prediction model according to Equation (22).

According to the definition of gray relational degree, $\xi_{0i} \in [0, 1]$. The accuracy of the prediction is accurate only when there is a gray correlation degree of 1 between them.

Step 3: Generate a combined forecast of primary energy consumption in China. The combined predicted value of primary energy consumption at time t is given by Equation (23):

$$\hat{x}_t = l_1 x_{1t} + l_2 x_{2t} \; t = 1, 2, \ldots, N \tag{23}$$

In Equation (23), l_1 is the weighted coefficient corresponding to the predicted value of the gray prediction model, l_2 is the weighting coefficient corresponding to the prediction value of the GRNN prediction model, which satisfies $l_1 + l_2 = 1$; Equation (23) shows that inequality (24) holds:

$$\min_{1\le i\le 2} x_{it} \le \hat{x}_t \le \max_{1\le i\le 2} x_{it} \; t = 1, 2, \ldots, N \tag{24}$$

Let e_t be the prediction error of the primary energy consumption at time t from the combined forecasting method, and according to $l_1 + l_2 = 1$, Equation (25) is as follows:

$$\begin{aligned} e_t &= x_t - \hat{x}_t = x_t - \sum_{i=1}^{2} l_i x_{it} \\ &= \sum_{i=1}^{2} l_i (x_t - x_{it}) = \sum_{i=1}^{2} l_i e_{it} \\ t &= 1, 2, \ldots N \end{aligned} \tag{25}$$

Letting ξ be the gray relational degree of the combined forecasting method, then the gray relational degree of the combined forecasting is given by Equation (26):

$$\xi = \frac{1}{N}\sum_{i=1}^{N} \frac{\min\limits_{1\le i\le 2}\min\limits_{1\le t\le N}|e_{it}| + \rho \max\limits_{1\le i\le 2}\max\limits_{1\le t\le N}|e_{it}|}{|\sum\limits_{i=1}^{2} l_i e_{it}| + \rho \min\limits_{1\le i\le 2}\min\limits_{1\le t\le N}|e_{it}|} \tag{26}$$

In Equation (26), the gray relational degree ξ of the combined forecasting method is a function of the weighting coefficient $L = (l_1, l_2)$ of each single forecasting model, so that ξ can be denoted as $\xi(L)$.

According to the gray relational theory, as the gray relational degree of the combined forecasting method increases, the combined forecasting model is more effective. If $\xi < \xi_{\min}$, the combined forecasting model is considered inferior forecasting; if $\xi_{\min} \le \xi \le \xi_{\max}$, the combined forecasting model is referred to as non-inferior combination forecasting; and if $\xi > \xi_{\max}$, the combined forecasting model is the optimal combined forecasting model.

2.2. Construction of the Forecast Model for Energy Consumption Structure in China

2.2.1. Energy Structure Prediction Based on the Markov Model

The evolution of primary energy consumption structure has its own changes and development laws. This evolutionary law provides the basis for our study of the future energy consumption structure. The main methods for predicting the energy consumption structure in existing research include the Markov forecasting model, and the energy and environment comprehensive policy evaluation model (Integrated Assessment Model, IAM model) [48]. Taking into account that the IAM model involves many factors, the data are not easy to obtain, the model is not generally suitable for cooperative research and development groups, and an individual research model is very difficult to establish. This paper uses the Markov forecasting model to predict the future energy consumption structure in China.

Step 1: Build a Markov model that predicts the primary energy consumption structure.

A represents the total primary energy consumption; for simplicity, it can be divided into four types of energy sources: coal, oil, natural gas and clean energy (water, nuclear, and wind electricity, etc.). At time n, the vector of the primary energy consumption structure is $S(n) = \{s_c(n), s_o(n), s_g(n), s_e(n)\}$, where $S_c(n)$, $S_o(n)$, $S_g(n)$, and $S_e(n)$ represent the shares of coal, oil, natural gas and clean energy, respectively, in the total primary energy consumption; the sum of their proportions is 1. We assume that the one-step transition probability matrix of the energy consumption structure from time n to time $n+1$ is:

$$P(n) = \begin{bmatrix} P_{c\to c}(n) & P_{c\to o}(n) & P_{c\to g}(n) & P_{c\to e}(n) \\ P_{o\to c}(n) & P_{o\to o}(n) & P_{o\to g}(n) & P_{o\to e}(n) \\ P_{g\to c}(n) & P_{g\to o}(n) & P_{g\to g}(n) & P_{g\to e}(n) \\ P_{e\to c}(n) & P_{e\to o}(n) & P_{e\to g}(n) & P_{e\to e}(n) \end{bmatrix} \tag{27}$$

In the probability matrix, every element is a positive number less than 1, and the sum of the probabilities in each row is always equal to 1. Here, the elements are classified according to the characteristics of the elements in the probability matrix.

First, the main diagonal elements of the matrix $P(n)$ are classified as the first category, referred to as the "retention probability elements." These elements represent the probability that various types of energy consumption continue to maintain the original ratio (for example:$P_{c\to c}(n)$ represents the probability that coal consumption will continue to maintain the original ratio from time n to time $n+1$). Second, the main diagonal line elements are classified as the second category, referred to as the "transition probability elements." These elements represent the proportion of such energy consumption to other types of energy consumption in terms of the transfer probability (for example:

$P_{c\to o}(n)$ represents the probability that the ratio of coal consumption to oil consumption transfers from time n to time $n+1$). Third, the column elements outside the main diagonal are classified as the third category, referred to as the "absorption probability elements." These elements represent the probable proportion of such energy consumption absorption of other types of energy consumption (for example, $P_{o\to c}(n)$ represents the probability that the percentage of coal consumption absorbs the oil consumption ratio from time n to time $n+1$).

Step 2: Determine the average transition probability matrix P.

To predict the future energy consumption structure, we must find the average transfer probability matrix based on the existing energy consumption structure. The specific procedure is as follows: first, calculate the primary transition probability matrix of the energy consumption structure in each year, and then average the transition matrix to obtain the average transition probability matrix.

Supposing that from the initial moment to moment m, the transition probability matrix for each step of the energy consumption structure is $P(1), P(2), \cdots, P(m)$, then the average transfer probability matrix is $P = [P(1) \cdot P(2) \cdots P(m)]^{1/m}$. According to the average transfer probability matrix P, the structure of primary energy consumption at the time $n+m$ can be predicted by Equation (28):

$$S(n+m) = S(n) \cdot P^m \tag{28}$$

Step 3: Determine the transition probability matrix for each step $P(n)$. To determine the average transition probability matrix P, the key lies in how to determine a transition probability matrix of the energy consumption structure $P(n)$. This paper uses the following four steps to calculate the value for each element in the matrix $P(n)$:

(I) Calculate the retention probability elements. If from time n to $n+1$, the proportion of energy consumption increases, the retention probability of this energy in the transition probability matrix is 1; if the proportion decreases, the retention probability is equal to the ratio of time $n+1$ to time n.

(II) Calculate the transition probability in the rows in which the element with a retention probability of 1 is located. If the retention probability of a row is 1, there is no possibility of transferring energy to other types of energy, and the sum of the elements in each row of the transition probability matrix has been set equal to 1; therefore, the probability of a row transition probability element is zero.

(III) Calculate the probability of absorption in the columns where the element with a retention probability of less than 1 is located. If the retention probability of a column is less than 1, the percentage of energy consumption represented by the column is reduced. There is no possibility for such energy consumption to absorb other types of energy. Therefore, the probability of absorption for this column is zero.

(IV) Calculate the nonzero transition probability in the rows in which the element with a retention probability of less than 1 is located. The retention probability of energy corresponding to a row of elements is less than 1, indicating the transfer of such energy consumption to other types of energy consumption from n to $n+1$. Using coal as an example, if $P_{c\to c}(n)$ is less than 1, the proportion of coal consumption decreases from n to $n+1$, and the transfer of coal consumption to the other three types of energy consumption occurs. The probability of coal consumption shifting to oil consumption, natural gas consumption, and clean energy consumption can be calculated according to Equations (29)–(31), respectively:

$$P_{c\to o}(n) = \frac{[1 - P_{c\to c}(n)] \times [s_o(n+1) - s_o(n)]}{[s_o(n+1) - s_o(n)] + [s_g(n+1) - s_g(n)] + [s_e(n+1) - s_e(n)]} \tag{29}$$

$$P_{c\to g}(n) = \frac{[1 - P_{c\to c}(n)] \times [s_g(n+1) - s_g(n)]}{[s_o(n+1) - s_o(n)] + [s_g(n+1) - s_g(n)] + [s_e(n+1) - s_e(n)]} \tag{30}$$

$$P_{c\to e}(n) = \frac{[1 - P_{c\to c}(n)] \times [s_e(n+1) - s_e(n)]}{[s_o(n+1) - s_o(n)] + [s_g(n+1) - s_g(n)] + [s_e(n+1) - s_e(n)]} \tag{31}$$

For other types of energy retention probabilities less than 1, based on the same principle, we can calculate the non-zero transition probabilities of such energy sources.

Based on the above steps, the transition probability matrix of the energy consumption structure from the initial moment to the moment m is obtained as $P(1), P(2), \cdots, P(m)$, the average transfer probability matrix P can be obtained, and then Equation (28) can be used to predict the future energy consumption structure.

2.2.2. Energy Structure Prediction Model Based on the External Cost Minimization of Carbon Emissions

In this section, we optimize the structure of energy consumption with the goal of minimizing the external costs of carbon emissions. Research has established that the factors of various types of energy carbon emissions are different, and the amount of carbon dioxide released by different types of energy combustion per unit mass is discrete, so that the external costs of carbon emissions consumed are not equal. Studies [49] have shown that external emissions of carbon dioxide cost approximately 20 dollars/ton, which, according to the current exchange rate, converts into 129.21 RMB/ton. Carbon emission factors of coal, oil and natural gas are 0.7476, 0.5825, and 0.4435, respectively. Burning a ton of carbon in oxygen releases five tons of carbon dioxide. Thus, the external CO_2 emission costs for various energy sources are shown in Table 3:

Table 3. CO_2 emission factors for various energy sources and external emission costs.

Specie	Coal	Oil	Natural Gas
CO_2 emission factor	2.744	2.138	1.628
External cost (RMB/tce)	354.81	276.46	210.49

We set the coal consumption x_1 at ten thousand tons, oil consumption x_2 at ten thousand tons, natural gas consumption x_3 at ten thousand tons, and the renewable energy consumption x_4 at ten thousand tons. According to the above analysis, the objective function $f(x)$ is set as follows:

$$f(x) = 354.81x_1 + 276.46x_2 + 210.49x_3 \tag{32}$$

$$\text{Total } CO_2 \text{ emission} = \sum Energy\ consumption \times Carbon\ emission\ factor \times \frac{44}{12} \tag{33}$$

$$\text{Carbon Intensity} = \frac{\text{Total } CO_2 \text{ emission}}{\text{GDP}} \tag{34}$$

According to China's energy long-term development strategy research and the 13th Five-Year Plan, the following constraints are set:

(1) Primary energy consumption is not greater than the predicted value:

$$x_1 + x_2 + x_3 + x_4 \le C_1 \tag{35}$$

(2) The total amount of CO_2 emissions is within the forecast range, where μ_i represents the corresponding emission factor of different types of energy:

$$\sum_{i=1}^{4} \mu_i x_i \le C_2 \tag{36}$$

(3) Various types of future energy structure changes in China are set in Table 4:

Table 4. The proportions of primary energy consumption for 2020 and 2030.

Year	Coal Ratio	Oil Ratio	Natural Gas Ratio	Clean Energy Ratio
2020	0.50–0.60	0.20–0.30	0.06–0.15	0.15–0.25
2030	0.40–0.50	0.25–0.30	0.10–0.20	0.20–0.25

(4) All types of energy are not less than zero:

$$x_i \geq 0 \quad (37)$$

(5) Of all types of energy, only coal, oil, and natural gas are considered to generate carbon dioxide during combustion, while clean energy releases very little carbon dioxide, which is not taken into account.

According to the above objective function and constraints, a non-linear programming model is constructed, and the relevant software is used to obtain the energy structure optimization results.

3. China's Energy Consumption Forecast Results

3.1. Result Analysis Based on the GM (1, 1) Model

This section uses MATLAB2017a (MathWorks, Natick, MA, USA) to realize the GM (1, 1) model, of which the MAPE is 0.0692, the residual standard deviation is 13971.35, actual standard deviation is 128779.08, and the posterior difference C calculated by Equation (15) is 0.1085, $C < 0.35$, which shows that the constructed gray forecasting model has a better forecasting effect. In addition, the probability of small residuals is calculated as P is 1, which shows that the model has a high goodness of fit and is suitable for predicting primary energy consumption in China.

Table 5 shows the forecasting results for the future energy consumption of China. The primary energy consumption forecast for 2017 will be 4815 million tons of standard coal, reaching 5786 million tons of standard coal in 2020 and 10,070 million tons of standard coal in 2030. This prediction shows that China's energy consumption for the future is high, but the prediction error of the gray prediction model is small and the accuracy is high, so the model itself is not a problem. The reason for the high forecast result is that the sample data predicted by this model are sourced from the primary energy consumption data for China from 1953 to 2016. This stage represents the process of China shifting from being a largely agricultural country to being a mature industrial country. The consumption of energy each year is considerable, with a high annual growth rate. This trend is further extended, resulting in a high growth rate of the predicted model values, which directly leads to a prediction of high energy consumption. The gray model is commonly used for short-term forecasting [50]. Therefore, it is reasonable to predict the economic variables through the gray model to supplement the later forecast data.

3.2. Result Analysis Based on the GRNN Model

In this paper, through cross-validation, the neural network is continuously trained and tested. When the radial basis function expansion speed is set to 0.56, the network error RMSE is the lowest and the function approximation is the best. Therefore, a smoothing factor of 0.56 was determined. The GRNN model was constructed, and the predicted values were inversely normalized to obtain the primary energy consumption forecast results. The MAPE for the forecast of primary energy consumption in the GRNN model from 1980 to 2016 is calculated as 0.0618. The goodness of fit of the model is high, so the model is suitable for the prediction of primary energy consumption in China.

In the following steps, the forecast variables of China are entered into the GRNN neural network by predicting the explanatory variables in turn, and the forecast is obtained for primary energy consumption in China from 2017 to 2030.

(1) The change in the energy price is a stochastic process. The mechanism of this change is complicated and affected by a combination of multiple factors. Therefore, for the trend forecast of energy price changes, we cannot simply assume that it changes at a fixed rate but rather use a measurement model for prediction. In this paper, we choose the ex-factory price index for coal as an alternative variable for energy price and predict the future price index through the ARIMA (2, 1, 2) model combined with the changing trend of the price index itself.

ARIMA (p, d, q) is a common and effective forecasting model that is widely used in time series forecasting [51]. In the model, p and q represent the order of autoregressive and moving average processes, respectively, and d is the degree of differencing. Through the Augmented Dickey-Fuller (ADF) test, the results showed that the original series is non-stationary (p-value = 0.1019). After taking the natural logarithm and performing a first-order difference, it becomes stationary (p-value = 0.000), so set d to 1. According to the autocorrelation graph (ACF) and the partial autocorrelation graph (PACF), there is no significant seasonal trend. To determine p and q, Akaike's information criteria (AIC) criterion was applied here. Through comparing the values of AIC, ARIMA (2, 1, 2) model is the best, with the smallest AIC of −2.7517, and the MAPE is 4.006%, which indicates the model's strong forecasting ability.

(2) The natural population growth rate in China has been declining since 1978. Building on recent literature regarding future population growth trends in China, this paper refers to the forecast results from the National Development Plan 2016–2030 [52] and predicts that in 2017, the total population will reach approximately 1.404 billion. By 2020, the total population will be 1.42 billion, and by 2030, it will reach 1.45 billion.

(3) With regard to the growth forecast for China's economy, combined with current researchers' studies [53–55], the commonly held view among experts is that from 1990 to 2010, China's economy will have grown at a very high rate, while it will grow at a medium rate from 2010 to 2030 and a low rate from 2030 to 2050. Therefore, China's future economic growth will occur at a 6.5% annual growth rate from 2011 to 2020 and at a 5.5% annual growth rate from 2021 to 2030, from which the total amount of China's GDP for 2017–2030 can be predicted.

(4) Residents' consumption levels are directly related to economic growth. Based on the research by Wang et al. [56], this paper predicts that there is a co-integration relationship between GDP and residents' consumption level; for every 1% increase in GDP, the consumer price index will increase by 0.679%. According to this co-integration relationship, the annual growth rate of consumer spending for 2011–2020 can be calculated as 4.41% and that for 2021–2030 as 3.735%.

(5) For the prediction of energy consumption in the industrial sector, this article refers to the method from the China Energy Economics Research Center at Xiamen University [57], and according to the historical trend for industrial energy efficiency, the industrial energy efficiency growth rate is set at 3% in 2010, decreasing by 0.5% every five years thereafter. Therefore, the industrial energy efficiency growth rate is 2% for 2016–2020, 1.5% for 2020–2025, and 1% for 2026–2030. From the ratio of industrial energy efficiency equal to the ratio of added value for the energy consumption of the industrial sector, the predicted value of the energy consumption for the industrial sector can be calculated.

(6) Chen's [58] research showed that the average annual growth rate of China's industrial added value as a portion of GDP was 0.4% beginning in the Ninth Five-Year Plan period. This article draws on this static calculation method, considering the recession of growth, which sets the annual average growth rate of industrial added value as a share of GDP for 2011–2020 as 0.3% and presents the annual average growth rate of 0.2% for 2021–2030.

From this process, we can calculate the forecast value for industrial added value from 2017 to 2030.

According to the above settings, the predicted values for the six explanatory variables are entered into the trained generalized regression neural network as the network input layer to predict the primary energy consumption for 2017–2030 in China. The results are shown in Table 5.

3.3. Result Analysis Based on the Combined Forecasting Model

The forecast processes for primary energy consumption in China in 2020 and 2030 are shown in Figure 2. China's primary energy consumption is predicted through the GM (1, 1) model and the GRNN model to obtain the predicted values x_{1t} and x_{2t}, and the gray relational method is used to calculate the corresponding weights l_{1t} and l_{2t} in the individual models. According to Equation (25), we can obtain the predicted value of a combined model for primary energy consumption at time t.

Due to the GRNN model, the data for the original variables in the input layer can be traced back to 1980 at the earliest. To improve the prediction accuracy of the model, the forecast values for primary energy consumption in the GM (1, 1) model and GRNN model from 1980 to 2016 are selected as the reference sequences and the gray relational degrees of the two reference sequences and original sequence are calculated, respectively. The gray relational degree of the GM (1, 1) model is calculated as $l_1 = 0.7165$, and the gray relational degree of the GRNN model is $l_2 = 0.7281$. Thus, the results of the combined forecasting are shown in Table 5.

The gray relational degree of the combined forecasting model ξ is 0.7368, and according to $\xi > \xi_{02} > \xi_{01}$, the combined forecasting model is considered the optimal combined forecasting model. Among the models, the MAPE of the GM (1, 1) model is 6.92%, and the MAPE of the GRNN model is 6.18%, while the MAPE of the combined forecasting model is 5.87%, the prediction accuracy is improved, and the combined forecasting method is superior to the single prediction method. Therefore, this paper uses the combined forecasting method to predict China's primary energy consumption from 2017 to 2030.

Table 5 shows the forecasting results of China's primary energy consumption in 2017–2030 under the GM (1, 1) model, the GRNN model and the combined model, respectively. It can be seen that the prediction results of GM (1, 1) model have a fast growth rate, while the prediction results of GRNN model are more robust. The forecasting results of the two models are relatively close in 2017. However, the differences between the predicted values of the two models become larger and larger. By 2030, the forecasting value of GM (1, 1) model is nearly two times more than that of GRNN model. The results are mainly determined by the characteristics of each model. The GM (1, 1) model has the advantages of small sample size, less parameter requirements, and simple calculation, etc. However, the GM (1, 1) model is more suitable for a smooth data sequence with exponential change, but for data with a jumping nature or a rapidly changing speed, the forecasting accuracy is not high [59]. The GRNN model has a strong nonlinear mapping ability and learning speed. The neural network usually converges to the optimal regression with large sample size aggregation, which is suitable for processing unstable data and lone-term prediction. However, the model often has high requirements on sample quantity and quality [60]. In view of the characteristics of the GM (1, 1) model and the GRNN model, this paper uses the combined model to forecast the energy consumption, so as to better make up for the defects of each model and to improve the prediction accuracy. According to Figure 2, the fitting of the original data shows that the MAPE under the combined model is smaller than that of the GM (1, 1) model and the GRNN model, which indicates that the prediction accuracy of the combined model proposed in this paper is improved compared with the traditional single forecasting model.

As shown in Table 5 and predicted with the combined forecasting model, China's primary energy consumption in 2020 will reach 5.06 billion tce, and the primary energy consumption in 2030 will reach 7.54 billion tce. Furthermore, the forecasting results of this study are compared with those of other studies, which are shown in Appendix B. In this paper, the forecasting result for 2020 is close to that of British Petroleum (BP), and the forecasting result for 2030 is close to the high economic growth scenario of the Energy Information Administration (EIA). The predictions of 2020 and 2030 are both less than that of the South Korea Scenario and the Baseline Scenario, which implies that the prediction results of those two scenarios may be high. The comparison further proves that our results are robust and reliable.

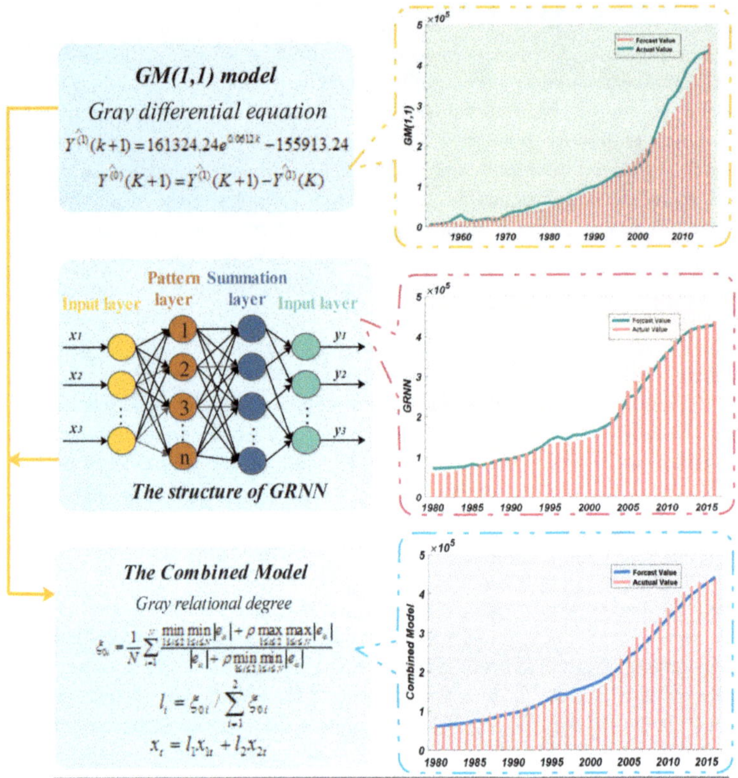

Model	GM(1,1)	GRNN	Combined Model
MAPE	6.92%	6.18%	5.87%

Figure 2. The forecasting process for the combined model.

Table 5. Prediction of primary energy consumption. (unit: 10^4 tce).

Year	GM (1, 1)	GRNN	The Combined Model
2017	481,509.07	428,574.45	455,252.80
2018	511,903.92	430,308.03	471,431.29
2019	544,217.41	431,775.91	488,444.95
2020	578,570.67	432,981.45	506,356.51
2021	615,092.45	433,770.76	525,154.52
2022	653,919.63	434,393.88	545,031.98
2023	695,197.75	434,867.40	566,070.49
2024	739,081.51	435,216.47	588,360.47
2025	785,735.40	435,467.43	611,997.90
2026	835,334.27	435,645.99	637,083.65
2027	888,064.03	435,768.62	663,719.59
2028	944,122.31	435,851.30	692,013.24
2029	1,003,719.22	435,906.06	722,076.46
2030	1,067,078.13	435,941.71	754,026.21

According to existing studies, China's primary energy consumption will continue to grow over the next 14 years. Adjusting the energy structure is a key factor in achieving the carbon intensity targets in 2020 and 2030.

4. China's Energy Consumption Structure Forecast Results

4.1. Different Scenario Settings

(1) Unconstrained Scenario (UCS)

The unconstrained scenario (UCS) is the scenario in which no specific measures are taken to reduce carbon intensity. At present, China is in the process of building a prosperous society in a comprehensive way. Industrialization is in a transition period, and it is heavily dependent on energy consumption. The coal-dominated energy consumption structure exerts tremendous ecological pressure on the environment. In recent years, due to the state's support and technological progress, the proportion of coal consumption has been gradually declining, the proportion of renewable energy consumption has been gradually rising, and the energy consumption structure has been gradually improving. This paper sets the unconstrained energy structure optimization scenario as a reference, which does not consider the national development plan and predicts the future development trend based on the natural evolution law, followed by the energy structure change.

(2) Policy-Constrained Scenario (PCS)

The policy-constrained scenario (PCS) is the scenario that optimizes the energy consumption structure according to energy policies. For a long time, China's primary energy consumption consisted of nearly 70% coal. To ease the pressure on carbon emissions, China launched the 13th Five-Year Plan in 2017 to clarify the guiding ideology, the basic principles, and the development goals for energy development. The plan proposes that non-fossil fuels account for 15% of primary energy consumption by 2020, the proportion of natural gas consumption reaches 10%, the proportion of coal is less than 62%, and that non-fossil fuels account for 20% of primary energy consumption by 2030. This paper sets this scenario, combined with the above energy development plan, as the policy-constrained scenario, and predicts the future energy consumption structure according to China's energy planning.

(3) Minimum external costs of carbon emissions scenario (MCS)

The minimum external costs of carbon emissions scenario (MCS) is a scenario that optimizes the energy structure with the objective of minimizing external costs. Given that carbon dioxide is the primary greenhouse gas, the external discharge of carbon dioxide has a negative impact on the environment and generates a certain cost. Therefore, from the cost perspective, this paper applies a minimum external cost of the carbon emissions scenario, which sets the minimum cost as the goal, determines the constraint conditions according to national energy policy planning, and constructs a non-linear programming model. By generating the non-linear programming model, we can obtain the results of energy structure optimization in this scenario.

4.2. Energy Structure Forecast Results

Based on the above three scenario settings, the energy structures in different scenarios are calculated separately.

(1) Energy structure prediction under the unconstrained scenario

In this paper, the energy structure is predicted with the Markov forecasting model. The energy structure variables from 2008 to 2015 are selected to calculate the average transfer probability matrix P (Appendix C).

Given the primary energy consumption structure and the average transition probability matrix P, we predict the energy consumption structure in the unconstrained scenario. Combined with the prediction for GDP and the total primary energy consumption (Table 5), we obtain the forecast results for the total energy consumption and the prediction of various types of energy in the future. The results are shown in Table 6.

In the unconstrained scenario, non-fossil energy will account for 14.1579% of the total energy consumption in 2020, natural gas will account for 7.57% in 2020, and non-fossil fuels will account for 17.4589% in 2030; these figures still lag behind the goals set in the 13th Five-Year Plan. Therefore, the energy structure under the policy-constrained scenario is adjusted as follows.

(2) Energy Structure Prediction under the Policy-Constrained Scenario

At present, coal accounts for the highest proportion of primary energy consumption in China. Compared to other energy sources, coal has low calorific value and high carbon emissions per unit. One of the principal ways to optimize energy structure is to reduce the proportion of coal and increase the proportion of renewable energy. Therefore, this paper assumes that in the future energy structure, the proportion of oil consumption will continue the same trend of adjustment. The increased ratio of clean energy and natural gas will be supplemented by a decrease in the coal proportion. Based on the above assumptions, the energy consumption structure optimization results can be calculated under the policy-constrained scenario. Table 6 shows that in this scenario, the primary energy consumption structure in 2020 will be 55.80% coal, 19.20% oil, 10% natural gas, and 15% clean energy; the primary energy consumption structure in 2030 will be 48.23% coal, 20.93% oil, 10.84% natural gas, and 20% clean energy.

(3) Energy Structure Prediction under the Minimum External Costs of a Carbon Emissions Scenario

From the cost perspective, this paper sets the minimum external cost of carbon emissions as a decision-making goal, and builds a non-linear programming model to predict the primary energy consumption structure. According to the analysis of the objective function and constraints in Section 2.2.2, we obtain the minimum scenario model of carbon emission costs in 2020 and the minimum scenario model of carbon emissions in 2030 (Appendix D).

Table 6 shows that the primary energy consumption structure in 2020 is 59% coal, 20% oil, 6% natural gas and 15% clean energy. The primary energy consumption structure in 2030 is 45% coal, 25% oil, 10% natural gas and 20% clean energy.

Table 6. China's primary energy consumption structure forecast results.

Year	2020			2030		
	UCS	PCS	MCS	UCS	PCS	MCS
Coal (10^4 tce)	299,115.43	282,541.36	298,750.34	382,843.23	363,683.43	339,311.79
Share (%)	59.07	55.80	59.00	50.77	48.23	45.00
Oil (10^4 tce)	97,226.02	97,226.02	101,271.30	157,828.24	157,828.24	188,506.55
Share (%)	19.20	19.20	20.00	20.93	20.93	25.00
Natural gas (10^4 tce)	38,325.11	50,635.65	30,381.39	81,709.30	81,709.30	75,402.62
Share (%)	7.57	10.00	6.00	10.84	10.84	10.00
Clean energy (10^4 tce)	71,689.45	75,953.48	75,953.48	131,644.68	150,805.24	150,805.24
Share (%)	14.16	15.00	15.00	17.46	20.00	20.00
Total (10^4 tce)	506,356.51	506,356.51	506,356.51	754,026.21	754,026.21	754,026.21

From the total primary energy consumption in China under different scenarios and their structures, the consumption levels of coal, oil, natural gas, and clean energy under various scenarios in 2020 and 2030 are calculated as shown in Figure 3. From Figure 3, we can observe the following. (1) In 2030, all types of primary energy consumption in China are higher than primary energy consumption in 2020, indicating that China's energy consumption will increase during this period. (2) In the same year, the coal consumption under the policy-constrained scenario is lower than that in the unconstrained scenario, indicating that at present, the energy consumption structure has a problem with an excessively high proportion of coal consumption. (3) From 2020 to 2030, under the unconstrained scenario, the proportion of coal consumption will decrease, and the proportions of oil, natural gas, and clean energy consumption will increase. Thus, China's energy structure has been optimized under natural evolution.

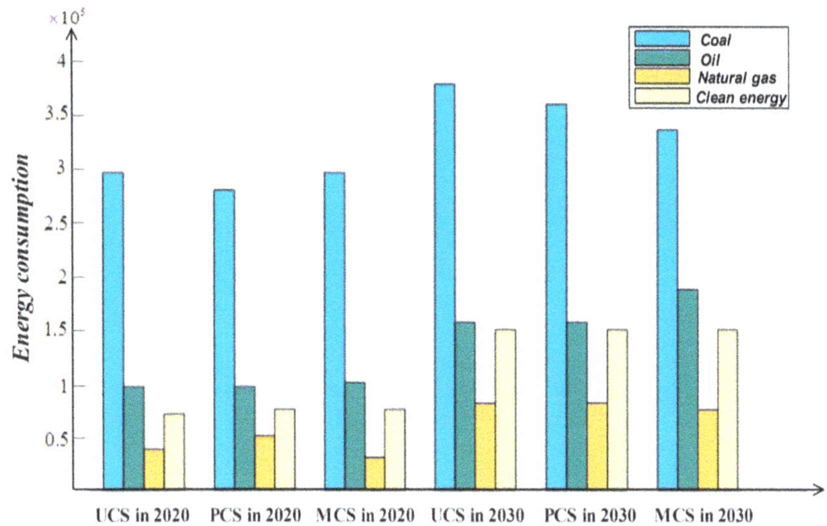

Figure 3. Energy consumption under different scenarios in China in 2020 and 2030.

5. Contribution Analysis of Energy Structure Optimization to Achieve the Carbon Intensity Targets

In 2005, China's GDP was 4,797.58 billion RMB (1980 constant price). By burning fossil fuel energy sources, the total amount of carbon dioxide emitted was 5.533 million tons, and the carbon intensity was 1.15 kg/RMB.

Based on the energy consumption in Table 6 and Equation (34), the total amount of carbon dioxide emitted from primary energy combustion in each scenario is estimated.

Given the prediction for China's national economy (GDP) in Section 3.2 of this paper and the carbon intensity in each scenario calculated with Equation (35), we can analyze the potential contribution of energy structure optimization to achieving the carbon intensity targets. The results are shown in Table 7 and Figure 4.

Table 7. Energy structure optimization results under various scenarios.

Year	2020			2030		
Scenario	UCS	PCS	MCS	UCS	PCS	MCS
GDP (billion)	164,048.47	164,048.47	164,048.47	280,218.49	280,218.49	280,218.49
Carbon emissions (10^4tce)	1,089,917	1,064,503	1,084,638.26	1,519,418	1,466,897	1,455,357.30
Carbon intensity (kg/RMB)	0.66	0.65	0.66	0.54	0.52	0.52
Degree of decline (%)	42.39	43.74	42.67	52.98	54.61	54.97

(1) Carbon Intensity Prediction Results in each Scenario

The carbon intensity prediction results in each scenario are shown in Table 7. Under natural evolution, the carbon intensity in 2017 is 0.74 kg/RMB. The carbon intensity will decline to 0.66 kg/RMB by 2020 under the unconstrained scenario, a total decrease of 10.81% from 2017, and the carbon intensity will decline to 0.54 kg/RMB by 2030, a total decrease of 27.03% from 2017. Under the policy-constrained scenario, the carbon intensity will decline to 0.65 kg/RMB by 2020, a total decrease of 12.16%, and the carbon intensity will decline to 0.52 kg/RMB in 2030, a total decrease of

29.07%. Additionally, under the minimum external cost scenario, the carbon intensity will decline to 0.66 kg/RMB by 2020, a total decrease of 11.63%. In 2030, the carbon intensity will decline to 0.52 kg/RMB, a total decrease of 29.63%.

(2) Contribution Analysis of Optimizing the Energy Structure to Realize the Carbon Intensity Targets

Based on the results, the contribution potential of optimizing the energy structure that drives the carbon intensity relative to the decrease in 2005, and the goal of achieving carbon intensity, are calculated. The contribution potential refers to the ratio of the decrease of carbon intensity in each energy structure optimization scenario relative to the target of the carbon intensity reduction. The carbon intensity target is deemed to be a range, so that the calculated contribution potential is also a range.

China set the target for carbon intensity to decline using 2005 as the base year, and the carbon intensity in 2005 was 1.15 kg/RMB. If the carbon intensity declines by 40–45% from 2005 to 2020, the carbon intensity must be reduced to 0.63–0.69 kg/RMB. If the carbon intensity in 2030 is 60–65% lower than that in 2005, the carbon intensity must decline to 0.40–0.46 kg/RMB.

Figure 4 comprehensively reflects the energy structure optimization results under various scenarios. The figure contains four graphs that reflect the predicted values of carbon intensity in each scenario for 2020 and 2030, and their potential contributions to achieving the carbon intensity targets. Among the graphs, Figure 4a,b provide the total amount of carbon dioxide emissions and carbon intensity results under the various scenarios in 2020 and 2030, respectively. Figure 4c shows the decrease in carbon intensity in each scenario between 2020 and 2030 compared to 2005. Figure 4d shows the potential contribution of the structure optimization of energy consumption to achieving the carbon intensity targets in the 2020 and 2030 scenarios.

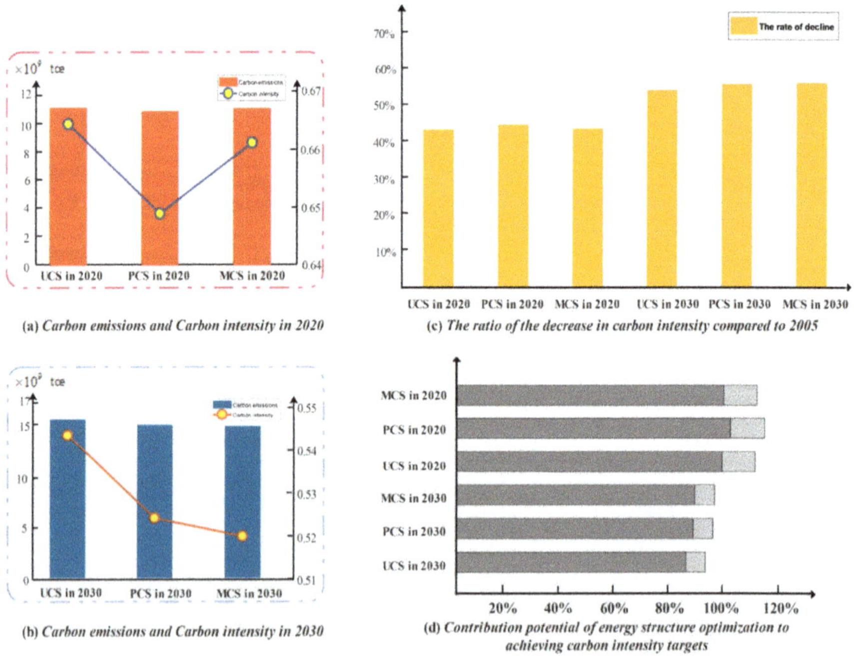

Figure 4. Energy structure optimization results under all scenarios.

Considering the results from Table 7, in 2020, the carbon intensity predicted by the unconstrained scenario is 0.66 kg/RMB, a decrease of 42.39% from 2005; the carbon intensity of the policy-constrained scenario is estimated to be 0.65 kg/RMB, a decrease of 43.74% from 2005; and the carbon intensity predicted by the minimum external costs scenario is 0.66 kg/RMB, a decrease of 42.67% from 2005. We can observe that the carbon intensity consequences predicted by the three scenarios are all within the target range; the energy structure optimization achieves 100% of the carbon intensity target in all scenarios, and the carbon intensity target for 2020 can be completely achieved.

In 2030, the carbon intensity predicted by the unconstrained scenario is 0.54 kg/RMB, which is a decrease of 52.98% from 2005, with a potential contribution of 81.51–88.31% to achieving the carbon intensity target. The carbon intensity of the policy-constrained scenario is 0.52 kg/RMB, which a decrease of 54.61% from 2005, with a potential contribution of 84.01–91.01%. The carbon intensity of the minimum external cost scenario is predicted to be 0.52 kg/RMB, which is 54.57% lower than that of 2005, with a potential contribution of 84.56–91.61%. To achieve the carbon intensity target for 2030, the carbon intensity must be reduced to 0.40–0.46 kg/RMB. We can observe that under the three scenarios, all failed to achieve the carbon intensity target of 2030 by relying on energy structure optimization; however, the contribution potential of energy structure optimization is greater than 80%. By further implementing other measures to save energy and reduce emissions, it is very possible to reach the goal of a 60–65% reduction in carbon intensity by 2030.

According to Table 6, the consumption proportion of natural gas is 10–11%, and that of oil is 20–25% in 2030. The proportion of oil is much higher than that of natural gas. It is known that natural gas is a relatively clean fossil energy. Under the same standard, the CO_2 produced by burning one ton of coal is 30% more than oil, and 70% more than natural gas. That is, natural gas emits the least amount of carbon dioxide when it generates the same amount of heat. According to the results of this paper, the predicted energy structure optimization fails to achieve the carbon intensity target by 2030. Therefore, as an efficient, clean, and high-quality fossil energy, increasing the proportion of natural gas consumption is one of the best choices to optimize the energy structure, improve energy efficiency, and achieve the carbon intensity target [61]. In addition, in the process of achieving energy-saving and emission reduction targets, natural gas has more cost advantages and technological advantages than new clean energy such as wind energy and nuclear energy. It is less restricted by natural conditions such as time and region. Natural gas has great potential in optimizing energy structure. At present, natural gas is a lagging energy in China, and the reason is that the price is on the high side. In the future, it will be necessary for the Chinese government to further strengthen the reform of natural gas price market, to change the situation in which resources such as natural gas exploration, to import channels and bargaining negotiations are highly concentrated in the three state-owned companies, to open up the natural gas market, and to encourage more market institutions to participate in. Higher market-oriented offshore LNG projects can also be used to increase the development and utilization of unconventional natural gas, in order to increase the consumption proportion of natural gas.

Through a comparison of the prediction results for the three energy structure optimization scenarios, the following conclusions are also obtained:

(1) The scenario with the lowest predicted carbon intensity in 2020 is the policy-constrained scenario. This scenario has the fewest total carbon emissions. Compared with the unconstrained scenario, CO_2 emissions will be reduced by 254.14 million tons in 2020 under the policy-constrained scenario, which is equivalent to 92.71 million tons of standard coal.
(2) The scenario with the lowest predicted carbon intensity in 2030 is the minimum external costs of the carbon emissions scenario. This scenario has the least total carbon emissions. Compared with the unconstrained scenario, CO_2 emissions will decrease by 640.607 million tons in 2030, which is equivalent to 233.7 million tons of standard coal.
(3) The comparison of different scenarios in the same year shows that the best energy structure optimization effect in 2020 occurs through the policy-constrained scenario. Coal consumption in this scenario accounts for 55.8%. The best energy structure optimization effect in 2030 occurs

through the minimum external costs of carbon emissions scenario, in which the coal consumption accounts for 45%. The above two scenarios are the scenarios with the lowest proportion of coal consumption in the same year. The reason for this result is perhaps that, compared with other types of energy, coal has low calorific value and high unit carbon emissions, and it is a poor-quality energy source. These characteristics indicate that reducing the proportion of coal consumption is critical to achieving the carbon intensity targets.

(4) The natural evolution of the energy consumption structure in 2020 will achieve the planned target of less than 62% of coal consumption, but it will not meet the target of 10% of natural gas consumption or 15% of non-fossil energy consumption. When the energy consumption structure evolves in 2030, the proportion of natural gas consumption will just reach the 2020 planning goal. Although non-fossil energy consumption accounts for more than 15% of total energy consumption, the target of 20% of consumption has not been achieved. All these scenarios show that China's current energy consumption transformation and upgrading process is slow, and that the goal is still arduous. It is necessary to further increase the energy conversion rate and upgrade energy technologies.

(5) Based on the results shown in Table 7, under the state of sustained economic growth, the carbon intensity in 2020 calculated under the three energy structure optimization scenarios will decrease by 42.39–43.74% compared with 2005, and the carbon intensity target of 2020 can be successfully realized. The calculated carbon intensity in 2030 will decrease by 52.98–54.97% compared with 2005, which is a gap of nearly 10% to achieve the carbon intensity target of 2030. An important reason for this result is that coal is still the most important source of energy consumption in the three scenarios, as shown in Figure 3. This indicates that the optimization of China's energy structure is not sufficient and needs to be further strengthened. To achieve the carbon intensity target by 2030, China needs to develop renewable energy vigorously, reduce the proportion of coal, replace fossil energy with non-fossil energy, and replace coal with natural gas in fossil energy, so as to promote low-carbon diversification of energy structure and to realize sustainable energy development.

6. Conclusions and Policy Suggestions

6.1. Conclusions

This paper studies China's primary energy consumption structure based on the carbon intensity targets for 2020 and 2030. First, primary energy consumption is predicted with a combined forecasting model. The single forecast is generated using the GM (1, 1) model and GRNN model, and the single model is weighted using the gray relational degree method to obtain the primary energy consumption for 2017–2030. Furthermore, we compare China's primary energy consumption predictions between this study and others, as shown in Appendix B. Second, from the perspectives of natural evolution, policy planning and cost, we set three scenarios to optimize the energy structure and obtain the prediction results. Finally, we analyze the energy structure and its contribution to achieving the carbon intensity targets in each scenario.

According to the above study, we can draw the following conclusions. (1) The estimated primary energy consumption is 5.06 and 7.54 billion tce in 2020 and 2030, respectively. Combined with the forecasting results of other studies, we can conclude that China's primary energy consumption will continue to grow over the next 14 years. (2) The carbon intensity target for 2020 can be achieved under the unconstrained scenario, policy-constrained scenario and minimum external costs of the carbon emissions scenario; among these scenarios, the predicted carbon intensity decreases the most and the carbon emissions are the lowest under the policy-constrained scenario. (3) The carbon intensity target will almost be attained in 2030 under the optimized energy structure in different scenarios, and so further emission reduction efforts are still required. In 2030, the predicted carbon intensity will decrease the most under the minimum external costs of the carbon emissions scenario, with a contribution potential of 84.56–91.61%. Under the unconstrained scenario, the predicted carbon intensity is 52.98% lower than that in 2005, with a potential contribution of 81.51–88.31%. Under the policy-constrained scenario, the predicted carbon intensity is 54.61% lower than that in 2005, with a contribution potential of 84.01–91.01%. (4) To achieve the carbon intensity targets, it is necessary to further develop clean energy and promote a change in the energy consumption structure from mainly coal in order to classify the consumption of coal, oil, natural gas, and clean energy.

6.2. Policy Suggestions

Based on the above conclusions, we propose the following policy suggestions:

(1) Control the primary energy consumption and reduce coal consumption.

According to the forecasting results of our study, China's primary energy consumption will continue to grow over the next 14 years. To achieve the carbon intensity targets in 2020 and 2030, controlling total energy consumption is a key measure. First, an overall national total primary energy consumption plan should be made, and the energy consumption targets should be assigned to each province according to equity principles and regional development strategies by taking full account of the economic development level and the industrial structure of each province. Second, the local government assessment systems should be improved, and the government and enterprises should be supervised to accomplish the established targets of energy conservation and emission reduction.

In addition, the forecasting results show that coal consumption will account for more than 45% of total energy consumption in 2030, which is still higher than the world average (30%). The excessive proportion of coal has put increasing pressure on the environment and contributed to climate change. To adjust the energy structure, total coal consumption should be controlled in key areas, such as the Pearl River Delta, Yangtze River Delta, and Beijing–Tianjin region, by gradually phasing out or upgrading coal-fired power stations, strictly limiting the growth of new heavy industries based on coal consumption, and promoting industrial equipment and technologies.

(2) Increase oil and natural gas consumption and develop renewable energy.

The proportion of oil and natural gas is relatively low in China's energy consumption structure. China has abundant oil resources, and given the monopoly of China's oil market and the pressure of large-scale oil imports to energy supply security, it is necessary to fully utilize the oil resources to meet basic demands. Meanwhile, the development of the natural gas industry must be accelerated, including the excavation of natural gas resources, introducing advanced technology and equipment, speeding up the construction of domestic natural gas pipelines, and strengthening the cooperation with neighboring countries regarding natural gas resources.

Based on the forecasting results of the energy consumption structure, it is clear that China's energy consumption is overly dependent on fossil fuels, which is not sustainable. Vigorously developing new types of clean energies, such as hydropower, wind, solar, and nuclear, is necessary in order to achieve a low-carbon economy. First, policy and financial support for renewable energies should be strengthened. Second, advanced foreign technologies should be introduced to the clean energy industry, and they should help clean energy enterprises to improve their technological level. Meanwhile, increasing the investment in Research and Development (R&D) in new energy fields is also significant to reduce the costs of renewable energy.

(3) Improve energy efficiency and develop carbon emissions reduction technology.

Given the conclusions of this study, energy structure optimization cannot fully achieve China's carbon intensity target by 2030. Thus, it is important to improve the energy efficiency and to develop corresponding technology. First, energy waste should be reduced, and the efficiency of energy deployment should be increased by further improving the reform of the energy price mechanism and linking domestic energy prices with international energy prices. Second, it is also important to learn from the advanced experience in emission reduction in developed countries, and to establish a sound legal system to supervise enterprises to reduce energy consumption and improve the efficiency of energy use. Third, the unified carbon trading market should be developed, and carbon taxes levied, in order to accelerate the green transformation of energy-intensive enterprises.

To develop carbon emissions reduction technology, financial support should be given to adopt environment-protecting technologies, and to formulate a sound low-carbon transition development evaluation system. Meanwhile, carbon capture, utilization, and storage (CCUS) is an important strategic choice for carbon reduction in China. Special funds and compensation mechanisms should be established for CCUS as soon as possible.

Author Contributions: Y.W. was mainly responsible for the writing of the full text. D.L. conceived and designed the study. P.S., L.H., and Y.Z. built the models of the paper.

Funding: This paper was supported by Humanities and Social Sciences Research Project of the Ministry of Education (18YJC910013), National Natural Science Foundation of China (71573034), Liaoning Social Science Fund (L17CTJ001, L17BJY042), China Postdoctoral Science Fund (2016M601318, 2017T100180), and the Research Project of Dongbei University of Finance and Economics (DUFE2017Q16), Natural Science Foundation of China (Nos. 71502026, 71872033).

Conflicts of Interest: The authors declare no conflict of interest.

Appendix A

Table A1. Summary of the aforementioned studies.

Author	Time	Research Topics	Methodologies	Main Results
Tan et al. [6]	1998–2008	Explore the driving forces for reducing China's CO_2 emission intensity	LMDI model	The decomposition results show that improvements in the energy intensity of power generation, electricity intensity of GDP, and energy intensity of GDP for other activities were mainly responsible for the success in reducing China's CO_2 emission intensity.
Huang et al. [7]	1996–2009	Carbon intensity factor decomposition	LMDI model	In the long run, optimizing the energy structure and the industrial structure is the fundamental way to reduce carbon intensity.
Xu et al. [8]	2001–2011	Decomposition analysis of factors affecting CO_2	PDA model and IDA model	Economic activity is the main cause of substantial increases in carbon dioxide emissions in China. Different driving factors affect carbon emissions in different provinces.
Su et al. [10]	1997–2000	China's regional emission embodiments	Input–output model	The paper combines the HEET approach and stepwise distribution analysis to allocate the carbon emission reduction targets.
Zhu et al. [11]	1980–2007	Decomposition of carbon emissions from energy consumption	Kaya model and the LMDI model	The economic output effect makes the greatest contribution to the carbon emissions from energy consumption in this stage in China. The population scale effect and the industrial structure effect are positive, and the energy intensity effect and energy structure effects are negative.
Raupach et al. [12]	2000–2004	The area and factor decomposition of carbon emissions	Kaya model	In 2000 to 2004, developing countries had a greater share of emissions growth than of emissions themselves.
Wang et al. [13]	1980–2010	Analyze the impact factors of CO_2 emissions in Guangdong province	STIPPRAT model	Population is the most important impact factor of CO_2 emissions. Industrialization level, urbanization level, energy consumption structure, service level, and GDP per capita are also significant impact factors.
Yue et al. [14]	2010–2020	Find the influencing factors and calculate CO_2 emissions	Expanded IPAT model and decomposition method	Rapid economic growth is the main determinant that results in increasing CO_2 emissions in Jiangsu province.
Liu et al. [15]	2008–2020	Investigate the energy consumption and CO_2 emissions trend	System dynamic analysis	The total energy consumption and CO_2 emissions in 2020 are 1.6 and 1.9 times higher than 2008 figures.
Parag Sen et al. [16]	2002–2013	Forecast energy consumption and GHG emission	ARIMA model	ARIMA (0,1,0) × (0,1,1) is the best model to forecast energy consumption.
Wang et al. [17]	2014–2020	Forecast Chinese carbon emissions	Non-linear grey model	Compared with other models, the TNGM (1, 2) has the smallest predicted error, and the prediction results can be obtained from this model.

Table A1. *Cont.*

Author	Time	Research Topics	Methodologies	Main Results
Nabavi-Pelesaraei et al. [18]	2013–2014	Forecast and sensitivity analysis of energy input and GHG emissions	Artificial neural networks model	The total energy consumption and energy output for kiwifruit production were 37.32 GJ ha^{-1} and 43.44 GJ ha^{-1}, respectively. The highest energy consumption among all energy sources was made by chemical fertilizers.
Ajay et al. [19]	2005–2050	Forecast China's carbon dioxide emission	The MESSAGE model and the Grantham Institute model	The total energy consumption and energy output for kiwifruit production were 37.32 GJ ha^{-1} and 43.44 GJ ha^{-1}, respectively. The highest energy consumption among all energy sources was by chemical fertilizers (mainly nitrogen, with 45.61%).
Yongrok et al. [20]	2001–2010	Estimate the energy efficiency, potential emission reductions and marginal abatement costs of energy related CO_2 emissions	DEA model	Most provinces in China are not performing energy efficiently, and Hebei and Shandong are the two provinces with the largest energy consumption and have profound energy use reduction potential.
Du et al. [21]	2006–2010	The estimation of the CO_2 emission efficiency and the potential emission reduction	Nonparametric metafrontier model	The potential CO_2 emission reduction is 1687 million tons, and over half of the potential emission reduction is caused by the technology gap.
Stern et al. [27]	2005–2020	Examine the carbon intensity targets of China and India in 2020	Stochastic frontier model	China is likely to need to adopt ambitious carbon mitigation policies in order to achieve its target; India's target is less ambitious.
Yi et al. [28]	2005–2020	Analyze the determining factors and diverse scenarios of CO_2 emissions intensity reduction	Extended Kaya model and scenario analysis	The carbon intensity target for 2020 is very likely to be achieved in China.
Xiao et al. [29]	2005–2020	Examine the carbon intensity targets of China in 2020	System dynamics model	The Chinese government can accomplish the target under carbon tax policy scenario and integrated policy scenario.
Yuan et al. [30]	2005–2020	Investigate clean energy targets	Scenario analysis	A 17% clean energy target can achieve the carbon intensity target of 2020 and exceed the IEA 450 ppm scenario.
Wang et al. [31]	2005–2020	Discuss low-carbon energy policies at the province level	Reactor model and scenario analysis	Due to their different energy structures, Chinese provinces should evaluate their own unique situations and how they relate to carbon emissions.
Yi et al. [32]	2005–2020	Allocate the CO_2 reduction target regionally	Index analysis and top-down model	Shanghai, Hebei, Shanxi, Shandong, Guangdong and Liaoning provinces may undertake greater burdens to achieve the 45% intensity reduction target by 2020.
Xu et al. [33]	2005–2030	Predict carbon emissions and carbon intensity	GM (1,1) model and STIPPRAT model	China's carbon intensity reduction targets in 2020 and 2030 can be met under current policies.
Zhang et al. [34]	1993–2035	Determine whether and how the carbon intensity targets will be realized from a sector-specific perspective	Dynamic Monte Carlo simulation and scenario analysis	It is very possible for the industrial sector to achieve the 2020 and 2030 intensity-reduction targets.

Appendix B

Table A2. Comparison of China's primary energy consumption predictions between this study and others. (unit: 10^4 tce).

Authors	Model	Scenarios	2020	2030
(Energy Information Administration) EIA [62]	World Energy Projections plus, WEPS+	High economic growth	619,384.40	738,768.70
		Low economic growth	605,657.20	679,284.50
		High oil price	609,401.00	731,281.20
		Low oil price	611,480.90	686,356.10
		Reference case	612,728.80	708,818.60
(British Petroleum) BP [63]	-	-	494,740.00	594,197.00
Shen et al. [64]	Exponential Smoothing (ES)	United States scenario	455,300.00	447,700.00
		United Kingdom scenario	419,100.00	432,800.00
		Germany scenario	454,200.00	429,400.00
		France scenario	518,900.00	564,300.00
		Japan scenario	458,300.00	577,300.00
		South Korea scenario	649,900.00	830,300.00
		Baseline scenario	617,100.00	892,500.00
Wang et al. [65]	Self-adaptive MVO-SVM	High economic growth	484,701.00	545,365.40
		Low economic growth	482,657.50	581,781.10
		High coal consumption	493,987.50	620,522.90
		Low coal consumption	467,364.20	553,733.10
		Reference case	483,929.90	565,624.80
Authors	Combined model	-	506,356.51	754,026.21

C.

Under the unconstrained scenario, the average transfer probability matrix P is:

$$P = \begin{bmatrix} 0.9813 & 0.0055 & 0.0036 & 0.0095 \\ 0.0035 & 0.9934 & 0.0024 & 0.00075 \\ 0 & 0 & 1 & 0 \\ 0.0083 & 0 & 0.005 & 0.9867 \end{bmatrix}$$

D.

I The non-linear programming model under the minimum external costs of carbon emissions scenario in 2020:

$$\min f(x) = 354.81x_1 + 276.46x_2 + 210.49x_3$$
$$s.t \begin{cases} x_1 + x_2 + x_3 + x_4 \le 506356.51 \\ (0.7476x_1 + 0.5825x_2 + 0.4435x_3) \times \frac{44}{12} \le 1089917 \\ 0.5 \le x_1 / \sum\limits_{i=1}^{4} x_i \le 0.6 \\ 0.2 \le x_2 / \sum\limits_{i=1}^{4} x_i \le 0.3 \\ 0.06 \le x_3 / \sum\limits_{i=1}^{4} x_i \le 0.15 \\ 0.15 \le x_4 / \sum\limits_{i=1}^{4} x_i \le 0.25 \end{cases}$$

II The non-linear programming model under the minimum external costs of carbon emissions scenario in 2030:

$$\min f(x) = 354.81x_1 + 276.46x_2 + 210.49x_3$$
$$s.t \begin{cases} x_1 + x_2 + x_3 + x_4 \le 754026.21 \\ (0.7476x_1 + 0.5825x_2 + 0.4435x_3) \times \frac{44}{12} \le 1519418 \\ 0.4 \le x_1 / \sum\limits_{i=1}^{4} x_i \le 0.5 \\ 0.25 \le x_2 / \sum\limits_{i=1}^{4} x_i \le 0.3 \\ 0.10 \le x_3 / \sum\limits_{i=1}^{4} x_i \le 0.20 \\ 0.2 \le x_4 / \sum\limits_{i=1}^{4} x_i \le 0.25 \end{cases}$$

References

1. Team, C.W.; Pachauri, R.K.; Meyer, L.A. Climate Change 2014: Synthesis Report. Contribution of Working Groups I, II and III to the Fifth Assessment Report of the Intergovernmental Panel on Climate Change. *J. Roman. Stud.* **2014**, *4*, 85–88.
2. Dong, F.; Li, X.H.; Long, R.Y.; Liu, X.Y. Regional carbon emission performance in China according to a stochastic frontier model. *Renew. Sustain. Energy Rev.* **2013**, *28*, 525–530. [CrossRef]
3. Nyakundi, M.M.; Jerald, F.; Wesley, B. An empirical analysis of the role of China's exports on CO_2 emissions. *Appl. Energy* **2013**, *104*, 258–267.
4. Liu, L.W.; Zong, H.J.; Zhao, E.D.; Chen, C.X.; Wang, J.Z. Can China realize its carbon emission reduction goal in 2020: From the perspective of thermal power development. *Appl. Energy* **2014**, *124*, 199–212. [CrossRef]
5. Chen, W.D.; Wu, F.Y.; Geng, W.X.; Yu, G.Y. Carbon emissions in China's industrial sectors. *Resour. Conserv. Recyl.* **2016**, *117*, 264–273. [CrossRef]
6. Tan, Z.F.; Li, L.; Wang, J.J.; Wang, J.H. Examining the driving forces for improving China's CO_2 emission intensity using the decomposing method. *Appl. Energy* **2011**, *88*, 4496–4504. [CrossRef]

7. Huang, F.; Jiang, K.S.; Lu, Y.Q.; Wu, Y. Factor Decomposition of Carbon Intensity in China: Based on LMDI Method. *Math. Pract. Theory* **2012**, *42*, 40–46.
8. Xu, G.Q.; Liu, Z.Y.; Jiang, Z.H. Decomposition Model and Empirical Study of Carbon Emissions for China,1995–2004. *China Popul. Resour. Environ.* **2016**, *16*, 158–161.
9. Wang, Y.; Zhou, Y.; Zhu, L.; Zhang, F.; Zhang, Y.C. Influencing Factors and Decoupling Elasticity of China's Transportation Carbon Emissions. *Eneriges* **2018**, *11*, 1157. [CrossRef]
10. Su, B.; Ang, B.W. Input–output analysis of CO2 emissions embodied in trade: A multi-region model for China. *Ecol. Econ.* **2014**, *114*, 377–384. [CrossRef]
11. Zhu, Q.; Peng, X.Z.; Lu, Z.M.; Wu, K.Y. Factors Decomposition and Empirical Analysis of Variations in Energy Carbon Emission in China. *Resour. Sci.* **2009**, *31*, 2072–2079.
12. Michael, R.R.; Gregg, M.; Philippe, C.; Corinne, L.Q.; Josep, G.C.; Gemot, K.; Christopher, B.F. Global and Regional Drivers of Accelerating CO_2 Emissions. *Proc. Natl. Acad. Sci. USA* **2007**, *104*, 10288–10293.
13. Wang, P.; Wu, W.S.; Zhu, B.Z.; Wei, Y.M. Examining the impact factors of energy-related CO_2 emissions using the STIRPAT model in Guangdong Province, China. *Appl. Energy* **2013**, *106*, 65–71. [CrossRef]
14. Yue, T.; Long, R.Y.; Chen, H.; Zhao, X. The optimal CO_2 emissions reduction path in Jiangsu province: An expanded IPAT approach. *Appl. Energy* **2013**, *112*, 1510–1517. [CrossRef]
15. Liu, X.; Mao, G.Z.; Ren, J.; Li, Y.M.; Guo, J.H.; Zhang, L. How might China achieve its 2020 emissions target? A scenario analysis of energy consumption and CO_2 emissions using the system dynamics model. *J. Clean. Prod.* **2015**, *103*, 401–410. [CrossRef]
16. Sen, P.; Roy, M.; Pal, P. Application of ARIMA for forecasting energy consumption and GHG emission: A case study of an Indian pig iron manufacturing organization. *Energy* **2016**, *116*, 1031–1038. [CrossRef]
17. Wang, Z.X.; Ye, D.J. Forecasting Chinese carbon emissions from fossil energy consumption using non-linear grey multivariable models. *J. Clean. Prod.* **2017**, *142*, 600–612. [CrossRef]
18. Pelesaraei, A.N.; Rafiee, S.; Bandbafha, H.H.; Shamshirband, S. Modeling energy consumption and greenhouse gas emissions for kiwifruit production using artificial neural networks. *J. Clean. Prod.* **2016**, *133*, 924–931. [CrossRef]
19. Gambhir, A.; Schulz, N.; Napp, T.; Tong, D.; Munuera, L.; Faist, M.; Riahi, K. A hybrid modelling approach to develop scenarios for China's carbon dioxide emissions to 2050. *Energy Policy* **2013**, *59*, 614–632. [CrossRef]
20. Choi, Y.; Zhang, N.; Zhou, P. Efficiency and abatement costs of energy-related CO_2 emissions in China: A slacks-based efficiency measure. *Appl. Energy* **2012**, *98*, 198–208. [CrossRef]
21. Du, K.R.; Huang, L.; Yu, K. Sources of the potential CO_2 emission reduction in China: A nonparametric metafrontier approach. *Appl. Energy* **2014**, *115*, 491–501. [CrossRef]
22. Xiang, N.; Xu, F.; Sha, J.H. Simulation Analysis of China's Energy and Industrial Structure Adjustment Potential to Achieve a Low-carbon Economy by 2020. *Sustainability* **2013**, *5*, 5081–5099. [CrossRef]
23. Ye, F.; Xie, X.X.; Zhang, L.; Hu, X.L. An Improved Grey Model and Scenario Analysis for Carbon Intensity Forecasting in the Pearl River Delta Region of China. *Energies* **2018**, *11*, 91. [CrossRef]
24. Li, J.Y.; Shi, J.F.; Li, J.C. Exploring Reduction Potential of Carbon Intensity Based on Back Propagation Neural Network and Scenario Analysis: A Case of Beijing, China. *Energies* **2016**, *9*, 615. [CrossRef]
25. Wang, P.; Zhu, B.Z. Estimating the Contribution of Industry Structure Adjustment to the Carbon Intensity Target: A Case of Guangdong. *Sustainability* **2016**, *8*, 355. [CrossRef]
26. Tang, D.C.; Ma, T.Y.; Li, Z.J.; Tang, J.X.; Bethel, B.J. Trend Prediction and Decomposed Driving Factors of Carbon Emissions in Jiangsu Province during 2015–2020. *Sustainability* **2016**, *8*, 1018. [CrossRef]
27. Stern, D.I.; Jotzo, F. How ambitious are China and India's emissions intensity targets? *Energy Policy* **2010**, *38*, 6776–6783. [CrossRef]
28. Yi, B.W.; Xu, J.H.; Fan, Y. Determining factors and diverse scenarios of CO_2 emissions intensity reduction to achieve the 40–45% target by 2020 in China—A historical and prospective analysis for the period 2005–2020. *J. Clean. Prod.* **2016**, *122*, 87–101. [CrossRef]
29. Xiao, B.W.; Niu, D.X.; Guo, X.D. Can China achieve its 2020 carbon intensity target? A scenario analysis based on system dynamics approach. *Ecol. Indic.* **2016**, *71*, 99–112. [CrossRef]
30. Yuan, J.H.; Xu, Y.; Zhang, X.P.; Hu, Z.; Xu, M. China's 2020 clean energy target: Consistency, pathways and policy implications. *Energy Policy* **2014**, *65*, 692–700. [CrossRef]

31. Wang, R.; Liu, W.J.; Xiao, L.S.; Liu, J.; Kao, W. Path towards achieving of China's 2020 carbon emission reduction target—A discussion of low-carbon energy policies at province level. *Energy Policy* **2011**, *39*, 2740–2747. [CrossRef]
32. Yi, W.J.; Zou, L.L.; Guo, J.; Wang, K.; Wei, Y.M. How can China reach its CO_2 intensity reduction targets by 2020? A regional allocation based on equity and development. *Energy Policy* **2011**, *39*, 2407–2415. [CrossRef]
33. Xu, L.; Chen, N.C.; Chen, Z.Q. Will China make a difference in its carbon intensity reduction targets by 2020 and 2030. *Appl. Energy* **2017**, *203*, 874–882. [CrossRef]
34. Zhang, X.; Zhao, X.R.; Jiang, Z.J.; Shao, S. How to achieve the 2030 CO_2 emission-reduction targets for China's industrial sector: Retrospective decomposition and prospective trajectories. *Glob. Environ. Chang.* **2014**, *44*, 83–97. [CrossRef]
35. Deng, J.L. *Gray Prediction and Gray Decision*, 1st ed.; Huazhong University of Science and Technology Press: Wuhan, China, 2002; pp. 48–65. ISBN 7-5609-2727-0.
36. Liu, S.F.; Yang, J.Y.; Wu, L.F. *Gray System Theory and Its Application*, 5th ed.; Science Press: Beijing, China, 2014; pp. 49–97. ISBN 978-703-027-3925.
37. CEInet Statistics Database. Available online: http://db.cei.cn/page/Login.aspx (accessed on 8 September 2018).
38. Specht, D.F. A general regression neural network. *IEEE Trans. Neural Netw. Learn. Syst.* **1991**, *2*, 568–576. [CrossRef] [PubMed]
39. Baker, B.D. Can flexible non-linear modeling tell us anything new about educational productivity? *Econ. Educ. Rev.* **2001**, *20*, 81–92. [CrossRef]
40. Wei, J.Y.; Ru, F. Forecasting the Traffic Volume by the Model of GRNN and Studying It's Realization. *Transp. Eng.* **2006**, *22*, 46–50.
41. Liu, L.F.; Yi, X.J. An estimation and forecast simulation of energy demand in China. *J. Shanghai Univ. Financ. Econ.* **2008**, *10*, 84–91.
42. Guo, J.E.; Chai, J.; Lv, Z.D. Application of path analysis and PLSR to forecast the energy resource demand in China. *Chin. J. Manag.* **2008**, *5*, 651.
43. Lin, B.Q. Structural changes, efficiency improvement and electricity demand forecasting. *Econ. Res.* **2003**, *5*, 57–65.
44. He, X.P.; Liu, X.Y.; Lin, Y.P. China's Electricity Demand Forecast under Urbanization Process. *Econ. Res. J.* **2009**, *1*, 118–130.
45. Yang, W.; Wang, C.J.; Jin, F.J.; Li, L.L. Decomposition of Energy Intensity Change in Industrial Sub-Sectors and Its Spatio-Temporal Variation in China. *J. Nat. Resour.* **2013**, *28*, 81–91.
46. China Statistical Yearbook. Available online: http://data.cnki.net/Yearbook/Single/N2006010755 (accessed on 6 August 2018).
47. Wang, Y.M. Research on the methods of combining forecasts based on correlativity. *Forecasting* **2002**, *21*, 58–62.
48. Liu, Y.Q.; Zhao, G.H. The evolution analysis of China's energy consumption structure in the constraints of energy-saving and carbon emissions reduction. *Econ. Probl.* **2015**, *1*, 27–33.
49. Wang, S. Research on Energy Structure Optimization in Jiangsu Province Subject to Low Carbon Targets. Master's Thesis, Suzhou University, Suzhou, China, 2016.
50. Li, C.H. Applying the Grey prediction model to the global integrated circuit industry. *Technol. Forecast. Soc. Chang.* **2003**, *70*, 563–574.
51. Li, H.M.; Wang, J.Z.; Lu, H.Y.; Guo, Z.H. Research and application of a combined model based on variable weight for short term wind speed forecasting. *Renew. Energy* **2018**, *116*, 669–684. [CrossRef]
52. Central People's Government of the People's Republic of China. Available online: http://www.gov.cn/zhengce/content/node_330.htm (accessed on 15 August 2018).
53. Zhang, L.F. The Research of the Model of Energy Supply and Demand and the Development Countermeasure of China. Ph.D. Thesis, Capital University of Economics and Business, Beijing, China, 2006.
54. Jiang, K.J.; Hu, X.L.; Zhuang, X.; Liu, Q. China's low-carbon scenarios and roadmap for 2050. *Sino-Glob. Energy* **2009**, *14*, 1–7.
55. Eichengreen, B.; Park, D.; Shin, K. When Fast Growing Economies Slow Down: International Evidence and Implications for the People's Republic of China. *Asian Econ. Pap.* **2012**, *11*, 42–87. [CrossRef]

56. Wang, F.; Feng, F.G. Contribution of improving energy mix to carbon intensity target in China: Potential assessment. *China Ind. Econ.* **2011**, *4*, 127–137.
57. Xiamen University Energy and Economics Research Center. *"The 12th Five Year Plan" Preliminary Research-Energy Development Strategy*; Xiamen University Energy and Economics Research Center: Xiamen, China, 2009.
58. Chen, J.G.; Huang, H.Q. *Research on the Issues of Industrialization and Industrial Modernization in China*, 1st ed.; Economic Management Press: Beijing, China, 2009; pp. 62–135. ISBN 9787509606896.
59. Ma, X.J.; Wang, Y.; Wang, C. Low-carbon development of China's thermal power industry based on an international comparison: Review, analysis and forecast. *Renew. Sustain. Energy Rev.* **2017**, *80*, 942–970. [CrossRef]
60. Bendu, H.; Deepak, B.B.V.L.; Murugan, S. Application of GRNN for the prediction of performance and exhaust emissions in HCCI engine using ethanol. *Energy Conv. Manag.* **2016**, *122*, 165–173. [CrossRef]
61. Wang, Y.; Duan, F.M.; Ma, X.J.; He, L.C. Carbon emissions efficiency in China: Key facts from regional and industrial sector. *J. Clean. Prod.* **2019**, *206*, 850–869. [CrossRef]
62. International Energy Outlook 2016 with Projections to 2040. Available online: https://www.eia.gov/ (accessed on 14 September 2018).
63. BP Statistical Review of World Energy Beyond Petroleum. Available online: http://www.bp.com/statisticalreview (accessed on 25 August 2018).
64. Shen, L.; Liu, L.T.; Wang, L.M.; Chen, F.N.; Zhang, C.; Shen, M.; Zhong, S. 2050 Energy Consumption Projection for China. *J. Nat. Resour.* **2015**, *30*, 361–373.
65. Wang, X.Y.; Luo, D.K.; Zhao, X.; Sun, Z. Estimates of energy consumption in China using a self-adaptive multi-verse optimizer-based support vector machine with rolling cross-validation. *Energy* **2018**, *152*, 539–548. [CrossRef]

energies

MDPI

Article

Quota Allocation for Carbon Emissions in China's Electric Power Industry Based Upon the Fairness Principle

Ming Meng, Lixue Wang * and Qu Chen

Department of Economics and Management, North China Electric Power University, Baoding 071003, China; 51851341@ncepu.edu.cn (M.M.); ncepucq@ncepu.cn (Q.C.)

* Correspondence: ylxblx@ncepu.cn; Tel.: +86-155-3025-8606

Received: 6 August 2018; Accepted: 25 August 2018; Published: 27 August 2018

Abstract: As an essential measure to mitigate the CO_2 emissions, China is constructing a nationwide carbon emission trading (CET) market. The electric power industry is the first sector that will be introduced into this market, but the quota allocation scheme, as the key foundation of market transactions, is still undetermined. This research employed the gross domestic product (GDP), energy consumption, and electric generation data of 30 provinces from 2001 to 2015, a hybrid trend forecasting model, and a three-indicator allocation model to measure the provincial quota allocation for carbon emissions in China's electric power sector. The conclusions drawn from the empirical analysis can be summarized as follows: (1) The carbon emission peak in China's electric power sector will appear in 2027, and peak emissions will be 3.63 billion tons, which will surpass the total carbon emissions of the European Union (EU) and approximately equal to 2/3 of the United States of America (USA). (2) The developed provinces that are supported by traditional industries should take more responsibility for carbon mitigation. (3) Nine provinces are expected to be the buyers in the CET market. These provinces are mostly located in eastern China, and account for approximately 63.65% of China's carbon emissions generated by the electric power sector. (4) The long-distance electric power transmission shifts the carbon emissions and then has an impact on the quotas allocation for carbon emissions. (5) The development and effective utilization of clean power generation will play a positive role for carbon mitigation in China's electric sector.

Keywords: quotas allocation; carbon emissions; electric power industry; fairness

1. Introduction

China's carbon emissions have grown rapidly in recent years, from 3511.84 million tons in 2001 to 9232.58 million tons in 2017, with an average annual growth rate of 6.23% [1]. China surpassed the United States of America (USA) in 2006 to be the largest CO_2 emitter in the world. At present, China's global CO_2 emission share is as high as 27.61%, much larger than the second (USA, 15.21%) and the third (India, 7.01%) largest emitters [1]. To control the global carbon emissions, and therefore mitigate the aggravation of the greenhouse effect, the European Union (EU), as the traditional core force in emission reduction, set a binding economy-wide domestic emission reductions target of at least 40% by 2030 compared to 1990 [2]. China has always maintained it will control its carbon emissions according to responsibility and capability. To cooperate with the international community to mitigate the global climate change, China pledged to achieve its peak CO_2 emissions before 2030 and to make every possible effort to peak earlier in its INDCs (Intended Nationally Determined Contributions) to the Paris Agreement [3]. Besides, accompanied pollutants emitted with CO_2 have been deteriorating China's local environments. More than 99% of the 500 largest cities in China do not satisfy the standards on air quality suggested by the WHO (World Health Organization), and 7 cities of the 10 most polluted cities worldwide are in China [4]. Therefore, China is facing tremendous pressure for carbon emission mitigation from both internationally and domestically.

Carbon emissions generated by energy are regarded as the prime contributor to the greenhouse effect, which is one of the most urgent issues around the world [5,6]. Due to resource endowments and excessive development of coal-fired power plants, thermal power generated more than 75% of the total electricity from the foundation of the People's Republic of China from 1949 to the present [7]. In 2015, the carbon emissions resulting from thermal power plants were approximately 4212.68 million tons in China (45.95% of China's total), which far exceeded the country emissions of India (2218.43), the Russian Federation (1483.18), Japan (1207.79), Germany (753.64), Canada (532.5), and many other major emitters [1]. Fortunately, China's power industry has great potential for carbon reduction. Chinese carbon emissions per unit of electricity were almost 627 g/kWh in 2015 [7–9], which is still far above that of the developed countries. As a major energy consumer and carbon emitter in China, the electric power industry is expected to be the main force for China's carbon mitigation, which is the first sector to be introduced into the carbon emission trading (CET) market in China.

CET is a market mechanism where the emitters can trade carbon emission rights as a commodity to promote global greenhouse gas emissions reduction [10,11]. Since 1997, when Kyoto proposed the concept of CET, it has been widely used and has achieved significant emissions reduction effects around the world. The European Union's carbon emissions trading system (EU-ETS), which was started in 2005, is the biggest CET system worldwide [12]. By 2015, there have been 17 CET systems spread across four continents, and the gross domestic product (GDP) of these systems has accounted for 40% of the global GDP. In December 2017, the NDRC (National Development and Reform Commission) of China formally announced the launch of the nationwide CET market in the electric sector. The total carbon emissions of CET in the electric industry will exceed 3 billion tons, outdistancing any other CET in the world [13]. There is no doubt that CET in China's power industry will play an essential role in carbon emissions reduction.

The quota allocation for carbon emissions is an important part for the successful operation of CET, which impact the price of carbon quotas and ensure the effectiveness of emissions reduction targets. It is well known that there are great differences in economic development levels, natural resource endowments, industry structure, and population across China. Therefore, the scientific and rational quotas allocation for carbon emissions is even more important to facilitate the establishment of the CET market in China's electric industry. However, the quota allocation scheme, as the key foundation of market transactions, is still undetermined. This research studies the quota allocation for carbon emissions in China's electric sector's CET from the perspective of capacity, responsibility, and potential for carbon mitigation, which can exactly reflect the economic development, industrial structure, and efficiency of carbon emissions, and then provide convincing evidence for quota allocation.

The traditional methods of carbon emission quota allocation can be defined as grandfathering [14,15], benchmarking [16,17], auction [18,19], and so on. In addition to these traditional methods, an index method based upon fairness and efficiency was developed into a single-index and a multiple-index decision model. The latter is of particular interest to scholars because of the adaptive ability to address multi-dimensional differences among objects [20]. Wen et al. designed a comprehensive indicator model from the perspective of the capacity, responsibility, and potential for carbon emissions reduction [21]. To exactly measure the carbon efficiency, Qin et al. introduced a multi-criteria model combined with a weighted Russell direction distance model [22].

Apart from the method mentioned above, data envelopment analysis (DEA) is also a conspicuous optimization method. For example, Zhou et al. presented a DEA model with multiple abatement indexes considering the difference of marginal abatement cost in Chinese cities [23]. Based upon a non-radial Zero Sum Gains Data Envelopment Analysis (ZSG-DEA) model, Miao et al. efficiently allocated the carbon emission quota among Chinese provinces [24]. Zhou et al. introduced a DEA model based on spatial-temporal allocation strategies [25], which considered the imbalanced economic development and the discrepancy between development and carbon emissions. In addition, nonlinear programming and game theoretic approach have been advocated to explore optimal allocation of carbon emission permits. Liu et al. minimized the cost of carbon reduction to optimize the allocation for carbon emission quota based on a novel nonlinear programming method [26]. Stackelberg game models were employed by Ren et al. to study the quotas allocation for carbon emissions between the manufacturer and the retailer from the perspective of production [27]. An and Lee constructed a Stackelberg framework on the basis of a newsvendor non-cooperative game, which makes the target of carbon reduction come true by adjusting the individual and overall carbon quotas [28].

The current research related to carbon emission quota allocation is mainly concerned with the allocation between countries or regions. Little research has focused on the quotas initial allocation for carbon emissions from the perspective of industry, especially the power industry with a heavy proportion of carbon emissions. This paper allocates carbon emission quota for the Chinese power industry in 30 provinces from 2016 to 2030, providing a reference for China's national CET. Considering the striking differences that exist in the aspects of economics, resource endowment, and historical emissions [29], this study selected per capita GDP, historical accumulated per capita accumulated CO_2 emissions, and CO_2 emissions per unit of electricity as indicators based upon capacity, responsibility, and potential for carbon mitigation, respectively, to allocate the CO_2 emission permission in the power industry.

This study carefully introduces a three indicators allocation model for CET in China's electric industry in Section 2. The data and the calculation of three indicators are included in Section 2. Section 3 describes the allocation results of the carbon emission quota. The detailed empirical analysis is discussed in Section 4, and then some policy suggestions are put forward. Section 5 summarizes conclusions of this paper.

2. Methods and Data

2.1. Methods

2.1.1. Hybrid Trend Forecasting Model for Carbon Emissions

To distribute the carbon quota in China's power sector, the total carbon emissions in the power sector (denoted as C) should be calculated first. The change of C in China is mainly influenced by three factors: carbon emissions per unit of electricity (denoted as H), electric consumption per unit of GDP (denoted as P), and GDP. The above three variables are closely related to C. H is the most efficient factor for carbon mitigation. The decrease of H means lower carbon emissions are required to meet the electric demand. With other variables remaining constant, the changes in P will affect the total electricity consumption and hence the C. The impact of GDP on C is similar to that of P. Higher GDP will increase electricity consumption and correspondingly the carbon emissions will increase if other variables remain unchanged. Furthermore, GDP, P, and H cover all the possible factors (compose C completely) that can be considered in energy policy making. Therefore, the relationship equation between C and GDP, P, and H is described as

$$\begin{aligned} C &= \mathrm{GDP} \times \mathrm{P} \times \mathrm{H} \\ &= \mathrm{GDP} \times \frac{\mathrm{E}}{\mathrm{GDP}} \times \frac{\mathrm{C}}{\mathrm{E}} \end{aligned}, \tag{1}$$

where C, GDP, and E are the carbon emissions in the Chinese electric power industry, the gross domestic product, and electric generation of China, respectively.

In order to allocate carbon emission quota in China's electric sector, a forecasting method must be employed to predict GDP, P, and H in Equation (1). Influenced by economic and political factors, the three explanatory variables related to carbon emissions do not show a single linear or exponential relationship. The grey models, such as GM (1,1) which is a trend extrapolate method generated by first-order accumulation of single variate, are popularly used to predict a small sample trend. Unfortunately, the representativeness of GM (1,1) on the exponential growth trend is very limited and cannot represent all the exponential growth trends. In other words, only when the time-series data has a steady growth rate will GM (1,1) can have a good fitting effect. Therefore, the grey prediction methods commonly used in homogeneous equation prediction are not suitable for this study. In order to forecast GDP, P, and H more exactly, this paper adopts a hybrid trend extrapolation model [30], which can simultaneously fit the linear trend, exponential trend, and the combination trend of the two.

The homogeneous exponential and the linear fitting equations to a time series $x(k)$, $k = 1, 2, \ldots, n$, can be denoted as $\hat{x}_{(k+1)} = a\hat{x}_{(k)}$ and $\hat{x}_{(k)} = ck + d$, respectively. Then, the hybrid trend extrapolation equation can be expressed as

$$\begin{cases} \hat{x}_{(k+1)} = \hat{\lambda}_1 \hat{x}_{(k)} + \hat{\lambda}_2 k + \hat{\lambda}_3,\ k = 1, 2, \cdots \\ \hat{x}_{(1)} = x_{(1)} + \hat{\lambda}_4, \end{cases}, \tag{2}$$

where $\hat{\lambda}_1$, $\hat{\lambda}_2$, $\hat{\lambda}_3$, and $\hat{\lambda}_4$ are the estimated parameters.

Firstly, the estimations of $\hat{\lambda}_1$, $\hat{\lambda}_2$, $\hat{\lambda}_3$ are calculated. Let $\hat{\boldsymbol{\lambda}} = \begin{bmatrix} \hat{\lambda}_1 \\ \hat{\lambda}_2 \\ \hat{\lambda}_3 \end{bmatrix}$, $\mathbf{B} = \begin{bmatrix} x_{(1)} & 1 & 1 \\ x_{(2)} & 2 & 1 \\ \cdots & \cdots & \cdots \\ x_{(n-1)} & n-1 & 1 \end{bmatrix}$ and $\mathbf{Y} = \begin{bmatrix} x_{(2)} \\ x_{(3)} \\ \cdots \\ x_{(n)} \end{bmatrix}$. Then, the estimation of $\hat{\lambda}$ by ordinary least squares (OLS) is

$$\hat{\lambda} = (\mathbf{B}'\mathbf{B})^{-1}\mathbf{B}\mathbf{Y}, \tag{3}$$

By employing the results of the former parameters and the actual value of $x_{(k)}$, the result of $\hat{\lambda}_4$ is calculated as follows:

$$\hat{\lambda}_4 = \frac{\sum\limits_{k=1}^{n-1} \left[x_{(k+1)} - \hat{\lambda}_1^k x_{(1)} - \hat{\lambda}_2 \sum\limits_{j=1}^{k} j\hat{\lambda}_1^{k-j} - \frac{1-\hat{\lambda}_1^k}{1-\hat{\lambda}_1}\hat{\lambda}_3 \right] \hat{\lambda}_1^k}{1 + \sum\limits_{k=1}^{n-1} (\hat{\lambda}_1^k)^2}, \tag{4}$$

Let $k = 1, 2, \ldots$, in Equation (2), and the match value and predicting results to $x(k)$ can be gained.

2.1.2. Principles and Indicators for Carbon Emission Quota Allocation Based on Equity

Based on the current studies, three indicators are selected from the perspectives of equity and efficiency. This research selects per capita GDP, accumulated per capita carbon emissions from the electric power sector, and CO_2 emissions per unit of electricity as indicators, which reflect the capacity, responsibility, and potential for carbon emissions reduction, respectively. Details of these criteria are presented in Table 1.

(1) Carbon reduction capacity.

The economically developed provinces can better afford and are more willing for carbon reduction. Wealthier regions have some essential conditions: sufficient funds, advanced technologies, and the relative optimized energy structure. In addition, the less-developed provinces are given priority to develop economy and narrow the gap of regional economic development. This means richer provinces with greater reduction ability will shoulder more responsibility for carbon emission-mitigation targets. This indicator is effective to measure the economic development of each province. The developed areas with higher per capita GDP will obtain fewer carbon emission quotas. Inversely, those regions with less per capita GDP will get more carbon emission quotas. This research employed the mean per capita GDP from 2011 to 2015 to quantify the economy of each province.

(2) Carbon reduction responsibility.

This criterion reflects the historical egalitarian principle, which cannot be ignored for equitable allocation for carbon emission quotas. The greenhouse gases emitted into the atmosphere cannot disappear immediately, and the capacity of the atmosphere is limited. In this sense, the developed provinces encroached on the future emission space of the developing provinces owing to the excessive carbon emissions during the historic economic development. Therefore, the areas that had high carbon emissions in the past should be given more responsibility for carbon mitigation and fewer carbon quotas. Conversely, the regions with low CO_2 emissions historically should get more carbon emission quotas. In this study, the per capita accumulated carbon emission from 2011 to 2015 from the electric industry is employed to quantify the responsibility for carbon reduction of each province.

(3) Carbon reduction potential.

Carbon reduction potential is based on the perspective of cost equity for carbon reduction, which reflects the efficiency of energy utilization. For the economically developed areas, the technology of energy saving and emission reduction is relatively mature with efficient energy consumption, high cost of carbon reduction, and less potential for carbon emission reduction. For the economically backward provinces, the technology is relatively imperfect with poor efficient energy utilization, less capital investment into emissions mitigation, and more space for carbon emissions reduction. To minimize social emission reduction costs, more carbon emission quotas should be assigned in developed regions with higher emission reduction costs and less in developing regions with lower emission reduction costs. This research employed the mean of CO_2 emissions per unit of electricity from 2011 to 2015 to quantify the potential for carbon mitigation of each province. The larger the indicator value, the lower the carbon quota should be and vice versa.

Table 1. Principles of carbon emission allocation.

Dimension	Principle	Interpretation	Indicator
Carbon reduction capacity	Economic development	The developed provinces should undertake more responsibility.	Per capita gross domestic product (GDP)
Carbon reduction responsibility	Historical responsibility	Provinces with more historical emissions should be more responsible for carbon mitigation.	Historical accumulated per capita carbon emissions
Carbon reduction potential	Carbon emission efficiency	Provinces with inefficient energy consumption should take more responsibility.	Carbon efficiency

2.1.3. Allocation Model for Carbon Emission Quota

Multi-criteria allocation models for carbon emission quota have received prominence among researchers because of their practicability. On the basis of the criteria introduced above, the allocation model for carbon emission quota can be written as follows:

$$Y_{i,t} = (W_{1i}X_{1i} + W_{2i}X_{2i} + W_{3i}X_{3i}) \times C_t, \tag{5}$$

$Y_{i,t}$ is the carbon quota of province i in year t and W_{1i}, W_{2i}, and W_{3i} are the weights of three indicators. X_{1i}, X_{2i}, and X_{3i} are the percentage of per capita GDP, per capita accumulated CO_2 emissions, and the mean of CO_2 emissions per unit of electricity for province i, respectively. Since the values of the three indicators are inversely related to their carbon quotas, this means that the bigger the value is, the smaller the corresponding allocation will be. Therefore, this paper employs the following formula to achieve.

$$z_{ji} = -\ln X_{ji}, j = 1, 2, 3, \tag{6}$$

$$Z_{ji} = \frac{z_{ji}}{\sum\limits_{i=1}^{i=30} z_{ji}}, j = 1, 2, 3, \tag{7}$$

In addition, the Equation (5) could be written as

$$Y_{i,t} = (W_{1i}Z_{1i} + W_{2i}Z_{2i} + W_{3i}Z_{3i}) \times C_t, \tag{8}$$

The entropy method was employed to calculate the weight of each indicator. Shannon introduced the definition of entropy firstly into information theory, and it has been extensively employed in many fields, such as engineering technology and social economy [31]. As a means to measure the disorder of the system, the basic idea of the entropy is to measure the objective weight based on the variability of the indicator [22]. In general, if the entropy of a certain index is smaller, it means that the greater the extent of variation of the index value, then the wealthier the information will be that is provided and it may play a more important role in the comprehensive evaluation and have greater weight. Inversely, the bigger the entropy of an indicator is, the smaller is the extent of variation of the indicator value, the poorer is the amount of information provided, the more insignificant is the role played in the system evaluation, and the smaller is the weight of the index. The steps can be concluded briefly as follows:

(1) Data normalization.

Since the measurement units of various indexes are not unified, it is necessary to normalize the data first. In other words, data normalization converts the absolute value of the index into a relative value so as to solve the homogeneity problem of different quality index values. In addition, the normalized processes of the positive and negative index value are different. The bigger value is better for positive indicators, but the smaller value is better for negative indicators. In addition, different algorithms should be employed to normalize the data for the two kinds of indexes, which are shown as follows:

Positive indicator:

$$x'_{ki} = \frac{x_{ki} - \min|x_k|}{\max|x_{ki}| - \min|x_k|} \ (k = 1, 2, 3; i = 1, 2, \ldots 30), \tag{9}$$

Negative indicator:

$$x'_{ki} = \frac{\max|x_k| - x_{ki}}{\max|x_k| - \min|x_k|} \ (k = 1, 2, \ldots 3; i = 1, 2, \ldots 30), \tag{10}$$

(2) Calculate the share of sample i in indicator k.

$$u_{ki} = \frac{x'_{ki}}{\sum\limits_{i=1}^{n} x'_{ki}}, \tag{11}$$

where $\sum\limits_{i=1}^{30} u_{ki} = 1$.

(3) Calculate the entropy value of the indicator k.

$$e_k = -(\ln 30)^{-1} \sum_{i=1}^{30} u_{ki} \ln(u_{ki}), \tag{12}$$

If $u_{ik} = 0$, then it can stipulate that $u_{ki} \ln(u_{ki}) = 0$.

(4) Calculate the weights of indicator.

$$W_k = \frac{1 - e_k}{\sum\limits_{k=1}^{3}(1 - e_k)}, \ (k = 1, 2, \ldots 3), \tag{13}$$

At this point, the weights of indexes in Equation (8) can be obtained.

2.2. Data

2.2.1. Measure of Carbon Emissions

Since the data of carbon emissions from the electric sector have never been provided in China, it had to be calculated through the actual data of energy consumption [32]. Considering the consistency of the statistical caliber and proportion of energy consumption in the 30 provinces from the power industry, this paper selected nine energy sources: natural gas, refinery gas, fuel oil, diesel, crude oil, coke oven gas, other washed coal, cleaned coal, and raw coal. The carbon emissions can be obtained as follows,

$$CO_2 = FC_i \times NCV_i \times CC_i \times OF_i \times \frac{44}{12}, \tag{14}$$

FC_i is the energy consumption (standard coal measurement). NCV_i is mean the net heating value factor [33]. CC_i is the carbon content of energy selected from Intergovernmental Panel on Climate Change (IPCC) [34], which were widely used to estimate carbon emissions. OF_i is carbon oxidation coefficient, which is generally 100%; $\frac{44}{12}$ is the ratio of mass between a CO_2 molecule and a C atom.

This paper employed the average value of "Middling" and "Peat" to represent the NCV of washed coal [35], because its composition is complex and uncertain. Similarly, the NCV of washed coal was replaced by the average NCV of "Oilfield Gas" and "Gasfield Gas".

2.2.2. Data

To forecast the total carbon emissions in China's electric industry, this research employed the annual data of GDP, energy consumption, and electricity generation in the Chinese power sector from 2001 to 2015. For the quota allocation to carbon emissions, the annual data of GDP, population, and energy consumption in the power industry of each province from 2011 to 2015 were collected from the *China Statistical Yearbook* (2012–2016). It should be noted that the price index was employed to reduce the impact of price fluctuation. In this article, energy consumption in thermal power represented the data of the electric industry. Energy consumption in the electric industry of each province and the total were, respectively, from the *China Energy Statistical Yearbook* (2012–2016) and the *China Statistical Yearbook* (2002–2016). Considering the data availability, this paper selected the related data of 30 provinces.

3. Results

3.1. The Results of Carbon Emission Trends

To obtain the trend of C in Equation (1), a hybrid extrapolation method was employed to forecast the values of GDP, P, and H, respectively. Using the data of GDP, P, and H, the parameters were estimated and the trend extrapolation equations for GDP, P, and H were calculated, respectively, as follows:

$$\begin{cases} G\hat{D}P_{(k+1)} = 0.4638G\hat{D}P_{(k)} + 347.7759k + 5721.8756, k = 1, 2, \cdots \\ G\hat{D}P_{(1)} = GDP_{(1)} - 121.0431 \end{cases}, \tag{15}$$

$$\begin{cases} \hat{P}_{(k+1)} = 0.4923\hat{P}_{(k)} + 0.006k + 0.0831, k = 1, 2, \cdots \\ \hat{P}_{(1)} = P_{(1)} - 0.0026 \end{cases}, \tag{16}$$

$$\begin{cases} \hat{H}_{(k+1)} = 0.3656\hat{H}_{(k)} - 8.9538k + 413.4115, k = 1, 2, \cdots \\ \hat{H}_{(1)} = H_{(1)} - 0.9802 \end{cases}, (k = 1, 2, \ldots m), \tag{17}$$

Let k = 1~14 and k = 15~29 in Equations (12)–(14). Then, the fitting and forecasting results of the GDP, P, and H in years 2001–2015 and 2016–2030 were obtained, respectively, as shown in Figure 1a–c.

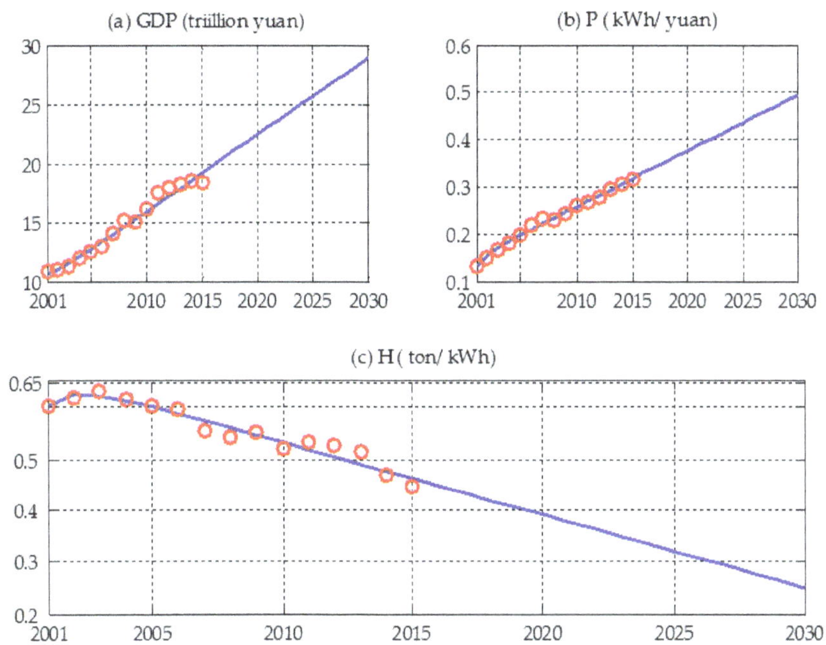

Figure 1. Fitting and forecasting results to the GDP, electric consumption per unit of GDP (P), and carbon emissions per unit of electricity (H).

Figure 1a–c show that the historical trend, fitting results, and the forecasting trend of GDP, P, and H, respectively. As shown in Figure 1a,b, the GDP and P will keep growing before 2030. For GDP, it will increase to 22.43 and 28.92 (eliminating the effects of inflation) trillion yuan in 2020 and 2030, respectively. For P, it will increase to 0.39 and 0.49 kWh/yuan in 2020 and 2030, respectively, which is equal to 2 and 2.5 times of US's, respectively, and the average annual growth rate is still strong. This means that the electricity consumption required for the growth unit GDP is increased. H measures the efficiency of electric generation. Contrary to GDP and P, the future trend of H is presenting a visible decreasing trend, which indicates the generation of electricity is becoming more efficient; this is conducive to realizing the sustainability development of China.

Based on the fitting and forecasting results to GDP, P, and H above and in Equation (1), the trend of carbon emissions in the electric industry was obtained, as is shown in Figure 2. The forecasting results show that the carbon emissions in the electric industry exhibit an increasing trend from 2016 to 2027 with the growth rate decreasing. The carbon emission peak in China's electric power industry will appear in 2027 and peak emissions are 3.63 billion tons, which will surpass the total carbon emissions of the EU and approximately equal 2/3 of the USA (their carbon emissions have remained stable in recent years). The results indicate that China may achieve their commitment at the Paris climate conference, which is to attain a peak on carbon emissions before 2030.

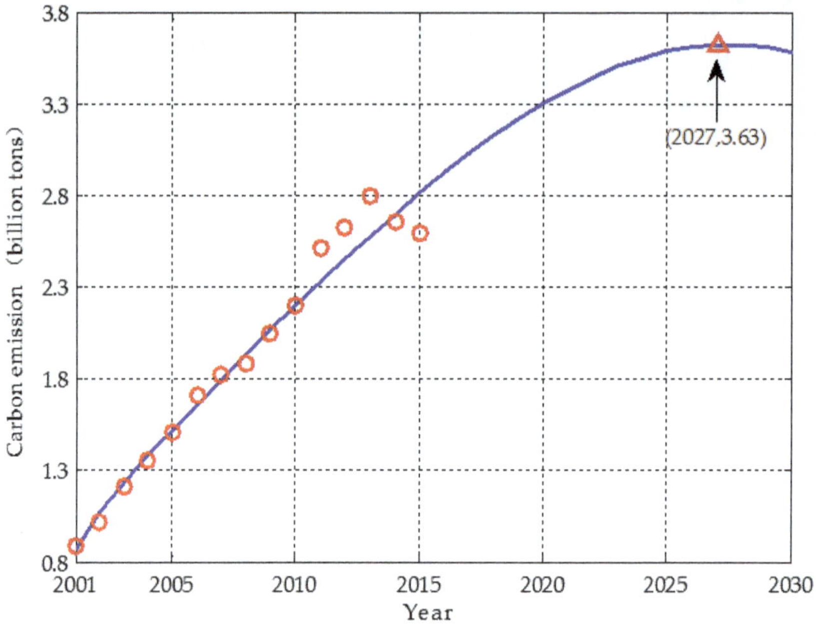

Figure 2. Fitting and forecasting results for the carbon emissions in the electric power industry.

3.2. Results of Quota Allocation to Carbon Emissions

The weights of indicators were obtained by Equations (9)–(13) and the results are listed in Table 2. For the three indicators, the proportion of carbon reduction responsibility is heaviest, which is high at 0.4011; the carbon reduction potential is smallest and the weight of carbon reduction capacity is in the middle.

Table 2. Weights for indicators.

	W_1	W_2	W_3
Value	0.3420	0.4011	0.2569

The indexes of CO_2 emission quota allocation of each province for the years 2016 to 2030 are listed in Table 3, which represent the share of CO_2 emission quota of each province in each year. For example, the index of Beijing is 0.0343, which indicates the quota from Beijing accounts for 3.43% of China's total in each year. The larger the value is, the higher the carbon emission quota will be, and the smaller the value, the lower the carbon emission quota will be. As can be seen from Table 3, Sichuan has the highest proportion at 4.12% and the smallest, 2.28%, is in Ningxia.

Table 3. Provincial shares for carbon emission quota.

Province	Share	Province	Share	Province	Share
Beijing	0.0343	Zhejiang	0.0323	Hainan	0.0274
Tianjin	0.0271	Anhui	0.0354	Chongqing	0.0366
Hebei	0.0351	Fujian	0.0330	Sichuan	0.0408
Shanxi	0.0304	Jiangxi	0.0374	Guizhou	0.0360
Inner Mongolia	0.0236	Shandong	0.0338	Yunnan	0.0412
Liaoning	0.0320	Henan	0.0361	Shaanxi	0.0333
Jilin	0.0316	Hubei	0.0322	Gansu	0.0353
Heilongjiang	0.0340	Hunan	0.0396	Qinghai	0.0329
Shanghai	0.0293	Guangdong	0.0357	Ningxia	0.0228
Jiangsu	0.0315	Guangxi	0.0394	Xinjiang	0.0298

Figure 3 shows the summated carbon emission quota from 2016 to 2030 on the three indicators and the total of the three components of the provinces examined in this paper. A larger carbon emission quota indicates a smaller emission reduction obligation, and a smaller carbon emission quota indicates more responsibility for emission reduction. For example, Yunnan has the largest carbon emission quota, indicating that it bears less emission reduction duty and can focus on developing the economy. The CO_2 emission quota of Inner Mongolia is the lowest, which indicates that the total scores of the comprehensive index on carbon reduction capacity, responsibilities for emission reduction, and potential for emission reduction are all high, therefore, Inner Mongolia must shoulder a heavier burden for emission reduction.

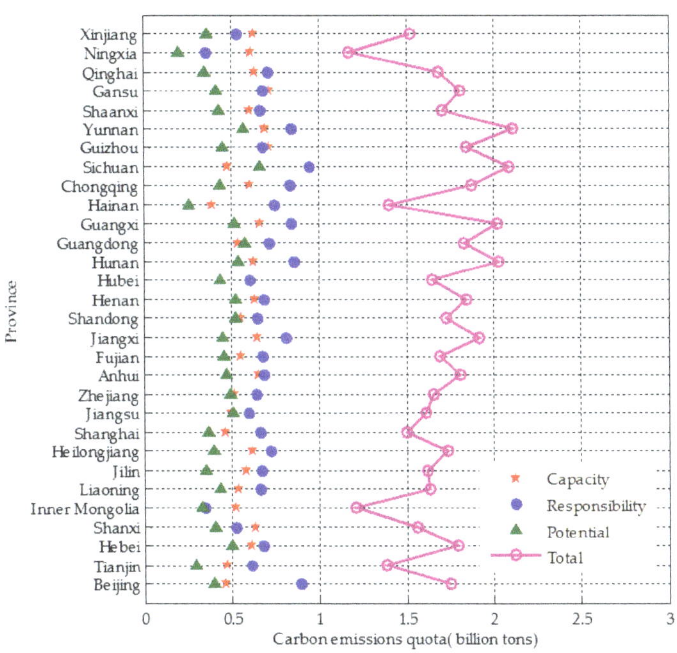

Figure 3. Carbon emission quota on indicators and the total for each province.

There are differences in economic development, historical carbon emissions, and power generation technologies among the provinces. Therefore, the carbon emission quota allocation obtained in terms of carbon emission reduction capacity, responsibilities, and potential also show great differences.

For carbon emission reduction capacity, Hainan, Beijing, Tianjin, and Shanghai have higher per capita GDP and a stronger ability for carbon mitigation, therefore, these provinces have lower carbon quotas for carbon emission reduction capacity; those provinces with low per capita GDP, such as Guizhou and Yunnan, have higher carbon emission quota allocation for their low carbon emission reduction capability. On the basis of responsibility for carbon mitigation, per capita accumulated carbon emissions of Xinjiang, Shanxi, Ningxia, and Inner Mongolia are high in the period of investigation, so they should be more responsible for the target of carbon reduction and accordingly, their carbon emission quotas are small; in contrast, the provinces where the per capita accumulated carbon emissions in the investigated period are low will take less carbon reduction responsibility and obtain more carbon quotas in this indicator. For the indicator of the potential for carbon emissions reduction, the indicator values of Ningxia, Hainan, Qinghai, Xinjiang, Tianjin, et al. are relatively high, which indicates that these provinces have more potential for carbon emission reduction, so they obtained lower carbon quotas in this aspect; the carbon emissions per unit of electricity in Sichuan and Guangdong are relatively high and the potential for carbon mitigation is smaller, therefore, the quota obtained from carbon reduction potential are higher.

4. Discussion

Further study of the relationship between carbon emission quotas and actual carbon emissions is essential for China's electric industry's CET market. To obtain the actual carbon emissions of each province from2016 to 2030, this study employed the mean of the carbon emission share from 2011 to 2015 as the actual carbon emission proportion from 2016 to 2030. In addition, the carbon quota remainder of each province can be defined as the difference between the total carbon quota and actual carbon emissions in 2016–2030, as demonstrated in Figure 4.

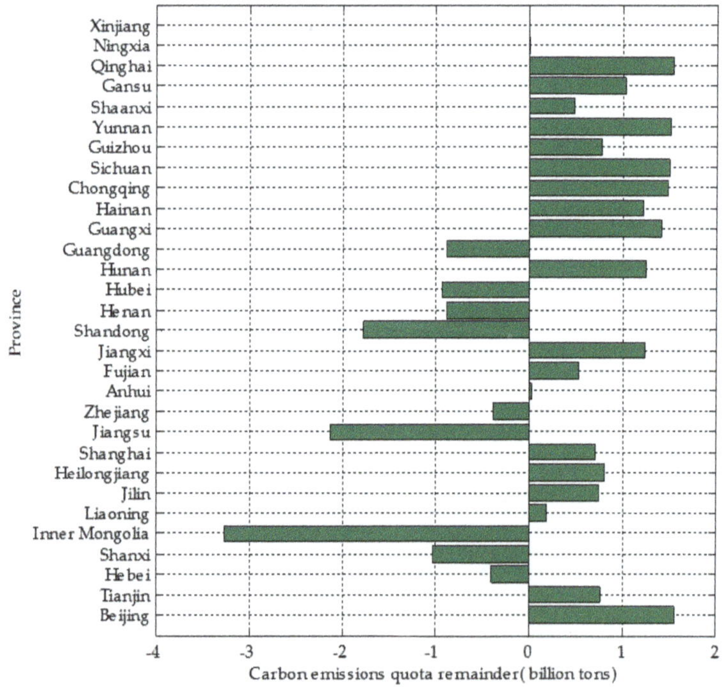

Figure 4. Carbon emission quota remainder of each province.

Based on the value of carbon quota remainder, these provinces can be divided into three types: surplus, balance, and loss. For the positive value of carbon quota remainder, the carbon quota is bigger than the actual carbon emissions, in other words, the carbon quota of these provinces has a surplus. The greater the value of the carbon quote remainder, the more surplus will be obtained, such as in Beijing, Tianjin, etc., so these areas are expected to become the sellers in the CET market. When the value of carbon quota is zero, the carbon quota is equal to the actual carbon emission so the carbon quota is in balance. Considering the development of the economy and technology, theses provinces have a certain self-regulation, hence, Anhui, Ningxia, and Xinjiang are considered to be in states of equilibrium. For the negative values of the carbon quota remainder, the carbon quota is less than actual carbon emission. The carbon quota of the province suffers a loss and the smaller the value, the greater the loss will be, such as in Inner Mongolia, Jiangsu, Shandong, etc., these areas are excepted to be the buyers in CET market. As shown in Figure 4, 70% of the investigated provinces will be in balance or have a surplus, but there are nine provinces that will have a loss: Guangdong, Hubei, Shandong, Zhejiang, Jiangsu, Inner Mongolia, Hebei, Henan, and Shanxi.

For the nine provinces, there are different features in economic development, industrial structure, and efficiency of electric generation and resource endowments, which lead to diverse reasons for their carbon emission quota being in a loss. The main reasons are summarized as follows:

(1) Economy is developed. The provinces with a developed economy, which have high per capita GDP, can provide more sufficient financial guarantees for the research technology to reduce carbon emissions than do those regions with less economic development. Hence, developed economies obtained lower carbon quota, such as those in Jiangsu, Zhejiang, Inner Mongolia, and Guangdong. Those provinces that are abundant in carbon quotas but have a high per capita GDP should also contribute to carbon reduction because they have stronger capacity for carbon reduction, for example, Hainan, Beijing, Shanghai, Tianjin, Sichuan, and Liaoning.

(2) Actual carbon emission is huge. These provinces are the major carbon emitters in China and account for approximately 63.65 percent of China's carbon emissions generated by the power sector. Of the nine provinces, the total actual carbon emissions will all exceed 200 million tons except Xinjiang.

(3) Electric generation is inefficient. More carbon emissions generated per unit power means that the carbon emissions reduction is a potential, such as in Inner Mongolia, Xinjiang, and Hubei. Correspondingly, these provinces should bear more duty, and the carbon quota by this province is small. It should be noted that there are some provinces with surplus carbon quotas but with inefficient electric generation, such as Ningxia, Hainan, Tianjin, Qinghai, Jilin, Shanghai, and Beijing, which will also be responsible for reducing emissions. For those provinces, research and introduction of advanced electric generation technology is essential to mitigate carbon emissions.

(4) The share of traditional industries is high. These provinces are supported by traditional industries. The contribution of the traditional industries output of nine provinces was more than 61% during 2011 to 2015, and for three provinces, Guangdong, Shandong, and Jiangsu, it was more than 10% [6]. In addition, the traditional industries are dependent on electricity consumption, which mainly includes mining, manufacturing, electricity, gas and water production and supply, and construction. This explains the high carbon emissions in these provinces. For these provinces, accelerating industrial transformation and upgrading will ease the pressure on carbon emissions mitigation.

(5) The carbon emissions are shifted. The long-distance electric power transmission shifts the carbon emissions, and then has an impact on the quotas allocation for carbon emissions. It is well known that there are is a great diversity in resource endowments in China. In order to alleviate the connection between energy shortage and economic development in the east, China carried out the "West-East electricity transmission project", transforming the rich energy resources into electric power resources in the west and transporting them to the east coastal areas. This policy is conducive to the transformation of western energy resources advantages into economic advantages while relieving the pressure on the environment and transportation. However, this policy put the electric power output provinces at a disadvantage in carbon emission quota allocation. Those provinces generated more

actual carbon emissions for electric power output and need to bear more carbon emission reduction duty, which means less carbon emission quota remainder than they should obtain. For example, Inner Mongolia and Shanxi, as sellers in the power market, provide a large amount of electricity for Beijing, Tianjin, and Hebei, leading to more carbon emissions, heavier carbon mitigation responsibility, and less carbon quota remainder. However, Beijing, Tianjin, and Hebei, as buyers in the power market, use part of the power with zero carbon emissions. This is equivalent to transferring carbon emissions from power buyers to the sellers and shifting the responsibility for carbon reduction too.

Figure 5 demonstrates the geographical location based on the carbon quota remainder. From Figure 5 it can be seen that the provinces in balance and in surplus are almost all located in the western, central, and northeast regions, while the provinces in loss are almost all in the east.

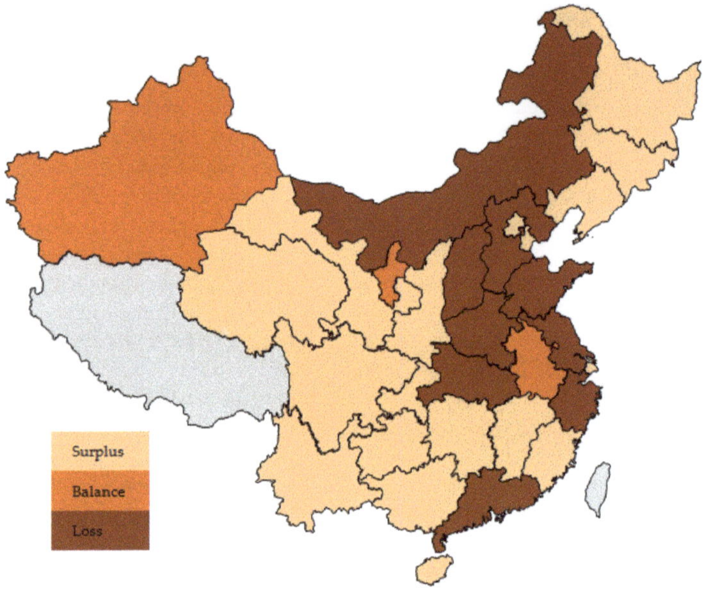

Figure 5. Geographical distribution of carbon quota remainder.

For the developed provinces in the east, the demands of electricity are enormous, but the electric generation is inefficient, so a large amount of carbon emission is inevitably produced. As the main force of carbon emission reduction, these provinces undertake more responsibility for emission reduction. This measure is conducive to research or introduction of advanced power generation technologies, restriction of the progress of carbon intensive enterprises, acceleration of industrial transformation and upgrading, and vigorous development of a low-carbon economy. The provinces with surplus carbon quota remainder located in the western and central regions lag behind in economic development. The CET market can bring economic benefits to these provinces and provide economic security for the introduction of advanced technologies to enhance electric generation efficiency.

5. Conclusions

As an essential measure to mitigate the CO_2 emissions, China is constructing a nationwide CET market. The electric power industry is the first sector which was introduced into this market, but the quota allocation scheme, as the key foundation of market transactions, is still undetermined. To guide the development of the CET market, this paper researched the issue of carbon quota allocation in the electric power industry of China's 30 provinces from 2016 to 2030.

This research employed the GDP and energy consumption and electric generation data of 30 provinces from 2001 to 2015 and a three-indicator allocation model to measure the provincial quota allocation for carbon emissions in China's electric sector. Furthermore, a hybrid trend forecasting model combined with a decomposition model was also used to predict the carbon emissions from the electric power sector. The conclusions drawn from empirical analyses are as follow: (1) The carbon emission peak in China's electric sector will appear in 2027 and peak emissions are 3.63 billion tons, which will surpass the total carbon emissions of the EU and approximately equal 2/3 of the USA. (2) The developed provinces driven by traditional industries will take more responsibility for carbon mitigation because of the excessive historical carbon emissions and high capacity for carbon reduction. (3) Nine provinces are expected to be the buyers in the CET market. These provinces are mostly located in eastern China, and account for approximately 63.65% of China's CO_2 emissions from the electric sector. (4) The long-distance electric power transmission shifts the carbon emissions, and then impacts the quotas allocation. (5) The development and effective utilization of clean power generation will play a positive role for carbon mitigation in China's electric sector, which will reduce the percentage of electricity generation from fossil fuels, and then cut down carbon emissions.

Based on the above analysis, the following policy recommendations are put forward:

(1) Advanced power generation technologies should be advocated. The fundamental way to achieve carbon emission reduction is to make the power generation efficient, regardless of the provinces with a carbon quota loss such as Inner Mongolia, or provinces with surplus carbon allowances, such as Tianjin, Liaoning, and Jilin.

(2) Financial support and transfer payment. The provinces with greater carbon reduction responsibility are mostly featured with single industrial structures and inefficient electrical generation. However, technological improvement is a long process and cannot be accomplished overnight. Hence, provinces devoting themselves to mitigating the carbon emissions should be incented economically. Particularly, the government should provide financial assistance to these regions. Furthermore, as both the production provinces and the consumption provinces benefit from the interprovincial transmission, they should share the newly added emission cost. The transfer payment led by the government is a feasible measure.

(3) Regional cooperation. Regional cooperation is an essential means to achieve emission reduction targets. Buyers in the CET market are likely to be sellers in the electricity market. Buyers of carbon quotas can reduce the pressure of emission reduction, while sellers can obtain advanced power generation technologies and economic benefits. Not only can provinces achieve the goal of emission reduction, but also help to narrow the economic gap between regions and coordinate common regional development.

(4) Actively develop clean power generation. China is vast and rich in clean energy. Hydropower, wind power, solar power, and other clean power resources are abundant, but the utilization rate is low at present. For example, the degree of hydropower development in China is about 10~18%, which is far below the world average level of 22% and the level of 50~100% in developed countries. Furthermore, the utilization of clean energy is insufficient. In 2016, the abandoned electricity generated by water, wind, and light reached up to 100 billion kWh. Actively developing and sufficiently utilizing clean energy can reduce the share of thermal power, which will be conducive to the achievement of China's carbon mitigation goal and the improvement of the ecological environment.

Author Contributions: M.M. designed this research and provided some professional suggestions. L.W. wrote this paper. Q.C. reviewed this paper and proposed some valuable ideas.

Funding: This study was supported by the National Natural Science Foundation of China (NSFC) (71471061) and the Fundamental Research Funds for the Central Universities (2017MS171).

Conflicts of Interest: The authors declare no conflict of interest.

References

1. Statistical Review of World Energy 2017. Available online: http://www.bp.com/statisticalreview (accessed on 20 June 2018).
2. Effort Sharing 2021–2030: Targets and Flexibilities. Available online: https://ec.europa.eu/clima/policies/effort/proposal_en (accessed on 20 August 2018).
3. The Intended Nationally Determined Contributions (INDCs). Available online: http://www.China.org.cn/chinese/China_key_words/2016-01/20/content_37622290.htm (accessed on 20 June 2018). (In Chinese)
4. Toward an Environmentally Sustainable Future: Country Environmental Analysis of the People's Republic of China. Available online: http://www.Adb.Org/publications/toward-environmentally-sustainable-future-country-environmental-analysis-prc (accessed on 20 June 2018).
5. Smol, J. Climate change: A planet in flux. *Nature* **2012**, *483*, S12–S15. [CrossRef] [PubMed]
6. Wachsmuth, D.; Cohen, DA.; Angelo, H. Expand the frontiers of urban sustainability. *Nature* **2016**, *536*, 391–393. [CrossRef] [PubMed]
7. National Bureau of Statistics of the People's Republic of China. *China Statistical Yearbook (2001–2015)*; China Statistics Press: Beijing, China, 2002–2016.
8. China Electricity Council. *Annual Development Report of China's Electric Power Industry 2016*; China Market Press: Beijing, China, 2016. (In Chinese)
9. Zhu, D. Analysis on advantages and disadvantages of initial allocation method for carbon emissions quota in electric power industry. *Manag. Chin. Power Enterp.* **2017**, *1*, 38–41. (In Chinese)
10. Guo, D.; Chen, H.; Long, R.; Ni, Y. An integrated measurement of household carbon emissions from a trading-oriented perspective: A case study of urban families in Xuzhou, China. *J. Clean. Prod.* **2018**, *188*, 613–624. [CrossRef]
11. Dai, H.; Xie, Y.; Liu, J.; Masui, T. Aligning renewable energy targets with carbon emissions trading to achieve China's INDCs: A general equilibrium assessment. *Renew. Sustain. Energy Rev.* **2018**, *82*, 4121–4131. [CrossRef]
12. Song, Y.; Liang, D.; Liu, T.; Song, X. How China's current carbon trading policy affects carbon price? An investigation of the Shanghai Emission Trading Scheme pilot. *J. Clean. Prod.* **2018**, *181*, 374–384. [CrossRef]
13. National Development and Reform Commission of China. *Nation Construction Program for Building a Carbon Emission Trading Market in Electric Power Industry*; National Development and Reform Commission of China: Beijing, China, 2017. (In Chinese)
14. Xu, J.; Yang, X.; Tao, Z. A tripartite equilibrium for carbon emission allowance allocation in the power-supply industry. *Energy Policy* **2015**, *82*, 62–80. [CrossRef]
15. Ji, J.; Zhang, Z.; Yang, L. Comparisons of initial carbon allowance allocation rules in an O2O retail supply chain with the cap-and-trade regulation. *Int. J. Prod. Econ.* **2017**, *187*, 68–84. [CrossRef]
16. Acquaye, A.; Genovese, A.; Barrett, J.; Koh, S.C.L. Benchmarking carbon emissions performance in supply chains. *Supply Chain Manag.* **2014**, *19*, 306–321. [CrossRef]
17. Dai, F.; Xiong, L.; Ma, D. How to set the allowance benchmarking for cement industry in China's carbon market: Marginal analysis and the case of the Hubei emission trading pilot. *Sustainability* **2017**, *9*, 322. [CrossRef]
18. Cong, R.; Wei, Y. Experimental comparison of impact of auction format on carbon allowance market. *Renew. Sustain. Energy Rev.* **2012**, *16*, 4148–4156. [CrossRef]
19. Liu, L.; Sun, X.; Chen, C.; Zhao, E. How will auctioning impact on the carbon emission abatement cost of electric power generation sector in China? *Appl. Energy* **2016**, *168*, 594–609. [CrossRef]
20. Zhou, P.; Wang, M. Carbon dioxide emissions allocation: A review. *Ecol. Econ.* **2016**, *125*, 47–59. [CrossRef]
21. Wen, J.; Zou, L.; Guo, J.; Wang, K.; Wei, Y. How can China reach its CO_2 intensity reduction targets by 2020? A regional allocation based on equity and development. *Energy Policy* **2011**, *39*, 2407–2415.
22. Qin, Q.; Liu, Y.; Li, X.; Li, H. A multi-criteria decision analysis model for carbon emission quota allocation in China's east coastal areas: Efficiency and equity. *J. Clean. Prod.* **2017**, *168*, 410–419. [CrossRef]
23. Zhou, Z.; Liu, C.; Zeng, X.; Jiang, Y.; Liu, W. Carbon emission performance evaluation and allocation in Chinese cities. *J. Clean. Prod.* **2018**, *172*, 1254–1272. [CrossRef]
24. Miao, Z.; Geng, Y.; Sheng, J. Efficient allocation of CO_2 emissions in China: A zero sum gains data envelopment model. *J. Clean. Prod.* **2016**, *112*, 4144–4150. [CrossRef]

25. Zhou, P.; Sun, Z.; Zhou, D. Optimal path for controlling CO_2 emissions in China: A perspective of efficiency analysis. *Energy Econ.* **2014**, *45*, 99–110. [CrossRef]
26. Liu, H.; Lin, B. Cost-based modelling of optimal emission quota allocation. *J. Clean. Prod.* **2017**, *149*, 472–484. [CrossRef]
27. Ren, J.; Bian, Y.; Xu, X.; He, P. Allocation of product-related carbon emission abatement target in a make-to-order supply chain. *Comput. Ind. Eng.* **2015**, *80*, 181–194. [CrossRef]
28. An, J.; Lee, J. A newsvendor non-cooperative game for efficient allocation of carbon emissions. *Sustainability* **2018**, *10*, 154. [CrossRef]
29. Yu, S.; Wei, Y.; Fan, J.; Zhang, X.; Wang, K. Exploring the regional characteristics of inter-provincial CO_2 emissions in China: An improved fuzzy clustering analysis based on particle swarm optimization. *Appl. Energy* **2012**, *92*, 552–562. [CrossRef]
30. Meng, M.; Niu, D.; Shang, W. A small-sample hybrid model for forecasting energy-related CO_2 emissions. *Energy* **2014**, *64*, 673–677. [CrossRef]
31. Liu, W.; Cai, W.; Cui, B.; Wang, M. Synthetic evaluation method of electronic visual display terminal visual performance based on the letter search task. In *Virtual, Augmented and Mixed Reality*; Shumaker, R., Lackey, S., Eds.; Springer International Publishing: Cham, Switzerland, 2015; Volume 9179, pp. 30–38.
32. National Bureau of Statistics of the People's Republic of China. *China Energy Statistical Yearbook (2011–2015)*; China Statistics Press: Beijing, China, 2012–2016.
33. GAQS (General Administration of Quality Supervision); IQPRC (Inspection and Quarantine of the People's Republic of China). *Standardization Administration of the People's Republic of China, General Principles for Calculation of Total Production Energy Consumption*; China Standard Press: Beijing, China, 2008. (In Chinese)
34. 2006 IPCC Guidelines for National Greenhouse Gas Inventories. Available online: http://www.ipcc-nggip.iges.or.jp/public/2006gl/vol2.html (accessed on 20 June 2018).
35. Meng, M.; Jing, K.; Mander, S. Scenario analysis of CO_2 emissions from China's electric power industry. *J. Clean. Prod.* **2017**, *142*, 3101–3108. [CrossRef]

energies

MDPI

Article

Optimization of Inventory Routing Problem in Refined Oil Logistics with the Perspective of Carbon Tax

Songyi Wang [1], Fengming Tao [1,2,*] and Yuhe Shi [3]

1 College of Mechanical Engineering, Chongqing University, Chongqing 400044, China; songyi_wang@cqu.edu.cn
2 School of Economics and Business Administration, Chongqing University, Chongqing 400044, China
3 School of Transportation and Logistics, Southwest Jiaotong University, Chengdu 610031, China; SHI681242@163.com
* Correspondence: taofengming@cqu.edu.cn; Tel.: +86-185-8070-7012

Received: 12 May 2018; Accepted: 28 May 2018; Published: 4 June 2018

Abstract: In order to solve the optimization problem of the refined oil distribution system from the perspectives of low-carbon and environmental protection, this paper focuses on the characteristics of the secondary distribution of refined oil and combines it with the integrated optimization concept of refined oil distribution network, where a low-carbon inventory routing problem (LCIRP) model is constructed with the minimum total costs as the objective function on the basis of considering carbon emissions. An adaptive genetic algorithm combined with greedy algorithm is designed to solve the model, and an example is given to verify the effectiveness of the algorithm. Then, this paper solves the model with two parts by introducing a practical numerical example: in the first part, the LCIRP models with different carbon tax values are solved, which verifies the effectiveness of the model and proves that carbon tax policies can effectively reduce the carbon emissions in the secondary distribution network of refined oil; in the second part, the LCIRP models with the different maximum load capacity of oil tank trucks are solved, which provides the economic and environmentally friendly distribution schemes for refined oil distribution enterprises under the premise of carbon tax policies and load limitation. Finally, the emission reduction proposals that take into account both economic and environmental benefits are given respectively from the aspect of government environmental protection agencies and from the aspect of refined oil distribution enterprises.

Keywords: refined oil distribution; inventory routing problem; hybrid genetic algorithm; carbon emissions; carbon tax

1. Introduction

As a special energy commodity, petroleum plays a decisive role in the national economy and people's livelihood in a country [1]. The distribution of refined oil refers to the entire logistics process of transporting refined oil from the oil refinery to consumers, which connects the oil refining company, distribution oil depots and sales outlets (gas stations). In addition, the distribution of refined oil can be divided into two stages: primary distribution and secondary distribution. A primary distribution refers to the process of transporting refined oil from the refinery to distribution oil depots, and secondary distribution refers to the process of transporting refined oil from a distribution oil depot to gas stations [2]. The secondary distribution of refined oil has the characteristics of small quantity and multiple batches. With the intensification of global market competition, the importance and necessity of a highly efficient distribution network integration strategy have begun to become prominent. How to achieve the reduction of transportation costs and the improvement of distribution efficiency in the inventory routing problem (IRP) of gas stations in a region has gradually become a core problem in corporate distribution management. At the same time, the combinatorial optimization research on the IRP of refined oil secondary distribution has also become the focus of many scholars.

In order to meet the needs of economic development, gas stations need a continuous supply of refined oil, but the transport of small quantity and multiple batches poses a series of threats to the ecological environment. The transportation of petrochemical products will generate a large amount of carbon emissions, which will lead to the increase of greenhouse gases, exacerbating air pollution and the greenhouse effect [3,4]. With the continuous promulgation of the carbon tax and other regulatory policies, the low-carbon transportation is imperative [5]. Therefore, it is necessary to optimize the design of the refined oil distribution network while considering the constraints of efficiency and environment, thus reaching a balanced state of economic efficiency and environmental benefits to achieve a win-win situation.

The remaining parts of this paper are organized as follows. In the next section, a literature review on the refined oil distribution optimization problem, as well as the balance between carbon reduction and cost optimization, is presented. Section 3 discusses the construction of the low-carbon IRP (LCIRP) model. The hybrid genetic algorithm is introduced to solve the model in Section 4. Section 5 gives a numerical experiment and results analysis. Finally, Section 6 concludes this paper and presents expectations for future work.

2. Literature Review

Since the main idea of the current research is to balance carbon emission reduction and cost optimization in IRP of refined oil logistics. We review the studies in two fields: refined oil distribution optimization and the balance between carbon reduction and cost optimization.

2.1. Refined Oil Distribution Optimization

Federgruen and Zipkin [6] first explicitly proposed the inventory routing problem (IRP) in 1984. They optimized the inventory routing scheme of the supply chain through scientifically making decisions about retailer's order quantity, order cycle, and route arrangement. In the same period, Golden et al. [7] also made a similar study. The ratio of the current inventory level to the storage capacity was used to describe the urgency of distribution demands, and a heuristic algorithm was designed to solve the problem.

With the improvement of the research level, the research of IRP has shown a trend of increasing complexity of the problem, and the depth and breadth of the problem have been extended. Avella et al. [8] studied the oil transportation problem with limited distribution resources (limited vehicles and limited drivers). A heuristic algorithm combined with the branch pricing algorithm was designed, which can be solved quickly. Moreover, they illustrated the effectiveness of the algorithm by multiple sets of real-world examples. Cornillier et al. [9–11] has been devoted to the study of IRP. They constructed mathematical models from the perspectives of multi-period distribution, distribution with time windows, multiple oil depot distribution and so on, and designed efficient algorithms to solve them. For the vehicle routing problem in the distribution of refined oil, Dai et al. [12] designed a human–computer interactively solving method based on a two-phase heuristic algorithm. Ma et al. [13] proposed a mathematical model with the shortest driving path and high full load rate as the optimization goals for the secondary distribution of refined oil, and the improved genetic algorithm was used to solve the model. Bocto et al. [14] designed several heuristic algorithms to solve the mathematical model for the problem of replenishment at gas stations, the simulation and real data were applied to verify the effectiveness of the algorithms respectively. Zhen et al. [15] studied the distribution of refined oil from the perspective of third-party logistics and introduced a combinatorial auction mechanism to construct a mixed-integer programming model with the objective of minimizing costs. A heuristic algorithm was devised to solve the problem. Popovic et al. [16] aimed at IRP in the refined oil logistics. An improved variable neighborhood search algorithm was introduced to solve the joint optimization model of the inventory and routing. In the case of known distribution routes and limited number of vehicles, Li et al. [2] introduced the concept of order neighborhood into refined oil distribution and established the optimization model of refined oil distribution route based on order neighborhood. The model was solved by an improved quantum genetic algorithm. Wang et al. [17] divided the secondary distribution of refined oil into two modes: one claimed that the different oil products required by a gas station was permitted to be separated and distributed by several vehicles, and the other must be entirely delivered by one single vehicle. According to the two modes, two kinds of distribution sub-models were derived on the basis of the basic model of the product oil multi-cabin distribution with time window, and they proposed an ant colony and tabu search hybrid algorithm (ACO-TS) to solve models.

2.2. The Balance between Carbon Emission Reduction and Cost Optimization

Environmental protection has been an issue of serious concern for many years [18]. In the perspectives of carbon tax and other series of carbon policies, many scholars pay attention to the trade-off between costs and carbon emissions in the optimization of low-carbon logistics.

Jabali et al. [19] studied the trade-off between carbon emissions and travel time in time-dependent vehicle routing problems and discussed the boundaries of carbon emissions in transportation. Elhedhli et al. [20] embedded carbon emission costs into the design costs of supply chain network, in which the calculation of carbon emissions was modeled by considering the vehicle weight. The test results indicate that the design scheme of supply chain network considering the carbon emission costs will has a great change. Wang et al. [21] proposed a multi-objective optimization model for the network design of green supply chain. The economic costs and carbon emissions in the network design of supply chain were taken into account, and the choice of facility location and transportation mode were determined by the intensity of emission reduction. Tsai et al. [22] presented a mixed activity-based costing decision model for green airline fleet planning. They analyzed the impact of carbon emissions on operating costs of aviation logistics under the constraints of the European Union Emissions Trading Scheme. Jaber et al. [23] studied the cooperation between buyers and sellers based on minimizing carbon emissions, and numerical analysis showed that carbon tax was a very effective way to control carbon emissions. The impacts of carbon regulatory mechanisms on replenishment decisions in a biofuel supply chain were analyzed by Palak et al. [24]. They employed the mathematical models to capture the trade-offs that exist between costs and emissions due to inventory and transportation.

In short, based on the above two parts of the analysis, there are many studies on combination optimization in refined oil logistics. At the same time, the research on the optimization design of low-carbon supply chains is abundant under carbon tax and other regulatory policies. However, there is no research on the low carbonation of IRP in refined oil logistics. In the existing refined oil logistics researches, whether it is the basic IRP or the variant problem of IRP, most of them only aim at the optimization of operating costs and rarely consider carbon emissions. However, with the gradual deepening of the concept of sustainable development and the implementation of a number of energy-saving and emission reduction policies, carbon emission in the optimization of refined oil distribution network has become a problem to be solved urgently. In view of this, this paper proposes a green and environmental protection model that considers carbon emissions in the optimization of refined oil logistics networks: the low-carbon inventory routing problem (LCIRP) model. An adaptive genetic algorithm combined with greedy algorithm is designed to solve the model. Finally, the effectiveness and feasibility of the model are verified by numerical experiments.

3. Model Formulation

3.1. Problem Description

The LCIRP model of refined oil distribution studied in this paper can be described as follows. An oil depot distributes refined oil to different gas stations through oil tank trucks. The location of each gas station is known, and a limited number of oil tank trucks start from the oil depot and return after the distribution tasks have been completed. Based on meeting the inventory requirements of gas stations and the constraint of the trucks load, the LCIRP model with the minimum comprehensive costs is constructed by considering the fixed costs, transportation costs, shortage costs, inventory costs, penalty costs, and carbon emission costs in the distribution process. Then, we hope to obtain the economic and environmental-friendly distribution plan under the premise of ensuring the delivery services are completed.

3.2. Parameters and Variables

According to the needs of building the model, this paper sets the following parameters and variables, as shown in Table 1.

Table 1. The meaning of parameters and variables.

Parameters and Variables	Meaning
V	The collection of nodes, 1, 2, ..., V is the serial number of gas stations, 0 represents oil depot, $V\{n\|n = 0, 1, 2, \ldots, \|V\|\}$
K	The collection of oil tank trucks owned by oil depot, $K\{k\|k = 1, 2, \ldots, \|K\|\}$
I_0	The initial inventory of oil depot
R	The actual amount of completed delivery of the oil depot
h_0	The inventory rates of oil depot (the inventory costs of unit cargoes for unit time)
M	The fixed cost of using the oil depot
C_0	The carbon tax on unit carbon emissions
Q_k	The maximum load allowed for oil tank truck k
v	The speed of oil tank trucks
r_i	The demand of gas station i
c_{ij}^k	The transportation costs for unit distance when the oil tank truck k from node i to node j
h_i	The inventory rates of gas station i, $h_0 < h_i$
C_k	The fixed cost of oil tank truck k
μ_1	The waiting costs for unit time when the oil tank truck arrives at gas station node in advance
μ_2	The penalty costs for unit time when the oil tank truck is late to gas station node
Q_{ij}^k	The weight of the loaded cargoes when the oil tank truck k travels between node i and node j
t_{si}	The time required for oil tank trucks to serve gas station i
d_{ij}	The transport distance from node i to node j
t_{ij}^k	The travel time of oil tank truck k from node i to node j
t_i^k	The time to arrive at the node i for oil tank truck k
b_i	The unit shortage cost for gas station i
I_i	The inventory of gas station i after distribution
CM	The carbon emissions from oil tank truck during the distribution process
$[ET_i, LT_i]$	Time window required by gas station i (namely, the range of time)
ρ	The fuel consumption of oil tank truck for unit distance
ρ_0	The fuel consumption for unit distance when oil tank truck is empty
e_0	The carbon emissions generated by unit fuel consumption
ρ^*	The fuel consumption for unit distance when oil tank truck is fully loaded
Y_k	$Y_k = 1$ represents oil tank truck k is used in oil depot, otherwise $Y_k = 0$
y_i^k	$y_i^k = 1$ represents the oil tank truck k serves the gas station i, otherwise $y_i^k = 0$
x_{ij}^k	$x_{ij}^k = 1$ represents the oil tank truck k passes through the road between node i and node j, otherwise $x_{ij}^k = 0$

3.3. Model Development

The LCIRP model of refined oil distribution constructed in this paper takes the total costs minimum as the objective function. Firstly, we need to analyze the sub-costs, and then the total costs of the IRP are determined by the various sub-costs.

3.3.1. Objective Function Analysis of Model

1. Fixed Costs

The fixed costs C_1 in the LCIRP model can be expressed as

$$C_1 = M + \sum_{k \in K} C_k Y_k \tag{1}$$

where C_1 contains the fixed costs of using oil depots and oil tank trucks, which mainly includes routine maintenance, depreciation expenses as well as labor costs for drivers and other employees.

2. Transportation Costs

Under the influence of fuel consumption, maintenance and other factors, the transportation costs are proportional to the travel mileage of the trucks. The transportation costs C_2 in the LCIRP model can be expressed as

$$C_2 = \sum_{k \in K} \sum_{i,j \in V} c_{ij}^k x_{ij}^k d_{ij} Y_k \tag{2}$$

3. Shortage Costs

The shortage costs of gas stations are proportional to the shortage volume in the current period.

$$C_3 = b_i \max\{0, r_i - I_i\} \tag{3}$$

4. Penalty Costs

When sudden events such as urban traffic jam, trucks breakdown and other emergencies occur, oil tank trucks may be unable to reach the destination within the time range required by gas stations (affected by inventory capacity and safety inventory of the gas stations, refined oil need to be sent to gas stations between the earliest delivery time and the latest delivery time). If oil tank trucks cannot arrive on time, it is necessary to pay a certain amount of penalty costs C_4.

$$C_4 = \sum_{k \in K} \sum_{i \in V \backslash \{0\}} \left(\mu_1 \max\left\{ ET_i - t_i^k, 0 \right\} + \mu_2 \max\left\{ t_i^k - LT_i, 0 \right\} \right) \tag{4}$$

where $\max\left\{ ET_i - t_j^k, 0 \right\}$ indicates the advance arrival time for oil tank truck k services gas station i; $\max\left\{ t_i^k - LT_i, 0 \right\}$ indicates the late time for oil tank truck k services gas station i.

5. Inventory Costs

Distribution activity occurs at the end of each distribution cycle and before the start of next cycle. The inventory costs of oil depot can be expressed as:

$$H_0 = h_0 (I_0 - R),\ R = \sum_{i \in V \backslash \{0\}} \left(r_i \sum_{j \in V} x_{ij}^k \right) \tag{5}$$

Due to the different inventory between the beginning of the delivery cycle and the end of the delivery cycle, the average inventory is used to calculate the inventory costs of the gas station i, which is denoted by H_i.

$$H_i = h_i\left(I_i - \frac{1}{2}\min\{r_i, I_i\}\right) \tag{6}$$

The inventory costs C_5 consist of two parts: the inventory costs of oil depot and the inventory costs of the gas station.

$$C_5 = h_0\left(I_0 - \sum_{i\in V\backslash\{0\}}\left(r_i \sum_{j\in V} x_{ij}^k\right)\right) + \sum_{i\in V\backslash\{0\}} h_i\left(I_i - \frac{1}{2}\min\{r_i, I_i\}\right) \tag{7}$$

6. Carbon Emission Costs

Carbon emissions during the distribution and transportation process of refined oil are mainly caused by fuel consumption.

The linear function formula of fuel consumption in per unit distance $\rho(X)$ is introduced in this paper.

$$\rho(X) = \rho_0 + \frac{\rho^* - \rho_0}{Q}X \tag{8}$$

Therefore, the carbon emissions can be expressed as follows:

$$CM = e_0\rho\left(Q_{ij}^k\right)d_{ij},\ (0 \le Q_{ij}^k \le Q_k) \tag{9}$$

The carbon emission costs of the LCIRP model C_6 shown as follows:

$$C_6 = C_0\sum_{k\in K}\sum_{i\in V}\sum_{j\in V} x_{ij}^k d_{ij}\left[e_0\rho\left(Q_{ij}^k\right)\right] \tag{10}$$

3.3.2. LCIRP Model Setting

Based on the analysis in Section 3.3.1, the LCIRP model of refined oil distribution constructed in this paper is as follows:

$$\begin{aligned}\min C = &\left(M + \sum_{k\in K} C_k Y_k\right) + \left(\sum_{k\in K}\sum_{i,j\in V} c_{ij}^k x_{ij}^k d_{ij} Y_k\right) \\ &+ (b_i \max\{0, r_i - I_i\}) \\ &+ \left(\sum_{k\in K}\sum_{i\in V\backslash\{0\}}\left(\mu_1 \max\left\{ET_i - t_i^k, 0\right\} + \mu_2 \max\left\{t_i^k - LT_i, 0\right\}\right)\right) \\ &+ h_0\left(I_0 - \sum_{i\in V\backslash\{0\}}\left(r_i \sum_{j\in V} x_{ij}^k\right)\right) + \sum_{i\in V\backslash\{0\}} h_i\left(I_i - \tfrac{1}{2}\min\{r_i, I_i\}\right) \\ &+ C_0 \sum_{k\in K}\sum_{i\in V}\sum_{j\in V} x_{ij}^k d_{ij}\left[e_0\rho\left(Q_{ij}^k\right)\right]\end{aligned} \tag{11}$$

Subject to

$$\sum_{k\in K}\sum_{j\in V\backslash\{0\}} x_{ij}^k = 1,\ \forall i \in V \tag{12}$$

$$\sum_{i\in V} x_{ij}^k = \sum_{i\in V} x_{ji}^k,\ \forall j \in V\backslash\{0\} \tag{13}$$

$$\sum_{j\in V\{0\}} x_{0j}^k = \sum_{i\in V\backslash\{0\}} x_{i0}^k = 1,\ \forall k \in K \tag{14}$$

$$\sum_{i \in V \backslash \{0\}} \left(r_i \sum_{j \in V} x_{ij}^k \right) \leq I_0, \forall k \in K \tag{15}$$

$$Q_{ij}^k \leq Q_k x_{ij}^k, \ i \neq j, \ \forall i \in V, \ \forall j \in V, \ \forall k \in K \tag{16}$$

$$x_{ij}^k + x_{ji}^k \leq 1, \ \forall i \in V \backslash \{0\}, \ \forall j \in V \backslash \{0\}, \ \forall k \in K \tag{17}$$

$$t_j^k = t_i^k + t_{si} + t_{ij}^k, \ \forall i \in V, \ \forall j \in V, \ \forall k \in K \tag{18}$$

The objective function of the model is to minimize the total costs, as shown in (11). Constraint (12) imposes the attention that there is only one oil tank truck provides delivery service for each gas station. Constraint (13) indicates the balance of oil tank trucks that entering and leaving each gas station. All oil tank trucks start from the oil depot and return to the oil depot after the distribution tasks are completed, and their operation is shown in (14). Constraint (15) represents that the delivery volume is limited by the remaining stock of the oil depot. Furthermore, the delivery volume is limited by the loading capacity of oil tank truck, as mentioned in (16). Constraint (17) is introduced to eliminate the sub-loops. The continuity of the travel time of oil tank trucks is emphasized in (18).

4. Algorithm Design

Genetic algorithm is a random parallel search algorithm based on the principles of natural selection and genetics, which is a more effective method for solving NP-Hard problem [25,26]. However, traditional genetic algorithm often has the disadvantages of slow convergence rate, easy falling into local optimum, and low optimization accuracy. Therefore, the goal of the continuous improvement of genetic algorithm is how to improve the convergence rate of the algorithm while ensuring the diversity of population, so that the optimization result is close to the optimal solution. In this paper, we propose an improved adaptive genetic algorithm combined with greedy algorithm (Hybrid Genetic Algorithm, HGA) to solve the LCIRP model. The following improvements have been made based on traditional genetic algorithms.

4.1. Combining Greedy Algorithm to Generate Initial Population

In traditional genetic algorithm, the initial population that is generated by a stochastic method has a low individual fitness, which will restrict the convergence rate of the algorithm to a certain extent. The greedy algorithm is used to optimize the initial individuals in this paper, which means the advantage of greedy algorithm in local optimization is utilized to generate new individuals. First, we select a gas station randomly and add it into individual. Then, the nearest gas station to the current gas station should be found through searching for all gas stations that have not been added to the individual, and it will be added into the individual as the current location. The initially optimized individual can be got by continuing searching the nearest gas station until all gas stations are added into the individual. The initial population generated by the greedy algorithm does not lose randomness, and its overall quality is improved, which helps to bring higher optimization speed. After we get the initial population, what we need to do is optimizing each initial individual by the genetic algorithm. In each iteration process, the fitness of the individuals should be calculated at first, and then the selection, crossover, and mutation operations are performed in sequence.

4.2. Adaptive Adjustment Mechanism of Crossover Probability

In the evolution of the traditional genetic algorithm, crossover probability p_c and mutation probability p_m are often set to fixed numbers [27]. However, parameters such as p_c and p_m are the key factors that affect the performance of the genetic algorithm [28]. Many researches have studied the adaptive calculation of parameters such as crossover probability and mutation probability of the genetic algorithm [27,29–31]. On the basis of these, the adaptive adjustment mechanisms are set up for p_c and p_m in this paper.

Crossover operation promotes the continuous updating of the population, and the size of p_c determines the updating rate of individuals. If the p_c is too large, it will destroy the excellent genetic modes. If the p_c is too small, the search speed of the algorithm will be slow, and the population will be difficult to evolve. In the early stages of evolution, in order to expand the overall search range and increase the updating speed of population, the value of p_c should be increased. In the later stages of evolution, the overall solution set of population tends to be stable, so p_c should be appropriately reduced to keep the superior gene structures preserved. In addition, the crossover operator can change or even destroy the gene structures. For individuals with poor fitness, it should be set higher p_c because more participation in crossover operation is benefit to its continuous optimization. Correspondingly, for individuals with higher fitness, the probability of crossover operation should be smaller in order to prevent the destruction of the gene structures. Based on the above analysis, we set the following adjustment mechanisms:

$$p_{ci} = \begin{cases} \max p_c - (\max p_c - \min p_c)\left(\frac{g}{2G} + \frac{f_i - \overline{f}}{2(f_{\max} - \overline{f})}\right), & f_i \geq \overline{f}; \\ \max p_c, & f_i < \overline{f}. \end{cases} \tag{19}$$

$$\max p_c = \begin{cases} 0.9, & g \leq G/4 \\ 0.8, & G/4 < g \leq 3G/4 \\ 0.7, & 3G/4 < g \leq G. \end{cases} \tag{20}$$

where p_{ci} is the probability that crossover operation occurs on the individual i, a higher crossover probability is given in the early evolution and the crossover probability is decreased in the later stages of evolution; G is the maximum iteration number during evolution; g is the current iteration number; the value of $\max p_c$ is related to the current iteration number, $\min p_c = 0.6$; f_i is the fitness of individual i, $f_{\max}$ is the maximum fitness value of all individuals, and $\overline{f}$ is the average fitness value of the current population.

4.3. Adaptive Adjustment Mechanism of Mutation Probability

The mutation of population will be affected by p_m, and the proper mutation of individuals can maintain the diversity of the population and prevent it from falling into a local optimum. However, if p_m is too large, the algorithm is similar to random search and loses the characteristic of genetic evolution. From the two aspects of genetic evolution number and the fitness values of individuals, an adjusting formula for mutation probability is established.

$$p_{mi} = \begin{cases} \min p_m + (\max p_m - \min p_m)\left(\frac{g}{2G} + \frac{f_i - \overline{f}}{2(f_{\max} - \overline{f})}\right), & f_i \geq \overline{f}; \\ \min p_m, & f_i < \overline{f}. \end{cases} \tag{21}$$

where p_{mi} is the probability that mutation occurs on the individual i. From the Formula (21), it can be seen that the smaller the fitness values of individuals, the less the possibility of mutation. With the increase in the iteration number, individuals tend to have a similar genetic structure, which is likely to fall into a local optimum. In order to avoid this situation, the mutation probability of individuals should be appropriately increased to encourage the emergence of new individuals and maintain individuals' diversity. We set $\max p_m = 0.005$, $\min p_m$ has been adjusted with the change of iteration number, which is shown as follows:

$$\min p_m = \begin{cases} 0.001, & g \leq G/4; \\ 0.002, & G/4 < g \leq 3G/4; \\ 0.003, & 3G/4 < g \leq G. \end{cases} \tag{22}$$

In the early stages of evolution, individuals are less likely to occur mutation; in the later stages of evolution, the higher mutation probability of individuals is conducive to expanding the search scope and jumping out of local optimum.

Based on the above analysis, the basic process of the improved algorithm designed in this paper is illustrated in Figure 1.

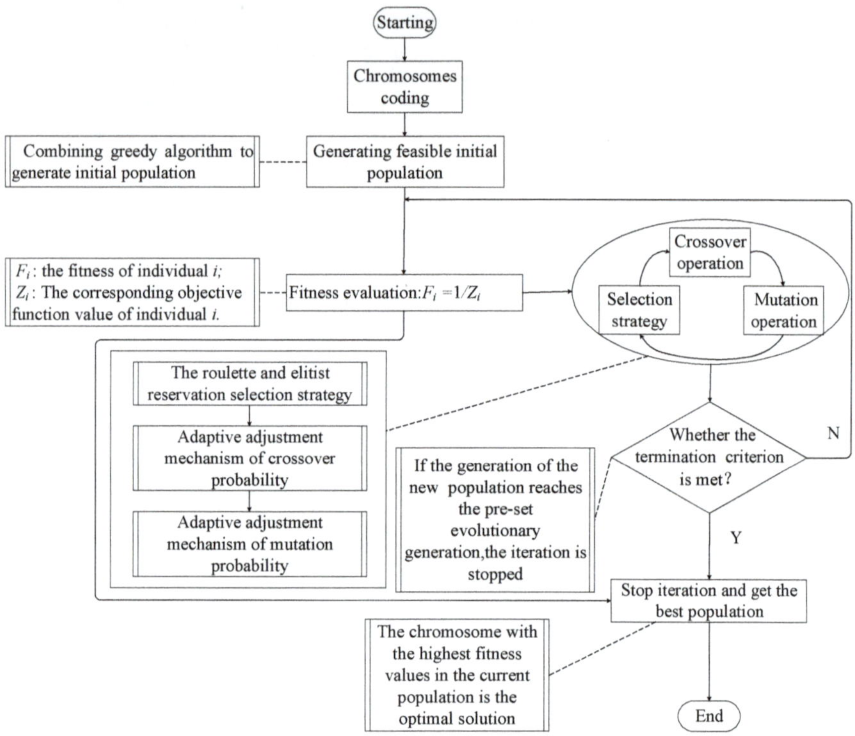

Figure 1. Basic process of hybrid genetic algorithm.

5. Experimental Design and Result Analysis

The example verification includes two parts: First, the HGA proposed in this paper is tested in Section 5.1 by using the typical VRP test question database. Second, in Section 5.2, the effectiveness of the LCIRP model is verified by an actual example of distribution problem in a refined oil logistics enterprise.

5.1. Algorithm Experiment

In this section, the typical test question database that is called Benchmark Problems [32] is used to verify the effectiveness of the proposed HGA. There are six test examples listed in the test question database (that is, R1, C1, RC1, R2, C2, and RC2), and this paper randomly selects two problems from each type of examples to form test data sets, which are solved by the traditional genetic algorithm (GA), the cycle evolutionary genetic algorithm (CEGA) [5], and HGA proposed in this paper. The results are shown in Table 2.

Table 2. Experimental results of HGA and other algorithms.

Examples	GA		CEGA		HGA	
	The Number of Trucks	Distance	The Number of Trucks	Distance	The Number of Trucks	Distance
R1-01	19	1701.92	18	1647.32	18	1613.56
R1-12	9	996.31	9	982.14	9	983.68
C1-04	10	881.23	10	831.58	10	824.78
C1-05	10	879.37	10	828.94	10	828.94
RC1-01	14	1789.36	14	1696.94	14	1634.29
RC1-08	10	1396.51	10	1139.82	10	1139.82
R2-02	6	1242.34	3	1191.70	4	1176.39
R2-10	4	1137.54	3	939.34	3	939.34
C2-01	3	643.57	3	591.56	3	591.56
C2-06	4	685.43	4	654.91	3	588.49
RC2-05	5	1517.68	4	1389.37	4	1297.19
RC2-07	4	1259.31	3	1061.14	3	1061.14

As can be seen from Table 2, when solving the above 12 examples, the results obtained by HGA are 100% better than the results obtained by GA, and nearly 91.7% are better than the results obtained by CEGA. Therefore, the HGA proposed in this paper is very competitive in solving the test examples.

5.2. Model Experiment

In this paper, the data of the refined oil distribution from an oil transportation company (hereinafter referred to as OTC) are introduced to verify the LCIRP model. The OTC needs to transport the refined oil in the oil depot to gas stations. The initial stock of the oil depot is 20 thousand cubic meters, and the oil depot coordinates are (54, 41). The demand for refined oil at 19 gas stations in a certain area should be met by the oil depot, the location of the gas stations and the information of the inventory and demand are shown in Table 3. All distribution tasks are carried out by road transportation, and all roads are non-forbidden roads. The service time for all gas stations is set as 30 min, and the parameters of oil tank truck are shown in Table 4. Other parameters are set as shown in Table 5 with referring to the literatures [33–35].

Table 3. Demand Information of gas stations.

Demand Point	x Coordinate	y Coordinate	Maximum Inventory (m^3)	Current Inventory (m^3)	Minimum Demand (m^3)	Consumption Rate (m^3/min)
1	91	71	32	11.52598	11.52598	0.024012
2	82	82	33	11.22246	11.22246	0.02338
3	10	7	16	7.62968	7.62968	0.015895
4	73	26	21	8.392063	8.392063	0.017483
5	7	6	21	7.830946	7.830946	0.016314
6	36	38	48	17.93179	17.93179	0.037358
7	64	31	30	10.20813	10.20813	0.021267
8	96	80	48	20.32358	20.32358	0.042341
9	23	32	49	20.03141	20.03141	0.042982
10	68	27	36	13.00829	13.00829	0.027101
11	20	66	39	13.17007	13.17007	0.027438
12	81	37	31	11.61286	11.61286	0.024193
13	6	14	32	12.32107	12.32107	0.025669
14	90	46	30	14.30075	14.30075	0.029793
15	79	74	32	11.34801	11.34801	0.023642
16	95	84	42	15.63903	15.63903	0.032581
17	25	6	14	4.683246	4.683246	0.009757
18	36	24	40	16.78833	16.78833	0.034976
19	25	22	48	21.44084	21.44084	0.044668

Table 4. The parameters of oil tank truck.

Parameter	Parameter Value	Parameter	Parameter Value
Outline dimension (mm)	12,350 × 2500 × 3850	Tank volume (m^3)	51
Total mass (kg)	40,000	Rated load capacity (kg)	33,400
The number of tires	12	Rated power (kw)	335
Swept volume (L)	4.0	Integrated fuel consumption (L/km)	0.4

Table 5. Model parameter settings.

Parameter	Parameter Value
M	200 CNY/d
h_0	32 CNY/t·d
h_i	50 CNY/t·d
C_k	100 CNY/d
b_i	1000 CNY/t
μ_1	30 CNY/h
μ_2	40 CNY/h
v	20 km/h
ρ_0	0.23 L/km
e_0	2.83 kg/L
ρ^*	0.45 L/km

5.2.1. Results with Different Carbon Tax Values

We set the value of carbon tax C_0 (unit: CNY/kg) from 0 to 25 in the model with referring to the literatures [5,36]. Each C_0 value is taken into the model to solve 10 times and the numerical value of the scheme with the optimal objective function is recorded, and then results can be obtained, as in Figures 2–4. Figure 2 is a line chart showing the changing trends of transportation costs, inventory costs, penalty costs, fixed costs, and carbon emission costs as carbon tax increases. It is possible to visually observe in detail the trend of each sub-cost changes as the carbon tax changes. Figure 3 is a stacked area chart showing the trend of total costs and each sub-cost changes as the carbon tax changes. The trends of total costs and carbon emissions change as carbon taxes increase, which are shown in Figure 4.

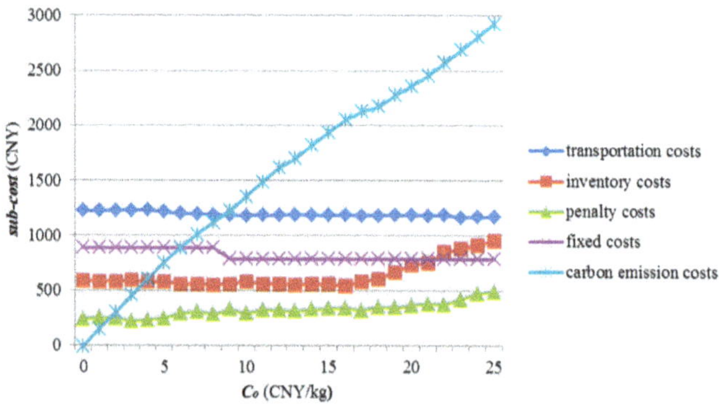

Figure 2. The line chart that shows the changing trends of each sub-cost as the carbon tax changes.

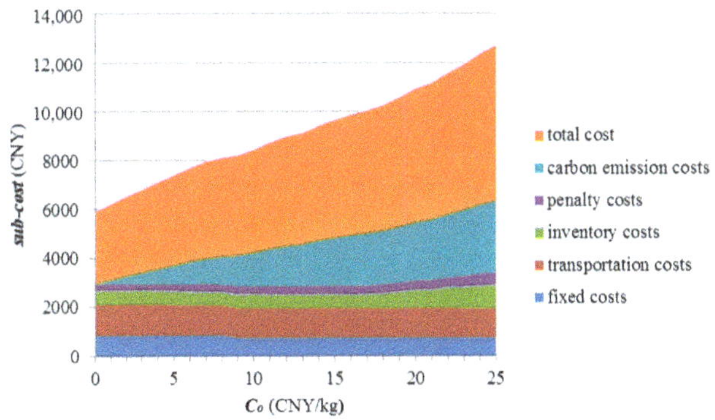

Figure 3. The stacked area chart that shows the changing trends of total cost and each sub-cost as the carbon tax changes.

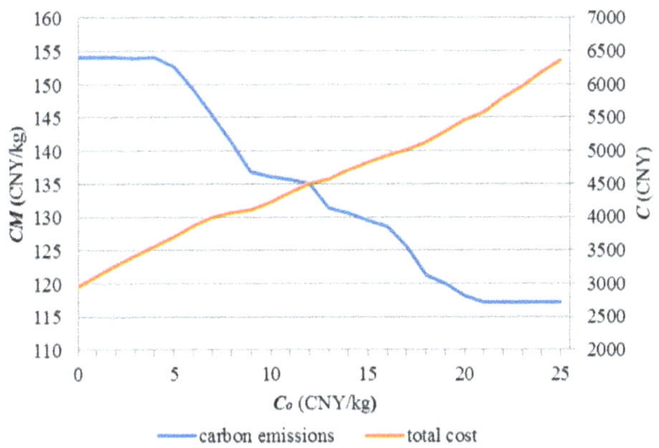

Figure 4. The line chart that shows the changing trends of total cost and carbon emissions as the carbon tax changes.

From the results in Figures 2–4, we can observe the following findings:

(1) All sub-costs in the distribution schemes have changed with the gradual increase of the carbon tax. The carbon emission costs, inventory costs, and penalty costs increase with the increase of carbon tax, of which the carbon emission costs change the most. However, the fixed costs and transportation costs decrease with the increase of carbon tax. According to Formula (2) and (10), it can be seen that both carbon emission costs and transportation costs are directly proportional to the transport distance. The model with the minimum total costs as the objective function will restrain the growth trend of carbon emission costs by optimizing the transport distance. At the same time, the reduction of the transport distance leads to the decrease of transportation costs. In addition, the Formula (10) shows that carbon emission costs are directly proportional to carbon emissions. In order to reduce the substantial increase in carbon emission costs, the model reduces the carbon emissions by optimizing the number of oil tank trucks, which also leads to a reduction in fixed costs.

(2) The proportion of each sub-cost in the total costs of distribution schemes has changed with the gradual increase of the carbon tax. The proportion of carbon emission costs in the total costs increases with the increase of the carbon tax, while the proportions of the remaining sub-costs in

the total cost decrease with the increase of the carbon tax. The Formula (10) shows that carbon emission costs are directly proportional to carbon tax value, and the substantial increase in the carbon emission costs contributes to the increase in the total costs. Therefore, the proportion of carbon emission costs in total costs continues to increase. Although the inventory costs and penalty costs increase with the increase of carbon tax, the total costs increase more greatly, so the proportions of the sub-costs in the total costs are reduced.

(3) With the increase of carbon tax, the overall carbon emissions are on a downward trend, and the total costs are on the rise. From Figure 4, we can see that when the carbon tax changes in three different ranges, the carbon emissions will have different trends. When $C_0 \in [0,4]$, carbon emissions remain unchanged because the carbon tax is too small at this time and the carbon emission costs account for a very small proportion of the total costs. When $C_0 \in (4,21)$, there is continuous reduction in carbon emissions because the proportion of carbon emission costs in total costs increases gradually with the increase of carbon tax value. The model with the minimum cost as the objective function will continue to be optimized and the dramatic increase in costs is controlled by reducing carbon emissions. When $C_0 \in [21,25]$, carbon emissions remain basically unchanged, the carbon tax value is very large at this time, and the carbon emission costs account for a large proportion of the total costs. The model with the minimum costs as the objective function has continuously optimized the carbon emissions to the minimum, so the carbon emissions remain unchanged. In summary, we can draw the conclusion that the carbon tax policy can reduce carbon emissions, but it will lead to the growth of distribution costs.

5.2.2. Results with the Different Maximum Load Capacity of Oil Tank Trucks

In this section, the carbon tax $C_0 = 12$ is taken as example to further analyze the impact of different maximum load capacity of oil tank trucks on carbon emissions and distribution schemes. The maximum load capacity of oil tank truck is set to 40, 50 and 60 (m^3), and then three groups of distribution schemes data are obtained respectively (each group is solved ten times, and the best schemes are selected), as shown in Table 6.

Table 6. Distribution schemes with the different maximum load capacity of oil tank trucks.

Load Capacity (m^3)	The Number of Truck	Distribution Scheme		Total Costs (CNY)	Carbon Emissions (kg)
40	8	0-14-1-15-0 0-12-2-0 0-6-3-5-0 0-19-11-0	0-8-16-0 0-7-17-13-0 0-10-9-0 0-4-18-0	5747.57	166.07
50	6	0-12-14-2-0 0-7-6-17-3-0 0-10-4-15-0	0-18-9-11-0 0-1-8-16-0 0-19-13-5-0	4517.73	135.24
60	5	0-19-3-5-0 0-14-11-13-0 0-1-8-16-2-0	0-6-9-18-17-0 0-7-10-4-12-15-0	4076.22	125.02

From the results in Table 6, we can observe the following findings:

(1) When the oil tank trucks with different maximum load capacity are selected for distribution, the number of trucks used in the distribution schemes will be changed. The delivery scheme needs to meet the minimum demand of each gas station to maintain the current inventory, and also to meet the potential demand time window of each gas station (the gas stations must be serviced before the current inventory is used up). Therefore, when the maximum load capacity of oil tank trucks is different, the model with the minimum costs as the objective function will choose different number of trucks. From Table 6, it can be seen that the greater the maximum load capacity of oil tank trucks, the smaller the number of trucks used.

(2) When the oil tank trucks with different maximum load capacity are selected for distribution, the total costs and carbon emissions will be affected. As shown in Table 6, when the oil tank trucks with the larger load capacity are used to distribution, the total costs of the distribution scheme are smaller as well as carbon emissions are relatively smaller.

5.3. Discussion and Analysis

For the inventory routing problem in the distribution process of refined oil, the LCIRP model is proposed in this paper to aim at the combination optimization of inventory and distribution. By taking the minimum total costs including carbon emission costs as the objective function, the distribution scheme that takes into account both economic and environmental benefits can be obtained, thereby achieving "green and low-carbon" energy distribution. However, from Section 5.2.1, we can see that the total costs of the distribution scheme have increased significantly with the increase of the carbon tax, which shows that is necessary to pay a certain economic costs for the realization of green and low-carbon logistics transportation under the trend of global warming. Through the analysis in Section 5.2.2, we know that under the development environment of energy-saving and emission-reduction, energy transport enterprises can use the delivery trucks with the appropriate load capacity to alleviate the increase in the total costs caused by the increase of carbon emission costs.

Government environmental protection agencies can adopt carbon tax policies and emission-reduction incentive policies to guide and encourage energy transportation enterprises to carry out energy conservation and emission reduction. Firstly, what they should do is to formulate carbon tax policies suitable for economic development in different industries, and strengthen supervision and monitoring of carbon emissions, thus urging enterprises to raise awareness of green and low-carbon. Secondly, the government can formulate a series of emission reduction incentive policies based on the current economic development status in different industries. The enterprises that have taken the initiative to save energy and reduce emissions should be subsidized, which will alleviate their pressure on the increase of operating costs caused by the carbon tax policies. This paper finds that carbon emissions remain basically unchanged when $C_0 \in [21, 25]$ in Section 5.2.1. It means that the carbon tax policy has limitation to reduce carbon emissions. In this situation, the government may use other ways such as subsidies to reduce carbon emissions. Thirdly, the government may think about supporting and promoting the development of low-carbon equipment. Meanwhile, the low-carbon operation companies with greater influence in the industries should be rewarded to form a demonstration effect of the low-carbon economy. Finally, the green and low-carbon consumption preferences need to be guided from the consumer side, which means that a "national low-carbon" consumption concept is established from the source of market demand.

Energy transport enterprises can adopt a series of appropriate optimization measures to balance economic and environmental benefits. First, it might make sense to introduce carbon emissions into the optimization of inventory routing problems for energy distribution and choose an optimization distribution scheme that takes into account both emission reduction and costs. Second, some measures can be taken to alleviate the pressure of costs increase brought by carbon regulatory policies, such as optimizing inventory management costs, distribution batch and distribution volume, choosing the type of transport truck reasonably (in line with the requirements of road weight restriction), and so on. Third, energy transport enterprises should respond to the low-carbon policies of the government and its industry associations and actively adopt low-carbon operation mode and transport equipment.

6. Conclusions

Simple inventory control and transportation strategy are classic problems studied over a long period of time in operational research. There are ideal solution algorithms for the models corresponding to most practical problems. However, there is little research on the joint optimization of inventory and transportation in refined oil logistics, especially IRP from a low-carbon perspective. With the aggravation of global warming and environmental pollution, low-carbon and environmentally friendly economies have become the development trends for the whole world. In the secondary distribution logistics network of refined oil with the characteristics of small quantity and multiple batches, countless gas stations require a continuous supply of refined oil, but energy transportation is accompanied by a large amount of carbon emissions. Under the background of carbon tax and other carbon regulatory policies that have been continuously enacted by governments around the world, it is necessary to optimize the secondary distribution network of refined oil while taking into account both economic and environmental benefits. In this paper, the LCIRP model is designed for inventory routing problem in the distribution process of refined oil, and an improved adaptive genetic algorithm combined with greedy algorithm is introduced to solve the model. Moreover, the numerical experiments are used to verify the effectiveness of the algorithm. Then, a practical example is introduced for simulation, and the effectiveness of the model is verified. The experimental results prove that the carbon tax policies can effectively reduce the carbon emissions in the distribution network of refined oil on the premise of paying a certain economic costs. In addition, the impact of the maximum load capacity of oil tank trucks on the distribution scheme is also analyzed.

However, in the actual distribution environment, the uncertainty of many parameters will affect the output of LCIRP model and then affect the optimal distribution scheme. In this paper, we develop our study from the perspective of the carbon tax policies. Carbon-constrained methods have a very significant contribution to emission reduction, but the implementation of mandatory carbon-constraints is hardly recognized by enterprises. For future research work, we will further study the emission-reduction optimization of refined oil distribution under various policies such as carbon supervision and carbon subsidy.

Author Contributions: S.W. was mainly responsible for the writing of the full text. S.W. implemented the experiments with the guidance of F.T. S.W. and Y.S. built the models of the paper.

Acknowledgments: This work is supported by the National Natural Science Foundation of China (No. 71571023).

Conflicts of Interest: The authors declare no conflict of interest.

References

1. Zeng, Y.; Cai, Y.; Huang, G.; Dai, J. A review on optimization modeling of energy systems planning and ghg emission mitigation under uncertainty. *Energies* **2011**, *4*, 1624–1656. [CrossRef]
2. Min, L.; Shaoquan, N.; Ling, Z.; Qiang, H. Vehicle routing problem with time windows of petroleum products distribution based on order neighborhood system. *Comput. Integr. Manuf. Syst.* **2015**, *21*, 2158–2169.
3. Wang, S.; Tao, F.; Shi, Y. Optimization of location-routing problem for cold chain logistics considering carbon footprint. *Int. J. Environ. Res. Public Health* **2018**, *15*, 86. [CrossRef] [PubMed]
4. Wang, Y.; Zhou, Y.; Zhu, L.; Zhang, F.; Zhang, Y. Influencing factors and decoupling elasticity of China's transportation carbon emissions. *Energies* **2018**, *11*, 1157. [CrossRef]
5. Wang, S.; Tao, F.; Shi, Y.; Wen, H. Optimization of vehicle routing problem with time windows for cold chain logistics based on carbon tax. *Sustainability* **2017**, *9*, 694. [CrossRef]
6. Federgruen, A.; Zipkin, P. A combined vehicle routing and inventory allocation problem. *Oper. Res.* **1984**, *32*, 1019–1037. [CrossRef]
7. Golden, B.; Assad, A.; Dahl, R. Analysis of a large-scale vehicle routing problem with an inventory component. *Large Scale Syst.* **1984**, *7*, 181–190.
8. Avella, P.; Boccia, M.; Sforza, A. Solving a fuel delivery problem by heuristic and exact approaches. *Eur. J. Oper. Res.* **2004**, *152*, 170–179. [CrossRef]
9. Cornillier, F. A heuristic for the multi-period petrol station replenishment problem. *Eur. J. Oper. Res.* **2008**, *191*, 295–305. [CrossRef]
10. Cornillier, F.; Laporte, G.; Boctor, F.F.; Renaud, J. The petrol station replenishment problem with time windows. *Comput. Oper. Res.* **2009**, *36*, 919–935. [CrossRef]
11. Cornillier, F.; Boctor, F.; Renaud, J. Heuristics for the multi-depot petrol station replenishment problem with time windows. *Eur. J. Oper. Res.* **2012**, *220*, 361–369. [CrossRef]

12. Dai, X.; Yao-Hua, Y.E.; Qin-Min, W.U.; Zhu, D.L. Interactively solving the vehicle routing problem for petroleum delivery. *J. Syst. Eng.* **2009**, *24*, 749–753.
13. Yi-Fei, M.A.; Sun, X.Y. Dispatching optimization model of second distribution of gasolin & diesel oil and solution based on genetic algorithm. *Oper. Res. Manag. Sci.* **2010**, *19*, 73–78.
14. Boctor, F.F.; Renaud, J.; Cornillier, F. Trip packing in petrol stations replenishment. *Omega* **2011**, *39*, 86–98. [CrossRef]
15. Zhen, F. Multi-period vehicle routing problem with recurring dynamic time windows. In Proceedings of the IEEE International Conference on Service Systems and Service Management, Tianjin, China, 25–27 June 2011.
16. Popović, D.; Vidović, M.; Radivojević, G. Variable neighborhood search heuristic for the inventory routing problem in fuel delivery. *Expert Syst. Appl.* **2012**, *39*, 13390–13398. [CrossRef]
17. Wang, X.P.; Zhan, H.X.; Sun, Z.L.; Gao, Y. Route optimization for the refined oil multi-compartment distribution based on ant colony and tabu search hybrid algorithm. *Syst. Eng. Theory Pract.* **2017**, *37*, 3215–3226.
18. Tsai, W.H.; Shen, Y.S.; Lee, P.L.; Chen, H.C.; Kuo, L.; Huang, C.C. Integrating information about the cost of carbon through activity-based costing. *J. Clean. Prod.* **2012**, *36*, 102–111. [CrossRef]
19. Jabali, O.; Woensel, T.V.; Kok, A.G.D. Analysis of travel times and co 2 emissions in time-dependent vehicle routing. *Prod. Oper. Manag.* **2012**, *21*, 1060–1074. [CrossRef]
20. Elhedhli, S.; Merrick, R. Green supply chain network design to reduce carbon emissions. *Transp. Res. Part D* **2012**, *17*, 370–379. [CrossRef]
21. Wang, F.; Lai, X.; Shi, N. A multi-objective optimization for green supply chain network design. *Decis. Support Syst.* **2011**, *51*, 262–269. [CrossRef]
22. Tsai, W.H.; Lee, K.C.; Liu, J.Y.; Lin, H.L.; Chou, Y.W.; Lin, S.J. A mixed activity-based costing decision model for green airline fleet planning under the constraints of the european union emissions trading scheme. *Energy* **2012**, *39*, 218–226. [CrossRef]
23. Jaber, M.Y.; Glock, C.H.; Saadany, A.M.A.E. Supply chain coordination with emissions reduction incentives. *Int. J. Prod. Res.* **2013**, *51*, 69–82. [CrossRef]
24. Palak, G.; Ekşioğlu, S.D.; Geunes, J. Analyzing the impacts of carbon regulatory mechanisms on supplier and mode selection decisions: An application to a biofuel supply chain. *Int. J. Prod. Econ.* **2014**, *154*, 198–216. [CrossRef]
25. Yuan, S.; Skinner, B.; Huang, S.; Liu, D. A new crossover approach for solving the multiple travelling salesmen problem using genetic algorithms. *Eur. J. Oper. Res.* **2013**, *228*, 72–82. [CrossRef]
26. Shima, T.; Rasmussen, S.J.; Sparks, A.G.; Passino, K.M. Multiple task assignments for cooperating uninhabited aerial vehicles using genetic algorithms. *Comput. Oper. Res.* **2006**, *33*, 3252–3269. [CrossRef]
27. Ying-Ying, Y.U.; Yan, C.; Tao-Ying, L.I. Improved genetic algorithm for solving tsp. *Control Decis.* **2014**, *29*, 1483–1488.
28. Liu, W.Y.; Lin, C.C.; Chiu, C.R.; Tsao, Y.S.; Wang, Q. Minimizing the carbon footprint for the time-dependent heterogeneous-fleet vehicle routing problem with alternative paths. *Sustainability* **2014**, *6*, 4658–4684. [CrossRef]
29. Vidal, T.; Crainic, T.G.; Gendreau, M.; Prins, C. A hybrid genetic algorithm with adaptive diversity management for a large class of vehicle routing problems with time-windows. *Comput. Oper. Res.* **2013**, *40*, 475–489. [CrossRef]
30. Srinivas, M.; Patnaik, L.M. Adaptive probabilities of crossover and mutation in genetic algorithms. *IEEE Trans. Syst. Man Cybern.* **2002**, *24*, 656–667. [CrossRef]
31. Dong, W.C.; Lee, Y.H.; Lee, T.Y.; Gen, M. An adaptive genetic algorithm for the time dependent inventory routing problem. *J. Intell. Manuf.* **2014**, *25*, 1025–1042.
32. Solomon, M.M. Algorithms for the vehicle routing and scheduling problems with time window constraints. *Oper. Res.* **1987**, *35*, 254–265. [CrossRef]
33. Mi, Z.; Meng, J.; Guan, D.; Shan, Y.; Song, M.; Wei, Y.M.; Liu, Z.; Hubacek, K. Chinese CO_2 emission flows have reversed since the global financial crisis. *Nat. Commun.* **2017**, *8*, 1712. [CrossRef] [PubMed]
34. Mi, Z.; Meng, J.; Guan, D.; Shan, Y.; Liu, Z.; Wang, Y.; Feng, K.; Wei, Y.M. Pattern changes in determinants of chinese emissions. *Environ. Res. Lett.* **2017**, *12*, 074003. [CrossRef]

35. Mi, Z.; Zhang, Y.; Guan, D.; Shan, Y.; Liu, Z.; Cong, R.; Yuan, X.C.; Wei, Y.M. Consumption-based emission accounting for chinese cities. *Appl. Energy* **2016**, *184*, 1073–1081. [CrossRef]
36. Mcausland, C.; Najjar, N. Carbon footprint taxes. *Environ. Resour. Econ.* **2015**, *61*, 37–70. [CrossRef]

energies

MDPI

Article

It is Worth Pondering Whether a Carbon Tax is Suitable for China's Agricultural-Related Sectors

Weiguo Fan [1,2], Zhicheng Gao [1,2], Nan Chen [3], Hejie Wei [1,2], Zihan Xu [1,2], Nachuan Lu [1,2], Xuechao Wang [4], Peng Zhang [1,2], Jiahui Ren [1,2], Sergio Ulgiati [5] and Xiaobin Dong [1,2,6,*]

1 State Key Laboratory of Earth Surface Processes and Resource Ecology, Faculty of Geographical Science, Beijing Normal University, Beijing 100875, China; fwgcnan@163.com (W.F.); 201721190004@mail.bnu.edu.cn (Z.G.); 201531190013@mail.bnu.edu.cn (H.W.); 201621190007@mail.bnu.edu.cn (Z.X.); 201521190007@mail.bnu.edu.cn (N.L.); 201731190014@mail.bnu.edu.cn (P.Z.); 201621190008@mail.bnu.edu.cn (J.R.)

2 College of Resources Science and Technology, Faculty of Geographical Science, Beijing Normal University, Beijing 100875, China

3 Department of Economics and Management, Baoding Vocational and Technical College, Baoding 071000, China; chennan@bvtc.com.cn

4 Sustainable Process Integration Laboratory—SPIL, NETME Centre, Faculty of Mechanical Engineering, Brno University of Technology—VUT Brno, Technická 2896/2, 616 69 Brno, Czech Republic; wang@fme.vutbr.cz

5 Department of Sciences for the Environment, Parthenope University, 80133 Napoli, Italy; sergio.ulgiati@uniparthenope.it

6 Joint Center for Global change and China Green Development, Beijing Normal University, Beijing 100875, China

* Correspondence: xbdong@bnu.edu.cn; Tel.: +86-010-5880-7058

Received: 11 July 2018; Accepted: 30 August 2018; Published: 31 August 2018

Abstract: Studying the characteristics, trends, and evolution of carbon emissions in agricultural related sectors is of great significance for rational formulation of carbon emission reduction policies. However, as an important carbon emission reduction policy, carbon tax has been controversial over whether or not it should be levied on China. Based on this consideration, this paper takes China's agricultural related sectors as an example and analyzes the degree of carbon tax on macro-environment, macroeconomy, and agricultural sectors during the period 2020–2050 by constructing a 3EAD-CGE (economy-energy-environmental-agricultural-dynamics Computable General Equilibrium) model. The results show that: (1) carbon tax has a time effect, specifically, the short-term effect is better than the long-term. (2) If the incremental rate of carbon tax is carried out alone, it will exert a great influence on the macroeconomy as well as on most of the agricultural related sectors. (3) If a carbon tax is introduced at the same time as indirect taxes are cut (proportionally), the policy will exert a negative impact on agriculture-related sectors that are subsidized. However, the policy will have a positive impact on those nonsubsidized sectors. Finally, based on the results, we put forward some suggestions that are more suitable for the introduction of a carbon tax in China's agricultural-related sectors.

Keywords: agricultural-related sectors; carbon emissions; carbon tax; China

1. Introduction

At present, China's economy has further developed. However, there is no denying that economic development has also brought various negative consequences, such as an energy crisis (from supply) and environmental pollution problems, which are inseparable from the current extensive economic development [1]. Because the energy and environmental problems are without borders, the extensive mode of economic development has brought huge challenges for sustainable global development, especially the global climate issue. More and more countries have realized that economic development accompanied by CO_2 emissions has caused a huge challenge to the biosphere [2]. China produces large amounts of CO_2; the environmental effects of CO_2 cannot be ignored. In order to deal with the challenges of global climate change, more and more countries are trying to create their own low carbon development strategies to minimize emissions of CO_2 at the national level [3].

As one of the key solutions to global climate change, low-carbon policies have gained more and more popularity among countries. Taking the Europe Union as an example, in order to build a low-carbon society, the EU has announced that by 2020, it will achieve the goal of reducing greenhouse gas emissions by 20% relative to 1990 [4]. In November 2013, the 19th Conference of the Parties to the United Nations Framework Convention on Climate Change and the 9th Conference of the Parties to the Kyoto Protocol were held in Warsaw, Poland, with topics focusing on the "Green Development Fund", which calls for developed countries to provide technology and capacity-building and financial support to developing countries to help them cope with climate change [5]. Developing countries, known for large CO_2 emissions, also expressed their strong willingness to cut CO_2 emissions [6]. In China, for example, a report on low-carbon planning for China's social development issued by China's National Development and Reform Commission points out that by 2020, the CO_2 emissions per unit of GDP will be reduced by 40% to 45% compared with 2005 [7]. In 2007, China's National Development Committee announced "China's National Climate Change Program". This program pointed out that China would start to impose a carbon tax on some enterprises in 2012. Despite the fact that China's current macro-policy has paved the way for its low-carbon development, because of historical reasons and technical limits, the increasing CO_2 emissions in China's economic development have been more and more questioned by the international community, and China is facing greater pressure because of its CO_2 emissions. If China refuses to assume responsibility for reducing emissions with the excuse of its right to development as a developing country, China will be confronted with unnecessary obstructions to its future economic development.

At present, theoretical and practical circles both at home and abroad usually make pollutants and CO_2 emissions in industrialization and urbanization a focus of attention, covering areas like general industry and the construction industry. It is undeniable, however, that large amounts of CO_2 are also produced in the agriculture-related sectors. Data released by the IPCC (Intergovernmental Panel on Climate Change) in 2007 showed that agricultural CO_2 emissions have become the world's second largest source of CO_2 [8]. Different data from FAOSTAT (Food and Agriculture Organization of the United Nations Statistics) in 2014 showed that agricultural CO_2 emissions in 2011 exceeded more than 10 billion tons of CO_2 equivalents, accounting for 14% of global CO_2 emissions [9]. As a traditionally agricultural country, China has a large rural population and a complex agricultural industry structure. The extensive agriculture, together with the large demand for agricultural products in other nonagricultural sectors, has resulted in huge CO_2 emissions from the agricultural sectors and has brought severe challenges to the environment in China. Based on relevant data from China's Statistical Yearbook [10] and China's input-output table in 2012 [11], and according to the specific input-output path of the intermediate production process of each sector, it can calculate CO_2 emissions in the main sectors of agriculture in China (including rice planting; wheat planting; corn planting; beans and potato planting; peanut, rapeseed, and sesame planting; cotton and hemp planting; tobacco planting; tea planting; fruit and other planting; animal husbandry; forestry; the fishery industry) due to the consumption of coal, oil, natural gas and electricity, which totals 26.94 million tons. The above data are only considered for CO_2 emissions in the intermediate production process due to coal, oil, and natural gas and electricity consumptions and do not take into account the final consumptions from households and other sources in agriculture such as fertilizers and pesticides. If they are all taken into account, CO_2 emissions will be even more significant. In addition, according to statistics, China's agricultural CO_2 emissions contribute 17% of the country's total CO_2 emissions, much higher than the contribution of China's transportation industry [12]. Moreover, in many European countries (not only in China), the extensive agriculture, together with the large demand for agricultural products in other nonagricultural sectors, has resulted in huge CO_2 emissions and brought severe challenges to the environment [13,14]. In this case, how to balance economic development and the sustainable development of agriculture has become particularly important.

CO_2 emissions reduction has become a hot issue in the international community, and in particular, the decoupling of CO_2 emissions from economic growth is receiving increasing attention. However, one fact cannot be ignored: at current rates and levels of development, the important roles that fossil fuels play in economic development will not be replaced in the near future, nor can the new carbon capture technology be widely adopted in the short term, owing to high costs and lack of supported techniques [15–17]. However, as one of the important carbon emission reduction measures, a carbon tax has been postulated for China, and it is worth considering whether the carbon tax is suitable for China. By analyzing and studying the carbon tax rates, we can find measures for decreasing adverse effects on economic development. Moreover, by levying a carbon tax on agriculture-related sectors, we want to find whether carbon tax is a feasible way to reduce carbon emissions in China.

In order to evaluate the influence of a carbon tax on China's macroeconomy and agriculture-related sectors, it is necessary to rely on a reasonable and effective model to simulate the structural effects of the carbon tax policy under the macroeconomic framework. In this regard, the Computable General Equilibrium (CGE) model provides the possibility of the above assumptions. The CGE model is a class of economic model and by taking each component of the national economy and every link of the economic cycle into a united framework, the model can simulate the final structural influence of the changes in energy and climate policies on the national economic sectors [18–20].

At present, the CGE model has been widely adopted for estimating the effectiveness of energy policy [21,22]. Mahmood and Marpaung [23] used a 20-sector CGE model to investigate the respective effects of a scenario with a carbon tax and a scenario with the cooperative implementation of a carbon tax and energy on the Pakistani economy. The results show that the impact of levying a carbon tax on the GDP is negative, but the impact on reducing emissions of pollutants is positive. Markandya et al. [24] used the CGE model to analyze and compare the differences between developing countries and developed countries in a trade-off between traditional economic development and low-carbon development and found that the adverse impact of the implementation of emissions reduction policies on developing countries is huge. Springmann et al. [25] explored the distribution of carbon quotas among different provinces in China. The results show that eastern China outsourced 14% of its own carbon quotas to central and western regions. Yan et al. [26] used the CGE model to study the environmental and economic effects of the carbon tax on China's net exports from a multiregional and multicommodity perspective. The simulation results show that China was lacking a driving force to reduce domestic carbon emissions. Carlos et al. [27] studied the impact of the carbon tax on China's smart-power-generation industry. The study shows that the effectiveness of the carbon tax policy is closely related to the variables that are not affected by policy makers' decisions, such as natural gas prices, the feasibility of the use of resources, etc. Yang et al. [28] studied the impact of the carbon market on China's environment and economy from the medium term to the long term by constructing a CGE model. The results show that the carbon market will have a positive impact on China's R&D investment.

From the existing literature, we can see that most scholars are focusing on exploring ways to control the reduction of CO_2 emissions and explore reduction degrees. Although many scholars focus on agricultural issues, such as the impacts of the export ban on the corn industry and the overall state of society [29] and the impacts of climate change on yields, production, and prices [30]. However, due to economic system and data constraints, the application of the CGE model to analyze the impacts of carbon tax on agricultural-related sectors is not systematic. Based on China's realities and the CGE model's theory and technology, this study builds a dynamic CGE model with a complex structure that reflects the energy-economy-environment system of Chinese agriculture-related sectors. The system simulates the impact of levying a carbon tax on the macro-environment, macroeconomic variables, and the agricultural-related sectors, so as to reveal the implementation effect of a carbon tax on agricultural related sectors.

2. Methods and Materials

2.1. Model Construction

As an energy reduction policy, carbon taxes achieve the ultimate goal of reducing fossil fuel consumptions and CO_2 emissions by levying tax on fossil fuel products such as coal, oil, and natural gas based on the proportion of carbon content. In order to simulate the impact of the carbon tax on macroeconomy and different agricultural-related sectors, an economy-energy-environmental-agricultural-dynamics CGE model (3EAD-CGE) is constructed in this article to comprehensively analyze the impact of the carbon tax on the production processes of Chinese agriculture-related sectors. In this regard, this study mainly involves 23 agriculture-related sectors, including rice planting; wheat planting; corn planting; beans and potato planting; peanut, rapeseed, and sesame planting; cotton and hemp planting; tobacco planting; tea planting; fruit and other planting; animal husbandry; the dairy industry; forestry; the fishery industry; the slaughtering and meat processing industry; the animal and plant oil processing industry; vegetables and other agricultural processing industries; the sugar products processing industry; the drinks and refined tea processing industry; the tobacco products processing industry; the feed processing industry; the textile industry; the leather products industry; and the wood products processing industry. The rest of the industries are combined into the manufacturing and mining industry, construction industry, transport industry, and service industry. The specific sector definition is shown in Table 1.

Table 1. Sector definition in this model.

Nr	Sector	Nr	Sector
1	Rice planting	15	Animal and plant oil processing industry
2	Wheat planting	16	Vegetables and other agricultural processing industry
3	Corn planting	17	Sugar products processing industry
4	Beans and potato planting	18	Drinks and refined tea processing industry
5	Peanut, rapeseed and sesame planting	19	Tobacco products processing industry
6	Cotton and hemp planting	20	Feed processing industry
7	Tobacco planting	21	Textile industry
8	Tea planting	22	Leather products industry
9	Fruit and other planting	23	Wood products processing industry
10	Animal husbandry	24	Manufacturing and mining industry
11	Dairy industry	25	Construction industry
12	Forestry	26	Transport industry
13	Fishery industry	27	Service industry
14	Slaughtering and meat processing industry		

The main economic entities in the 3EAD-CGE model include households, enterprises, and the government. The model is responsive to population, GDP, and capital recursive growth.

The 3EAD-CGE model normally consists of five modules, including the production block, the market block, the income block, the expenditure block, and the overall balance block. In this paper, in order to reflect the influence of carbon tax on the macroenvironment, the macroeconomy, and the agricultural sectors, we choose the above blocks as a whole to analyze.

The relationship between the specific blocks is shown in Figure 1.

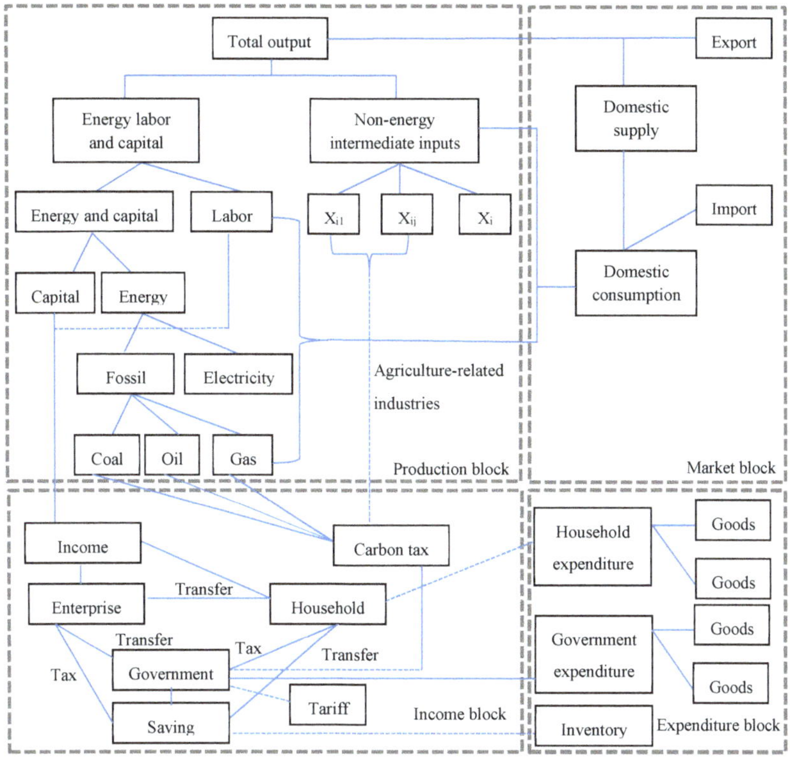

Figure 1. General structure of the economy-energy-environmental-agricultural-dynamics CGE model (3EAD-CGE) model in Chinese agriculture-related sectors.

In general, in the production block, the total output is decomposed in the form of the CES total output function, and corresponding to the market block, the total output consists of two parts, one for export and the other for domestic supply. The domestic supply plus the import component constitutes the total domestic consumption. In the income block, the labor and capital from the production block constitute the source of income. The income is divided into enterprise income, household income, and government income. The three parts are linked through transfer payment and taxation. Corresponding to the income block, the expenditure block mainly describes the expenditure status of enterprises, household, and the government and the expenditure is equal to the income. The specific relationship of each part is described below.

2.1.1. Production Block

In the model, we assume that the agriculture-related sectors are perfectly competitive enterprises and that scale returns remain unchanged. Based on this, we construct a multilevel nested constant elasticity of substitution (CES) production function to describe the substitutability of different factors of production. The production block is divided into five levels. In the first level, the inputs of the aggregation of intermediate commodities and the aggregation of capital-energy-labor are transformed into the total output in the form of the CES total output function. The first-level module formulas are as follows:

$$QA_{i(t)} = A_i{}^q{}_{(t)}[a_i{}^q{}_{(t)}QKEL_{i(t)}{}^{\rho_i{}^q} + (1 - a_i{}^q{}_{(t)})QINTA_{i(t)}{}^{\rho_i{}^q}]^{1/\rho_i{}^q} \quad (1)$$

$$PKEL_{i(t)}/PINTA_{i_{(t)}} = a_i{}^q{}_{(t)}/(1 - a_i{}^q{}_{(t)})\left[QINTA_{i(t)}/QKEL_{i(t)}\right]^{1-\rho_i{}^q} \quad (2)$$

In time point t, where $QA_{i(t)}$ denotes the output in i sector, $A^q_{i\,(t)}$ is the size parameter under the i department output, $a_i{}^q{}_{(t)}$ is the share parameters of the output of capital-energy-labor in department i, $QKEL_{i(t)}$ is the demand for finished products of capital-energy-labor in department i, $\rho_i{}^q$ is the substitution parameter between intermediate inputs and capital-energy-labor in department i, $QINTA_{i(t)}$ is the requirements of intermediate inputs in department i, $PKEL_{(t)}$ is the price of finished products of capital-energy-labor in department i, $PINTA_{i(t)}$ is the price of intermediate inputs in department i. In the second level, the aggregation of capital-energy-labor is decomposed into the synthesis beam of capital-energy and the labor force. The second-level module formulas are as follows:

$$QKEL_{i(t)} = A_i{}^{kel}{}_{(t)}[a_i{}^{ke}{}_{(t)}QKE_{i(t)}{}^{\rho_i{}^{kel}} + (1 - a_i{}^{ke}{}_{(t)})QLD_{i(t)}{}^{\rho_i{}^{kel}}]^{1/\rho_i{}^{kel}} \quad (3)$$

$$PKE_{i(t)}/WL_{(t)} = a_i{}^{ke}{}_{(t)}/(1 - a_i{}^{ke}{}_{(t)})\left[QLD_{i(t)}/QKE_{i(t)}\right]^{1-\rho_i{}^{kel}} \quad (4)$$

In time point t, where $A_i{}^{kel}{}_{(t)}$ denotes the size parameter for finished products of capital-energy labor in department i, $a_i{}^{ke}{}_{(t)}$ is the share parameters of the output of capital-energy in department i, $QKE_{i(t)}$ is the demand for finished products of capital-energy in department i, $QLD_{i(t)}$ is the demand for labor in department i, $\rho_i{}^{kel}$ is the substitution parameter between labor and products of capital-energy in department i, $PKE_{i(t)}$ is the price of finished products of capital-energy in department i, $WL_{(t)}$ is the average price of labor.

For the nonenergy intermediate inputs in the second level, the Leontief production function is used to represent the intermediate input requirements and prices. The specific formulas are as follows:

$$QINT_{ne,i(t)} = a_{ne,i(t)} \cdot QINTA_{i(t)} \quad (5)$$

$$PINT_{ne,i(t)} = \sum_{ne} a_{ne,i(t)} \cdot PC_{ne(t)} \quad (6)$$

In time point t, where $QINT_{ne,i(t)}$ denotes the requirements of intermediate inputs in ne by produce unit i, $a_{ne,i(t)}$ is the direct consumption coefficient of the intermediate input, $PINT_{ne,i(t)}$ is the price of intermediate inputs in ne by produce unit i, $PC_{ne(t)}$ is the price of intermediate inputs in ne.

In the third level, the synthesis beam of capital-energy is further broken down into two parts: capital and energy. The specific formulas for the module are as follows:

$$QKE_{i(t)} = A_i{}^{ke}{}_{(t)}[a_i{}^{e}{}_{(t)}QED_{i(t)}{}^{\rho_i{}^{ke}} + (1 - a_i{}^{e}{}_{(t)})QKD_{i(t)}{}^{\rho_i{}^{ke}}]^{1/\rho_i{}^{ke}} \quad (7)$$

$$PED_{i(t)}/PK_{i(t)} = a_i{}^{e}{}_{(t)}/(1 - a_i{}^{e}{}_{(t)})\left[QKD_{i(t)}/QED_{i(t)}\right]^{1-\rho_i{}^{ke}} \quad (8)$$

In time point t, where $A_i{}^{ke}{}_{(t)}$ denotes the size parameter for finished products of capital-energy in department i, $a_i{}^{e}{}_{(t)}$ is the share parameters of the energy of capital-energy in department i, $QED_{i(t)}$ is the demand for energy in department i, ρ_i^{ke} is the substitution parameter between labor and energy in department i, $QKD_{i(t)}$ is the demand for capital in department i, $PED_{i(t)}$ is the price of finished products of energy in department i, $PK_{i(t)}$ is the price of capital input in department i.

In the fourth level, the synthesis beam of energy is further decomposed into fossil energy and electricity in the form of the CES function. The specific formulas for the module are as follows:

$$QED_{i(t)} = A_i{}^{e}{}_{(t)}[a_i{}^{el}{}_{(t)}QELD_{i(t)}{}^{\rho_i{}^{elf}} + (1 - a_i{}^{el}{}_{(t)})QFD_{i(t)}{}^{\rho_i{}^{elf}}]^{1/\rho_i{}^{elf}} \quad (9)$$

$$PELD_{i(t)}/PFD_{i(t)} = a_i{}^{el}{}_{(t)}/(1 - a_i{}^{el}{}_{(t)})\left[QFD_{i(t)}/QELD_{i(t)}\right]^{1-\rho_i{}^{elf}} \quad (10)$$

In time point t, where $A_i{}^e{}_{(t)}$ denotes the size parameter for finished products of energy in department i, $QELD_{i(t)}$ is the demand for thermal power-energy in department i, $\rho_i{}^{elf}$ is the substitution parameter between thermal power and the composite beam of fossil energy in department i, $QFD_{i(t)}$ is the demand for the composite beam of fossil energy in department i, $PELD_{i(t)}$ is the price of finished products of thermal power-energy in department i, $PFD_{i(t)}$ is the price of the composite beam of fossil energy in department i.

In the last level, the synthesis beam of fossil energy is further decomposed into coal, oil, and natural gas in the form of the CES function. The specific formulas for the module are as follows:

$$QFD_{i(t)} = A_i{}^f{}_{(t)}[a_i{}^c{}_{(t)}QCD_{i(t)}{}^{\rho_i{}^{cog}} + a_i{}^o{}_{(t)}QOD_{i(t)}{}^{\rho_i{}^{cog}} + (1 - a_i{}^c{}_{(t)} - a_i{}^o{}_{(t)})QGSD_{i(t)}{}^{\rho_i{}^{cog}}]^{1/\rho_i{}^{cog}} \quad (11)$$

$$PCD_{i(t)}/POD_{i(t)} = (a_i{}^c{}_{(t)}/a_i{}^o{}_{(t)}) \cdot \left[QOD_{i(t)}/QCD_{i(t)}\right]^{1-\rho_i{}^{cog}} \quad (12)$$

$$PCD_{i(t)}/PGSD_{i(t)} = a_i{}^c{}_{(t)}/(1 - a_i{}^c{}_{(t)} - a_i{}^o{}_{(t)})\left[QGSD_{i(t)}/QCD_{i(t)}\right]^{1-\rho_i{}^{cog}} \quad (13)$$

In time point t, where $A_i{}^f{}_{(t)}$ denotes the size parameter of the composite beam of fossil energy in department i, $a_i{}^c{}_{(t)}$ is the share parameters of coal in department i, $a_i{}^o{}_{(t)}$ is the share parameters of oil in department i, $QCD_{i(t)}$ is the demand for coal in department i, $QOD_{i(t)}$ is the demand for oil in department i, $QGSD_{i(t)}$ is the demand for natural gas in department i, $\rho_i{}^{cog}$ is the substitution parameter of coal, oil, and natural gas in department i, $PCD_{i(t)}$ is the price of coal in department i, $POD_{i(t)}$ is the price of oil in department i, $PGSD_{i(t)}$ is the price of natural gas in department i.

In the part of Section 2.1.1, in order to facilitate the calculation, all the input units of demand are 10,000 Yuan. For example, the units of $QCD_{i(t)}$, $QOD_{i(t)}$, and $QGSD_{i(t)}$ are 10,000 Yuan. In addition, the above demands a calculated in one year.

2.1.2. Market Block

In the 3EAD-CGE model, assuming that the market clears, the number of commodities supplied on the market $QC_{i(t)}$ is the sum of the quantity of imports $QM_{i(t)}$ and the quantity of goods produced domestically $QAM_{i(t)}$ in the CES form. Under Ammington's condition, the quantity of the products supplied by enterprises to the market $QC_{i(t)}$ can maximize the profits of the enterprises. The specific formulas for the module are as follows:

$$QC_{i(t)} = A_i^{am}{}_{(t)}[a_i{}^m{}_{(t)}QM_{i(t)}{}^{\rho_i{}^{am}} + (1 - a_i{}^m{}_{(t)})QAM_{i(t)}{}^{\rho_i{}^{am}{}_{(t)}}]^{1/\rho_i{}^{am}} \quad (14)$$

$$PM_{i(t)}/PAM_{i(t)} = a_i{}^m{}_{(t)}/(1 - a_i{}^m{}_{(t)})\left[QAM_{i(t)}/QM_{i(t)}\right]^{1-\rho_i{}^{am}} \quad (15)$$

In time point t, where $QC_{i(t)}$ denotes total supply of goods i in the domestic market, $A_i{}^{am}{}_{(t)}$ is the size parameter of the supply of goods i in the domestic market, $QM_{i(t)}$ is the quantity of imported goods i, $QAM_{i(t)}$ is the quantity of the synthetic products in the domestic market, $\rho_i{}^{am}$ is the substitution parameter of the synthetic products and imported goods i, $PM_{i(t)}$ is the price of imported goods i, $PAM_{i(t)}$ is the price of the synthetic products i in domestic market.

In Equation (15), the domestic price of imported goods is determined by the world exchange rate $EXR_{(t)}$ and the international price $\overline{pwm_{i(t)}}$, which includes the import duty rate $\tau_i{}^m{}_{(t)}$. The specific formula for the module is as follows:

$$PM_{i(t)} = \left(1 + \tau_i{}^m{}_{(t)}\right) \cdot \overline{pwm_{i(t)}} \cdot EXR_{(t)} \quad (16)$$

Similarly, the total output of the sector $QA_{i(t)}$ is the sum of the domestic market supply and exports. Under Ammington's condition, the number of products that enterprises supply to the market can minimize the cost of the enterprises. The specific formulas for the module are as follows:

$$QA_{i(t)} = A_i{}^t{}_{(t)}[a_i{}^t{}_{(t)}QE_{i(t)}{}^{\rho_i{}^t} + (1 - a_i{}^t{}_{(t)})QDS_{i(t)}{}^{\rho_i{}^t}]^{1/\rho_i{}^t} \tag{17}$$

$$PE_{i(t)}/PDS_{i(t)} = a_i{}^t{}_{(t)}/(1 - a_i{}^t{}_{(t)})\left[QDS_{i(t)}/QE_{i(t)}\right]^{\rho_i{}^t - 1} \tag{18}$$

In time point t, where $A_i{}^t{}_{(t)}$ denotes the size parameter of producing goods i in domestic market, $a_i{}^t{}_{(t)}$ is the share parameters of export goods i, $QE_{i(t)}$ is the quantity of goods i produced domestically for export, $QDS_{i(t)}$ is the quantity of goods i produced domestically and used domestically, $\rho_i{}^t$ is the substitution parameter of domestic sales and exported goods i, $PE_{i(t)}$ is the price of goods i produced domestically for export, $PDS_{i(t)}$ the price of goods i produced domestically and used domestically.

In Equation (18), the domestic price of the exported commodity is determined by the world exchange rate $EXR_{(t)}$ and the international price $\overline{pwe_{i(t)}}$, which includes the import duty rate $\tau_i{}^e{}_{(t)}$. The specific formula for the module is as follows:

$$PE_{i(t)} = \left(1 + \tau_i{}^e{}_{(t)}\right) \cdot \overline{pwe_{i(t)}} \cdot EXR_{(t)} \tag{19}$$

Similar to Section 2.1.1, in this part, in order to facilitate the calculation, the units of all kinds of quantity of goods are 10,000 Yuan, such as the $QC_{i(t)}$, $QM_{i(t)}$, and $QAM_{i(t)}$, etc.

2.1.3. Income Block

The income block mainly describes the income distributions of the households, the enterprises, and the government. The specific formulas for the module are as follows:

$$YH_{(t)} = \sum\left(WL_{(t)} \cdot QLD_{i(t)} + shif_{hk(t)} \cdot PK_{i(t)} \cdot QKD_{i(t)}\right) + \overline{TR_{ent(t)}} + \overline{TR_{gov(t)}} \tag{20}$$

$$YENT_{(t)} = \sum PK_{i(t)} \cdot shif_{entk(t)} \cdot QKD_{i(t)} + \overline{TR_{entgov(t)}} \tag{21}$$

$$YG_{(t)} = INDTAX_{(t)} + YH_{(t)} \cdot \eta + YENT_{(t)} \cdot v + CTAX_{(t)} + TARIFF_{(t)} \tag{22}$$

$$INDTAX_{(t)} = \sum_i \tau_i \cdot PA_{i(t)} \cdot QA_{i(t)} \tag{23}$$

In time point t, where $YH_{(t)}$ denotes the households' income, $\overline{TR_{ent(t)}}$ is the transfer payment for households from enterprises, $\overline{TR_{gov(t)}}$ is the transfer payment for households from government, $YENT_{(t)}$ is the enterprise income, $shif_{entk(t)}$ is the share parameters of income distribution to the enterprise from capital, $shif_{hk(t)}$ is the share parameters of income distribution to the household from capital, $\overline{TR_{entgov(t)}}$ is the transfer payments to the enterprise from government, $YG_{(t)}$ is government revenue, $INDTAX_{(t)}$ is the production of indirect tax income, τ_i is the production tax rate of goods i, η is the rate of households' income taxes, v is the rate of enterprise income taxes, $CTAX_{(t)}$ is carbon tax, $TARIFF_{(t)}$ is tariffs.

In this study, we follow the assumption that direct and indirect taxes are defined as fixed shares in the model. In addition, the assumption that government financial revenues and expenditures are balanced is applied in this model.

Similar to Section 2.1.1, in this part, in order to facilitate the calculation, the units involved in the income are all 10,000 Yuan, such as the $YH_{(t)}$, $\overline{TR_{ent(t)}}$, and $\overline{TR_{gov(t)}}$, etc.

2.1.4. Expenditure Block

Corresponding to the income block, the expenditure block mainly describes the operations of the market economy in the process of the households, the enterprises, and the government spending and saving. The relationship between savings, income, and expenditure is as follows:

$$EH_{(t)} = mpc_{(t)} \cdot (1 - \eta) YH_{(t)} / PC_{i(t)} \tag{24}$$

$$HS_{(t)} = YH_{(t)} - EH_{(t)} \tag{25}$$

$$EENH_{(t)} = \overline{TR_{ent(t)}} + v \cdot YENT_{(t)} \tag{26}$$

$$ENTS_{(t)} = YENT_{(t)} - EENT_{(t)} \tag{27}$$

$$EG_{(t)} = \sum_i PC_{i(t)} \cdot QG_{i(t)} + \overline{TR_{entgov(t)}} + \overline{TR_{gov(t)}} + ESUB_{(t)} \tag{28}$$

$$ESUB_{(t)} = \sum_i esub_{i(t)} \cdot \overline{pwe_{i(t)}} \cdot EXR_{(t)} \cdot QE_{i(t)} \tag{29}$$

$$GS_{(t)} = YG_{(t)} - EG_{(t)} \tag{30}$$

$$TS_{(t)} = HS_{(t)} + ENTS_{(t)} + GS_{(t)} \tag{31}$$

In time point t, where $EH_{(t)}$ denotes households' consumption, $mpc_{(t)}$ is residents' marginal propensity to consume, $HS_{(t)}$ is household savings, $ENTS_{(t)}$ is enterprise savings, $EENT_{(t)}$ is enterprise spending, $GS_{(t)}$ is government savings, $EG_{(t)}$ is government spending, $INV_{(t)}$ is total investment, $TS_{(t)}$ is total savings, $FINV_{(t)}$ is foreign investment, $esub_{i(t)}$ is households' marginal propensity to consume, $ESUB_{(t)}$ is export tax rebate.

Similar to Section 2.1.1, in this part, in order to facilitate the calculation, the units involved in the expenditure are all 10,000 Yuan, such as the $EH_{(t)}$, $EG_{(t)}$, and $EENT_{(t)}$, etc.

2.1.5. Closed Module

For the balance of payments, the exchange rate of the general equilibrium is determined according to Equations (32)–(34). The specific formula is as follows:

$$\sum_i PE_i \cdot QE_i = \sum_i PM_i \cdot QM_i + FINV \tag{32}$$

$$\sum_i QKD_{i(t)} = \sum_i QKS_{i(t)} \tag{33}$$

$$\sum_i QLD_{i(t)} = \sum_i QLS_{i(t)} \tag{34}$$

where $QKS_{i(t)}$ donates the capital supply of industry i in period t, $QLS_{i(t)}$ is the size of the labor force in t of department i.

For the savings-investment closure, this paper uses neoclassical closure rule, that is, savings decide investments. All the savings in the economic entities will be transformed into investment and the specific formula is shown as follows:

$$INV_{(t)} = TS_{(t)} \tag{35}$$

For the zero profit condition, this paper uses the unit product price equal to unit sales price to determine the level of production activities in a balanced state.

2.2. Dataset

Based on the basic principles of the input-output table of China in 2012 [11] combined with the China Statistical Yearbook [10] and the China Financial Yearbook [31], which are authorized by the Ministry of Finance, the National Bureau of Statistics, and other related central ministries and commissions as well as information from social research, this article has compiled a social accounting matrix (SAM) of Chinese agriculture-related sectors, which is coordinated with the 3AED-CGE model. Because the input-output table of China in 2012 includes 139 sectors, for the purposes of this analysis, the agricultural industry is decomposed and 23 agriculture-related sectors are selected based on this decomposition. The rest of the industries are combined into the manufacturing and mining industry, construction industry, transport industry, and service industry.

CO_2 emissions are mainly reflected in the production block. Based on social accounting matrix which can reflect the input-output situation of different agriculture-related sectors in China, and on the proportion of other sectors which consume in the intermediate production process of different agricultural-related sectors accounting for the total output of the sector, CO_2 emissions of agricultural-related sectors in the intermediate production process can be calculated due to the consumption amount of coal, oil, natural gas, and electricity by other sectors. Correspondingly, in the production structure, the carbon tax is levied on the basis of CO_2 emissions of agriculture-related sectors. The result of the carbon tax will affect a variety of energy substitutions as well as energy and capital substitutions. As the scope of carbon tax levied mainly due to the combustion of fossil fuels caused by CO_2 emissions, based on this, this paper does not consider other CO_2 sources such as pesticide, fertilizer, and agricultural film.

Moreover, the State Council Development Research Center of China pointed out that China's economy will experience a weak resurgence in 2013 with hope for year-on-year growth of 8.1%. From 2013 to the next decade, China's economy will change to a medium growth at a rate of 6% to 8% [32]. According to the above analysis, from 2013 to 2020, China's GDP growth rates are set at 7.55%. From 2021 to 2050, the GDP growth rate refers to the research of Wang et al. [33]. For the population, according to the estimate in the Research Report of Chinese Population Development Strategy [34], the population will increase by 8 million annually in China. Based on 135,404 million people in China in 2012, it is possible to roughly estimate the growth rate of China's population by 2050. Specifically, China's GDP and population growth rates from 2012 to 2050 are shown in Table 2.

Table 2. Predicted values of China's GDP and Population Growth Rates in 2012–2050.

Period	GDP	Population
2012	518,942	135,404
2013–2020	7.55%	0.58%
2021–2025	7.35%	0.56%
2026–2030	6.85%	0.54%
2031–2035	6.35%	0.53%
2036–2040	5.50%	0.51%
2041–2045	5.21%	0.50%
2046–2050	4.86%	0.49%

Note: The unit of GDP is 100 million RMB, and the unit of population is 10,000.

The 3AED-CGE model uses a recursive dynamic mechanism, which can be used to solve the equilibrium solution in each period from the base year to 2050. The specific formulas are as follows:

$$QKS_{i(t+1)} = QKS_{i(t)} \cdot (1 - \kappa_i) + ID_{i(t)} \tag{36}$$

$$QLS_{i(t+1)} = QLS_{i(t)} \cdot (1 + \lambda_i) \tag{37}$$

$QKS_{i(t+1)}$ represents the capital supply of industry i in period $t + 1$, and κ_i represents the current depreciation rate of industry i. $ID_{i(t)}$ is the current investment of industry i. $QLS_{i(t+1)}$ represents the size of the labor force in $t + 1$ of department i, and λ_i is the labor growth rate.

A small number of the parameters in each function in the 3AED-CGE model are exogenously determined and estimated. Such as the substitution parameters which are reflecting the substitution between different factors. Most of the parameter values are obtained by substituting data from the SAM in the base year and then inversely deducing the values of unknown parameters such as the size parameters which are reflecting the efficiency of the overall use of society, the share parameters which are reflecting the contribution of factors in the production process, etc. For the size parameters, according to the Equations (1), (3), (7), (9), (11), (14) and (17), the size parameters of the respective variables can be obtained. For example, the size parameter ${A_i}^q{}_{(t)}$ can be obtained in the following formula converted by Equation (1):

$$ {A_i}^q{}_{(t)} = QA_{i(t)} / [{a_i}^q{}_{(t)} QKEL_{i(t)}{}^{{\rho_i}^q} + (1 - {a_i}^q{}_{(t)}) QINTA_{i(t)}{}^{{\rho_i}^q}]^{1/{\rho_i}^q} \quad (38) $$

The remaining size parameters can also be obtained in a similar manner.

For the share parameters, the values can be obtained according to the Equations (2), (4), (8), (10), (12), (13), (15) and (18). For example, the share parameter ${a_i}^q{}_{(t)}$ can be obtained in the following formula converted by Equation (2):

$$ {a_i}^q{}_{(t)} = PKEL_{i(t)} \cdot QKEL_{i(t)}{}^{1-{\rho_i}^q} / [PINTA_{i(t)} \cdot QINTA_{i(t)}{}^{1-{\rho_i}^q} + PKEL_{i(t)} \cdot QKEL_{i(t)}{}^{1-{\rho_i}^q}] \quad (39) $$

The remaining share parameters can also be obtained in a similar manner. The derivation of the size parameters and share parameters can be found in Pan [35].

For the substitution parameters that are exogenously determined and estimated, this study mainly refers to the results of Pan [35] and Dai et al. [36] who think that the substitution parameters are homogenous between different departments. The specific values are as follows: ${\rho_i}^q$ is -0.11, ${\rho_i}^{kel}$ is -0.43 ρ_i^{ke} is -0.43, ${\rho_i}^{elf}$ is -0.43, ${\rho_i}^{cog}$ is -0.43, ${\rho_i}^{am}$ is -0.25, and ${\rho_i}^{t}$ is -0.25.

2.3. Scenarios Design

Since Finland first implemented a carbon tax policy in 1990, the policy of a carbon tax as a response to global warming has been adopted by some Western countries. Carbon tax policy implementation has been brewing in China. The task force of the National Development and Reform Commission and the Ministry of Finance has pointed out that 2012 would be the appropriate time to launch a carbon tax. In addition, due to the most recent year of China's input-output table being 2012, and from the practical significance of this study and the data acquisition point of view, in this analysis, the carbon tax base period is set in 2012. Through the 3EAD-CGE model, during 2012–2050, a carbon tax on CO_2 from the consumption of fossil fuels in the production processes of agriculture-related sectors and from the consumption of energy products in the production processes is simulated.

China has made an ambitious commitment to cut down its carbon dioxide emissions of GDP per unit from 40% to 45% by 2020 compared with that in 2005. Accordingly, CO_2 emissions come from the fossil energy consumption when enterprises produce their products. And all sectors should take emissions reduction into account and then decide the appropriate activity level. Agricultural-related sectors are no exception. According to the above analysis, the carbon tax may be an effective means of reducing emissions. Internationally, most countries impose a fixed carbon tax rate, and the fixed carbon tax rate is usually lower in the early period of introduction and then is gradually increased. However, carbon tax levied in China has not really been implemented, and the tax rate is still no basis to determine the level of carbon tax to be imposed. In this regard, referring to the international carbon tax precedent, Li et al. [37] whose research introduces carbon tax into China, Meng and Pham [38] whose research introduces carbon tax into the tourism sector in Australia, and Meng [39] whose research focuses on the impact of carbon tax on electricity sector, this article sets up six different scenarios to impose a carbon tax on agriculture-related sectors. Among them, according to the international general experience, the carbon tax should be set gradually in accordance with the low to high, and it is not reasonable to set the carbon tax too high, otherwise the impact on the macroeconomy will be more significant. Scenario 1 is the business as usual (BAU) scenario, in which China does not implement a carbon tax policy between 2012 and 2050. Scenario 2, scenario 3, and scenario 4 can be referred to as CT20, CT30, and CT40 based on the carbon tax rates 20 yuan/ton, 30 yuan/ton, and 40 yuan/ton, respectively. In these scenarios, starting in 2012, the carbon tax rate is increased by 4.33%, 3.22%, and 2.44% per year, respectively, until they each reach 100 yuan/ton in 2050. Scenario 5 and scenario 6, which can be referred to as CTB40 and CTB60, implement balanced tax rates. In scenario 5 and 6, the carbon tax rates are assumed to be 40 yuan/ton and 60 yuan/ton, the indirect production tax is reduced proportionally so that the total tax burden is unchanged. The specific scenarios are shown in Table 3.

Table 3. The design of different scenarios of imposed carbon tax.

Scenarios	Increasing Rate			Balanced Rate	
Tax rate	20	30	40	40	60
BAU	−	−	−	−	−
CT20	+	−	−	−	−
CT30	−	+	−	−	−
CT40	−	−	+	−	−
CTB40	−	−	−	+	−
CTB60	−	−	−	−	+

Note: "+" indicates the given scenario contains the corresponding policy, and "−" indicates that the corresponding measure is not considered.

3. Results

3.1. The Changes in Macro-Environment

At present, the world economy is in transition to a low-carbon economy. In this regard, China is also responding positively. In September 2007, China's National Development and Reform Commission (NDRC) issued the Medium and Long-term Plan of Renewable Energy Source Development to raise the weight of renewable energy to 10%, as a proportion of total primary energy consumption in 2010, and further to 15% in 2020, which is expected to contribute greatly to the objectives of saving about 5–6 billion tons of CO_2 emissions by 2020 [40]. However, despite this situation, China's current energy structure is still unbalanced compared with the world level, and China's unbalanced structure results in a series of problems, especially regarding the CO_2 emissions that may hinder sustainable development [41]. This kind of situation is also reflected in agriculture-related sectors. Controlling agricultural CO_2 emissions may not only have a positive impact on the sustainable development of agriculture itself, but it may also make a corresponding contribution to the development of low-carbon agriculture. Based on the GDP forecast in China from 2012 to 2050, this section uses carbon intensity as an indicator to compare the CO_2 emission values of a unit of GDP under different scenarios. Because the indicator of carbon intensity can not only reflect economic condition, but reveal the environmental effects with the development of economy. The results are shown in Figure 2.

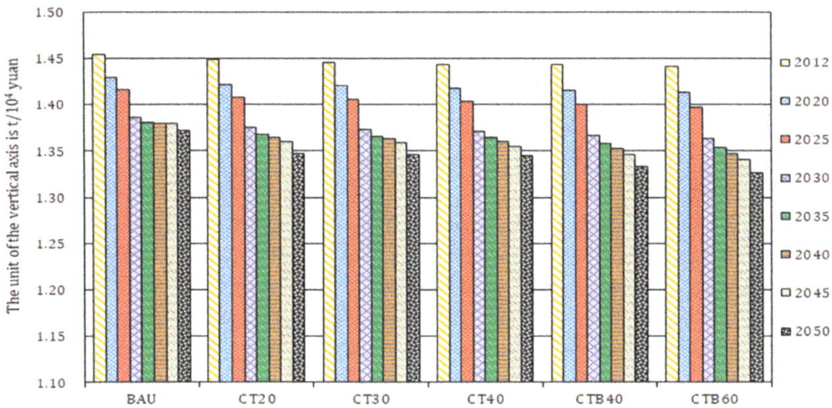

Figure 2. The trend in China's carbon intensity change under different scenarios.

In general, if agriculture-related sectors do not impose a carbon tax in the future, carbon intensity will decrease from 1.454 $t/10^4$ yuan in 2012 to 1.372 $t/10^4$ yuan in 2050, for a total decrease of 5.640%. However, according to the simulation results, the average annual growth rate of CO_2 emissions in the period from 2012 to 2050 is 6.172%, given the rapid growth rate of GDP, which is due to the absence of carbon emission reduction measures. It can be seen from Figure 2 that after the carbon tax is levied, the reduction in the carbon intensity is obviously relative to the scenario without a carbon tax. In the CTB40 and CTB50 scenarios, the reduction of the carbon intensity is more obvious than in the CT20, CT30, and CT40 scenarios. This is mainly because increasing the carbon tax while reducing the indirect production tax in the same proportion will decrease product prices, increase consumer demand for products, enlarge the scale of reproduction, and increase GDP compared to the BAU scenario. At the same time, compared with CT20, CT30, and CT40, the extent of the impact on carbon emissions is greater than the extent of the impact on GDP, so the changes in carbon intensity are more obvious than in the previous three scenarios. In addition, Figure 2 shows that the impact of a carbon tax on China's carbon intensity gradually weakens as time passes by. Taking CT30 as an example, from 2012 to 2030, the value of carbon intensity varies from 1.446 $t/10^4$ yuan to 1.373 $t/10^4$ yuan, and the magnitude of the decrease is 5.048%. From 2030 to 2050, the value of carbon intensity varies from 1.373 $t/10^4$ yuan to 1.346 $t/10^4$ yuan, and the magnitude of the decrease is 1.966%, which is similar to other scenarios. This suggests that the application of a carbon tax policy to reduce CO_2 emissions in agriculture-related sectors has an effect in the short term.

3.2. The Changes in Macroeconomy

The impact of the carbon tax on China's carbon intensity is mainly through the impacts on GDP and on carbon emissions. Although the effect of reducing the carbon intensity is significant, as an important energy-saving policy, the feasibility of the carbon tax policy depends not only on its emission reduction effect but also on its economic costs. Considering this situation, the six indicators of GDP, CO_2 emissions, households' consumption, households' welfare, government expenditure, and investment are selected to reflect the impact of a carbon tax on the macroeconomy. The welfare of the households is mainly measured based on the extent of change in the income level of the households relative to BAU scenario. The specific results are shown in Figure 3 and Table 4.

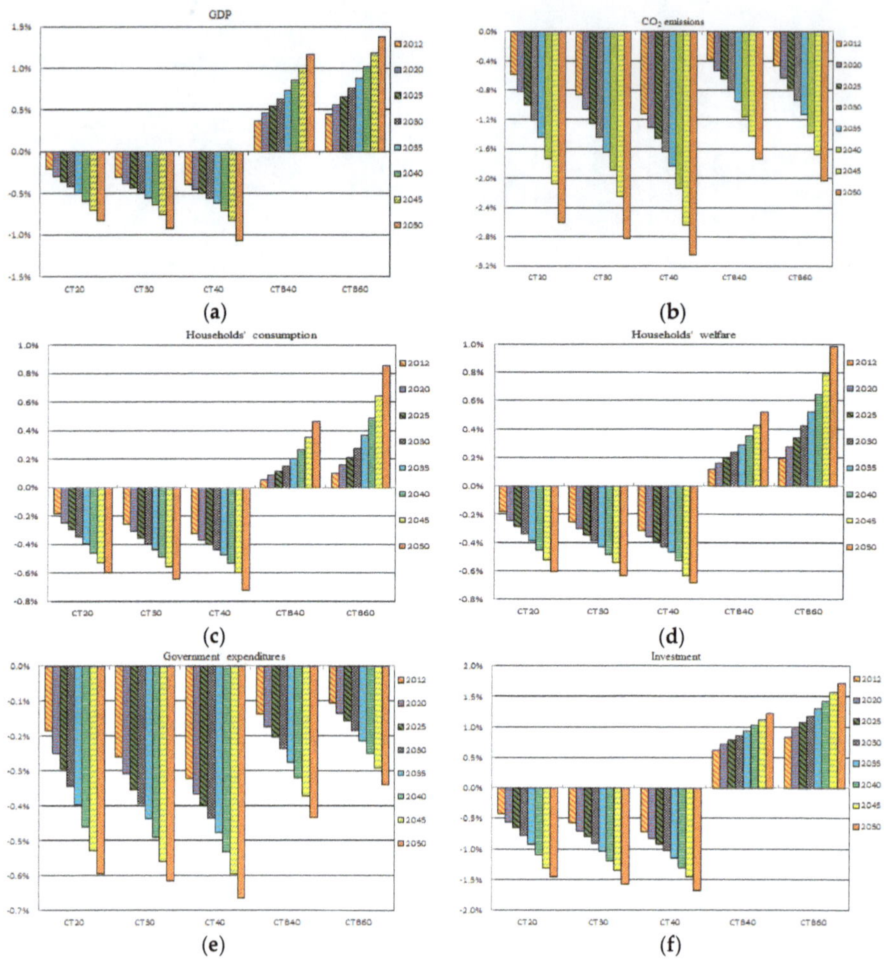

Figure 3. The trend of China's macroeconomy under different scenarios (compared with the BAU scenario).

The results of the simulation show that carbon tax has different effects on key macroeconomic variables in different scenarios. Overall, from the trend point of view, the impacts of carbon tax on CO_2 emissions and government expenditures are similar. With the increase in taxation, CO_2 reduction will be more and more obvious. In addition, emissions and expenditures of CTB40 and CTB60 are lower than those of CT60, but there is still a decreasing trend in CTB40 and CTB60. Moreover, the imposition of carbon tax on agricultural-related sectors will have the same impact trend on GDP, households' consumption, households' welfare and investment, which take effect by the impact of market supply and demand. Specifically, from 2012 to 2050, compared with BAU, the percentage changes in government expenditures and CO_2 emissions show reverse bias, and the percent deviation gradually decreases. As shown in Table 4, in the CT20 scenario, in 2012 and 2025, compared with BAU, CO_2 emissions are 0.580% and 0.998% lower, respectively, and the magnitude of decrease is 72.069% compared with BAU. In 2025 and 2035, CO_2 emissions are 0.998% and 1.443% lower, respectively, and the magnitude of decrease is 44.589%, which is in line with the above-mentioned finding that a carbon tax has a positive effect on reducing CO_2 emissions and improving energy efficiency in the short term. In addition, the CT20, CT40, and CT60 scenarios reverse the bias of percentage changes in GDP, households' consumption, households' welfare, and investment compared with BAU. However, the CTB40 and CTB60 scenarios show a positive deviation in above indicators compared with BAU. This suggests that if the carbon tax is applied alone, though the effect on CO_2 emissions will be significant, the overall macroeconomy will be sluggish, and the market demand will be depressed. As a result, the level of investment consumption will fall. When imposing a carbon tax and reducing the indirect production tax at the same proportion, due to economic expansion, the domestic market demand for products will increase and the level of investment in goods will also increase, thus stimulating consumer spending and leading to a substantial increase in GDP, as well as welfare.

Table 4. The changes in macroeconomic indicators under different scenarios between 2012–2050 (%) (compared with the BAU scenario).

Scenarios	Year	GDP	CO_2 Emissions	Households' Consumption	Households' Welfare	Government Expenditures	Investment
CT20	2012	−0.215	−0.580	−0.184	−0.180	−0.184	−0.419
	2025	−0.358	−0.998	−0.294	−0.287	−0.294	−0.643
	2035	−0.500	−1.443	−0.398	−0.389	−0.396	−0.908
	2050	−0.828	−2.597	−0.598	−0.604	−0.594	−1.443
CT30	2012	−0.311	−0.862	−0.259	−0.254	−0.259	−0.562
	2025	−0.439	−1.250	−0.353	−0.346	−0.353	−0.790
	2035	−0.561	−1.643	−0.438	−0.430	−0.436	−1.022
	2050	−0.930	−2.821	−0.643	−0.633	−0.613	−1.562
CT40	2012	−0.397	−1.119	−0.322	−0.315	−0.322	−0.711
	2025	−0.503	−1.460	−0.399	−0.391	−0.398	−0.912
	2035	−0.621	−1.832	−0.475	−0.469	−0.475	−1.132
	2050	−1.072	−3.043	−0.723	−0.682	−0.662	−1.671
CTB40	2012	0.360	−0.384	0.056	0.118	−0.136	0.631
	2025	0.538	−0.643	0.116	0.196	−0.202	0.793
	2035	0.732	−0.956	0.202	0.291	−0.274	0.945
	2050	1.162	−1.733	0.465	0.523	−0.432	1.229
CTB60	2012	0.437	−0.460	0.101	0.197	−0.105	0.842
	2025	0.647	−0.765	0.210	0.342	−0.157	1.077
	2035	0.874	−1.131	0.368	0.523	−0.214	1.301
	2050	1.373	−2.033	0.854	0.989	−0.339	1.729

Note: all the values are percentage variations; these also apply to all subsequent tables.

By comparing the CT20, CT30, and CT40 scenarios, with the change of carbon tax rate ranging from 20 yuan/t to 40 yuan/t, the effect of a carbon tax on macroeconomic variables is becoming more and more obvious. Taking 2050 as an example, compared with BAU, the GDP in the CT20, CT30, and CT40 scenarios is 0.828%, 0.930%, and 1.072% lower, respectively. Carbon emissions are 2.597%, 2.821%, and 3.043% lower, respectively. Households' consumption is 0.598%, 0.643%, and 0.723% lower, respectively. The trend of households' welfare, government expenditure and investment are the same as the above indicators. This indicates that as tax rates increase, the degree of impact on macroeconomic variables is more serious. Although a high tax rate will help decrease the carbon intensity, promote energy efficiency, and restructure industries as well as reduce emissions, which are in line with the purpose of the carbon tax levied, it will also cause enormous pressure on the macroeconomy and will ultimately lead to economic depression. Therefore, the objective of a carbon tax policy is to determine the optimal carbon tax with the aim of the maximum reduction in emissions and the least negative impact on the economy. On the contrary, in the balanced tax scenarios of CTB40 and CTB60, increasing the carbon tax will expand the scale of production. The higher the carbon tax is, the lower the production tax is, and the more positive the impact on the economy.

3.3. The Impact of Carbon Tax on Agriculture-Related Sectors

This article mainly focuses on the impact of a carbon tax policy on agriculture-related sectors. Therefore, the results of other departments such as the manufacturing and mining industry, construction industry, transport industry, and service industry are not listed. On the analysis of the macro-environment and the macroeconomic effects of a carbon tax levied on the production process in agriculture-related sectors, this article has elaborated on the positive effect of a carbon tax on emission reduction and the negative effect of a carbon tax on restraining economic development, and it further considers the impact of a carbon tax on the various industries of agriculture. Specifically, 23 different industries related to agricultural activities are considered. As the level of the return on capital will have a direct effect on farmers' willingness to invest, it will directly determine the future direction of the flow of farmers' funds with a strong incentive orientation. Moreover, the output levels and income levels in agricultural industries are major concerns of the government and farmers, which are related to the livelihood of farmers. These indicators are also key factors to maintain social stability. Therefore, the effects of the different scenarios on the various agriculture-related sectors are analyzed from those three perspectives: the return on capital, the output level, and the income level.

3.3.1. The Changes in the Return on Capital of Agriculture-Related Sectors

As shown in Figure 4, levying a carbon tax on the production process in agriculture-related sectors will have a significant impact on the return on capital of agriculture-related sectors compared with levying a noncarbon tax. However, different scenarios can generate different positive and negative offsets. As shown in Tables 5–7, the five agriculture-related sectors that experience the greatest impact on the return on capital when a carbon tax is levied on the production process are fruit and other planting, forestry, rice planting, the dairy industry, and corn planting. The five industries with the least impact on the return on capital are the leather products industry, tea planting, tobacco planting, cotton, hemp planting, and the textile industry. The main reason for the polarization of the impact on capital return is the difference in carbon emissions between the different industries. A carbon tax on agriculture-related sectors with more carbon emissions will increase the input costs much more than in agriculture-related sectors with less carbon emissions.

By comparing different scenarios according to the positive and negative offsets of the return on capital, the agriculture-related sectors can be roughly divided into two categories. In the first category, the carbon tax has a negative impact on agriculture-related sectors. This category includes rice planting; wheat planting; corn planting; beans and potato planting; peanut, rapeseed, and sesame planting; cotton and hemp planting; tobacco planting; tea planting; fruit and other planting; animal husbandry; forestry; and the fishery industry for a total of 12 industries. In the second category, the CT20, CT30, and CT40 scenarios have negative impacts on the return on capital in the agriculture-related sectors, but the CTB40 and CTB60 scenarios have positive impacts on the return on capital. This category includes 11 agriculture-related industries other than the above 12 industries. The main reason may be that the first category of agriculture-related sectors includes basic agricultural industries, in which the cycle of congenital production is long and supply regulations lag behind changes in market regulations, and the provision of food cannot be substituted. If a carbon tax is levied on those subsidized industries while the indirect production tax is reduced at the same ratio, for the first category of subsidized industries, since these industries are already subsidized by the government, they cannot enjoy the benefits of reducing the indirect taxation. Instead, it will increase the production costs of these subsidized industries. Moreover, it is not conducive to the production of capital accumulation, and will eventually lead to a lower return on capital. For the second category of nonsubsidized industries, the benefits of reducing indirect tax can be mainly applied to these nonsubsidized industries such as the dairy industry, the slaughter and meat processing industry, etc. As a result, these nonsubsidized industries will expand their production scale to lower input costs, improve revenue, and increase the input-output level, ultimately improving the return on capital.

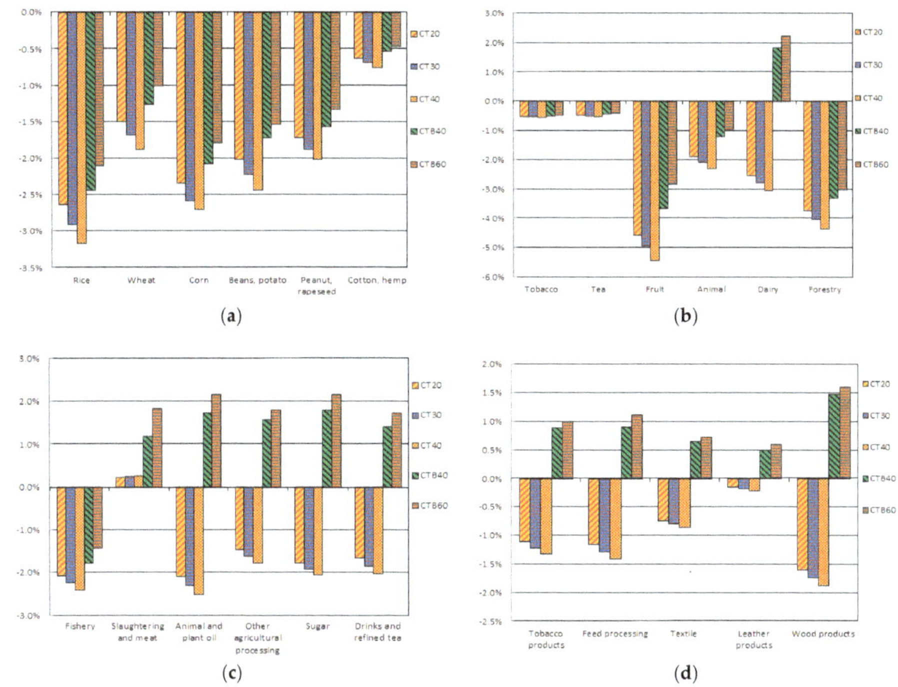

Figure 4. The change in the return on capital in agriculture-related sectors under different scenarios in 2050 (compared with the BAU scenario).

Table 5. The changes in the return on capital in agriculture-related sectors under different scenarios compared with the BAU scenario (%) (from rice to tea).

Scenarios	Year	Rice	Wheat	Corn	Beans, Potato	Peanut, Rapeseed, Sesame	Cotton, Hemp	Tobacco	Tea
CT20	2012	−0.645	−0.474	−0.602	−0.577	−0.532	−0.329	−0.287	−0.215
	2030	−1.261	−0.862	−1.142	−1.067	−0.971	−0.548	−0.503	−0.289
	2050	−2.635	−1.500	−2.337	−2.014	−1.729	−0.632	−0.515	−0.467
CT30	2012	−0.922	−0.661	−0.854	−0.805	−0.743	−0.477	−0.354	−0.242
	2030	−1.485	−0.983	−1.423	−1.230	−1.111	−0.577	−0.530	−0.308
	2050	−2.911	−1.683	−2.582	−2.223	−1.884	−0.642	−0.531	−0.492
CT40	2012	−1.171	−0.811	−1.014	−0.999	−0.913	−0.534	−0.421	−0.280
	2030	−1.685	−1.085	−1.618	−1.373	−1.228	−0.595	−0.555	−0.324
	2050	−3.175	−1.879	−2.701	−2.443	−2.020	−0.660	−0.550	−0.520
CTB40	2012	−0.503	−0.312	−0.521	−0.493	−0.473	−0.302	−0.243	−0.193
	2030	−1.062	−0.605	−1.004	−0.893	−0.834	−0.416	−0.339	−0.282
	2050	−2.441	−1.265	−2.081	−1.728	−1.564	−0.592	−0.492	−0.432
CTB60	2012	−0.382	−0.283	−0.413	−0.400	−0.398	−0.273	−0.220	−0.169
	2030	−0.857	−0.517	−0.828	−0.759	−0.705	−0.388	−0.316	−0.252
	2050	−2.104	−1.010	−1.795	−1.543	−1.332	−0.572	−0.473	−0.392

Table 6. The changes in the return on capital in agriculture-related sectors under different scenarios compared with the BAU scenario (%) (from fruit to vegetables and other processing).

Scenarios	Year	Fruit	Animal Husbandry	Dairy Industry	Forestry	Fishery Industry	Slaughtering and Meat Processing	Animal and Plant Oil Processing	Vegetables and Other Processing
CT20	2012	−1.164	−0.638	−0.801	−1.252	−0.857	0.106	−0.526	−0.451
	2030	−2.299	−1.063	−1.350	−2.145	−1.367	0.125	−1.043	−0.798
	2050	−4.552	−1.862	−2.506	−3.714	−2.070	0.235	−2.079	−1.458
CT30	2012	−1.681	−0.849	−1.062	−1.671	−1.112	0.111	−0.766	−0.616
	2030	−2.697	−1.194	−1.535	−2.435	−1.510	0.150	−1.222	−0.912
	2050	−4.911	−2.064	−2.754	−4.020	−2.220	0.246	−2.281	−1.606
CT40	2012	−2.137	−1.008	−1.274	−2.023	−1.304	0.115	−0.970	−0.751
	2030	−3.046	−1.308	−1.699	−2.693	−1.626	0.169	−1.379	−1.009
	2050	−5.431	−2.275	−3.015	−4.353	−2.391	0.261	−2.490	−1.772
CTB40	2012	−0.954	−0.463	0.402	−1.015	−0.554	0.432	0.701	0.521
	2030	−1.802	−0.728	0.828	−1.772	−0.959	0.699	1.076	0.877
	2050	−3.656	−1.203	1.846	−3.289	−1.763	1.192	1.732	1.563
CTB60	2012	−0.684	−0.310	0.621	−0.732	−0.438	0.633	0.845	0.660
	2030	−1.336	−0.530	1.139	−1.426	−0.765	1.047	1.319	1.059
	2050	−2.813	−0.963	2.232	−2.993	−1.421	1.832	2.163	1.790

Table 7. The changes in the return on capital in agriculture-related sectors under different scenarios compared with the BAU scenario (%) (from sugar to wood products).

Scenarios	Year	Sugar	Drinks and Refined Tea	Tobacco Products	Feed processing Industry	Textile	Leather	Wood Products
CT20	2012	−0.580	−0.393	−0.283	−0.248	−0.148	−0.027	−0.417
	2030	−1.017	−0.788	−0.442	−0.529	−0.346	−0.054	−0.824
	2050	−1.773	−1.652	−1.110	−1.160	−0.744	−0.150	−1.605
CT30	2012	−0.790	−0.570	−0.335	−0.372	−0.246	−0.033	−0.599
	2030	−1.154	−0.931	−0.519	−0.634	−0.422	−0.063	−0.966
	2050	−1.910	−1.831	−1.220	−1.280	−0.801	−0.182	−1.737
CT40	2012	−0.959	−0.730	−0.412	−0.487	−0.323	−0.038	−0.765
	2030	−1.271	−1.059	−0.597	−0.726	−0.485	−0.073	−1.088
	2050	−2.052	−2.010	−1.320	−1.411	−0.859	−0.221	−1.871
CTB40	2012	0.743	0.443	0.220	0.260	0.167	0.084	0.491
	2030	1.128	0.767	0.427	0.469	0.317	0.194	0.826
	2050	1.794	1.413	0.892	0.902	0.644	0.493	1.473
CTB60	2012	0.833	0.521	0.264	0.294	0.222	0.102	0.600
	2030	1.308	0.918	0.495	0.552	0.387	0.236	0.956
	2050	2.160	1.724	0.995	1.114	0.721	0.598	1.604

3.3.2. The Changes in the Output Level of Agriculture-Related Sectors

As shown in Figure 5, a carbon tax on agriculture-related sectors will have a significant impact on the output level compared with the non-introduction of a carbon tax, but the impacts vary depending on the specific agriculture-related industry. At the same time, different scenarios generate different positive and negative offsets. As shown in Tables 8–10, the five agriculture-related sectors with the greatest negative impact on the levels of output are fruit and other planting, the animal and plant oil processing industry, forestry, the wood products processing industry, and the sugar products processing industry. The five agriculture-related sectors with the least negative impact on the levels of output are tobacco planting, the textile industry, cotton and hemp planting, tea planting, and the fishery industry. In contrast, the impact of the introduction of a carbon tax on agriculture-related sectors that are subsidized is generally smaller than the impact on nonsubsidized agriculture-related sectors.

In addition, Figure 5 shows that, in the CT20, CT40, and CT60 scenarios, although the slaughter and meat processing industry and the leather products industry have positive deviations from the BAU levels of output, the other 21 agriculture-related sectors show different degrees of negative deviation in the levels of output. This may be due to taxation leading to increased costs for agriculture-related sectors that emit CO_2 into the air, ultimately leading to a rise in the output prices of their agricultural products. In order to avoid the cost of the carbon tax, agriculture-related sectors such as the slaughter and meat processing industry and the leather products industry will promote agricultural CO_2 emissions reduction through technological innovation or changing the mode of production, thus will lead to an increase in output levels of those sectors. However, in other agriculture-related sectors may reduce CO_2 emissions by reducing the scale of production. The former are called technology-intensive industries, and the latter are known as labor-intensive industries. That means that the more labor-intensive agriculture-related sectors are forced, by a carbon tax, to reduce their production scale, resulting in a negative shift in the levels of output in various agriculture-related sectors.

Finally, by comparing the different scenarios shown in Figure 5 and in Tables 8–10, it can be seen that the impacts of the different scenarios with offsets, which are positive or negative relative to the output level, are different. The result is the same as with the impact on the return on capital of agriculture-related sectors. The main difference is between subsidized and nonsubsidized industries. It also shows that for subsidized agriculture-related sectors, introducing a carbon tax at the same time as indirect taxes are cut in proportion is not conducive to expanding production.

Table 8. The changes in outputs in agriculture-related sectors under different scenarios compared with the BAU scenario (%) (from rice to tea).

Scenarios	Year	Rice	Wheat	Corn	Beans, Potato	Peanut, Rapeseed, Sesame	Cotton, Hemp	Tobacco	Tea
CT20	2012	−0.159	−0.114	−0.139	−0.136	−0.127	−0.049	−0.037	−0.020
	2030	−0.317	−0.187	−0.296	−0.248	−0.220	−0.064	−0.044	−0.044
	2050	−0.636	−0.249	−0.616	−0.390	−0.286	−0.102	−0.093	−0.094
CT30	2012	−0.231	−0.155	−0.211	−0.191	−0.176	−0.054	−0.039	−0.030
	2030	−0.372	−0.200	−0.356	−0.280	−0.243	−0.071	−0.048	−0.053
	2050	−0.702	−0.278	−0.676	−0.411	−0.288	−0.111	−0.101	−0.103
CT40	2012	−0.294	−0.180	−0.272	−0.234	−0.210	−0.060	−0.042	−0.040
	2030	−0.420	−0.207	−0.411	−0.305	−0.258	−0.078	−0.052	−0.061
	2050	−0.777	−0.309	−0.737	−0.432	−0.292	−0.119	−0.110	−0.113
CTB40	2012	−0.114	−0.090	−0.097	−0.090	−0.091	−0.035	−0.027	−0.015
	2030	−0.213	−0.121	−0.189	−0.149	−0.130	−0.048	−0.040	−0.029
	2050	−0.426	−0.167	−0.396	−0.261	−0.191	−0.069	−0.062	−0.060
CTB60	2012	−0.084	−0.073	−0.077	−0.072	−0.067	−0.026	−0.020	−0.013
	2030	−0.168	−0.094	−0.150	−0.116	−0.096	−0.035	−0.030	−0.024
	2050	−0.365	−0.125	−0.315	−0.196	−0.144	−0.048	−0.047	−0.048

Table 9. The changes in outputs in agriculture-related sectors under different scenarios compared with the BAU scenario (%) (from fruit to vegetables and other processing).

Scenarios	Year	Fruit	Animal Husbandry	Dairy Industry	Forestry	Fishery Industry	Slaughtering and Meat Processing	Animal and Plant Oil Processing	Vegetables and Other Processing
CT20	2012	−0.268	−0.128	−0.159	−0.287	−0.110	0.100	−0.230	−0.159
	2030	−0.598	−0.221	−0.294	−0.554	−0.191	0.277	−0.510	−0.309
	2050	−1.313	−0.347	−0.548	−1.015	−0.266	1.008	−1.124	−0.551
CT30	2012	−0.412	−0.175	−0.223	−0.409	−0.151	0.154	−0.355	−0.230
	2030	−0.722	−0.246	−0.338	−0.644	−0.212	0.376	−0.613	−0.358
	2050	−1.438	−0.369	−0.591	−1.086	−0.274	1.164	−1.237	−0.589
CT40	2012	−0.548	−0.210	−0.275	−0.516	−0.181	0.240	−0.468	−0.289
	2030	−0.832	−0.267	−0.376	−0.720	−0.227	0.473	−0.704	−0.398
	2050	−1.564	−0.392	−0.641	−1.160	−0.287	1.323	−1.358	−0.629
CTB40	2012	−0.193	−0.856	0.044	−0.220	0.150	0.392	0.310	0.215
	2030	−0.395	−0.433	0.129	−0.374	0.150	0.778	0.681	0.408
	2050	−0.879	−0.203	0.424	−0.674	0.331	1.665	1.632	0.830
CTB60	2012	−0.141	−0.646	0.072	−0.157	0.188	0.513	0.393	0.272
	2030	−0.293	−0.351	0.200	−0.273	0.188	0.928	0.862	0.528
	2050	−0.661	−0.178	0.620	−0.503	0.450	1.793	2.063	1.103

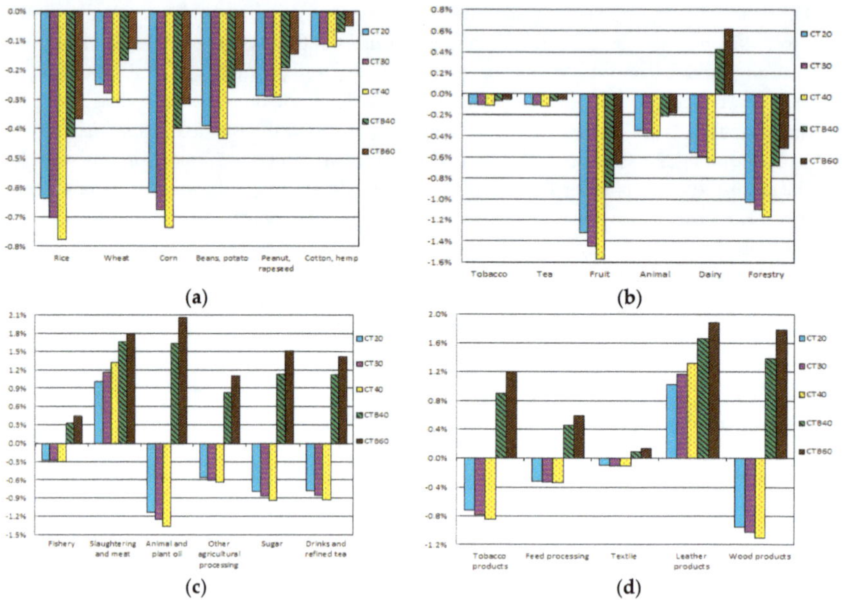

Figure 5. The changes in the outputs in agriculture-related sectors under different scenarios (compared with the BAU scenario).

Table 10. The changes in outputs in agriculture-related sectors under different scenarios compared with the BAU scenario (%) (from sugar to wood products).

Scenarios	Year	Sugar	Drinks and Refined Tea	Tobacco Products	Feed Processing Industry	Textile	Leather	Wood Products
CT20	2012	−0.186	−0.198	−0.151	−0.122	−0.040	0.107	−0.263
	2030	−0.396	−0.393	−0.337	−0.217	−0.055	0.407	−0.517
	2050	−0.787	−0.773	−0.721	−0.315	−0.095	1.017	−0.957
CT30	2012	−0.282	−0.289	−0.243	−0.171	−0.046	0.166	−0.379
	2030	−0.469	−0.460	−0.406	−0.242	−0.061	0.546	−0.587
	2050	−0.855	−0.841	−0.781	−0.328	−0.101	1.164	−1.027
CT40	2012	−0.366	−0.366	−0.310	−0.206	−0.052	0.256	−0.481
	2030	−0.531	−0.518	−0.470	−0.260	−0.068	0.680	−0.632
	2050	−0.928	−0.912	−0.842	−0.341	−0.108	1.316	−1.109
CTB40	2012	0.260	0.290	0.190	0.150	0.041	0.452	0.360
	2030	0.522	0.551	0.396	0.255	0.061	0.837	0.683
	2050	1.131	1.123	0.893	0.460	0.095	1.661	1.391
CTB60	2012	0.304	0.351	0.230	0.191	0.050	0.580	0.449
	2030	0.649	0.680	0.504	0.325	0.080	1.015	0.865
	2050	1.506	1.419	1.201	0.590	0.134	1.890	1.791

3.3.3. The Changes in the Income Level of Households in Various Industries of Agriculture

It can be seen from Figure 6 that, subject to the decline in the rate of return on capital and the level of output, the imposition of a carbon tax on agriculture-related sectors will significantly change the income levels of the various industries in agriculture. Tables 11–13 show that the five industries with the greatest negative impacts on the income level after introduction of the carbon tax in agricultural production are the animal and plant oil processing industry, the sugar products processing industry, forestry, the wood products processing industry, and the drinks and refined tea processing industry. In the CT20 scenario in 2020, the corresponding income levels are 1.577%, 1.166%, 1.159%, 1.124%, and 0.952% lower, respectively, compared with BAU, whereas the corresponding rates of return on capital are 2.079%, 1.773%, 3.714%, 1.605%, and 1.652% lower, respectively, compared with BAU, and the corresponding reductions of the output level are 1.124%, 0.787%, 1.015%, 0.957%, and 0.713%, respectively, compared with BAU, which are higher than those of the same index compared with BAU. This indicates that the level of income is closely related to the rate of the return on capital and the level of output.

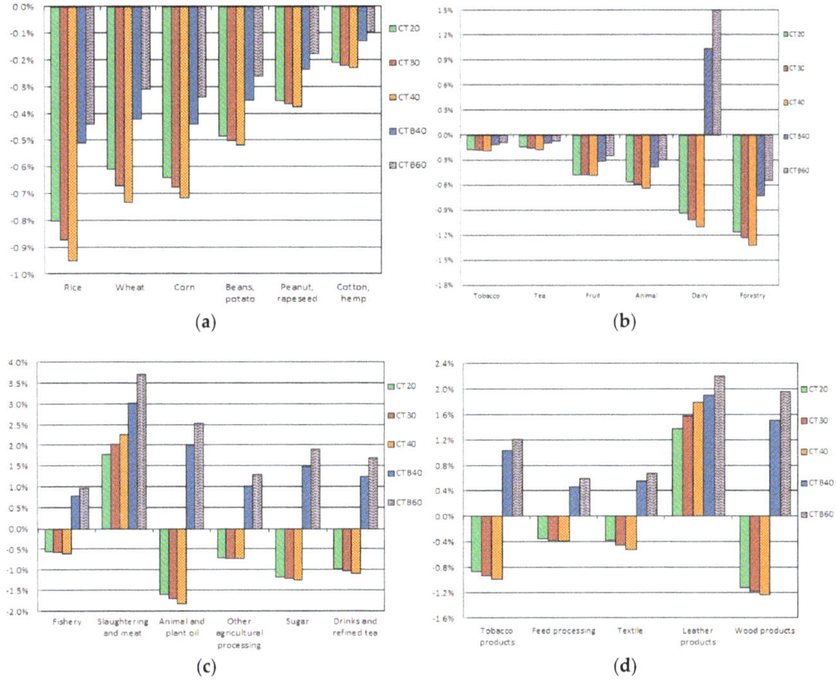

Figure 6. The changes in the households' income in agriculture-related sectors under different scenarios (compared with the BAU scenario).

Table 11. The changes in the households' income in agriculture-related sectors under different scenarios compared with the BAU scenario (%) (from rice to tea).

Scenarios	Year	Rice	Wheat	Corn	Beans, Potato	Peanut, Rapeseed, Sesame	Cotton, Hemp	Tobacco	Tea
CT20	2012	−0.216	−0.149	−0.179	−0.189	−0.171	−0.092	−0.073	−0.032
	2030	−0.417	−0.235	−0.272	−0.330	−0.288	−0.145	−0.109	−0.065
	2050	−0.803	−0.611	−0.641	−0.485	−0.354	−0.210	−0.170	−0.142
CT30	2012	−0.310	−0.198	−0.229	−0.260	−0.234	−0.131	−0.106	−0.047
	2030	−0.486	−0.248	−0.305	−0.368	−0.313	−0.159	−0.117	−0.076
	2050	−0.874	−0.671	−0.678	−0.501	−0.365	−0.221	−0.179	−0.156
CT40	2012	−0.389	−0.228	−0.259	−0.313	−0.276	−0.169	−0.136	−0.061
	2030	−0.545	−0.256	−0.333	−0.397	−0.330	−0.173	−0.125	−0.086
	2050	−0.951	−0.732	−0.718	−0.518	−0.376	−0.232	−0.190	−0.171
CTB40	2012	−0.160	−0.107	−0.136	−0.141	−0.123	−0.066	−0.053	−0.024
	2030	−0.278	−0.205	−0.237	−0.217	−0.168	−0.091	−0.075	−0.046
	2050	−0.511	−0.421	−0.439	−0.350	−0.237	−0.129	−0.110	−0.095
CTB60	2012	−0.120	−0.079	−0.105	−0.100	−0.090	−0.049	−0.039	−0.017
	2030	−0.223	−0.150	−0.183	−0.157	−0.124	−0.067	−0.056	−0.034
	2050	−0.441	−0.307	−0.341	−0.260	−0.178	−0.097	−0.086	−0.072

Table 12. The changes in the households' income in agriculture-related sectors under different scenarios compared with the BAU scenario (%) (from fruit to vegetables and other processing).

Scenarios	Year	Fruit	Animal Husbandry	Dairy Industry	Forestry	Fishery Industry	Slaughtering and Meat Processing	Animal and Plant Oil Processing	Vegetables and Other Processing
CT20	2012	−0.215	−0.213	−0.276	−0.339	−0.223	0.221	−0.484	−0.400
	2030	−0.362	−0.363	−0.504	−0.645	−0.387	0.549	−0.907	−0.609
	2050	−0.476	−0.561	−0.939	−1.159	−0.550	1.782	−1.577	−0.704
CT30	2012	−0.290	−0.290	−0.384	−0.480	−0.305	0.318	−0.692	−0.518
	2030	−0.398	−0.403	−0.579	−0.747	−0.430	0.732	−1.037	−0.647
	2050	−0.482	−0.596	−1.020	−1.235	−0.566	2.030	−1.690	−0.709
CT40	2012	−0.345	−0.345	−0.473	−0.602	−0.367	0.403	−0.852	−0.590
	2030	−0.423	−0.436	−0.644	−0.833	−0.463	0.904	−1.145	−0.671
	2050	−0.490	−0.639	−1.103	−1.326	−0.586	2.280	−1.810	−0.716
CTB40	2012	−0.161	−0.166	0.350	−0.260	0.286	0.514	0.653	0.551
	2030	−0.222	−0.247	0.585	−0.424	0.465	1.187	1.112	0.739
	2050	−0.319	−0.384	1.034	−0.730	0.799	3.012	2.010	1.022
CTB60	2012	−0.120	−0.122	0.421	−0.180	0.341	0.661	0.827	0.659
	2030	−0.169	−0.187	0.767	−0.306	0.560	1.498	1.407	0.907
	2050	−0.246	−0.299	1.494	−0.550	0.973	3.722	2.538	1.292

Table 13. The changes in households' income in agriculture-related sectors under different scenarios compared with the BAU scenario (%) (from sugar to wood products).

Scenarios	Year	Sugar	Drinks and Refined Tea	Tobacco Products	Feed Processing Industry	Textile	Leather	Wood Products
CT20	2012	−0.543	−0.335	−0.301	−0.173	−0.069	0.150	−0.406
	2030	−0.879	−0.589	−0.545	−0.276	−0.136	0.407	−0.727
	2050	−1.166	−0.952	−0.871	−0.351	−0.381	1.382	−1.124
CT30	2012	−0.723	−0.463	−0.413	−0.232	−0.091	0.231	0.562
	2030	−0.958	−0.662	−0.622	−0.293	−0.177	0.546	−0.819
	2050	−1.199	−1.013	−0.932	−0.371	−0.451	1.583	−1.175
CT40	2012	−0.843	−0.558	−0.513	−0.267	−0.120	0.355	−0.686
	2030	−1.015	−0.722	−0.682	−0.303	−0.212	0.680	−0.891
	2050	−1.235	−1.082	−0.995	−0.392	−0.517	1.792	−1.231
CTB40	2012	0.604	0.452	0.389	0.210	0.093	0.562	0.510
	2030	0.927	0.731	0.618	0.305	0.217	1.002	0.852
	2050	1.493	1.247	1.033	0.460	0.554	1.903	1.507
CTB60	2012	0.826	0.560	0.494	0.260	0.118	0.693	0.661
	2030	1.228	0.948	0.755	0.384	0.270	1.199	1.106
	2050	1.907	1.702	1.211	0.592	0.676	2.203	1.960

At the same time, compared with the other scenarios, we find that in the CT20, CT40, and CT60 scenarios, although the income levels of the slaughter and meat processing industry and the leather products industry show a positive deviation, the income levels of the other 21 agriculture-related sectors show different degrees of negative deviation. In addition, in the CTB40 and CTB60 scenarios, the level of income will increase compared with BAU in nonsubsidized agricultural industries, which may be related to a reduction in the production tax resulting in the expansion of the scale of production, increasing production and stimulating consumption.

4. Discussions

4.1. Policy Implications

Through the above analysis, we can see that in order to achieve agricultural CO_2 emissions reductions, we cannot blindly imitate foreign advanced concepts and successful practices. On the contrary, we should combine these concepts with the specific circumstances of the various regions of

China, according to local conditions, to explore suitable paths and methods for China's agricultural CO_2 emissions. Specific practices are as follows:

(1) Find the optimal combination of carbon tax policies.

The simulation results show that levying a carbon tax at the same time as cutting indirect taxes in proportion can reduce the negative impact on agriculture-related sectors. Therefore, the government should determine the carbon tax levied and at the same time set a reasonable carbon tax levy and increase the implementation of transfer payments and other supporting measures to ensure that energy-saving emission reduction measures are effective.

(2) Consider the time effect of the carbon tax.

Given that the effect of the carbon tax in the short term is stronger than in the long term, in the short term, a carbon tax can be the main measure to reduce emissions. In the long term, a carbon tax is as a supplemental tool, and other energy-saving emission reduction methods are the main measures.

(3) Actively explore ways to reduce CO_2 emissions in agriculture-related sectors.

As the results of this study show, due to the natural fragility of China's agriculture-related sectors, a carbon tax would impose a serious impact on China's economy of agriculture-related sectors. There is an urgent need to explore other ways to solve the problem of agricultural CO_2 emissions reductions, such as improving energy efficiency or using alternative energy sources, etc., in order to minimize the intensity of the impact on agriculture-related sectors.

As the scope of carbon tax levied mainly due to the combustion of fossil fuels caused by CO_2 emissions, this paper does not consider other CO_2 sources such as pesticide and fertilizer. Finally, the relevant policy recommendations based on the results mainly reflect the impact of the carbon tax levied on the intermediate production process of agricultural-related sectors. Therefore, when considering limiting CO_2 emissions in agricultural-related sectors, on the one hand, tax means can be useful; on the other hand, reducing the intensity of agricultural chemicals, and gradually increasing the input proportion of organic fertilizer, biological pesticides, and other low-carbon green production can limit CO_2 emissions effectively.

4.2. Future Prospects

Although the carbon tax is considered to be one of the most efficient policy means for reducing emissions, it has been controversial about how exactly the carbon tax is levied. Due to the limitations of research conditions, research level, and time, this paper has some of the shortcomings. In the future, efforts and improvements can be made in the following aspects.

(1) Introduce carbon trading mechanism to make the scenarios more colorful under the premise of the combination of the carbon tax policy and the carbon trading mechanism.
(2) Further improve the accuracy of the data. Most of the parameters in this model refer to the existing research results. In future research, more detailed data surveys should be conducted to calibrate the parameters more accurately and should be more in line with the characteristics of the real economy.
(3) Strengthen the model scenarios design to improve the degree of discrimination of carbon tax impacts between different scenarios.
(4) Update the basic data in a timely manner to improve the accuracy of the results.

5. Conclusions

As a basic industry in China, agriculture has faced more and more problems from CO_2 emissions. Therefore, under the premise of ensuring the stability of China's agriculture-related economics, formulating effective policy mechanisms to reduce agricultural CO_2 emissions is a serious problem.

Based on this situation, this article analyzes the degree of influence of agricultural CO_2 emissions on the macroenvironment, on macroeconomic variables, and on the specific agriculture-related sectors of the 3EAD-CGE model. The conclusions are as follows:

(1) Whether the short-term and long-term effects of carbon taxes on agriculture-related sectors would be significant if the carbon tax was simply implemented at an incremental rate rather than compensating for the negative effects of carbon taxes. This study provides detailed data that supports the views of most people against the imposition of a carbon tax. From the simulation results, we can see that the capacity in China's agriculture-related sectors to deal with CO_2 emissions is very fragile. The conclusion of this analysis also verifies why the Chinese government did not adopt the relevant agencies' recommendations to levy a carbon tax in 2012.

(2) A policy levying a carbon tax at the same time as cutting indirect taxes in proportion will exert a negative impact on agriculture-related sectors that are subsidized, but the impact is gentler than carrying out a carbon tax alone. From the results of the current model simulations, scenario CTB60 can achieve the best result. However, such a policy is still not conducive to capital accumulation and capital expansion in agriculture-related sectors that are subsidized. In the future development of carbon tax policy, we must carefully consider its impact on the subsidized industries to ensure the output does not decline below China's base crop yield red line.

Author Contributions: Conceptualization, W.F. and X.D.; Data curation, W.F., Z.G., H.W., Z.X., and S.U.; Formal analysis, W.F., N.C., H.W., N.L., X.W., and P.Z.; Funding acquisition, X.D.; Investigation, Z.G., N.C., H.W., Z.X., and J.R.; Methodology, W.F., N.C., H.W., Z.X., N.L., X.W., P.Z., and S.U.; Project administration, X.D.; Resources, X.D.; Supervision, X.D.; Validation, X.D.; Visualization, W.F.; Writing—original draft, W.F.; Writing—review & editing, W.F. and X.D.

Funding: This research was funded by the National Natural Science Foundation of China (41671531, 41271549), International S&T Cooperation Program of China (YS2017YFGH000562), the Key Project of the National Societal Science Foundation of China (15ZDB163), Science &Technology Supporting Program of China (2012BAD14B03), and the EU project "Sustainable Process Integration Laboratory—SPIL" (No. CZ.02.1.01/0.0/0.0/15_003/0000456) funded by EU "CZ Operational Programme Research, Development and Education".

Conflicts of Interest: The authors declare no conflicts of interest.

References

1. Deng, M.; Li, W.; Hu, Y. Decomposing Industrial Energy-Related CO_2 Emissions in Yunnan Province, China: Switching to Low-Carbon Economic Growth. *Energies* **2016**, *9*, 23. [CrossRef]
2. Vaissiere, A.C.; Levrel, H.; Pioch, S.; Carlier, A. Biodiversity offsets for offshore wind farm projects: The current situation in Europe. *Mar. Policy* **2014**, *48*, 172–183. [CrossRef]
3. Guan, D.B.; Barker, T. Low-carbon development in the least developed region: A case study of Guangyuan, Sichuan province, southwest China. *Nat. Hazards* **2012**, *62*, 243–254. [CrossRef]
4. Liobikiene, G.; Butkus, M. The European Union possibilities to achieve targets of Europe 2020 and Paris agreement climate policy. *Renew. Energy* **2017**, *106*, 298–309. [CrossRef]
5. Lu, J.; Fan, W.; Meng, M. Empirical Research on China's Carbon Productivity Decomposition Model Based on Multi-Dimensional Factors. *Energies* **2015**, *8*, 3093–3117. [CrossRef]
6. Gao, C.K.; Wang, D.; Zhao, B.H.; Chen, S.; Qin, W. Analyzing and forecasting CO2 emission reduction in China's steel industry. *Front. Earth Sci.* **2015**, *9*, 105–113. [CrossRef]
7. China Daily. China Likely to Levy Carbon Tax Around 2012[EB/OL]. Available online: http://www.chinadaily.com.cn/china/2010-05/11/content_9834525.htm (accessed on 6 September 2010).
8. IPCC. *Climate Change 2007: Impacts, Adaptation and Vulnerability*; Cambridge University Press: New York, NY, USA, 2007.
9. Food and Agriculture Organization of the United Nations (FAOSTAT). FAO Statistical Yearbook 2013[EB/OL]. Available online: http://www.fao.org/economic/ess/ess-publications/ess-yearbook/en/ (accessed on 21 August 2015).
10. National Bureau of Statistics (NBS). *China Statistical Yearbook 2013*; China Statistics Press: Beijing, China, 2015.
11. National Bureau of Statistics (NBS). *Input-output Tables of China 2012*; China Statistics Press: Beijing, China, 2015.

12. Dong, H.M.; Li, Y.E.; Tao, X.P.; Li, N.; Zhu, Z.P. China greenhouse gas emissions from agricultural activities and its mitigation strategy. *Trans. Chin. Soc. Agric. Eng.* **2008**, *24*, 269–273.
13. Iodice, P.; Senatore, A. Atmospheric pollution from point and diffuse sources in a National Interest Priority Site located in Italy. *Energy Environ.* **2016**, *27*, 586–596. [CrossRef]
14. Abagnale, C.; Cardone, M.; Iodice, P.; Strano, S.; Terzo, M.; Vorraro, G. Power requirements and environmental impact of a pedelec. a case study based on real-life applications. *Environ. Impact Assess. Rev.* **2015**, *53*, 1–7. [CrossRef]
15. Armaroli, N.; Balzani, V. Towards an electricity-powered world. *Energy Environ. Sci.* **2014**, *4*, 3193–3222. [CrossRef]
16. Charles, D. Stimulus gives DOE billions for carbon-capture projects. *Science* **2009**, *323*, 1158. [CrossRef] [PubMed]
17. Iodice, P.; Langella, G.; Amoresano, A. Ethanol in gasoline fuel blends: Effect on fuel consumption and engine out emissions of SI engines in cold operating conditions. *Appl. Therm. Eng.* **2018**, *130*, 1081–1089. [CrossRef]
18. Kitwiwattanachai, A.; Nelson, D.; Reed, G. Quantitative impacts of alternative East Asia free trade areas: A computable general equilibrium (CGE) assessment. *J. Policy Model* **2010**, *32*, 286–301. [CrossRef]
19. Bor, Y.J.; Huang, Y. Energy taxation and the double dividend effect in Taiwan's energy conservation policy-an empirical study using a computable general equilibrium model. *Energy Policy* **2010**, *38*, 2086–2100. [CrossRef]
20. Hermeling, C.; Löschel, A.; Mennel, T. A new robustness analysis for climate policy evaluations: A CGE application for the EU 2020 targets. *Energy Policy* **2013**, *55*, 27–35. [CrossRef]
21. Mittal, S.; Dai, H.C.; Fujimori, S.; Masui, T. Bridging greenhouse gas emissions and renewable energy deployment target: Comparative assessment of China and India. *Appl. Energy* **2016**, *166*, 301–313. [CrossRef]
22. Qi, T.Y.; Winchester, N.; Karplus, V.J.; Zhang, D.; Zhang, X.L. An analysis of China's climate policy using the China-in-Global Energy Model. *Econ. Model.* **2016**, *52*, 650–660. [CrossRef]
23. Mahmood, A.; Marpaung, O.P. Carbon pricing and energy efficiency improvement—Why to miss the interaction for developing economies? An illustrative CGE based application to the Pakistan case. *Energy Policy* **2014**, *67*, 87–103. [CrossRef]
24. Markandya, A.; Antimiani, A.; Costantini, V.; Martini, C.; Palma, A.; Tommasino, M.C. Analyzing Trade-offs in International Climate Policy Options: The Case of the Green Climate Fund. *World Dev.* **2015**, *74*, 93–107. [CrossRef]
25. Springmann, M.; Zhang, D.; Karplus, V.J. Consumption-Based Adjustment of Emissions-Intensity Targets: An Economic Analysis for China's Provinces. *Environ. Resour. Econ.* **2015**, *61*, 615–640. [CrossRef]
26. Dong, Y.; Ishikawa, M.; Hagiwara, T. Economic and environmental impact analysis of carbon tariffs on Chinese exports. *Energy Econ.* **2015**, *50*, 80–95. [CrossRef]
27. Carlos, B.; Luis, G.; Manuel, D.; Rodrigo, F.; Gonzalo, G.; Rodrigo, P.; Catalina, R. The Impact of a Carbon Tax on the Chilean Electricity Generation Sector. *Energies* **2015**, *8*, 2674–2700.
28. Yang, L.; Yao, Y.F.; Zhang, J.T.; Zhang, X.; McAlinden, K.J. A CGE analysis of carbon market impact on CO_2 emission reduction in China: A technology-led approach. *Nat. Hazards* **2016**, *81*, 1107–1128. [CrossRef]
29. Diao, X.S.; Kennedy, A. Economywide Impact of Maize Export Bans on Agricultural Growth and Household Welfare in Tanzania: A Dynamic Computable General Equilibrium Model Analysis. *Dev. Policy Rev.* **2016**, *34*, 101–134. [CrossRef]
30. Wiebe, K.; Lotze-Campen, H.; Sands, R.; Tabeau, A.; van der Mensbrugghe, D.; Biewald, A.; Bodirsky, B.; Islam, S.; Kavallari, A.; Mason-D'Croz, D.; et al. Climate Change Impacts on Agriculture in 2050 under a Range of Plausible Socioeconomic and Emissions Scenarios. Available online: http://iopscience.iop.org/article/10.1088/1748-9326/10/8/085010 (accessed on 1 September 2015).
31. Editorial board of China's financial yearbook. *Finance Yearbook of China 2013*; China State Finance Magazine: Beijing, China, 2014.
32. Liu, S.J. The Survey of China's per Capita Income in 2013[EB/OL]. Available online: http://www.chinairn.com/news/20130327/095807960.html (accessed on 27 March 2013).
33. Wang, K.; Wang, C.; Chen, J. Analysis of the economic impact of different Chinese climate policy options based on a CGE model incorporating endogenous technological change. *Energy Policy* **2013**, *37*, 2930–2940. [CrossRef]

34. China's Population Development Strategy Research Group, 2007[EB/OL]. Available online: http://www.gov.cn/gzdt/2007-01/11/content_493677.htm\T1\textgreater{} (accessed on 11 January 2007).
35. Pan, H.R. *Computable General Equilibrium Model Primary Tutorial*, 1st ed.; China Population Publishing House: Beijing, China, 2016.
36. Dai, H.C.; Masui, T.; Matsuoka, Y.; Fujimori, S. Assessment of China's climate commitment and non-fossil energy plan towards 2020 using hybrid AIM/CGE model. *Energy Policy* **2011** *39*, 2875–2887. [CrossRef]
37. Li, W.; Li, H.; Sun, S. China's Low-Carbon Scenario Analysis of CO_2 Mitigation Measures towards 2050 Using a Hybrid AIM/CGE Model. *Energies* **2015**, *8*, 3529–3555. [CrossRef]
38. Meng, S.; Pham, T. The impact of the Australian carbon tax on the tourism industry. *Tourism Econ.* **2017**, *23*, 506–522. [CrossRef]
39. Meng, S. How may a carbon tax transform Australian electricity industry? A CGE analysis. *Appl. Econ.* **2014**, *46*, 796–812. [CrossRef]
40. Zhang, D.H.; Wang, J.Q.; Lin, Y.G.; Si, Y.L.; Huang, C.; Yang, J.; Huang, B.; Li, W. Present situation and future prospect of renewable energy in China. *Renew. Sustain. Energy Rev.* **2017**, *76*, 865–871. [CrossRef]
41. Zhao, X.; Luo, D.K. Driving force of rising renewable energy in China: Environment, regulation and employment. *Renew. Sustain. Energy Rev.* **2017**, *68*, 48–56. [CrossRef]

energies

MDPI

Article

A Green Quality Management Decision Model with Carbon Tax and Capacity Expansion under Activity-Based Costing (ABC)—A Case Study in the Tire Manufacturing Industry

Wen-Hsien Tsai

Department of Business Administration, National Central University, Jhongli, Taoyuan 32001, Taiwan; whtsai@mgt.ncu.edu.tw; Tel.: +886-3-4267247

Received: 22 May 2018; Accepted: 12 July 2018; Published: 16 July 2018

Abstract: Issues related to global environmental protection are highly important. Under the global trend of energy saving and carbon reduction, in order to lower the carbon emissions of products or services offered by enterprises, the Taiwanese government aims to control carbon emissions by constructing a carbon tax system and mandating enterprises to pay a carbon tax. The collection of a carbon tax can minimize the total social environmental cost and increase the efficiency of carbon reduction; the need to control the green quality cost can serve as a criterion of green management decision-making. This study aimed to reorganize carbon emissions in different stages of production in order to lower the total carbon emissions of products. Activity-based costing (ABC) was adopted to assess green quality management and production cost. The optimal green quality production portfolio was selected via a mathematical programming model to focus on the expansion of productivity and outsourcing strategy in order to effectively lessen the harmful effects on the environment and maximize profits. Besides academic contributions, the findings of this study could serve as a reference to enterprises on assessing the effects of carbon emissions, carbon taxes, and environmental management on production decision-making.

Keywords: carbon emissions; carbon tax; activity-based costing (ABC); capacity expansion; green quality management; product-mix decision model; mathematical programming

1. Introduction

Issues regarding climate change have received increased attention in recent years. The misuse of natural resources by human and industrial activities affect the balance of the life cycle sustained by Earth and cause abnormal weather and global warming. It is the goal for every nation to stop exhausting greenhouse gases (GHG) for the Earth to provide a sustainable growing environment. In May 1992, the United Nations Framework Convention on Climate Change (UNFCCC) was adopted. It aimed to control the concentration of greenhouse gases in the atmosphere to "a level that would prevent dangerous anthropogenic interference with the climate system" [1]. Developed economies took responsibility for environmental protection through forming international organizations or signing protocols such as the Copenhagen Accord [2], the Kyoto Protocol [3], and the Paris Agreement [1]. Developed economies use these agreements as self-discipline or criterions for countries around the world to balance economic growth and environment protection. In addition, they recognize that these agreements will significantly reduce the risks and impact of climate change and meet the objective of "making finance flows consistent with a pathway towards low greenhouse gas emissions and climate-resilient development" [1].

Green Quality Management mainly aims at eco-efficiency in the complete product life cycle, including design, raw material selection, process improvement, prevention and treatment of pollution, after-sales services and waste recycling, and processing. Quality cost measurement is the first step of quality cost management. Several researchers have proposed various methods to measure the cost of quality (COQ), which includes prevention, appraisal, internal failure, and external failure costs [4]. The Activity-Based Costing (ABC) approach has been widely used in measuring COQ [5–9] because many COQ elements are activity-related costs [9]. ABC uses a two-stage cost assignment to compute the costs of activities in the first stage and the costs of the objects in the second stage [10]. Thus, activity costs can be achieved in the first stage and sources of activity costs can be traced in the second stage of ABC cost assignment. ABC divides activities into value-added (VA) and non-value-added (NVA) activities. The purpose of green quality management is to reduce the workload of VA activities and to finally eliminate the execution of NVA activities related to COQ.

The theory of constraints (TOC), proposed by Goldratt in the book titled "The Goal" [11], aims to identify the constraints and managing them. There are several constraints in business operations, and TOC uses five steps to identify and eliminate the bottlenecks one by one to increase business performance. One method is capacity expansion with a fixed capital investment to increase the possible sales and profits.

Carbon taxes aim to reduce carbon emissions, and they offer a potentially cost-effective means of reducing greenhouse gas emissions [12]. Carbon taxes are also an important policy tool for environmental protection to prompt companies to pursue optimal environmental management under tax considerations [13–15]. Currently, carbon taxes are assessed differently from country to country due to variances in environmental policies, so companies must incorporate them in their production and pricing decisions [16,17].

In view of the above discussion, the purpose of this paper was to propose a green quality management decision model with the consideration of carbon taxes and capacity expansion under ABC. This study took the tire industry as an example to illustrate the application of the model presented in this paper. In addition to considering factors such as carbon taxes, capacity resource limitations, and market demand, this study also took the emissions reduction of pollution during the production process into account. This study combined models of TOC, ABC, green quality management and capacity expansion and used mix-integer programming (MIP) to find the optimal product mix and its resource consumption analysis as a basis for green decision-making. In addition, this paper also discussed how the production plan, derived from the green quality management decision model, can be executed with the help of Industry 4.0 techniques. The remainder of this paper is organized into four sections. Section 2 details the literature about the concepts of environmental management under intelligent manufacturing and the production flow in the Tire industry. Section 3 presents the research method. The green quality management decision model is developed in Section 4. A numerical example is used to demonstrate how to solve these models with MIP and sensitivity analysis under the consideration of carbon taxes and capacity expansion in Section 5. Finally, discussions and conclusions are presented in Sections 6 and 7, respectively.

2. Green Quality Management in the Tire Manufacturing Industry

2.1. Green Quality Cost Management

The purpose of green quality cost management is to find, identify, and quantify the direct and indirect environment-related costs in production so as to help improve the environmental performance of enterprises, realize cost control, evaluate investments in equipment and technology of clean production and pollution prevention and treatment, and develop better processes and green products according to the concept of environmental protection and clean production [6]. Further, it may provide valuable information for decision makers in terms of product structure, retention, and pricing strategies, etc. Green quality cost data can provide important indicators to evaluate the environment quality of enterprises, and can also play an essential part in determining whether the green quality cost management system is effective or not. Objective standards are provided to determine whether the green quality management cost stays aligned with the purpose of environmental protection. Green quality management cost consists of the prevention cost, appraisal cost, internal failure, and external failure cost [9].

The preventive activities of green production quality management are intended to establish a mechanism of enhancing green quality awareness and preventing the impact on the environment. It reduces the negative impact of products during the life cycle on the environment. The prevention cost refers to all expenses incurred in this process and mainly includes green quality planning and renovation projects, green brainstorming activities, the analysis of green process capacity, and green product design, etc. The prevention cost consists of relevant expenses incurred when preventing environmental pollution during production and service activities [18].

Appraisal activity costs refer to all expenses incurred in determining whether the prevention measures achieve the anticipated goals or outcomes. From the design of the product life cycle to the first delivery of goods, the evaluation and assessment activities mainly focus on those which fail to satisfy the green standards. The appraisal cost assures product specifications and quality in the production process based on the quality standards. The appraisal cost includes the necessary expenses on green quality standards tests such as environment check and testing expenses, the purchase of materials and process examination fees, as well as quality checks and diagnostic products tests, etc. [9].

Internal failure activity costs refer to the costs and losses incurred when the products have defects and fail to meet the green quality standards after production but before delivery [19]. Internal failure costs include the cost of reproduction, losses from waste, damaged products, failure analysis, and reexamination due to the failure of meeting the green standards and quality standards [19,20].

External failure activity costs refer to the costs and losses incurred when the products have defects and fail to meet the green quality standards after delivery. External failure costs include the cost of after-sale service and warranties, any penalties and compensation incurred due to the failure of observing relevant laws and regulations in the process, any unexpected expenses from the disposal of waste, and the reduction of operating income resulting from negative public images regarding environmental protection, etc. [21].

2.2. Production Flow in the Tire Industry

Figure 1 shows the tire production flow chart. There are four main activities, as shown below:

(1) Mixing and Component preparation: First, natural rubber, processed oil, carbon black, and other chemicals are mixed into a compound using a Banbury mixer, which is then squeezed into a thin sheet using extruders. Other components include fabric calendar transferred from textiles and fabric, and wire calendar, steel calendar, bead, and belt transferred from steel.

(2) Tire building: The tire building machine pre-shapes radial tires into a form very close to their final dimensions to make sure the many components are in their proper positions before the tire goes into the mold.

(3) Curing press: The curing press is where tires attain their final shape and tread pattern. Hot molds shape and vulcanize the tire. The molds are then engraved with the tread pattern and sidewall marking.

(4) Tire inspection: Trained inspectors perform a visual inspection to check the tire surfaces. A balance inspection and force variation are then carried out to check uniformity. Finally, some tires are sampled from the production line and X-rayed to detect any hidden weaknesses or internal failures. In addition, quality control engineers regularly perform cut sections and study details of the tire construction that affect performance, ride, or safety [22].

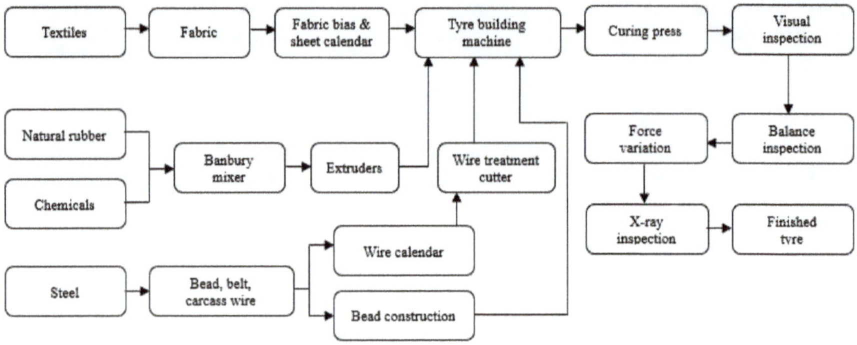

Figure 1. Tire production flow chart.

2.3. *Shop Flow Control and ABC Costing with the Help of Industry 4.0*

2.3.1. Origin and Meaning of Industry 4.0

The era of Industry 4.0 is coming for the global manufacturing industry. This is the biggest change since the third industrial revolution in 1970. Automated production has been upgraded to the level of intelligent manufacturing [23], which is also called the fourth industrial revolution. The concept of Industry 4.0 was first mentioned at the Hannover Fair in 2011. It means to apply the Internet of Things and system service applications into the production process to become a cyber-physical production system [24]. It vertically integrates production factors such as materials, human resources, equipment, and factories to form an intelligent manufacturing network and, at the same time, horizontally integrates the value chains of supply vendors and customers as a value creation network and reduces pollution and greenhouse gas emissions during production as much as possible [25].

Industry 4.0 places emphasis on its products' influences on the environment by taking advantages of different concepts of a smart factory, which include green production, intelligent production and urban production. Meanwhile, big data is adopted to analyze corporate conditions and resources to guide enterprises with environmentally friendly guidelines during their development of innovative green technology [26] so that they can make efficient use of resources, reduce costs, and relieve the impact of operating activities on the environment in current and expanding production.

Under Industry 4.0, virtual models based on the orders of customized tires are produced according to the advice of existing experts and product design and processing knowledge on the cloud through the Internet [27]. Then, the physical products are put into production based on the model and a production schedule is made. A virtual production system control workshop is built and robotic arms take the place of a traditional workforce in the production of tires by applying automatic assembly and sensors/actuators in the virtual workshop to the production process [28]. The production data are sent to the cloud database as digital knowledge accumulation.

2.3.2. Collecting Data for ABC Costing Using Industry 4.0 Techniques

In the past, enterprises were in pursuit of high quality with the most appropriate costs in the shortest production cycle and lead time, and they made dramatic investments in the establishment of mechanisms and production management systems, as well as automation. Although they might have had knowledge of existing problems through various information dashboards, they failed to make full use of the databases of those systems and the valuable information in the systems. In order to integrate the goals of multiple systems and keep those goals in line with their corporate mission, enterprises need to combine data mining technology and the intelligent analysis methods of specific fields and identify key opportunities for breakthroughs from massive amounts of data [29]. As Industry 4.0

and the era of the Internet of Things approach [30], with the help of wider usage of information extraction technology through the Internet and cloud computing, integration of both upstream and downstream information can be easily realized in combination with an automatized factory system. Intelligent manufacturing will be a niche market that can cater to customers' demands in the future [31]. In addition, sensors/actuators, high-performance computing systems, and the production control of intelligent factories will manage to detect various available production capacities in a short time and arrange configurations correspondingly and effectively, which avoids inadequate production.

Technically speaking, intelligent factories emphasize mobile sensors/actuators and artificial intelligent self-organized machine learning [26]. At the same time, they give full play to the mature computing virtual technology to accomplish the integration of manufacturing and service in the vertical activities of corporate entities and further improve the support of relevant activities of enterprises.

Big data for production emphasizes the digitalization of all information, from product design to after-sales service. With the necessary information for cost analysis available on the cloud system, and by combining the theory of constraint with the expansion of production capacity and the system of fundamental operating costs, enterprises can oversee the resource consumption and carbon emissions of each activity. Through virtual design and analysis of the virtual factory, real-time cost analysis results and a review of the carbon footprint can be provided before production, which can reduce the number of potentially damaged products, reproduced products, and carbon emissions during the production of physical products.

In summary, Industry 4.0 uses cyber-physical systems (CPS) and the Internet of Things (IoT) [13] to link all the elements in the production system. Then, the system will real-timely collect and monitor the activity data of all the elements and send out intelligent responses to various problems that may arise in the manufacturing spots according to the real-time analysis results of big data and cloud computing. Finally, the production process can be flexibly adjusted or set up differently on the basis of specific customer needs to attain the objectives of mass customization and customer satisfaction. The activity cost calculation and activity improvement of Activity-Based Costing (ABC) can be easily achieved since all the elements in the manufacturing systems can be integrated and monitored under Industry 4.0.

3. Research Method

3.1. TOC and Capacity Expansion

First proposed by Goldratt [11], TOC is mainly used to resolve short term product mix and processing bottleneck resource problems. It is an improved method of the conventional cost accounting system that hinders the effective output of a company, arguing that all organizations have restrictions that will affect the activities of the entire organization. Paraskevopoulos, Karakitsos and Rustem [32] held that the expansion of production capacity should also be included in the production plan when uncertainties emerge in the production environment so that the production capacity can be adjusted if there are any changes in the external environment. They found that the bottleneck factors determine the level of product mix flexibility, and that bottleneck improvements make a better contribution than non-bottleneck improvements.

Targeted at the present operating process and production level, the theory of constraints explores the best management method. However, for decision-making about long-term production, the production capacity of each activity must be specifically planned. When confronted with gaps in the production capacity, enterprises should adopt different strategies to expand production capacity in order to meet demands, such as recruiting new employees, purchasing more equipment, outsourcing, working overtime, or renting equipment. Aghezzaf [33] proposed that production capacity does not merely mean the production capacity of the equipment but also the available workforce in each period, which should be given high priority and taken into consideration to develop a series of efficient algorithms for human resource schemes based on strategies. When the production capacity of human resources is calculated, the results should be transferred into the production capacity for both regular time and overtime. Kok [34] indicated that when demands surpass the available production capacity, two strategies of production capacity, including postponing and outsourcing, should be utilized to build a mathematical model. Then, the enterprises can choose the best strategy with lower costs and other relevant parameters between the two strategies.

3.2. Green Quality Cost Measurement under ABC

Kaplan and Cooper first proposed ABC in 1988 to overcome the distortion of the conventional cost accounting system [35]. ABC uses a two-stage cost assignment method and considers resource drivers and activity drivers as the bases of cost assignment, as shown in Figure 2. ABC assumes that cost objects (e.g., products, product lines, processes, customers, channels, and markets, etc.) create the need for activities, and that activities create the need for resources. Accordingly, ABC uses a two-stage cost assignment method to assign resource costs to cost objects. In the first stage, resource costs are assigned to various activities by using resource drivers. Resource drivers are the factors that approximate the consumption of resources by the activities. The appropriate resource drivers are determined according to the analysis of the correlation between resources and activity costs [36]. Each type of resource traced to an activity becomes a cost element of an activity cost pool. Thus, an activity cost pool is the total cost associated with an activity. An activity center is composed of related activities that are usually clustered by function or process. It is possible to create activity centers by various ways according to different information needs [9].

In the second stage, each activity cost pool is distributed to the cost objects using an adequate activity driver which is used to measure the consumption of activities by the cost objects [10]. If the cost objects are products, then the total cost of a specific product can be calculated by adding the direct material cost and the direct labor cost, as well as the costs of various activities assigned to that product [9]. At the second stage, the appropriate activity drivers are determined according to different activity levels (e.g., unit-level, batch-level, process-level, product-level, or facility-level) in order to summarize the cost databases before attributing costs to the cost objects, thus obtaining reliable cost assignment information [10,36]. Unit-level activities are performed one time for one unit of product, e.g., 100 percent inspections, machining, or finishing, etc. Batch-level activities are performed one time for a batch of products, e.g., sampling inspections, set-up, or scheduling, etc. Process-level activities are performed to sustain the operations of a process, e.g., the various tasks executed in a process. Product-level activities are performed to benefit all units of a particular product, e.g., product design, design verification, or review, etc. Facility-level activities are performed to sustain the manufacturing facility, e.g., plant guards and management, or zero defect programs, etc. [9].

Kaplan and Cooper argued that ABC emphasizes the relationship between the activities and the consumption of resources; therefore, it can provide product decision-making information and help managers with decisions regarding product design, pricing, product mix, marketing, and process improvements [35]. ABC provides information in the recognition of activities that cause poor quality by classifying all activities as either value-added (VA) or non-value-added (NVA) activities. According to this categorization, VA activities are those that contribute to the value (increase the quality and effectiveness of use) of the product/services delivered. As NVA activities do not contribute to the value of the product/services, their elimination decreases the related costs and has no effect on the value of the products/services [9,37]. As ABC can provide an effective and complete estimation of product costs, enterprises may use ABC and calculate the costs of products to improve the product mix based on costs and determine the pricing strategies [38]. By connecting corporate strategies with the decision-making of the operating management, managers can make full use of the valuable activities of their companies. With regard to the activities that cannot fulfill the goals of enterprises, reforms or changes can be made. Rectifications can be immediately performed when the project is found to deviate from the original plan to ensure the accomplishment of investment plans. ABC can effectively trace each activity of the costs and then categorize the costs to corresponding products according to the causes of the activities so as to provide a reasonable sharing method and improve the production performance measuring method. Particularly, when an enterprise has complicated categories of products or processes, ABC can provide assistance with managers to learn the resource consumption of different products and processes and offer better cost management [16].

In this paper, the activities of ABC model were classified into green quality-related activities and green quality-unrelated activities. A COQ activity center was created for green quality-related activities. Within the COQ activity center, four nested activity centers were established, i.e., prevention, appraisal, internal failure, and external failure activity centers [9]. For the purpose of improvement, regardless of being green quality-related or quality-unrelated activities, all activities should be identified as VA or NVA activities, as shown in Figure 2.

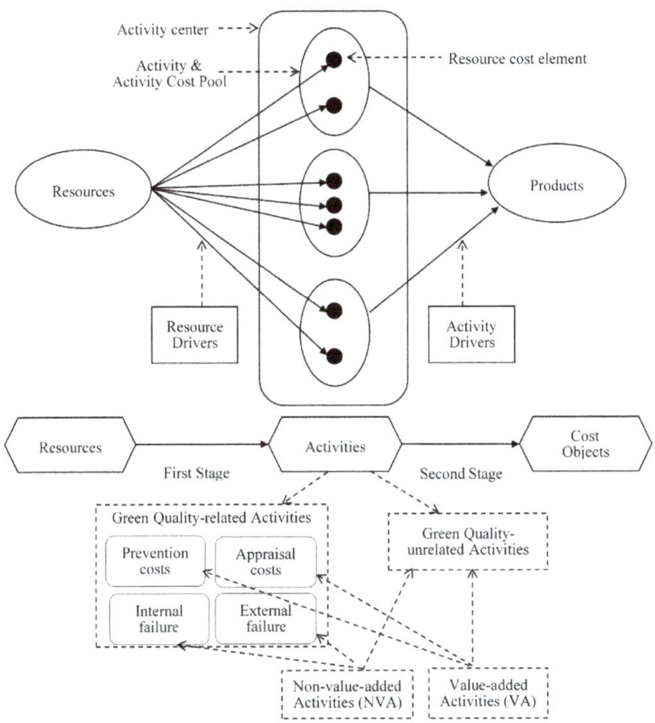

Figure 2. Cost Assignment of ABC's two-stage model.

3.3. Carbon Tax Functions

The objective of a carbon tax is to reduce the harmful and unfavorable levels of carbon dioxide emissions, thereby decelerating climate change and its negative effects on the environment and human health [39]. A carbon tax imposes a tax on each unit of carbon emissions and gives companies an incentive to reduce pollution whenever doing so will cost less than paying the tax [40]. In the literature, "carbon tax" [12,13,15,17,41] and "emission tax" [42–44] are used interchangeably, although some argue that a carbon tax simply puts a price on carbon dioxide emissions while an emissions tax puts a price on all pollution emissions, not just carbon dioxide, including methane, carbon monoxide, nitrous oxide, or fluorinated gases, etc. [45]. Some authors use "carbon emission tax" [46–48]. However, most authors mean to levy on all pollution emissions which are converted to a carbon equivalent [49]. This paper also adopted the concept of all pollution emissions and the carbon equivalent.

This study intended to explore the effects of carbon tax levies on production mix decisions. Figure 3 is the carbon tax function with a fixed tax rate, which has been adopted by many countries presently. However, if the carbon tax level is too low, companies may want to pay the tax and continue to pollute. Thus, the government may use carbon tax functions that have a full progressive tax rate without a threshold (Figure 4) [50] or with a threshold (Figure 5). In these two carbon tax functions, the higher a company's carbon emission quantity, the higher the carbon tax rate. It will have aggressive effect on carbon emission reduction. However, considering small companies with very low levels of production and high levels of carbon pollution, the government may use the carbon tax system with a threshold, as shown in Figure 5, where a company will have a carbon tax exemption when its carbon emission quantity is under threshold quantity [51]. In this paper, a carbon tax system with progressive tax rates and a threshold was used for illustration. On the other hand, the government should help companies reduce carbon emissions by providing subsidies for carbon emission control equipment or other methods. A carbon tax can be a regressive tax, in that it may directly or indirectly affect low-income groups disproportionately [41]. The latter allows for different rates at different carbon levels and can still create a meaningful impact in carbon reduction [52].

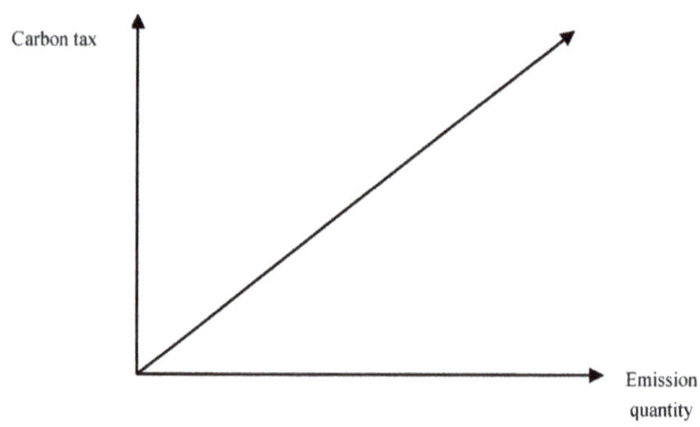

Figure 3. The fixed tax rate.

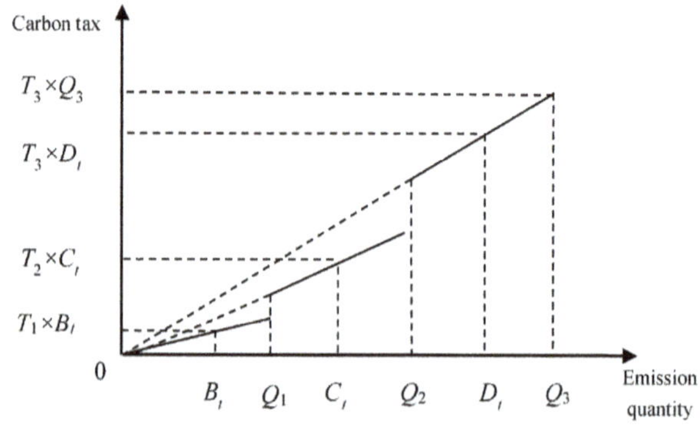

Figure 4. The full progressive tax rate without a threshold.

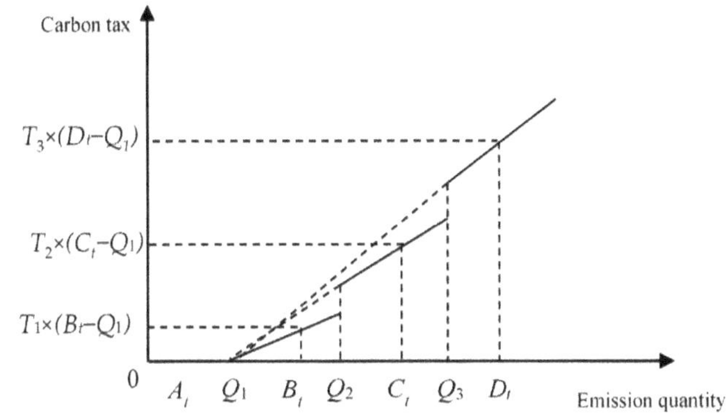

Figure 5. The full progressive tax rate with a threshold.

4. Model Formulation

4.1. Assumptions

The green quality management decision model under ABC presented in this paper had several assumptions. First, assume that a multi-product company had various levels of activities including unit-level, batch-level, product-level, and facility-level activities. The company had determined the resource drivers and activity drivers used in ABC analysis. Second, the company also had calculated the unit cost of each activity based on the actual running activity cost per activity driver and used in the model. Third, the facility-level activity cost was a common fixed cost. Fourth, direct labor resources can be expanded by using additional shifts or overtime work with higher wage rates. Fifth, machine hour resources also can be expanded by renting more machines. Sixth, the products' selling prices were constant within the relevant range of the analysis.

4.2. Green Quality Management Decision Model under ABC

According to the assumptions described above, this paper presented a general model that incorporated material purchase discounts, capacity expansions, capacity constraints, waste disposal, and carbon tax expenditures to determine an optimal product-mix decision.

4.2.1. Notations

The following notations were used in this paper:

π: The company's maximized profit

p_i: The unit selling price of product i

q_i: The production quantity of product i

L_r: The unit cost of the rth material, without a purchase discount

Ld_r: The unit cost of the rth material, with the first purchase discount rate

Ldd_r: The unit cost of the rth material, with the second purchase discount rate

M_r: The quantity of the rth material, without a purchase discount

Md_r: The quantity of the rth material, with the first purchase discount rate

Mdd_r: The quantity of the rth material, with the second purchase discount rate

d_{ir}: The requirement of the rth material for product i

X_r: The upper limit of the rth material quantity with the second purchase discount rate; it is also the available quantity of the rth material

W_r: The upper limit of the rth material quantity with the first purchase discount rate

TD_r: The upper limit of the rth material quantity without a purchase discount

ND_r: a 0–1 variable. $ND_r = 1$ means that the purchase quantity of the rth material falls within the first range of the rth material purchase quantity which dissatisfies the threshold for a discount; otherwise, $ND_r = 0$

SD_r: a 0–1 variable. $SD_r = 1$ means that the purchase quantity of the rth material falls within the second range of the rth material purchase quantity which will have the first purchase discount rate; otherwise, $SD_r = 0$

OD_r: a 0–1 variable. $OD_r = 1$ means that the quantity of the rth material falls within the third range of the rth material purchase quantity which will have the second purchase discount rate; otherwise, $OD_r = 0$

WC_1: Total direct labor cost when labor hour is below the total normal labor hour, WH_1

WC_2: Total direct labor cost in WH_2

WC_3: Total direct labor cost in WH_3

TWH: Total direct labor hour

WH_1: The available normal direct labor hour

WH_2: The upper limit of labor hour with the first overtime wage rate

WH_3: The upper limit of labor hour with the second overtime wage rate

(η_1, η_2): An SOS1 (special ordered set of type 1) set of 0–1 variables, within which exactly one variable must be non-zero

$(\alpha_0, \alpha_1, \alpha_2)$: An SOS2 (special ordered set of type 2) set of non-negative variables, within which at most two adjacent variables, in the order given to the set, can be non-zero

t_{oi}: Time of rework mix processing of product i

t_{ci}: Time to shift each batch of product i to rework mix department

$t_{\delta i}$: Time to maintain building tires

$t_{\tau i}$: Time for the inspection of product i

BA_i: Batches of product i in rework mix processing

δ_i: Batches of product i in maintenance

s_i: Quantity of each maintenance of product i

R_i: Demand of product i

τ_i: 0–1 variable. If product i is not produced, it is 0; otherwise, it is 1

σ_o: Capacity of rework mix processing department

σ_δ: Capacity of maintenance department

σ_τ: Capacity of inspection department

WA_1: Total waste disposal cost in WQ_1

WA_2: Total waste disposal cost in WQ_2

WA_3: Total waste disposal cost in WQ_3

w_i: The average waste disposal quantity for one unit of product i

WAQ: The total quantity of waste disposal; $WAQ = \sum_{i=1}^{n} w_i q_i$

WQ_1: The upper limit of waste disposal quantity with the first waste disposal cost rate

WQ_2: The upper limit of waste disposal quantity with the second waste disposal cost rate

WQ_3: The upper limit of waste disposal quantity with the third waste disposal cost rate

(μ_1, μ_2, μ_3): An SOS1 (special ordered set of type 1) set of 0–1 variables, within which exactly one variable must be non-zero

$(\beta_0, \beta_1, \beta_2, \beta_3)$: An SOS2 (special ordered set of type 2) set of non-negative variables, within which at most two adjacent variables, in the order given to the set, can be non-zero.

TA_i: The carbon emission quantity per unit of product i

TCQ: The company's total carbon emission quantity; $TCQ = \sum_{i=1}^{n} TA_i q_i$

T_1: The carbon tax rate when TCQ falls within the first taxable range of carbon emission quantity

T_2: The carbon tax rate when TCQ falls within the second taxable range of carbon emission quantity

T_3: The carbon tax rate when TCQ falls within the third taxable range of carbon emission quantity

A_t: The company's total carbon emission quantity when TCQ falls within the range of carbon emission quantity without carbon tax

B_t: The company's total carbon emission quantity when TCQ falls within the first taxable range of carbon emission quantity

C_t: The company's total carbon emission quantity when TCQ falls within the second taxable range of carbon emission quantity

D_t: The company's total carbon emission quantity when TCQ falls within the third taxable range of carbon emission quantity

Q_1: The upper limit of carbon emission quantity without carbon tax

Q_2: The upper limit of carbon emission quantity with the first carbon tax rate

Q_3: The upper limit of carbon emission quantity with the second carbon tax rate

G_1: 0–1 variable. $G_1 = 1$ means that TCQ falls within the range of carbon emission quantity without carbon tax; otherwise, $G_1 = 0$

G_2: 0–1 variable. $G_2 = 1$ means that TCQ falls within the first taxable range of carbon emission quantity; otherwise, $G_2 = 0$

G_3: 0–1 variable. $G_3 = 1$ means that TCQ falls within the second taxable range of carbon emission quantity; otherwise, $G_3 = 0$

G_4: 0–1 variable. $G_4 = 1$ means that TCQ falls within the third taxable range of carbon emission quantity; otherwise, $G_4 = 0$

(G_1, G_2, G_3, G_4): An SOS1 set of 0–1 variables, within which exactly one variable must be one, which is a set of 0–1 indicator variables

H_i The carbon emission quantity for one unit of product i

MC_k: Total machine cost when the machine hour is expanded to kth level of machine resource, MA_K

MA_K: The machine hour when the machine hour is expanded to kth level of machine resource

$(\theta_{i1}, \theta_{i2}, \ldots, \theta_{ik})$: An SOS1 set of 0–1 variables, within which exactly one variable must be one, which is a set of 0–1 variables indicating which level of machine hours the machine hour resource is expended to.

4.2.2. The Objective Function

The objective of the model is to maximize the total profit of the company, as shown in Equation (1):

$$\begin{aligned} MAX\pi = & \sum_{i=1}^{n} p_i q_i - \sum_{r=1}^{s} (L_r M_r + Ld_r Md_r + Ldd_r Mdd_r) - [WC_1 + (WC_2 - WC_1)\alpha_1 + (WC_3 - WC_1)\alpha_2] \\ & - \sum_{i=1}^{n} (t_{oi} + t_{ci}) k_o BA_i - \sum_{i=1}^{n} (t_{\delta i}\delta_i + t_{pi} q_i) k_\delta - \sum_{i=1}^{n} (t_{\tau i} k_\tau \tau_i) - (WA_1\beta_1 + WA_2\beta_2 + WA_3\beta_3) \\ & - [T_1(B_t - Q_1) + T_2(C_t - Q_1) + T_3(D_t - Q_1)] - \sum_{k=0}^{t} MC_k \theta_k - FC \end{aligned} \quad (1)$$

$MAX\pi$ = Total revenue − Direct material cost − Direct labor cost − Rework cost − Maintenance cost − Inspection cost − Waste disposal cost − Carbon tax − Machine cost − Other fixed cost, where $\sum_{i=1}^{n} p_i q_i$ is the total revenue, $\sum_{r=1}^{s} (L_r M_r + Ld_r Md_r + Ldd_r Mdd_r)$ is the total direct material cost, $[WC_1 + (WC_2 - WC_1)\alpha_1 + (WC_3 - WC_1)\alpha_2]$ is the total direct labor cost, $\sum_{i=1}^{n} (t_{oi} + t_{ci}) k_o o_i$ is the total rework cost, $\sum_{i=1}^{n} (t_{\delta i}\delta_i + t_{pi} q_i) k_\delta$ is the total maintenance cost, $\sum_{i=1}^{n} (t_{\tau i} k_\tau \tau_i)$ is the total inspection cost, $(WA_1\beta_1 + WA_2\beta_2 + WA_3\beta_3)$ is the waste disposal cost, $[T_1(B_t - Q_1) + T_2(C_t - Q_1) + T_3(D_t - Q_1)]$ is the total carbon tax, $\sum_{k=0}^{t} MC_k \theta_k$ is the total machine cost, and FC is the total other fixed cost. These costs and their associated constraints are described in the following sub-sections.

4.2.3. Total Direct Material Cost

According to the assumption for attaining a discount on the purchase of material, the term $\sum_{r=1}^{s}(L_rM_r + Ld_rMd_r + Ldd_rMdd_r)$ in Equation (1) represent the total direct material costs with $(r \in D)$ and without $(r \in D)$ a purchase discount. The material cost function with a purchase quantity discount is shown in Figure 6 [53–56]. There are three ranges of material purchase quantities, $[0,TD_r]$, $(TD_r,W_r]$, and $(W_r,X_r]$, in which the unit costs are L_r, Ld_r, and Ldd_r, respectively, and $L_r < Ld_r < Ldd_r$. The constraints associated with the material cost are shown in Equations (2)–(10).

Direct material quantity constraints:

$$\sum_{i=1}^{n} d_{ir}q_i \leq X_r;\ r = 1,2,\ldots,s; r \notin D \tag{2}$$

$$\sum_{i=1}^{n} d_{ir}q_i \leq M_r + Md_r + Mdd_r;\ r = 1,2,\ldots,s; r \in D \tag{3}$$

$$M_r \geq 0;\ r = 1,2,\ldots,s \tag{4}$$

$$M_r \leq TD_rND_r;\ r = 1,2,\ldots,s; r \in D \tag{5}$$

$$Md_r \geq TD_rSD_r;\ r = 1,2,\ldots,s; r \in D \tag{6}$$

$$Md_r \leq W_rSD_r;\ r = 1,2,\ldots,s; r \in D \tag{7}$$

$$Mdd_r \geq W_rOD_r;\ r = 1,2,\ldots,s; r \in D \tag{8}$$

$$Mdd_r \leq X_rOD_r;\ r = 1,2,\ldots,s; r \in D \tag{9}$$

$$ND_r + SD_r + OD_r = 1;\ r = 1,2,\ldots,s; r \in D \tag{10}$$

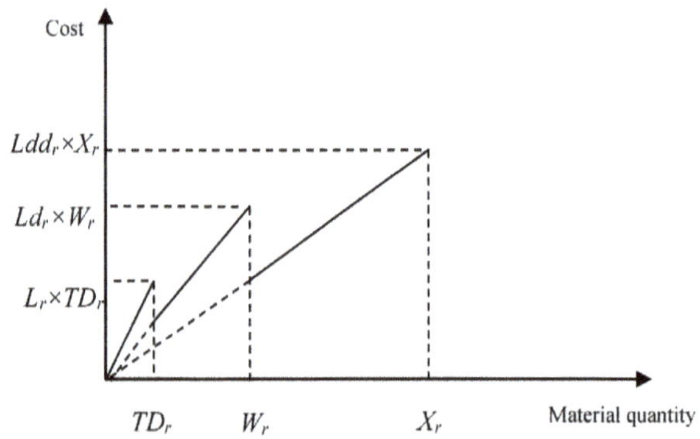

Figure 6. Direct material cost.

Equation (2) is the direct material quantity constraint for materials without a purchase discount $(r \in D)$. Equations (3)–(10) are the constraints for the materials with a purchase discount$(r \in D)$. M_r, MD_r, and Mdd_r represent the variables of the material purchase quantity in the three ranges of Figure 5. In addition, $ND_r + SD_r + OD_r = 1$, and ND_r, SD_r, OD_r are 0–1 indicator variables which are used to indicate which range the material purchase quantity is in for the optimal solution. For example, if $ND_r = 1$, then $SD_r = OD_r = 0$ from Equation (10), which indicates the material purchase quantity is in first range of Figure 5. Then, $0 \leq M_r \leq TD_r$ from Equations (4) and (5), $Md_r = Mdd_r = 0$ from Equations (6)–(9), and $\sum_{i=1}^{n} d_{ir}q_i \leq M_r$ from Equation (3). Therefore, the total material cost is L_rM_r for the material with a purchase discount.

4.2.4. Total Direct Labor Cost

In Equation (1) $WC_1 + (WC_2 - WC_1)\alpha_1 + (WC_3 - WC_1)\alpha_2$ represents the total direct labor costs. It is assumed that the direct labor cost for production can be expanded using overtime work, additional night shifts and the hiring of temporary workers as shown in Figure 7. In Figure 7, the available normal direct labor hour is WH_1, with the fixed cost WC_1 used no matter how many labor hours the company uses. The labor resources can be expanded to WH_2 and WH_3 with two different higher wage rates. The total additional labor costs are represented by $(WC_2 - WC_1)\alpha_1 + (WC_3 - WC_1)\alpha_2$, and the associated constraints are shown in Equations (11)–(16).

Direct labor hour constraints:

$$TWH \leq WH_1 + (WH_2 - WH_1)\alpha_1 + (WH_3 - WH_1)\alpha_2 \tag{11}$$

$$\alpha_0 - \eta_1 \leq 0 \tag{12}$$

$$\alpha_1 - \eta_1 - \eta_2 \leq 0 \tag{13}$$

$$\alpha_2 - \eta_2 \leq 0 \tag{14}$$

$$\alpha_0 + \alpha_1 + \alpha_2 = 1 \tag{15}$$

$$\eta_1 + \eta_2 = 1 \tag{16}$$

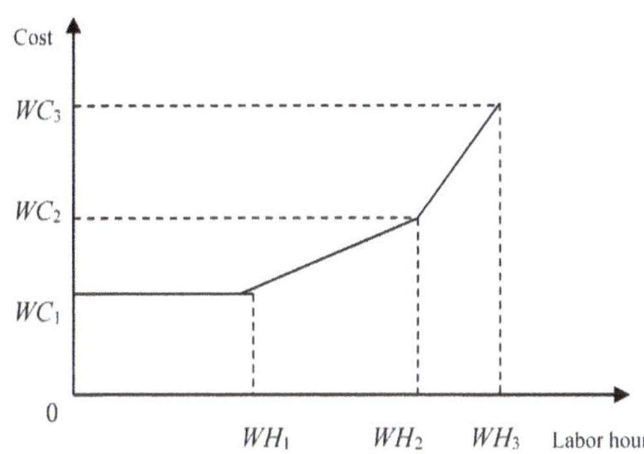

Figure 7. Direct labor cost.

In Equations (11)–(16), (η_1, η_2) is an SOS1 set of 0–1 variables, within which exactly one variable must be one; $(\alpha_0, \alpha_1, \alpha_2)$ is an SOS2 set of non-negative variables, within which at most two adjacent variables in the order can be non-zero.

For example, if $\eta_1 = 1$, then $\eta_2 = 0$ from Equation (16), $\alpha_0 \leq 1, \alpha_1 \leq 1, \alpha_2 = 0$ from Equations (12)–(14), and $\alpha_0 + \alpha_1 = 1$ from Equation (15). Thus, the total labor hours (TWH) needed is $WH_1 + (WH_2 - WH_1)\alpha_1$, and the total labor cost is $TWC = WC_1 + (WC_2 - WC_1)\alpha_1$. This means that the production will require overtime work in the first overtime work range of Figure 7, and the final (TWH, TWC) will fall within the second segment of Figure 7, where (TWH, TWC) is the linear combination of (WH_1, WC_1) and (WH_2, WC_2). In addition, if $\eta_2 = 1$, then $\eta_1 = 0$ from Equation (16), $\alpha_0 = 0, \alpha_1 \leq 1, \alpha_2 \leq 1$ from Equations (12)–(14), and $\alpha_1 + \alpha_2 = 1$ from Equation (15). Thus, the total labor hour needed is $WH_1 + (WH_2 - WH_1)\alpha_1 + (WH_3 - WH_1)\alpha_2$ and total labor cost is $WC_1 + (WC_2 - WC_1)\alpha_1 + (WC_3 - WC_1)\alpha_2$. This indicates that the production will require overtime work in the second overtime work range of Figure 7, and the final (TWH, TWC) will fall within the second segment of Figure 6, where (TWH, TWC) is the linear combination of (WH_2, WC_2) and

(WH_3, WC_3). When $\eta_2 = 1$ and $\alpha_0 = 1$, then $\eta_1 = \alpha_1 = \alpha_2 = 0$ from Equations (12)–(16), the total direct labor hours is $TWH \leq WH_1$ from Equation (11) and the total labor cost is the fixed cost WC_1.

4.2.5. Total Rework Cost

In Equation (1), $\sum_{i=1}^{n} (t_{oi}+t_{ci})k_o BA_i$ represents the total rework mix processing costs, and the associated constraints are shown in Equation (17). t_{oi} is the time of rework mix processing and t_{ci} is the time to shift each batch of product i to the rework mix department [57–59]. After mixing, the rubber charge is dropped into a chute and fed by an extruding screw into a roller die. Alternatively, the batch can be dropped onto an open rubber mill system. k_0 is the unit labor/hour cost of the rework mix processing department. Rework mix processing refers to applying mechanical work to the ingredients in order to blend them into a homogeneous substance. BA_i is the batches of rework mix processing and σ_o is the rework mix processing department production capacity.

Rework constraints:

$$\sum_{i=1}^{n}(t_{oi}BA_i + t_{ci}BA_i) \leq \sigma_o;\ i = 1,2,\ldots,n \quad (17)$$

4.2.6. Total Maintenance Cost

In Equation (1), $\sum_{i=1}^{n} (t_{\delta i}\delta_i + t_{pi}q_i)k_\delta$ represents the tire building maintenance costs, and the associated constraints are shown in Equations (18) and (19). $t_{\delta i}$ is the time to maintain building tires, t_{pi} is the time to package tires, k_δ is the unit labor/hour costs of the maintenance tire building department, δ_i is the amount of maintenance, and s_i is the quantity of each maintenance activity. Equation (18) is the constraint concerning the product quantity and Equation (19) is the constraint concerning the capacity of the maintenance department.

Maintenance constraints:

$$q_i = s_i\delta_i;\ i = 1,2,\ldots,n \quad (18)$$

$$\sum_{i=1}^{n}(t_{\delta i}\delta_i + t_{pi}q_i) \leq \sigma_\delta;\ i = 1,2,\ldots,n \quad (19)$$

4.2.7. Total Inspection Cost

After the tire has been cured, large commercial truck/bus tires, as well as some passenger and light truck tires, are inspected by X-ray or magnetic induction based inspection machines. Tire balance measurement is a test where the tire is automatically placed on wheel halves, rotated at a high speed, and measured for imbalance. In Equation (1), $\sum_{i=1}^{n} (t_{\tau i}k_\tau \tau_i)$ represents the tire inspection cost, and the associated constraints are shown in Equations (20) and (21). k_τ is the unit tire inspection hourly cost, $t_{\tau i}$ is the inspection time for product i, and σ_τ is the tire inspection capacity. Equation (20) is the constraint for tire inspection hour capacity, and Equation (21) is the constraint for product demand. τ_i is a 0–1 variable, in other words, if a tire is not produced, then $\tau_i = 0$ will result in $q_i = 0$ from Equation (21); conversely, if it is produced, then $\tau_i = 1$ will result in $0 \leq q_i \leq R_i$ from Equation (21).

Inspection constraints:

$$\sum_{i=1}^{n}(t_{\tau i}\tau_i) \leq \sigma_\tau;\ i = 1,2,\ldots,n \quad (20)$$

$$0 \leq q_i \leq R_i\tau_i;\ i = 1,2,\ldots,n \quad (21)$$

4.2.8. Total Waste Disposal Cost

The term $(WA_1\beta_1 + WA_2\beta_2 + WA_3\beta_3)$ in Equation (1) represents the total waste disposal cost of tire production. It is assumed that the total waste disposal cost function is shown in Figure 8, which is a piecewise linear function composed of three segments with different disposal rates. The constraints associated with the company's total waste disposal quantity (WAQ) are as shown in Equations (22)–(38). In Equation (22), WAQ represents the total waste disposal quantity, and $WAQ = \sum_{i=1}^{n} w_i q_i$, where w_i is the average waste disposal quantity for one unit of product i.

In Equations (23)–(28), (μ_1, μ_2, μ_3) is an SOS1 set of 0–1 variables, within which exactly one variable must be one; $(\beta_0, \beta_1, \beta_2, \beta_3)$ is an SOS2 set of non-negative variables, within which no more than two adjacent variables to the order in the set are non-zero. (μ_1, μ_2, μ_3) is a set of indicator variables that indicate which range the company's total waste disposal quantity falls within. If $\mu_1 = 1$, then $\mu_2 = \mu_3 = 0$ from Equation (28), $\beta_0 \leq 1, \beta_1 \leq 1$, and $\beta_2 = \beta_3 = 0$ from Equations (23)–(26). Thus, the company's total waste disposal quantity falls within the first range, $[0, WQ_1]$, i.e., $WAQ = WQ_1\beta_1$ from Equation (22), and the total waste disposal cost is $WA_1\beta_1$. Similarly, If $\mu_2 = 1$, then the company's total waste disposal quantity falls within the second range, $[WQ_1, WQ_2]$; If $\mu_3 = 1$, then the company's total waste disposal quantity falls within the third range, $[WQ_2, WQ_3]$.

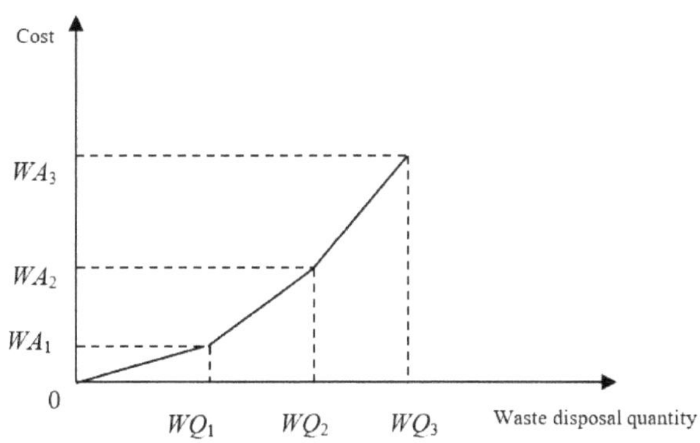

Figure 8. Waste disposal cost.

Waste disposal quantity constraints:

$$WAQ = \sum_{i=1}^{n} w_i q_i = WQ_1\beta_1 + WQ_2\beta_2 + WQ_3\beta_3 \tag{22}$$

$$\beta_0 - \mu_1 \leq 0 \tag{23}$$

$$\beta_1 - \mu_1 - \mu_2 \leq 0 \tag{24}$$

$$\beta_2 - \mu_2 - \mu_3 \leq 0 \tag{25}$$

$$\beta_3 - \mu_3 \leq 0 \tag{26}$$

$$\beta_0 + \beta_1 + \beta_2 + \beta_3 = 1 \tag{27}$$

$$\mu_1 + \mu_2 + \mu_3 = 1 \tag{28}$$

4.2.9. Carbon Tax Expenditure

The terms $[T_1(B_t - Q_1) + T_2(C_t - Q_1) + T_3(D_t - Q_1)]$ in Equation (1) represent the total carbon tax incurred during tire production. It is assumed that the carbon tax function is shown in Figure 4 (replicated for easy reading), which has the full progressive tax rates with a threshold. The constraints associated with the carbon tax function in Figure 4 are shown in Equations (29)–(37).

Carbon tax constraints:

$$TCQ = \sum_{i=1}^{n} TA_i q_i = A_t + B_t + C_t + D_t \quad (29)$$

$$A_t \geq 0 \quad (30)$$

$$A_t \leq G_1 Q_1 \quad (31)$$

$$B_t > G_2 Q_1 \quad (32)$$

$$B_t \leq G_2 Q_2 \quad (33)$$

$$C_t > G_3 Q_2 \quad (34)$$

$$C_t \leq G_3 Q_3 \quad (35)$$

$$D_t > G_4 Q_3 \quad (36)$$

$$G_1 + G_2 + G_3 + G_4 = 1 \quad (37)$$

The tire industry is a high emission industry, and the carbon tax is expected to be efficient in terms of the downscaling of production and reduced emissions. A higher rate is imposed on higher carbon contents, and all carbon content levels are accompanied with corresponding progressive rates. The full progressive tax system reports better results in carbon reductions. In Figure 4, the carbon tax rate $T_1 < T_2 < T_3$. In Equation (29), the company's total carbon emission quantity is $TCQ = \sum_{i=1}^{n} TA_i q_i$ where TA_i is the carbon emission quantity per unit of product i; and A_t, B_t, C_t, D_t are the variables of TCQ when it falls within the first, second, third, and fourth range of carbon emission quantity shown in Figure 4, respectively. Also, B_t, C_t, D_t are the variables of TCQ when it falls within the first, second, and third taxable range of carbon emission quantity whose carbon tax rates are T_1, T_2, and T_3, respectively. In Equation (37), (G_1, G_2, G_3, G_4) is an SOS1 set of 0–1 variables, within which exactly one variable must be one, which is a set of 0–1 indicator variables. If $G_1 = 1$, then $G_2 = G_3 = G_4 = 0$ from Equation (37), $0 \leq A_t \leq G_1 Q_1$ from Equations (30) and (31), and $B_t = C_t = D_t = 0$ from Equations (32)–(36). Thus, TCQ falls within the first range of Figure 4, in which the carbon tax is zero. Similarly, TCQ falls within the second, third, or fourth range of carbon emission quantities in Figure 4 when $G_2 = 1$, $G_3 = 1$, or $G_4 = 1$.

4.2.10. Total Machine Cost

As shown in Figure 9, this paper assumes that the machine costs are regarded as fixed costs with stepwise function for various machine hour ranges. The total machine cost is MC_0 under the current capacity of MA_{i0} machine hours. If tire production requires an expansion of machine hours to $MA_{i1}, MA_{i2}, \ldots$, or MA_{ik}, the total machine cost will increase to $MC_{i1}, M_{i2}, \ldots$, or MC_{ik}, respectively. According to the assumption, the term $\sum_{k=0}^{t} MC_{ik}\theta_{ik}$ in Equation (1) represents the total machine costs, and the associated constraints are shown in Equations (38) and (39). $(\theta_{i1}, \theta_{i2}, \ldots, \theta_{ik})$ is an SOS1 set of 0–1 variables, within which exactly one variable must be one, which is a set of 0–1 indicator variables. If $\theta_{ik} = 1$, then the machine hours will be expanded to $MA_{ik}\theta_{ik}$, i.e., $H_i q_i \leq MA_k \theta_k$, where H_i is the carbon emission quantity for one unit of product i. Thus, the machine cost will be $MC_{ik}\theta_{ik}$.

Stepwise machine hour constraints:

$$H_i q_i \leq \sum_{k=0}^{t} MA_{ik}\theta_{ik};\ i = 1, 2, \ldots, n \quad (38)$$

$$\sum_{k=0}^{t} \theta_{ik} = 1;\ i = 1, 2, \ldots, n \quad (39)$$

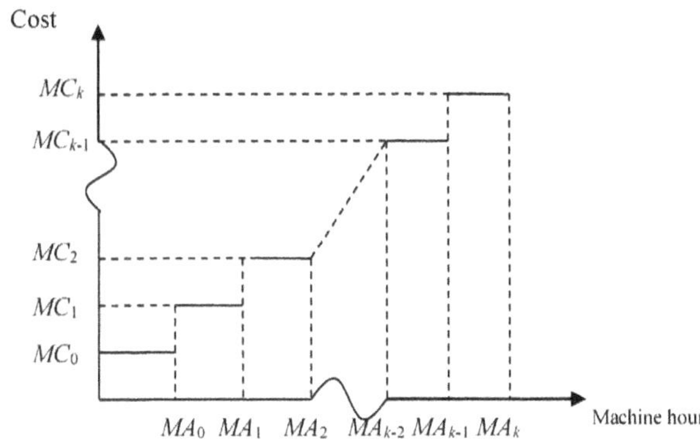

Figure 9. Stepwise process-level activity cost.

4.2.11. The Complete Model

The complete green quality management decision model included the objective function, i.e., Equation (1), and Equation (38) constraints associated with various costs, i.e., Equations (2)–(39). This model was used for the analyses of four scenarios for carbon tax and capacity expansion:

- Scenario 1: Current Capacity without Carbon Tax. The complete model for Scenario 1 included the objective function, Equation (1), and 28 constraints, Equations (2)–(28), and Equation (38), in which Equation (38) is $H_i q_i \leq MA_{i0}$.
- Scenario 2: Capacity Expansion without Carbon Tax. The complete model for Scenario 2 included the objective function, Equation (1), and Equation (29) constraints, Equations (2)–(28), and Equations (38) and (39).
- Scenario 3: Current Capacity with Carbon Tax. The complete model for Scenario 3 included the objective function, Equation (1), and Equation (37) constraints, Equations (2)–(38), in which Equation (38) is $H_i q_i \leq MA_{i0}$.
- Scenario 4: Capacity Expansion with Carbon Tax. The complete model for Scenario 4 included the objective function, Equation (1), and Equation (38) constraints, Equations (2)–(39).

5. Numerical Illustration

In this section, a numerical example was used to illustrate the application of the model.

5.1. Data and Description of a Numerical Example

In this section, a numerical example in Company X was used to illustrate the concepts described in the previous section. First, this research obtained the optimal product-mix decision under current capacity status. Then, this research considered the capacity expansion. Finally, this research analyzed the optimal product-mix decision under current capacity and capacity expansion with a carbon tax. It was assumed that Product 1, Product 2 and Product 3 could be sold for NT $4000, NT $6000, and NT $7500 per unit.

The production process of Company X is shown in Figure 1. Company X needed the following main activities in producing tire products: two unit-level activities, tire building and mixing (U = {1,2}); two batch-level activities, component preparation and curing (B = {3,4}); and one product-level activity, product design (P = {5}). These activities would need direct material and direct labor hours. Three process-level activities using machine 1, machine 2, and machine 3 were used for product 1, product 2, and product 3, respectively. The company's rework mix processing and extruder control departments were mainly used for rework processing and inspection of each batch of chemicals to the mixing department. On average, each rework mix processing took about 15 min for Product 1, Product 2, and Product 3, and rework mixing natural rubber and chemicals required 50 min, 55 min, and 60 min, respectively, to transfer Product 1, Product 2, and Product 3 to the tire building department. It took about 50 min for each order in the maintenance and pressing department, and it took eight minutes for the pressing of each unit of tires. It required 150 h, 300 h, and 250 h to process each batch of Product 1, Product 2, and Product 3 for inspection, respectively, and the inspection costs were NT $702,500. In the second stage of the ABC cost assignment view, the activity costs were traced to the cost objects. ABC uses activity drivers to measure the consumption of activities by the cost objects. In this example, products were used as the cost objects. This could trace COQ-related and COQ-unrelated costs to the products, as shown in Table 1.

Table 1. Activity analysis in accordance with COQ and ABC.

Activities	Activity Drivers	ABC Categories	COQ Scheme
Direct labor	Labor hours	VA	-
Machine 1	Machine hours	VA	-
Machine 2	Machine hours	VA	-
Machine 3	Machine hours	VA	-
Marketing, Plant guard & management	Labor hours	VA	-
Waste disposal	Number of disposal	NVA	Internal failure
Carbon emission	Carbon emission quantities	NVA	External failure
Rework	Labor hours	NVA	Internal failure
Inspection	Inspection hours	VA	Appraisal
Maintenance	Machine hours	VA	Prevention

VA: Value-added; NVA: Non-value-added.

Compounding is the activity of bringing together natural rubber, process oil, carbon black, accelerators, and other additives. The curing press is the activity where tires attain their final shape and tread pattern. Inspection did not just stop at the surface; some tires were sampled from the production line and X-rayed to detect any hidden weaknesses or internal failures. The data of the activity cost for each process and the available capacity are also presented in Table 2.

Table 2. Data for the numerical example.

Panel A: Production Information					Product 1	Product 2	Product 3	Available Capacity
Maximum Demand					70,000	40,000	50,000	-
Selling Price					$4000	$6000	$7500	-
Quantity of Batch			BA_i		10	4	4	-
Quantity of Maintenance			s_i		5	2	2	-
Direct Material Constraint								
Cost/Unit	L_1 = $50/unit	Ld_1 = $40/unit		Ldd_1 = $30/unit	6	8	10	X_1 = 30,000
	L_2 = $40/unit	-		-	2.5	3	4	X_2 = 10,000
Material Quantities	TD_1 = 10,000	W_1 = 20,000		-	-	-	-	-
Direct Labour Constraint								
Cost	WC_1 = $600,000	WC_2 = $1,100,000		WC_3 = $2,400,000	-	-	-	-
Labour Hours	WH_1 = 3000	WH_2 = 5000		WH_3 = 8000	2	2	3.5	WH_3 = 8000
Wage Rate	WR_1 = $200/h	WR_2 = $250/h		WR_3 = $300/h	-	-	-	-
Waste Disposal Constraint								
Cost	WA_1 = $400,000	WA_2 = $900,000		WA_2 = $2,000,000	-	-	-	-
Disposal Quantity	WQ_1 = 2000	WQ_2 = 3000		WQ_3 = 5000	2	3	3	WQ_3 = 5000
Disposal Rate	DR_1 = $200/h	DR_2 = $300/h		DR_3 = $400/h	-	-	-	-
Time Driver								
Inspection Hours			$t_{\tau i}$		150	300	250	-
Machine Hours			H_i		2	2	3	
Carbon Emission Constraint								
Cost/Unit	T_1 = $40/unit	T_2 = $50/unit		T_3 = $60/unit	-	-	-	-
Upper Limit of Carbon Emission Quantity	Q_1 = 10,000	Q_2 = 12,000		Q_3 = 15,000	20	40	50	Q_3 = 15,000

Process-level activity	Machine 1	Machine 2	Machine 3
Current capacity			
Cost	MC_{10} = $200,000	MC_{20} = $300,000	MC_{30} = $720,000
Machine hours	MA_{10} = 2000	MA_{20} = 3000	MA_{30} = 6000
Capacity expansion 1			
Cost	MC_{11} = $450,000	MC_{21} = $600,000	MC_{31} = $1,200,000
Machine hours	MA_{11} = 3000	MA_{21} = 4000	MA_{31} = 8000
Capacity expansion 2			
Cost	MC_{12} = $1,000,000	MC_{22} = $1,200,000	MC_{32} = $1,800,000
Machine hours	MA_{12} = 5000	MA_{22} = 6000	MA_{32} = 9000

Panel B: Resources consumed	Rework	Maintenance	Inspection
Resources	$100,000	$1,230,000	$702,500
Capacity (hours)	$\sigma_o = 250$	$\sigma_\delta = 4100$	$\sigma_\tau = 1000$
Per hour/rate	k_0 = $400/h	k_δ = $300/h	k_τ = $702.5/h

5.2. Four Scenario Analyses

5.2.1. Scenario 1: Current Capacity without Carbon Tax

The mathematical programming model for current capacity without carbon tax is shown in Table A1, and the optimal solution is shown in Table 3. The optimal product mix was to produce 580 units, 204 units and 408 units of Product 1, Product 2 and Product 3, respectively. It required 10,000 pounds of rubber and 3694 pounds of chemical materials, and it used 1160 h, 408 h and 1224 h for Machine 1, Machine 2, and Machine 3, respectively. A total of 2996 units of waste was generated, 250 h were spent on rework, 700 h were spent on inspection, and 511 h were spent on maintenance. For this scenario, the company gained the maximum profit of NT $2,391,377.

5.2.2. Scenario 2: Capacity Expansion without Carbon Tax

The mathematical programming model for capacity expansion without carbon tax is shown in Table A2, and the optimal solution is shown in Table 4. The optimal product mix was to produce 2300 units of Product 1 only. It required 13,800 pounds of rubber and 5750 pounds of chemical materials, and it used 4600 h for Machine 1. A total of 4600 units of waste was generated, 249 h were spent on rework, 150 h were spent on inspection, and 690 h were spent on maintenance. For this scenario, the company gained the maximum profit of NT $2,685,958.

5.2.3. Scenario 3: Current Capacity with Carbon Tax

The mathematical programming model for current capacity with carbon tax is shown in Table A3, and the optimal solution is shown in Table 5. The optimal product mix was to produce 1000 units and 284 units of Product 1 and Product 3, respectively. It required 10,000 pounds of rubber and 3636 pounds of chemical materials, and it used 2000 h and 852 h for Machine 1 and Machine 3, respectively. A total of 2852 units of waste was generated, 197 h were spent on rework, 450 h were spent on inspection, and 456 h were spent on maintenance. For this scenario, the company gained the maximum profit of NT $715,867 and paid a carbon tax of NT $1,452,000.

5.2.4. Scenario 4: Capacity Expansion with Carbon Tax

The mathematical programming model for capacity expansion with carbon tax is shown in Table A4, and the optimal solution is shown in Table 6. The optimal product mix was to produce 1500 units of Product 1 only. It required 10,000 pounds of rubber and 3750 pounds of chemical materials, and it used 3000 h for Machine 1. A total of 3000 units of waste was generated, 163 h were spent on rework, 150 h were spent on inspection, and 450 h were spent on maintenance. For this scenario, the company gained the maximum profit of NT $774,625 and paid a carbon tax NT $1,200,000.

6. Discussion of Results

From Scenario 1 to Scenario 3, the production quantity of Product 1 increased from 580 units to 2300 units, and the production quantities of Product 2 and Product 3 decreased to zero. However, the total profit increased from $2,391,377 to $2,695,958. In Scenario 3, the optimal product mix was to produce 2300 units of Product 1, which was determined by the revenue, resource consumptions, resource expansion costs, the carbon emission quantity of the three products, and the resources available, etc. Considering the scale of economy or satisfying the present customers' need for each product, this study added the constraints of minimum production quantities for products.

Table 3. Optimal solution for current capacity without carbon tax (Scenario 1).

Optimal Product Mix Solution for Current Capacity without Carbon Tax (Scenario 1)
$q_1 = 580$ $q_2 = 204$ $q_3 = 408$ $M_{r1} = 0$ $Md_{r1} = 10,000$ $Mdd_{r1} = 0$ $M_{r2} = 3694$ $\alpha_0 = 1$ $\alpha_1 = 0$ $\alpha_2 = 0$ $\beta_0 = 0$ $\beta_1 = 0$ $\beta_2 = 1$ $\beta_3 = 0$ $BA_1 = 58$ $BA_2 = 51$ $BA_3 = 102$ $\delta_1 = 116$ $\delta_2 = 102$ $\delta_3 = 204$ $\tau_1 = 1$ $\tau_2 = 1$ $\tau_3 = 1$ $ND = 0$ $SD = 1$ $OD = 0$ $\eta_1 = 1$ $\eta_2 = 0$ $\mu_1 = 0$ $\mu_2 = 1$ $\mu_3 = 0$

The mathematical programming model for current capacity without carbon tax is shown in Table A1.

Table 4. Optimal solution for capacity expansion without carbon tax (Scenario 2).

Optimal Product Mix Solution for Capacity Expansion without Carbon Tax (Scenario 2)
$q_1 = 2300$ $q_2 = 0$ $q_3 = 0$ $M_{r1} = 0$ $Md_{r1} = 13,800$ $Mdd_{r1} = 0$ $M_{r2} = 5750$ $\alpha_0 = 0$ $\alpha_1 = 1$ $\alpha_2 = 0$ $\beta_0 = 0$ $\beta_1 = 0$ $\beta_2 = 0$ $\beta_3 = 1$ $BA_1 = 230$ $BA_2 = 0$ $BA_3 = 0$ $\delta_1 = 460$ $\delta_2 = 0$ $\delta_3 = 0$ $\tau_1 = 1$ $\tau_2 = 0$ $\tau_3 = 0$ $ND = 0$ $SD = 1$ $OD = 0$ $\eta_1 = 0$ $\eta_2 = 1$ $\mu_1 = 0$ $\mu_2 = 0$ $\mu_3 = 1$ $\theta_{10} = 0$ $\theta_{11} = 0$ $\theta_{12} = 1$ $\theta_{20} = 1$ $\theta_{21} = 0$ $\theta_{22} = 0$ $\theta_{30} = 1$ $\theta_{31} = 0$ $\theta_{32} = 0$

The mathematical programming model is shown in Table A2.

Table 5. Optimal solution for current capacity with carbon tax (Scenario 3).

Optimal Product Mix Solution for Current Capacity with Carbon Tax (Scenario 3)
$q_1 = 1000$ $q_2 = 0$ $q_3 = 284$ $M_{r1} = 0$ $Md_{r1} = 10,000$ $Mdd_{r1} = 0$ $M_{r2} = 3636$ $\alpha_0 = 1$ $\alpha_1 = 0$ $\alpha_2 = 0$ $\beta_0 = 0$ $\beta_1 = 0$ $\beta_2 = 1$ $\beta_3 = 0$ $BA_1 = 100$ $BA_2 = 0$ $BA_3 = 71$ $\delta_1 = 200$ $\delta_2 = 0$ $\delta_3 = 142$ $\tau_1 = 1$ $\tau_2 = 0$ $\tau_3 = 1$ $ND = 0$ $SD = 1$ $OD = 0$ $\eta_1 = 1$ $\eta_2 = 0$ $\mu_1 = 0$ $\mu_2 = 1$ $\mu_3 = 0$ $At = 0$ $Bt = 0$ $Ct = 0$ $Dt = 34,200$ $G_1 = 0$ $G_2 = 0$ $G_3 = 0$ $G_4 = 1$

The mathematical programming model for current capacity with carbon tax is shown in Table A3.

Table 6. Optimal solution for capacity expansion with carbon tax (Scenario 4).

Optimal Product Mix Solution for Capacity Expansion with Carbon Tax (Scenario 4)
$q_1 = 1500$ $q_2 = 0$ $q_3 = 0$ $M_{r1} = 0$ $Md_{r1} = 10,000$ $Mdd_{r1} = 0$ $M_{r2} = 3750$ $\alpha_0 = 1$ $\alpha_1 = 0$ $\alpha_2 = 0$ $\beta_0 = 0$ $\beta_1 = 0$ $\beta_2 = 1$ $\beta_3 = 0$ $BA_1 = 150$ $BA_2 = 0$ $BA_3 = 0$ $\delta_1 = 300$ $\delta_2 = 0$ $\delta_3 = 0$ $\tau_1 = 1$ $\tau_2 = 0$ $\tau_3 = 0$ $ND = 0$ $SD = 1$ $OD = 0$ $\eta_1 = 1$ $\eta_2 = 0$ $\mu_1 = 0$ $\mu_2 = 1$ $\mu_3 = 0$ $\theta_{10} = 0$ $\theta_{11} = 1$ $\theta_{12} = 0$ $\theta_{20} = 1$ $\theta_{21} = 0$ $\theta_{22} = 0$ $\theta_{30} = 1$ $\theta_{31} = 0$ $\theta_{32} = 0$ $At = 0$ $Bt = 0$ $Ct = 0$ $Dt = 30,000$ $G_1 = 0$ $G_2 = 0$ $G_3 = 0$ $G_4 = 1$

The mathematical programming model for capacity expansion with carbon tax is shown in Table A4.

Similarly, from Scenario 2 to Scenario 4, the production quantity of Product 1 increased from 1000 units to 1500 units, the production quantities of Product 3 decreased to zero, and the total profit increased from $715,867 to $774,625. The total revenue in the capacity expansion decreased by 130,000 (6,130,000–6,000,000) compared to the original total revenue. Product 1 increased by 500 units, Product 3 decreased by 284 units, and the company's total carbon tax decreased by 252,000 (1,452,000–1,200,000). The quantity of Product 1 increased but the company's total carbon tax did not increase accordingly, showing that the carbon emission coefficient of Product 1 was lower than that of the other products. This built-in tax flexibility could help firms to minimize their compliance costs over time.

The green quality management (GQM) report, shown in Table 7, included the data of the optimal product-mix, resources consumed, profit, COQ cost, and value-added and non-value-added activity costs for the four scenarios illustrated in this paper. This report provided several considerable advantages to the company. First, with the use of ABC, the activities and associated costs were classified as quality-related or quality-unrelated and VA or NVA in terms of activities. This classification could allow management to detect opportunities for cost reduction and the elimination of NVA activities which could be used in cost management [9]. The applied method was allowed to produce optimally and redirect the productive capacity towards products that save environmental costs and change, modify or redesign the production system from the environmental and quality cost point of view (such as the introduction of low carbon technologies or fuel sources, for example). Second, the measurement and reporting of GQM could provide opportunities which could be used by the organization as long-term performance measurements. Finally, the GQM report presented the cost of each product by activity and quality basis, which could be used to compare the contribution of each product to the company.

Table 7. Green quality management report for comparison of the four decision scenarios.

		Current Capacity		Capacity Expansion	
		Without Carbon Tax (Scenario 1)	With Carbon Tax (Scenario 2)	Without Carbon Tax (Scenario 3)	With Carbon Tax (Scenario 4)
Panel A: Production-mix	Product A1	580	1000	2300	1500
	Product A2	204	0	0	0
	Product A3	408	284	0	0
Panel B: Resources Consumed	Direct material 1	10,000	10,000	13,800	10,000
	Direct material 2	3694	3636	5750	3750
	Direct labor hours	2996	2994	4600	3000
	Machine 1 (hours)	1160	2000	4600	3000
	Machine 2 (hours)	408	0	0	0
	Machine 3 (hours)	1224	852	0	0
	Waste disposal	2996	2852	4600	3000
	Carbon emission	-	34,200	-	30,000
	Rework	250	197	249	163
	Inspection	700	450	150	150
	Maintenance	511	456	690	450
Panel C: Profit	Revenue	6,604,000	6,130,000	9,200,000	6,000,000
	Direct material 1	400,000	400,000	552,000	400,000
	Direct material 2	147,760	145,440	230,000	150,000
	Direct labor (VA)	600,000	600,000	1,100,000	600,000
	Machine 1 (VA)	200,000	200,000	1,000,000	450,000
	Machine 2 (VA)	300,000	300,000	300,000	300,000
	Machine 3 (VA)	720,000	720,000	720,000	720,000
	Waste disposal cost (NVA, Internal failure)	900,000	900,000	2,000,000	900,000
	Carbon tax (NVA, External failure)	-	1,452,000	-	1,200,000
	Rework cost (NVA, Internal failure)	99,933	78,833	99,667	65,000
	Inspection cost (VA, Appraisal)	491,750	281,000	105,375	105,375
	Maintenance cost (VA, Prevention)	153,180	136,860	207,000	135,000
	Marketing, Plant guard & management (VA)	200,000	200,000	200,000	200,000
	Income based on resources consumed	2,391,377	715,867	2,685,958	774,625
Panel D: COQ Report	Total product cost	4,012,623	5,214,133	6,314,042	5,025,375
	Total activity cost	3,664,863	4,868,693	5,732,042	4,675,375
	Total VA cost	2,664,930	2,437,860	3,632,375	2,510,375
	Total NVA cost	999,933	2,430,833	2,099,667	2,165,000
	Total COQ cost	1,644,863	2,848,693	2,412,042	2,405,375
Data Sources		Table 3	Table 5	Table 4	Table 6

7. Conclusions

Enterprises expand capacity for profit while governments levy carbon emissions and tax standards for environmental protection, and it is necessary to take these two factors into consideration in current enterprise development. As global environmental awareness has become increasingly valued, enterprises also seek environmental protection while at the time gaining profit [9]. In addition to government pressure to reduce environmental pollution, people are forcing the government to formulate strict environmental regulations and levy fines for environmental pollution. In this way, it is hoped that enterprises can reduce pollution via environmental management [29].

Capacity expansion concerns whether an enterprise can achieve sustainable growth, increase sales volume, and even improve enterprise competitiveness. Therefore, how to apply capacity is an important and key factor of enterprises' sustainable activities and ability to maintain competitiveness. The flexible use of capacity strategies to improve competitiveness is also a goal jointly pursued by all enterprises. As a result, how to make good use of capacity to intensify enterprises' competitiveness and advantages is an important issue.

The green quality management decision model proposed in this paper has the following features:

- It uses mathematical programming to simultaneously consider material purchase discounts, capacity expansions, capacity constraints, waste disposal, and carbon tax expenditures in order to determine an optimal product-mix decision.
- It uses Activity-Based Costing to consider various levels of activities in order to accurately measure the cost of activities, identify the costs of quality, and identify the costs of value-added and non-value-added activities, which can indicate the possible benefits of improving or eliminating non-value-added activities.
- It successfully formulates the various cost functions in the mathematical programming, such as material quantity discounts, the labor cost with high overtime rates, the piecewise linear waste disposal cost function, the machine cost with capacity expansion, and the carbon tax with full progressive tax rates and a threshold (used in the illustration). In such a way, companies can incorporate various resource expansions into the mathematical programming model to alleviate the workload of post-optimal analysis.

Additionally, there are two main contributions of this study:

- First, it can provide decision-makers in tire production with the decisions and actions needed to respond to carbon tax policies, and it can provide environmental policy-makers with strategic thinking in policy making.
- Second, it can help enterprises plan new practices for emission reduction and energy saving by combining Activity-based Costing Method and adopting Mathematical Programming Model analysis to create a win-win production method for environmental protection and enterprise profit.

Funding: This research was funded by the Ministry of Science and Technology of Taiwan under Grant No. MOST104-2410-H-008-045 and MOST106-2410-H-008-020-MY3.

Acknowledgments: The author is extremely grateful to the energies journal editorial team and reviewers who provided valuable comments for improving the quality of this article. The author also would like to thank the Ministry of Science and Technology of Taiwan for financial support of this research under Grant No. MOST104-2410-H-008-045 and MOST106-2410-H-008-020-MY3.

Conflicts of Interest: The author declares no conflict of interest.

Appendix A

Table A1. The quality management decision model for current capacity without carbon tax (Scenario 1).

$$Max\ \pi = 4000 \times q_1 + 6000 \times q_2 + 7500 \times q_3 - 50 \times M_{r1} - 40 \times Md_{r1} - 30 \times Mdd_{r1} - 40 \times M_{r2} - 600{,}000 - 500{,}000 \times \alpha_1 - 1{,}800{,}000 \times \alpha_2$$
$$-15 \times 400 \times (BA_1 + BA_2 + BA_3) \div 60 - 400 \times (50 \times BA_1 + 55 \times BA_2 + 60 \times BA_3) \div 60 - 300 \times (50 \times \delta_1 + 50 \times \delta_2 + 50 \times \delta_3) \div 60 - 300 \times (8 \times q_1 + 8 \times q_2 + 8 \times q_3) \div 60$$
$$-702.5 \times (150 \times \tau_1 + 300 \times \tau_2 + 250 \times \tau_3) - 400{,}000 \times \beta_1 - 900{,}000 \times \beta_2 + 200{,}000 \times \beta_3 - 200{,}000 - 300{,}000 - 720{,}000 - 200{,}000$$

Subject to—Direct Material

$$6 \times q_1 + 8 \times q_2 + 10 \times q_3 - M_{r1} - Md_{r1} - Mdd_{r1} \le 0$$
$$2.5 \times q_1 + 3 \times q_2 + 4 \times q_3 - M_{r2} \le 0$$
$$ND + SD + OD = 1M_{r2} \le 10{,}000M_{r2} \ge 0$$
$$M_{r1} \le 10{,}000 \times NDMd_{r1} \le 20{,}000 \times SDMdd_{r1} \le 30{,}000 \times OD$$
$$Md_{r1} \ge 10{,}000 \times SDMdd_{r1} \ge 20{,}000 \times OD$$

Subject to—Direct Labour

$$2 \times q_1 + 2 \times q_2 + 3.5 \times q_3 - 3000 - 2000 \times \alpha_1 - 5000 \times \alpha_2 \le 0$$
$$\alpha_0 - \eta_1 \le 0\alpha_1 - \eta_1 - \eta_2 \le 0\alpha_2 - \eta_2 \le 0$$
$$\eta_1 + \eta_2 = 1\alpha_0 + \alpha_1 + \alpha_2 = 1$$

Subject to—Machine Hour

$$2 \times q_1 - 2000 \le 02 \times q_2 - 3000 \le 03 \times q_3 - 6000 \le 0$$

Subject to—Inspection

$$150 \times \tau_1 + 300 \times \tau_2 + 250 \times \tau_3 \le 1000$$
$$q_1 \le 7000 \times \tau_1 q_2 \le 4000 \times \tau_2 q_3 \le 5000 \times \tau_3$$

Subject to—Waste Disposal

$$2 \times q_1 + 3 \times q_2 + 3 \times q_3 - 2000\beta_1 - 3000 \times \beta_2 - 5000 \times \beta_3 \le 0$$
$$\beta_0 - \mu_1 \le 0\beta_1 - \mu_1 - \mu_2 \le 0\beta_2 - \mu_2 - \mu_3 \le 0\beta_3 - \mu_3 \le 0$$
$$\beta_0 + \beta_1 + \beta_2 + \beta_3 = 1\mu_1 + \mu_2 + \mu_3 = 1$$

Subject to—Rework

$$15 \times BA_1 + 15 \times BA_2 + 15 \times BA_3 + 50 \times BA_1 + 55 \times BA_2 + 60 \times BA_3 \le 250 \times 60$$
$$q_1 = BA_1 \times 10q_2 = BA_2 \times 4q_3 = BA_3 \times 4$$

Subject to—Maintenance

$$50\delta_1 + 50\delta_2 + 50\delta_3 + 8q_1 + 8q_2 + 8q_3 \le 4100 \times 60$$
$$q_1 = 5 \times \delta_1 q_2 = 2 \times \delta_2 q_3 = 2 \times \delta_3$$

Table A2. The quality management decision model for capacity expansion without carbon tax (Scenario 2).

$Max\ \pi$	$= 4000 \times q_1 + 6000 \times q_2 + 7500 \times q_3 - 50 \times M_{r1} - 40 \times Md_{r1} - 30 \times Mdd_{r1} - 40 \times M_{r2} - 600{,}000 - 500{,}000 \times \alpha_1 - 1{,}800{,}000 \times \alpha_2$ $-15 \times 400 \times (BA_1 + BA_2 + BA_3) \div 60 - 400 \times (50 \times BA_1 + 55 \times BA_2 + 60 \times BA_3) \div 60 - 300 \times (50 \times \delta_1 + 50 \times \delta_2 + 50 \times \delta_3) \div 60 - 300 \times (8 \times q_1 + 8 \times q_2 + 8 \times q_3) \div 60$ $-702.5 \times (150 \times \tau_1 + 300 \times \tau_2 + 250 \times \tau_3) - 400{,}000 \times \beta_1 - 900{,}000 \times \beta_2 + 200{,}000 \times \beta_3$ $-200{,}000 \times \theta_{10} - 300{,}000 \times \theta_{20} - 720{,}000 \times \theta_{30} - 450{,}000 \times \theta_{11} - 600{,}000 \times \theta_{21} - 1{,}200{,}000 \times \theta_{31} - 1{,}000{,}000 \times \theta_{12} - 1{,}200{,}000 \times \theta_{22} - 1{,}800{,}000 \times \theta_{32} - 200{,}000$
Subject to—Direct Material $6 \times q_1 + 8 \times q_2 + 10 \times q_3 - M_{r1} - Md_{r1} - Mdd_{r1} \leq 0$ $2.5 \times q_1 + 3 \times q_2 + 4 \times q_3 - M_{r2} \leq 0$ $ND + SD + OD = 1 M_{r2} \leq 10{,}000 M_{r2} \geq 0$ $M_{r1} \leq 10{,}000 \times ND Md_{r1} \leq 20{,}000 \times SD Mdd_{r1} \leq 30{,}000 \times OD$ $Md_{r1} \geq 10{,}000 \times SD Mdd_{r1} \geq 20{,}000 \times OD$	Subject to—Direct Labour $2 \times q_1 + 2 \times q_2 + 3.5 \times q_3 - 3000 - 2000 \times \alpha_1 - 5000 \times \alpha_2 \leq 0$ $\alpha_0 - \eta_1 \leq 0 \alpha_1 - \eta_1 - \eta_2 \leq 0 \alpha_2 - \eta_2 \leq 0$ $\eta_1 + \eta_2 = 1 \alpha_0 + \alpha_1 + \alpha_2 = 1$ Subject to—Rework $15 \times BA_1 + 15 \times BA_2 + 15 \times BA_3 + 50 \times BA_1 + 55 \times BA_2 + 60 \times BA_3 \leq 250 \times 60$ $q_1 = BA_1 \times 10 q_2 = BA_2 \times 4 q_3 = BA_3 \times 4$
Subject to—Inspection $150 \times \tau_1 + 300 \times \tau_2 + 250 \times \tau_3 \leq 1000$ $q_1 \leq 7000 \times \tau_1 q_2 \leq 4000 \times \tau_2 q_3 \leq 5000 \times \tau_3$ Subject to—Waste Disposal $2 \times q_1 + 3 \times q_2 + 3 \times q_3 - 2000 \beta_1 - 3000 \times \beta_2 - 5000 \times \beta_3 \leq 0$ $\beta_0 - \mu_1 \leq 0 \beta_1 - \mu_1 - \mu_2 \leq 0 \beta_2 - \mu_2 - \mu_3 \leq 0 \beta_3 - \mu_3 \leq 0$ $\beta_0 + \beta_1 + \beta_2 + \beta_3 = 1 \mu_1 + \mu_2 + \mu_3 = 1$	Subject to—Stepwise Machine Hour $2 \times q_1 - 2000 \times \theta_{10} - 3000 \times \theta_{11} - 5000 \times \theta_{12} \leq 0 \theta_{10} + \theta_{11} + \theta_{12} = 1$ $2 \times q_2 - 3000 \times \theta_{20} - 4000 \times \theta_{21} - 6000 \times \theta_{22} \leq 0 \theta_{20} + \theta_{21} + \theta_{22} = 1$ $3 \times q_3 - 6000 \times \theta_{30} - 8000 \times \theta_{31} - 9000 \times \theta_{32} \leq 0 \theta_{30} + \theta_{31} + \theta_{32} = 1$ Subject to—Maintenance $50\delta_1 + 50\delta_2 + 50\delta_3 + 8q_1 + 8q_2 + 8q_3 \leq 4100 \times 60$ $q_1 = 5 \times \delta_1 q_2 = 2 \times \delta_2 q_3 = 2 \times \delta_3$

Table A3. The quality management decision model for current capacity with carbon tax (Scenario 3).

$Max\ \pi$	$= 4000 \times q_1 + 6000 \times q_2 + 7500 \times q_3 - 50 \times M_{r1} - 40 \times Md_{r1} - 30 \times Mdd_{r1} - 40 \times M_{r2} - 600,000 - 500,000 \times \alpha_1 - 1,800,000 \times \alpha_2$ $-15 \times 400 \times (BA_1 + BA_2 + BA_3) \div 60 - 400 \times (50 \times BA_1 + 55 \times BA_2 + 60 \times BA_3) \div 60 - 300 \times (50 \times \delta_1 + 50 \times \delta_2 + 50 \times \delta_3) \div 60 - 300 \times (8 \times q_1 + 8 \times q_2 + 8 \times q_3) \div 60$ $-702.5 \times (150 \times \tau_1 + 300 \times \tau_2 + 250 \times \tau_3) - 400,000 \times \beta_1 - 900,000 \times \beta_2 + 200,000 \times \beta_3 - 40 \times (Bt - 10,000) - 50 \times (Ct - 10,000) - 60 \times (Dt - 10,000)$ $-200,000 - 300,000 - 720,000 - 200,000$
Subject to—Direct Material $6 \times q_1 + 8 \times q_2 + 10 \times q_3 - M_{r1} - Md_{r1} - Mdd_{r1} \leq 0$ $2.5 \times q_1 + 3 \times q_2 + 4 \times q_3 - M_{r2} \leq 0$ $ND + SD + OD = 1 M_{r2} \leq 10,000 M_{r2} \geq 0$ $M_{r1} \leq 10,000 \times ND Md_{r1} \leq 20,000 \times SD Mdd_{r1} \leq 30,000 \times OD$ $Md_{r1} \geq 10,000 \times SD Mdd_{r1} \geq 20,000 \times OD$	Subject to—Direct Labour $2 \times q_1 + 2 \times q_2 + 3.5 \times q_3 - 3000 - 2000 \times \alpha_1 - 5000 \times \alpha_2 \leq 0$ $\alpha_0 - \eta_1 \leq 0 \alpha_1 - \eta_1 - \eta_2 \leq 0 \alpha_2 - \eta_2 \leq 0$ $\eta_1 + \eta_2 = 1 \alpha_0 + \alpha_1 + \alpha_2 = 1$ Subject to—Machine hour $2 \times q_1 - 2000 \leq 02 \times q_2 - 3000 \leq 03 \times q_3 - 6000 \leq 0$
Subject to—Waste Disposal $2 \times q_1 + 3 \times q_2 + 3 \times q_3 - 2000\beta_1 - 3000 \times \beta_2 - 5000 \times \beta_3 \leq 0$ $\beta_0 - \mu_1 \leq 0 \beta_1 - \mu_1 - \mu_2 \leq 0 \beta_2 - \mu_2 - \mu_3 \leq 0 \beta_3 - \mu_3 \leq 0$ $\beta_0 + \beta_1 + \beta_2 + \beta_3 = 1 \mu_1 + \mu_2 + \mu_3 = 1$ Subject to—Carbon Tax $20 \times q_1 + 40 \times q_2 + 50 \times q_3 \leq At + Bt + Ct + Dt$ $At \geq 0 At \leq G_1 \times 10,000 Bt \geq G_2 \times 10,000 Bt \leq G_2 \times 12,000$ $Ct \geq G_3 \times 12,000 Ct \leq G_3 \times 15,000 Dt \geq G_4 \times 15,000$ $G_1 + G_2 + G_3 + G_4 = 1$	Subject to—Rework $15 \times BA_1 + 15 \times BA_2 + 15 \times BA_3 + 50 \times BA_1 + 55 \times BA_2 + 60 \times BA_3 \leq 250 \times 60$ $q_1 = BA_1 \times 10 q_2 = BA_2 \times 4 q_3 = BA_3 \times 4$ Subject to—Maintenance $50\delta_1 + 50\delta_2 + 50\delta_3 + 8q_1 + 8q_2 + 8q_3 \leq 4100 \times 60$ $q_1 = 5 \times \delta_1 q_2 = 2 \times \delta_2 q_3 = 2 \times \delta_3$ Subject to—Inspection $150 \times \tau_1 + 300 \times \tau_2 + 250 \times \tau_3 \leq 1000$ $q_1 \leq 7000 \times \tau_1 q_2 \leq 4000 \times \tau_2 q_3 \leq 5000 \times \tau_3$

Table A4. The quality management decision model for capacity expansion with carbon tax (Scenario 4).

$Max\ \pi$	$= 4000 \times q_1 + 6000 \times q_2 + 7500 \times q_3 - 50 \times M_{r1} - 40 \times Md_{r1} - 30 \times Mdd_{r1} - 40 \times M_{r2} - 600,000 - 500,000 \times \alpha_1 - 1,800,000 \times \alpha_2 - 15 \times 400 \times (BA_1 + BA_2 + BA_3) \div 60$ $-400 \times (50 \times BA_1 + 55 \times BA_2 + 60 \times BA_3) \div 60 - 300 \times (50 \times \delta_1 + 50 \times \delta_2 + 50 \times \delta_3) \div 60 - 300 \times (8 \times q_1 + 8 \times q_2 + 8 \times q_3) \div 60 - 702.5 \times (150 \times \tau_1 + 300 \times \tau_2 + 250 \times \tau_3)$ $-400,000 \times \beta_1 - 900,000 \times \beta_2 - 200,000\beta_3 - 40 \times (Bt - 10,000) - 50 \times (Ct - 10,000) - 60 \times (Dt - 10,000) - 200,000 \times \theta_{10} - 300,000 \times \theta_{20} - 720,000 \times \theta_{30}$ $-450,000 \times \theta_{11} - 600,000 \times \theta_{21} - 1,200,000 \times \theta_{31} - 1,000,000 \times \theta_{12} - 1,200,000 \times \theta_{22} - 1,800,000 \times \theta_{32} - 200,000$	
	Subject to—Direct Material $6 \times q_1 + 8 \times q_2 + 10 \times q_3 - M_{r1} - Md_{r1} - Mdd_{r1} \leq 0$ $2.5 \times q_1 + 3 \times q_2 + 4 \times q_3 - M_{r2} \leq 0$ $ND + SD + OD = 1M_{r2} \leq 10,000M_{r2} \geq 0$ $M_{r1} \leq 10,000 \times NDMd_{r1} \leq 20,000 \times SDMdd_{r1} \leq 30,000 \times OD$ $Md_{r1} \geq 10,000 \times SDMdd_{r1} \geq 20,000 \times OD$	Subject to—Direct Labour $2 \times q_1 + 2 \times q_2 + 3.5 \times q_3 - 3000 - 2000 \times \alpha_1 - 5000 \times \alpha_2 \leq 0$ $\alpha_0 - \eta_1 \leq 0\alpha_1 - \eta_1 - \eta_2 \leq 0\alpha_2 - \eta_2 \leq 0$ $\eta_1 + \eta_2 = 1\alpha_0 + \alpha_1 + \alpha_2 = 1$ Subject to—Rework $15 \times BA_1 + 15 \times BA_2 + 15 \times BA_3 + 50 \times BA_1 + 55 \times BA_2 + 60 \times BA_3 \leq 250 \times 60$ $q_1 = BA_1 \times 10q_2 = BA_2 \times 4q_3 = BA_3 \times 4$
	Subject to—Inspection $150 \times \tau_1 + 300 \times \tau_2 + 250 \times \tau_3 \leq 1000$ $q_1 \leq 7000 \times \tau_1 q_2 \leq 4000 \times \tau_2 q_3 \leq 5000 \times \tau_3$ Subject to—Waste disposal $2 \times q_1 + 3 \times q_2 + 3 \times q_3 - 2000\beta_1 - 3000 \times \beta_2 - 5000 \times \beta_3 \leq 0$ $\beta_0 - \mu_1 \leq 0\beta_1 - \mu_1 - \mu_2 \leq 0\beta_2 - \mu_2 - \mu_3 \leq 0\beta_3 - \mu_3 \leq 0$ $\beta_0 + \beta_1 + \beta_2 + \beta_3 = 1\mu_1 + \mu_2 + \mu_3 = 1$ Subject to—Shipping $50\delta_1 + 50\delta_2 + 50\delta_3 + 8q_1 + 8q_2 + 8q_3 \leq 4100 \times 60$ $q_1 = 5 \times \delta_1 q_2 = 2 \times \delta_2 q_3 = 2 \times \delta_3$	Subject to—Stepwise Machine Hour $2 \times q_1 - 2000 \times \theta_{10} - 3000 \times \theta_{11} - 5000 \times \theta_{12} \leq 0\theta_{10} + \theta_{11} + \theta_{12} = 1$ $2 \times q_2 - 3000 \times \theta_{20} - 4000 \times \theta_{21} - 6000 \times \theta_{22} \leq 0\theta_{20} + \theta_{21} + \theta_{22} = 1$ $3 \times q_3 - 6000 \times \theta_{30} - 8000 \times \theta_{31} - 9000 \times \theta_{32} \leq 0\theta_{30} + \theta_{31} + \theta_{32} = 1$ Subject to—Carbon Tax $20 \times q_1 + 40 \times q_2 + 50 \times q_3 \leq At + Bt + Ct + Dt$ $At \geq 0At \leq G_1 \times 10,000Bt \geq G_2 \times 10,000Bt \leq G_2 \times 12,000$ $Ct \geq G_3 \times 12,000Ct \leq G_3 \times 15,000Dt \geq G_4 \times 15,000$ $G_1 + G_2 + G_3 + G_4 = 1$

References

1. Report of the Conference of the Parties on Its Twenty-First Session. 2015. Available online: http://unfccc.int/resource/docs/2015/cop21/eng/10a01.pdf. (accessed on 11 July 2018).
2. Hunter, D. Implications of the Copenhagen Accord for Global Climate Governance. *Sustain. Dev. Law Policy* **2010**, *10*, 4–15, 56–57.
3. Abas, N.; Kalair, A.R.; Khan, N.; Haider, A.; Saleem, Z.; Saleem, M.S. Natural and synthetic refrigerants, global warming: A review. *Renew. Sustain. Energy Rev.* **2018**, *90*, 557–569. [CrossRef]
4. Douiri, L.; Jabri, A.; El Barkany, A. Models for optimization of supply chain network design integrating the cost of quality: A literature review. *Am. J. Ind. Bus. Manag.* **2016**, *6*, 860–876. [CrossRef]
5. García-Pastor, A.; Guirao, B.; López-Lambas, M.E. Quality cost in bus operations based on activity-based costing. *Proc. Inst. Civ. Eng. Transp.* **2016**, *169*, 107–117. [CrossRef]
6. Ittner, C.D. Activity-based costing concepts for quality improvement. *Eur. Manag. J.* **1999**, *17*, 492–500. [CrossRef]
7. Khataie, A.H.; Bulgak, A.A. A cost of quality decision support model for lean manufacturing: Activity-based costing application. *Int. J. Qual. Reliab. Manag.* **2013**, *30*, 751–764. [CrossRef]
8. Özkan, S.; Karaibrahimoğlu, Y.Z. Activity-based costing approach in the measurement of cost of quality in SMEs: A case study. *Total Qual. Manag. Bus. Excell.* **2013**, *24*, 420–431. [CrossRef]
9. Tsai, W.-H. Quality cost measurement under activity-based costing. *Int. J. Qual. Reliab. Manag.* **1998**, *15*, 719–752. [CrossRef]
10. Turney, P.B.B. *Common Cents: The ABC Performance Breakthrough—How to Succeed with Activity-Based Costing*; McGraw-Hill Education: New York, NY, USA, 2005.
11. Goldratt, E.M.; Cox, J. *The Goal: Excellence in Manufacturing*; North River Press: Great Barrington, MA, USA, 1984.
12. Yi, Y.; Li, J. Cost-Sharing Contracts for Energy Saving and Emissions Reduction of a Supply Chain under the Conditions of Government Subsidies and a Carbon Tax. *Sustainability* **2018**, *10*, 895.
13. Baranzini, A.; Carattini, S. Effectiveness, earmarking and labeling: testing the acceptability of carbon taxes with survey data. *Environ. Econ. Policy Stud.* **2017**, *19*, 197–227. [CrossRef]
14. Bhupendra, K.V.; Sangle, S. Pollution Prevention Strategy: A Study of Indian Firms. *J. Clean. Prod.* **2016**, *133*, 795–802. [CrossRef]
15. Tsai, W.-H.; Yang, C.-H.; Huang, C.-T.; Wu, Y.-Y. The Impact of the Carbon Tax Policy on Green Building Strategy. *J. Environ. Plan. Manag.* **2017**, *60*, 1412–1438. [CrossRef]
16. Tsai, W.-H.; Chang, J.-C.; Hsieh, C.-L.; Tsaur, T.-S.; Wang, C.-W. Sustainability Concept in Decision-Making: Carbon Tax Consideration for Joint Product Mix Decision. *Sustainability* **2016**, *8*, 1232. [CrossRef]
17. Revoredo-Giha, C.; Chalmers, N.; Akaichi, F. Simulating the Impact of Carbon Taxes on Greenhouse Gas Emission and Nutrition in the UK. *Sustainability* **2018**, *10*, 134. [CrossRef]
18. Chopra, A.; Garg, D. Behavior patterns of quality cost categories. *TQM J.* **2011**, *23*, 510–515. [CrossRef]
19. Chopra, S.; Wu, P.-J. Eco-activities and operating performance in the computer and electronics industry. *Eur. J. Oper. Res.* **2016**, *248*, 971–981. [CrossRef]
20. Grottke, M.; Schleich, B. Cost Optimality in Testing and Rejuvenation. In Proceedings of the 23rd IEEE International Symposium on Software Reliability Engineering Workshops, Dallas, TX, USA, 27–30 November 2012; pp. 259–264.
21. Tye, L.H.; Halim, H.A.; Ramayah, T. An exploratory study on cost of quality implementation in Malaysia: The case of Penang manufacturing firms. *Total Qual. Manag. Bus. Excell.* **2011**, *22*, 1299–1315. [CrossRef]
22. Trongkaew, P.; Utistham, T.; Reubroycharoen, P.; Hinchiranan, N. Photocatalytic Desulfurization of Waste Tire Pyrolysis Oil. *Energies* **2011**, *4*, 1880–1896. [CrossRef]
23. Yu, H.; Solvang, W.D. Enhancing the Competitiveness of manufacturers through Small-scale Intelligent Manufacturing System (SIMS): A Supply Chain Perspective. In Proceedings of the 6th International Conference on Industrial Technology and Management (ICITM), Cambridge, UK, 7–10 March 2017. [CrossRef]
24. Blanchet, M.; Rinn, T.; Thaden, G.V.; Thieulloy, G.D. INDUSTRY 4.0 The New Industrial Revolution: How Europe Will Succeed. Available online: http://www.iberglobal.com/files/Roland_Berger_Industry.pdf (accessed on 11 July 2018).

25. Erol, S.; Schumacher, A.; Sihn, W. Strategic guidance towards Industry 4.0—A three-stage process model. In Proceedings of the International Conference on Competitive Manufacturing, Stellenbosch, South Africa, 27–29 January 2016.
26. Zhong, R.Y.; Xu, X.; Klotz, E.; Newman, S.T. Intelligent Manufacturing in the Context of Industry 4.0: A Review. *Engineering* **2017**, *3*, 616–630. [CrossRef]
27. Liu, Y.; Xu, X. Industry 4.0 and cloud manufacturing: A comparative analysis. *J. Manuf. Sci. Eng.* **2016**, *139*. [CrossRef]
28. Talkhestani, B.A.; Jazdi, N.; Schlögl, W.; Weyrich, M. A concept in synchronization of virtual production system with real factory based on anchor-point method. *Procedia CIRP* **2018**, *67*, 13–17. [CrossRef]
29. Oviroh, P.; Jen, T.-C. The Energy Cost Analysis of Hybrid Systems and Diesel Generators in Powering Selected Base Transceiver Station Locations in Nigeria. *Energies* **2018**, *11*, 687. [CrossRef]
30. Wan, J.; Tang, S.; Shu, Z.; Li, D.; Wang, S.; Imran, M.; Vasilakos, A.V. Software-Defined Industrial Internet of Things in the Context of Industry 4.0. *IEEE Sensors J.* **2016**, *16*, 7373–7380. [CrossRef]
31. Yang, Y.; Li, X.; Yang, Z.; Wei, Q.; Wang, N.; Wang, L. The Application of Cyber Physical System for Thermal Power Plants: Data-Driven Modeling. *Energies* **2018**, *11*, 690. [CrossRef]
32. Paraskevopoulos, D.; Karakitsos, E.; Rustem, B. Robust capacity planning under uncertainty. *Manag. Sci.* **1991**, *37*, 787–800. [CrossRef]
33. Aghezzaf, B.; Hachimi, M. Generalized invexity and duality in multiobjective programming problems. *J. Glob. Optim.* **2000**, *18*, 91–101. [CrossRef]
34. De Kok, T.G. Capacity allocation and outsourcing in a process industry. *Int. J. Prod. Econ.* **2000**, *68*, 229–239. [CrossRef]
35. Kaplan, R.S. *Cost and Effect: Using Integrated Cost Systems to Drive Profitability and Performance*; Harvard Business School Press: Boston, MA, USA, 1998.
36. Huang, S.-Y.; Chen, H.-J.; Chiu, A.-A.; Chen, C.-P. The application of the theory of constraints and activity-based costing to business excellence: The case of automotive electronics manufacture firms. *Total Qual. Manag. Bus. Excell.* **2014**, *25*, 532–545. [CrossRef]
37. Tsai, W.-H.; Hsu, J.-L.; Chen, C.-H.; Chou, Y.-W.; Lin, S.-J.; Lin, W.-R. Application of ABC in hot spring country inn. *Int. J. Manag. Enterp. Dev.* **2010**, *8*, 152–174. [CrossRef]
38. Oh, S.-C.; Hildreth, A.J. Decisions on Energy Demand Response Option Contracts in Smart Grids Based on Activity-Based Costing and Stochastic Programming. *Energies* **2013**, *6*, 425–443. [CrossRef]
39. Nordhaus, W.D. *Carbon taxes to move toward fiscal sustainability. The Economists' Voice 2.0: The Financial Crisis, Health Care Reform, and More*; Edlin, A.S., Stiglitz, J.E., Eds.; Columbia University Press: New York, NY, USA, 2012; pp. 208–214.
40. Carbon Tax vs Cap-and-Trade: Which Is Better? Available online: https://www.theguardian.com/environment/2013/jan/31/carbon-tax-cap-and-trade. (accessed on 11 July 2018).
41. Descateaux, P.; Astudillo, M.F.; Amor, M.B. Assessing the life cycle environmental benefits of renewable distributed generation in a context of carbon taxes: The case of the Northeastern American market. *Renew. Sustain. Energy Rev.* **2016**, *53*, 1178–1189. [CrossRef]
42. Ouchida, Y.; Goto, D. Environmental research joint ventures and time-consistent emission tax: Endogenous choice of R&D formation. *Econ. Model.* **2016**, *55*, 179–188.
43. Lambertini, L.; Poyago-Theotoky, J.; Tampieri, A. Cournot competition and "green" innovation: An inverted-U relationship. *Energy Econ.* **2017**, *68*, 116–123. [CrossRef]
44. Lee, S.H.; Xu, L. Endogenous timing in private and mixed duopolies with emission taxes. *J. Econ.* **2018**, *124*, 175–201. [CrossRef]
45. The Difference between a Carbon Tax and an Emissions Tax. Available online: https://www.emissionstax.org/what/carbon-tax/ (accessed on 14 July 2018).
46. Wang, M.; Liu, K.; Choi, T.-M.; Yue, X. Effects of carbon emission taxes on transportation mode selections and social welfare. *IEEE Trans. Syst. Man Cybern.* **2015**, *45*, 1413–1423. [CrossRef]
47. Wang, X.; Zhu, Y.; Sun, H.; Jia, F. Production decisions of new and remanufactured products: Implications for low carbon emission economy. *J. Clean. Prod.* **2018**, *171*, 1225–1243. [CrossRef]
48. Yang, L.; Hu, X.; Fang, L. Carbon emissions tax policy of urban road traffic and its application in Panjin, China. *PLoS ONE* **2018**, *13*, e0196762. [CrossRef] [PubMed]

49. Liu, Y.; Xing, J.; Li, Y.; Wang, Y.; Wang, L.; Zheng, B.; Tao, D. Effect of carbon equivalent on thermal and mechanical properties of compacted graphite cast iron. *Int. J. Mater. Res.* **2016**, *31*, 2516–2523. [CrossRef]
50. Wu, B.; Huang, W.; Liu, P. Carbon reduction strategies based on an NW small-world network with a progressive carbon tax. *Sustainability* **2017**, *9*, 1747. [CrossRef]
51. Böhringer, C. Carbon taxes with exemptions in an open economy: A general equilibrium analysis of the German tax initiative. *J. Econ. Manag.* **1997**, *32*, 189–203. [CrossRef]
52. Zheng, J.; Qiao, H.; Wang, S. The Effect of Carbon Tax in Aviation Industry on the Multilateral Simulation Game. *Sustainability* **2017**, *9*, 1247. [CrossRef]
53. Pereira, V.; Costa, H.G. A literature review on lot size with quantity discounts: 1995–2013. *J. Model. Manag.* **2015**, *10*, 341–359. [CrossRef]
54. Alfares, H.K.; Ghaithan, A.M. Inventory and pricing model with price-dependent demand, time-varying holding cost, and quantity discounts. *Comput. Ind. Eng.* **2016**, *94*, 170–177. [CrossRef]
55. Rezaee, M.J.; Yousefi, S.; Hayati, J. A multi-objective model for closed-loop supply chain optimization and efficient supplier selection in a competitive environment considering quantity discount policy. *J. Ind. Eng. Int.* **2017**, *13*, 199–213. [CrossRef]
56. Shahsavar, A.; Zoraghi, N.; Abbasi, B. Integration of resource investment problem with quantity discount problem in material ordering for minimizing resource costs of projects. *Oper. Res.* **2018**, *18*, 1–28. [CrossRef]
57. Demeere, N.; Stouthuysen, K.; Roodhooft, F. Time-driven activity-based costing in an outpatient clinic environment: development, relevance and managerial impact. *Health Policy* **2009**, *92*, 296–304. [CrossRef] [PubMed]
58. Szychta, A. Time-driven activity-based costing in service industries. *Social Sci.* **2010**, *1*, 49–60.
59. Öker, F.; Adıgüzel, H. Time-driven activity-based costing: An implementation in a manufacturing company. *J. Corp. Acc. Financ.* **2016**, *27*, 39–56. [CrossRef]

energies

MDPI

Article

Green Production Planning and Control for the Textile Industry by Using Mathematical Programming and Industry 4.0 Techniques

Wen-Hsien Tsai

Department of Business Administration, National Central University, 300, Jhongda Rd., Jhongli, Taoyuan 32001, Taiwan; whtsai@mgt.ncu.edu.tw; Tel.: +886-3-4267247; Fax: +886-3-4222891

Received: 4 July 2018; Accepted: 3 August 2018; Published: 9 August 2018

Abstract: The textile industry is one of the world's major sources of industrial pollution, and related environmental issues are becoming an ever greater concern. This paper considers the environmental issues of carbon emissions, energy recycling, and waste reuse, and uses a mathematical programming model with Activity-Based Costing (ABC) and the Theory of Constraints (TOC) to achieve profit maximization. This paper discusses the combination of mathematical programming and Industry 4.0 techniques to achieve the purpose of green production planning and control for the textile industry in the new era. The mathematical programming model is used to determine the optimal product mix under various production constraints, while Industry 4.0 techniques are used to control the production progress to achieve the planning targets. With the help of an Industry 4.0 real-time sensor and detection system, it can achieve the purposes of recycling waste, reducing carbon emission, saving energy and cost, and finally achieving a maximization of profit. The main contributions of this research are using mathematical programming approach to formulate the decision model with ABC cost data and TOC constraints for the textile companies and clarifying the relation between mathematical programming models and Industry 4.0 techniques. Managers in the textile companies can apply this decision model to achieve the optimal product-mix under various constraints and to evaluate the effect on profit of carbon emissions, energy recycling, waste reuse, and material quantity discount.

Keywords: activity-based costing (ABC); mathematical programming; textile industry; green manufacturing; Industry 4.0; carbon emissions

1. Introduction

The traditional textile industry has always been labor-intensive and highly polluting [1]. In the past, the textile industry production process was very complicated. Many detailed actions needed to be executed; this caused a production trend of large volume but less variety [2]. Now people are paying more attention to product quality and unique requirements due to technology and people's living habits change. As a result, the manufacturing industry has moved toward customized production [3,4]. Technology has also been strengthened in response to such changes. Recently, many manufactures have been influenced by Industry 4.0. They are not only optimizing the manufacturing processes, but also effectively controlling industrial pollution with the assistance of data maintenance and monitoring [5,6].

The Industry 4.0 smart manufacturing concept was first presented at the Hanover Fair (Germany) in 2011, where it received great attention from different industry sectors in various nations [7,8]. Some researchers claim that the main essentials of Industry 4.0 are the integration of system components and the digitalization of manufacturing/service operations [9,10]. However, most companies in the textile industry are hesitant to introduce Industry 4.0 because of serious concerns about the uncertain financial benefits and lack of professional knowledge. Thus, a Textile Learning Factory 4.0 was set up and open in March 2017 by the Institut für Textiltechnik der RWTH Aachen University in Aachen, Germany [9]. The Textile Learning Factory 4.0 is a building with two implementation levels: Level 1 demonstrates the current state operation (lean production) and Level 2 demonstrates the future state operation (Industry 4.0) for the textile industry. The purpose of the factory is to provide a real-life demonstration and learning environment to teach textile companies how to plan the digital transformation to Industry 4.0. The factory is also used as a base for piloting new digital solutions for the textile industry. This may promote the applications of Industry 4.0 in textile companies around the world.

Industry 4.0 will use various sensor systems to real-timely monitor and collect the operations data of production systems and return immediate responses to various problems that may arise during production by using the results of real-timely analyzing big data. Finally, mass customization can be achieved by fine-tuning or adjusting the production process differently with the customer needs [7,11]. Industry 4.0 has been applied in various industries, including textile [9,10,12,13] and other process industries [8,14,15]. This research focuses on the textile industry.

In response to the current situation, this research uses the Activity-Based Costing (ABC) method to enhance the accuracy of cost estimates [16], in conjunction with the Theory of Constraints (TOC) to consider the possible constraints of production and sales, to achieve maximum profit under various constraints [17]. ABC and TOC are combined to deal with the problem of short-term operations and long-term cost management [18,19]. ABC uses two-stage cost assignment to compute the accurate costs of cost objects. ABC calculates the costs of activities in the first stage and the costs of the cost objects in the second stage [20]. Thus, textile processes/activity costs can be achieved in the first stage. TOC uses five steps to identify and eliminate the bottlenecks one by one to increase business performance. TOC can be used in the mathematical programming models for various kinds of problem [21–27]. In this research, TOC is used to form the various production and sales constraints. In summary, the mathematical programming model with ABC costs and TOC constraints is used for production planning to derive maximum profit [28–30]. In the literature, the mathematical programming approach has been applied in the textile industry for inventory [31], scheduling [32], and product-mix problems [33,34].

In this paper, Industry 4.0, activity-based costing and environmental issues are combined to improve the efficiency of green production. This paper considers the environmental issues of carbon emissions, energy recycling, and waste reuse, and uses a mathematical programming model with Activity-Based Costing (ABC) and the Theory of Constraints (TOC) to achieve profit maximization. This paper discusses the combination of mathematical programming and Industry 4.0 techniques to achieve the purpose of green production planning and control for the textile industry in the new era. The mathematical programming model is used to determine the optimal product mix under various production constraints, while Industry 4.0 techniques are used to control the production progress to achieve the planning targets. With the help of an Industry 4.0 real-time sensor and detection system, it can achieve the purposes of recycling waste, reducing carbon emission, saving energy and cost, and finally achieving a maximization of profit. With the approach proposed in this paper, corporations can simultaneously control the impact on the environment and ensure their profitability.

The remainder of this paper is organized into five sections. Section 2 describes the research background of this research. Section 3 develops the green production planning model under ABC. An illustrative case study is presented in Section 4 to demonstrate how to apply the model proposed in this paper and to conduct the sensitivity analysis of the direct material discount. Manufacturing execution system for production control under Industry 4.0 in the textile industry is explained in Section 5. Finally, discussions and conclusions are presented in Sections 6 and 7, respectively.

2. Research Background

2.1. Brief of Industry 4.0

From the eighteenth century up to today, there have been four industrial revolutions. The first industrial revolution started with the invention of the steam engine and led to a manual-production industry. The second included the use of electrical energy and assembly lines for mass production. The third was marked by the emergence of the computer, IT systems and automation. The fourth industrial revolution is called Industry 4.0; it combines the new technologies of Cyber-Physical Systems (CPS), Internet of Things (IoT), big data and cloud computing to enhance production efficiency [35,36]. Industry 4.0 is a new developmental stage rather than an extension of the third industrial revolution. It is especially suited to mass customization production. By means of CPS and IOT, machines can communicate with each other and send data in real time to systems and people [37,38].

Manufacturing operations have been transformed from traditional production into intelligent production. Industry 4.0 faces increased competitiveness and more complex challenges in dealing with the status of demographic changes, production processes, resources and the environment [39,40]. Hence, Industry 4.0 will solve problems with the use of CPS and IoT. Enterprises need vertical integration by the use of networks. CPS can connect resources and products through large data analysis and intelligent sensing technology to automate the monitoring of plant inventory, demand and equipment failure, and is well suited for maintenance management. At the same time, all the processing stages of the production process are recorded, including product variability, order modification, quality instability or machine failure, and other factors. As a result, material wear can be more effectively controlled, and waste in the production process will be reduced, thus increasing efficiency. However, enterprises also need horizontal integration of the production process, from purchasing, production and even sales, or from suppliers to the company, and finally to the customer. It is not only applicable to the production stage, but also extends to commodity development, subscriptions, plans, assembly and distribution, to ensure quality, time control, risk, price and environmental protection; in so doing, other factors in the entire value chain of each link are immediately controlled [5,38,41].

For the textile industry, Küsters et al. claimed that there are the following Industry 4.0 applications: (1) Status monitoring for real-time process parameters; (2) Product shadow for tracking the product production cycle; (3) Digital assistant system for instructing operators; (4) Digital performance management for problem solving by providing the digital Key Performance Indicator (KPI) board; (5) Automated Guided Vehicles (AGV) for material handling; (6) YET for analyzing Yield, Energy, and Throughput; (7) iCycle time for real-time line balancing; (8) Real-time WIP for real-time tracking of the Work-In-Process; (9) Predictive maintenance for breakdown prevention by the analysis of Big Data; (10) Smart routing and work station setup for self-adjusting work stations; (11) In-line Quality Control (QC) and adaptive machining by using the parameter or image analysis of Big Data; (12) Human-machine collaboration; (13) 3D printing for tailored production components; and so on [9].

2.2. The Influence of Industry 4.0 and Green Production on the Textile Industry

The textile industry is one of the major sources of environmental pollution in the industrial world [42]. The main environmental problems are those attributable to carbon emissions and the discharge of untreated wastewater. Recently, there has been an increased awareness of the need for environmental protection and green production. It is therefore obligatory that the textile production process must conform to the required standards of emissions.

The textile industry's Industry 4.0 requires the digital factory. Its core technology is Cyber-Physical Systems (CPS). This is the real equipment's computer software mapping. Every step of the production process is compiled in the virtual world to establish a high simulation of digitalism in the virtual and real worlds; this process will be simulated and optimized on the materials, products and factories in the real physical world. Next, the Internet of Things (IoT) is used to gather information by sensor, Programmable Logic Controller (PLC), visual equipment, etc., to transfer information through the

Internet to accomplish Human-Computer Interaction (HCT) [43–45]. Then, using the Internet to transmit information, people or machines can be operate based on the data collected for analysis and decision-making, with objects controlled by connection to the Internet for the implementation of the tasks and decision-making; objects connected with the Internet can even communicate and cooperate with each other directly.

The textile industry is a labor-intensive industry, and its complex manufacturing processes and long supply chain characteristics mean that large variety and small volume customized production requires an "intelligent" factory [46,47]. Industry 4.0, as it relates to the textile industry, has three characteristics: production network, Cyber-Physical Systems (CPS) and Internet of Things (IoT). In the smart factory, CPS is the real equipment's software mapping in the computer. Every step of the production process may be compiled in the virtual world to establish high level digital simulation of in the virtual and real world. The process will simulate and optimize the materials, products and factories of the real physical world. Manufacture execution systems (MES) will manage the master data, situation and maintenance records of equipment, like helping suppliers in the production value chain obtain and exchange real-time production information, and ensure that all of the components supplied by the supplier will arrive at the right time in the right order [48,49].

Industry 4.0 is more than machine automation; it involves the whole production process. First, there is real-time monitoring, digitizing and the setting up of the cyber-physical systems for the plant's machinery. The manufacturer sets the sensors on the machines and concatenates all of them via a network to continuously collect production data and send the data to the enterprise resource planning system (ERP) [50–54]. Then there is the gathering and organizing of data to be sent to the cloud platform to analyze the big data [37,55]. Cloud platforms serve just like human vertebra; they are vertical integrations of systems from upstream to downstream. According to customer demand, material supply may be changed via machine-to-machine conversation and real-time status monitoring [55–57]. In the past, production was limited by the extant technology. Since different lengths of cloth consume almost the same amount of dye and water, the product yield is less and the cost is higher. While it was very difficult to achieve large variety and small volume manufacturing, the smart factory solves the problem of the cost of customized production, matching the cost of mass production [58].

3. Green Production Planning Model under ABC by Using Mathematical Programming

3.1. A Production Process for a Typical Textile Company

The traditional textile industry is a labor-intensive. Every process needs a lot of workers to complete the product [46,47]. The product is a result of yarn spinning, weaving, dyeing and finishing. There are three main processes involved in fabric production as shown in Figure 1. The first step is yarn production made from fibers by drawing and twisting bobbins together and then spinning it to turn in into yarn. The next step is the weaving. The weaving machine requires two sets of the yarn: warp and weft set, and the two sets of yarn are then joined together by weaving. The fabric made from the looms without any further processing is called greige fabric. Next, the fresh greige fabric is stained, but as it still contains impurities, it must be washed. The fabric is then dyed in a high temperature, high pressure dyeing machine. Finally, the fabric is given a special tactile impression and function, like smooth, suede, waterproof, wrinkle-free, non-felting, flame retardant, etc. Currently, as a result of the development of science and technology, enterprises can vertically integrate all aspects of the textile industry, so that the quantity and quality can be self-controlled.

In addition, dyeing machines require high heat, often produced by coal in the dyeing and finishing process, but burning coal is harmful to the environment, A plant will mix the waste cinder with cement to produce the light texture and good permeability of ecological bricks (eco-brick). The production of eco-bricks is not only beneficial to the environment, but also saves costs by avoiding the expenses incurred by shipping out the cinders.

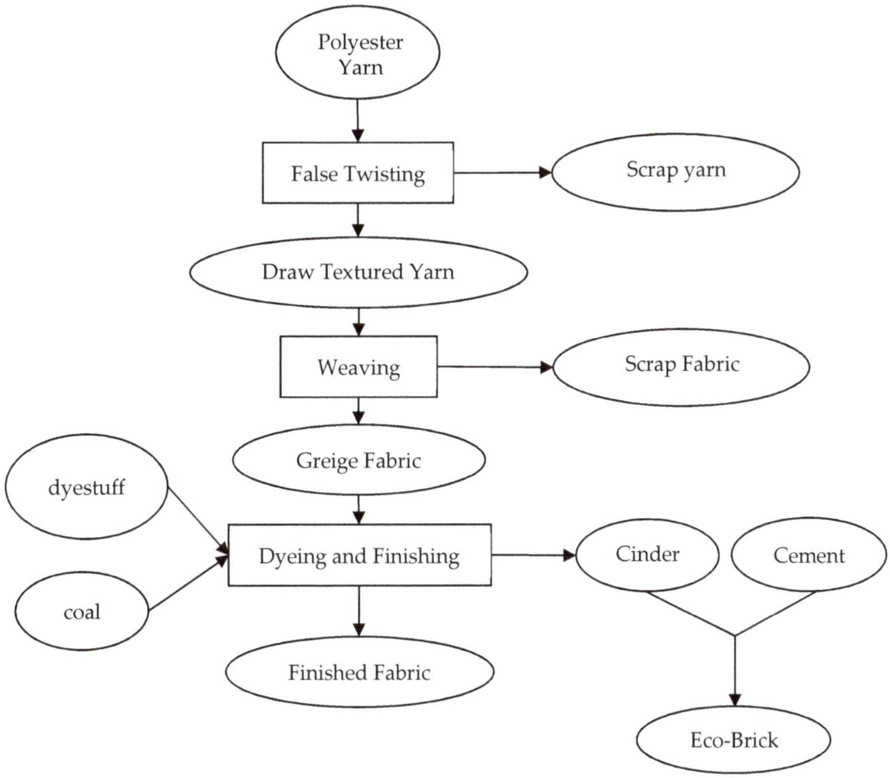

Figure 1. The process of producing textiles.

3.2. Assumptions

To develop vertical integration following the Industry 4.0 decision model and without loss of generality (WLOG), this paper discuss the classic textile production process in the following. One unit material processed in the process of twisting will produce e_1 units of draw textured yarn and $(1 - e_1)$ units of scrap include false yarn. Drawn textured yarn can not only be sold, but may be further processed; scrap yarn will be sold at a lower price. After further processing by weaving, it will produce e_2 units of greige fabric and $(1 - e_2)$ units of scrap fabric. Greige fabric can be sold or further processed. The final process is that of dyeing and finishing, and then the finished fabric is produced.

Except for assumptions of the production process, the green production of the Industry 4.0 decision model purposed has the following assumptions. First, the activities are categorized as unit-level and batch-level; resource drivers and activity drivers connect with ABC and Theory of Constraints (TOC) [23–25]. Second, the model does not involve capacity expansion and outsourcing of resources and materials, such as machine hours. Third, the labor working hours can be enhanced through overtime work with higher wage rates conforming to government regulations. Fourth, when the material quantity exceeds the threshold quantities, the purchaser obtains a piecewise price discount for all material resources. Fifth, the cost of CO_2 emissions is dependent on emission quantities that are taxed at different rates as a piecewise variable cost. Sixth, water and heat resulting from energy recycling save costs. Seventh, fixed costs include machine costs, depreciation and other required production costs.

3.3. Objective Function

The objective function of the green production planning model under ABC and Industry 4.0 is as follows:

Maximize π = Total Revenue of main product + Revenue of by product − Total material cost − Total direct labor cost − Carbon tax − Energy recycling cost saving − Total other fixed cost

$$\pi = [P_1X_{11}+P_2X_{21}+P_3X_{22}+\beta_1(1-e_1)M+\beta_2(1-e_2)X_{12}+\beta_3e_3X_{22}] - [(C_1M+C_2m_{22d})+(C_3m_{22c}+C_4m_{22m})] - [L_0+\eta_1 r_{ot}(Q_1-G_0)] - [\delta_1 r_{c1}(TC_1-GC_0)+\delta_2 r_{c2}(TC_2-GC_0)]+(C_5RE_h+C_6RE_w) - F, \quad (1)$$

where

P_i	The unit selling price of i product
X_{11}	The selling quantities of drawn textured yarn
X_{12}	The future processing quantities of drawn textured yarn
X_{21}	The selling quantities of greige fabric
X_{22}	The selling quantities of finished fabric
β_i	The unit selling price of i by-product
e_1	The input-output coefficient from POY to drawn textured yarn and scrap yarn
e_2	The input-output coefficient from textured yarn to greige fabric and scrap fabric
e_3	The input-output coefficient of Eco-Brick
C_1, C_2, C_3, C_4	The unit material cost of M, m_{22d}, m_{22c} and m_{22m}
C_5, C_6	The unit recycle saving cost of heat and water
M	The material quantities of POY
m_{22d}	The material quantities of dye
m_{22c}	The material quantities of cinder
m_{22m}	The material quantities of cement
L_0	The direct labor cost of available normal hours
η_0, η_1	Special ordered sets of type 1 (SOS1) of 0/1 variables, where only one variable will be 1
r_{ot}	The wage rate of the total direct labor hours in overtime work situations
Q_1	The overtime working hours (extend working hours to 2 hours per worker)
G_0	The limit of the normal direct labor hour
$\delta_0, \delta_1, \delta_2$	Special ordered sets of type 1 (SOS1) of 0/1 variables, where only one variable will be 1
r_{c1}	The carbon tax rate of available normal CO_2 emission quantities
r_{c2}	The carbon tax rate that produces excessive CO_2 emission quantities
TC_1	The total CO_2 emission quantities in normal production
TC_2	The total CO_2 emission quantities that produces excessive CO_2 emission quantities
GC_0	No charge for CO_2 emission quantities
RE_h, RE_w	Green energy recycling saving quantities of heat and water
F	Fixed cost

3.4. Unit-Level Direct Labor Cost Function

Assume that overtime work can expand the direct labor resources, and labor is used for handling the material and products. The total cost function of direct labor is the piecewise linear function, as shown in Figure 2. The available normal direct labor hours are G_0 and the direct labor hours can be expanded to G_1 with the total direct labor cost respectively being L_0 and $L_0+m_1G_1$ at G_0 and G_1. As well as the handling and setup, direct labor is also needed to transfer the products to the next plant. The setup of direct labor hours incudes the time required to replace material or reset programs during each batch start in the dyeing process, like the setup of dyestuff. The total direct labor cost is $L_0+\eta_1 r_{ot}(Q_1-G_0)$ in Equation (1), and the associated constraints are shown in Equations (2)–(5):

$$l_1M+l_2X_{12}+l_3X_{22}+\mu_{01}B_{01}+\mu_{12}B_{12}+\mu_{10}B_{10}+\mu_{23}B_{23}+\mu_{20}B_{20}+\mu_{30}B_{30}+\omega\ BS = Q_0+Q_1 \quad (2)$$

$$0 \le Q_0 \le \eta_0 G_0, \quad (3)$$

$$\eta_1 G_0 < Q_1 \leq \eta_1 G_1, \quad (4)$$

$$\eta_0 + \eta_1 = 1, \quad (5)$$

where:

l_1	The direct labor hours required by M
l_2	The direct labor hours required by X_{12}
l_3	The direct labor hours required by X_{22}
$\mu_{s,e}$	The direct labor hours of the handling of each batch-level activity from start to finish
$B_{s,e}$	The product of shipping from start to finish in terms of quantities needed at each batch-level
ω	Each batch of direct labor hours of setup
BS	The setup product in terms of quantities of each batch-level
Q_0	The total direct labor hours in the normal working hour
Q_1	The overtime working hours (extend working hours to 2 h per worker)
G_0	The limit of the normal direct labor hour
G_1	The limit of the overtime work direct labor hour
η_0, η_1	Special ordered sets of type 1 (SOS1) of 0/1 variables.

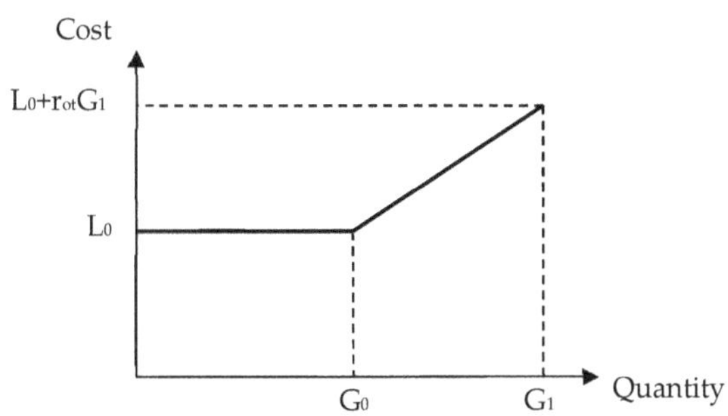

Figure 2. Direct labor cost function.

3.5. Batch-Level Activity Cost Function for Material Handling and Setup Activities

Batch-level activity in the production includes material and products handling from the start to finish; when the goods are finished, they will be shipped back to the plant awaiting sale. The other activity is the material dyestuff setup:

$$M \leq \sigma_{01} B_{01}, \quad (6)$$

$$X_{12} \leq \sigma_{12} B_{12}, \quad (7)$$

$$X_{11} + (1 - e_1)M \leq \sigma_{10} B_{10}, \quad (8)$$

$$X_{22} \leq \sigma_{23} B_{23}, \quad (9)$$

$$X_{21} + (1 - e_2)X_{12} \leq \sigma_{20} B_{20}, \quad (10)$$

$$X_{22} + e_3 X_{22} \leq \sigma_{30} B_{30}, \quad (11)$$

$$X_{22} \leq \lambda\, BS, \quad (12)$$

where

$\sigma_{s,e}$	The quantity of handling of each batch-level activity from start to finish
λ	The quantity of setup of batch-level activity

3.6. Carbon Tax Function

Global warming is an important issue. Reducing the greenhouse effect is a primary concern, and one of the main factors affecting the greenhouse effect is that of carbon dioxide emissions [1,59–61]. The textile industry is the one of the major industries guilty of excessive CO_2 emissions in global manufacturing, especially in the dyeing and finishing process [62–64]. Assume that CO_2 emissions are taxed at different rates as a piecewise variable cost as shown in Figure 3. When CO_2 emissions are well controlled according to particular criteria, there is free carbon tax in TC_0. The standard of the CO_2 emissions is TC_1, and the tax will be higher than the standard when CO_2 emissions exceed GC_1 and become TC_2. The total carbon tax cost is $\delta_1 r_{c1}(TC_1 - GC_0) + \delta_2 r_{c2}(TC_2 - GC_0)$ as shown in Equation (1), and the associated constraints are Equations (13)–(17):

$$q_c X_{22} = TC_0 + TC_1 + TC_2, \tag{13}$$

$$0 \leq TC_0 \leq \delta_0 GC_0, \tag{14}$$

$$\delta_1 GC_0 < TC_1 \leq \delta_1 GC_1, \tag{15}$$

$$\delta_2 GC_2 < TC_2, \tag{16}$$

$$\delta_0 + \delta_1 + \delta_2 = 1, \tag{17}$$

where:

q_c	The total CO_2 emission quantities in the dyeing and finishing process
TC_0	The total CO_2 emission quantities where there is no charge in standard range
TC_1	The total CO_2 emission quantities in normal production
TC_2	The total CO_2 emission quantities that produces excessive CO_2 emission quantities
GC_0	No charge for CO_2 emission quantities
GC_1	Carbon tax cost of normal capacity
GC_2	Carbon tax cost for over emission capacity
δ_0, δ_1, δ_2	Special ordered sets of type 1 (SOS1). The summation of 0/1 variables is 1

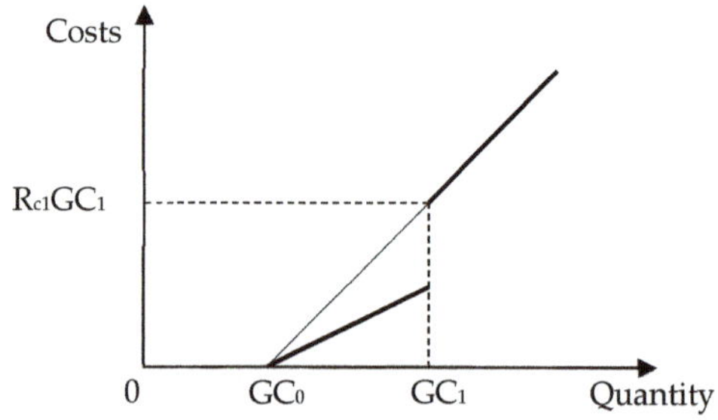

Figure 3. Direct Carbon tax function.

3.7. Energy Recycling

The textile industry, during the dyeing and finishing processes, needs to generate a lot of heat. It generally uses heat from burning coal, steam boilers, and medium heat boilers and heating furnaces. The use of its high temperature and air heat transfer can enhance the combustion air temperature, and reduce the exhaust temperature to achieve waste heat recovery purposes [65,66]. However, the textile industry is also a heavy consumer of water [42]. It is therefore evident that water recycling and reuse is also an important wastewater treatment issue [67]. Heat and water are used mainly in the dyeing and finishing process [68]; thus, energy recycling of heat and water will vary with the production quantity of the finished fabric. The energy recycling cost saving for heat and water are C_5RE_h and C_6RE_w, respectively, in Equation (1), and the associated constraints are Equations (18) and (19):

$$RE_h = \rho_1 \times X_{22} \tag{18}$$

$$RE_w = \rho_2 \times X_{22} \tag{19}$$

where:

ρ_1 The relation coefficient between energy recycling of heat and X_{22}
ρ_2 The relation coefficient between energy recycling of water and X_{22}

3.8. Input-Output Relationship

The amount of material input and product output differs because the material suffers loss in the production process, such as in the weaving process. Fabric is made of yarn, and some scrap from the fabric articles will remain that is the input-output coefficient. X_{12} is the quantity of yarn, e_2X_{12} is the quantity of fabric and $(1-e_2)X_{12}$ is the quantity of scrap fabric, as below:

$$X_1 - e_1M = 0, \tag{20}$$

$$X_1 = X_{11} + X_{12}, \tag{21}$$

$$X_2 - e_2X_{12} = 0, \tag{22}$$

$$X_2 = X_{21} + X_{22}, \tag{23}$$

$$M = e_1M + (1-e_1)M, \tag{24}$$

$$X_{12} = e_2X_{12} + (1-e_2)X_{12} = 0, \tag{25}$$

$$BP_1 = (1-e_1)M, \tag{26}$$

$$BP_2 = (1-e_2)X_{12}, \tag{27}$$

$$BP_3 = e_3 \times X_{22}, \tag{28}$$

$$m_{22d} = \vartheta_1 \times X_{22}, \tag{29}$$

$$m_{22c} = \vartheta_2 \times X_{22}, \tag{30}$$

$$m_{cin} = \vartheta_3 \times m_{22c}, \tag{31}$$

$$m_{cem} = \vartheta_4 \times m_{cin}, \tag{32}$$

where

BP_i The quantities of i byproduct
ϑ_1 The relation coefficient between m_{22d} and X_{22}
ϑ_2 The relation coefficient between m_{22c} and X_{22}
ϑ_3 The relation coefficient between m_{cin} and X_{22}
ϑ_4 The relation coefficient between m_{cem} and X_{22}

Textile manufacturing process with the symbols of quantity variables mentioned in Equations (20)–(32) is shown as Figure 4.

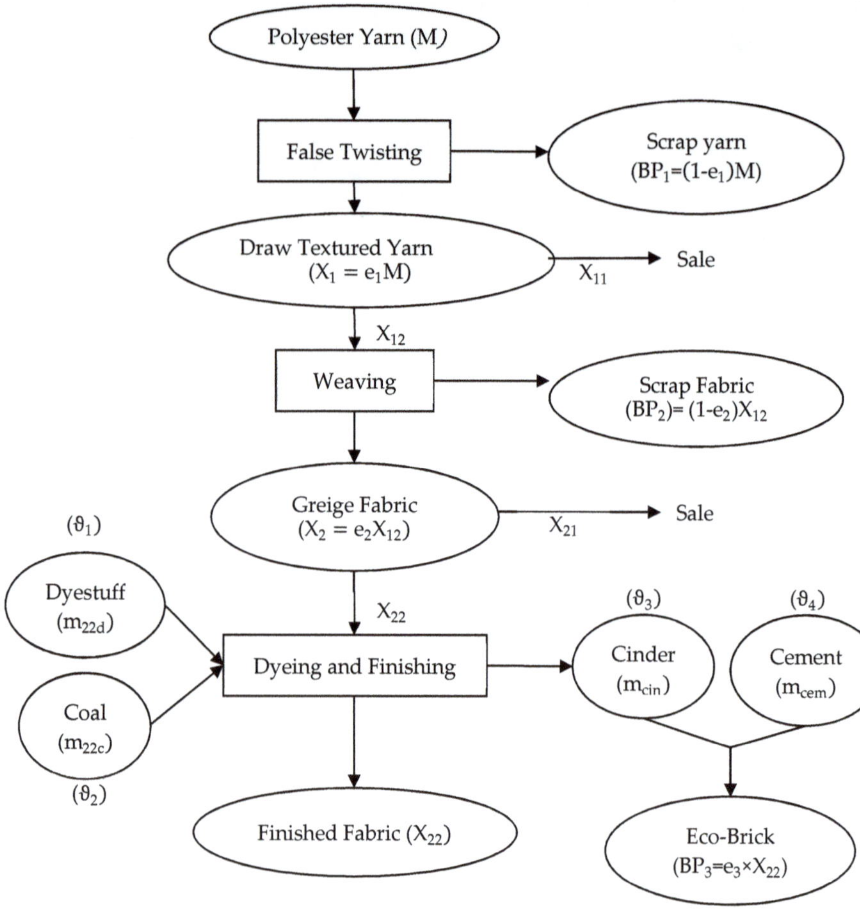

Figure 4. Textile manufacturing process with the symbols of quantity variables.

3.9. Other Sale and Production Constraints

There are three kinds of machines to generate products. A twisting machine is used to false twist the yarn, a loom is used to weave the fabric, and a dyeing and finishing machine is used to dye the fabric and give the fabric other special qualities. Those are limiting machines in so far as their efficiency is limited by the material, operator, and machine maintenance time [69]. Therefore, producing a product requires limited machine hours, including twisting machine hours (h_1), weaving machine hours (h_2) and dyeing and finishing machine hours (h_{22}), as expressed in the following Equations (23)–(25):

Machine hour constraints:

$$h_1 M \leq H_1, \tag{33}$$

$$h_2 X_{12} \leq H_2, \tag{34}$$

$$h_{22} X_{22} \leq H_{22}, \tag{35}$$

where:

h_1	the resources of machine hours of false twisting
h_2	the resources of machine hours of weaving
h_{22}	the resources of machine hours of dyeing and finishing
H_1	The limited machine hours of false twisting
H_2	The limited machine hours of weaving
H_{22}	The limited machine hours of dyeing and finishing

4. Illustrative Case

4.1. Illustrative Data and Optimal Decision Analysis

The model assumes that the textile factory is a vertically integrated plant that uses one material to produce three different products, which are similar to the joint-product model. After the production of false twist yarns with a raw yarn, part of the drawn textured yarn is sold while the rest continues to be processed into greige fabric. Part of the embryo fabric is sold, the rest undergoes dyeing and finishing procedures, and finally a high price finished fabric is produced. The associated costs in this model include: (1) Unit-level activity: material costs and director labor cost, (2) Batch-level activity: material and product handling and setup of the dyeing process, (3) environment-level activity: carbon tax and energy recycling cost, (4) Fixed cost: To more effectively trace and identify cost in the production, our model considers related production costs which include depreciation, land, plant and equipment as fixed cost to accurately evaluate production. This model can also be used to help corporations find the optimal solution to effectively decrease production cost by means of an optimal product portfolio and resource allocation.

The production data for production planning model for the production process of Figure 1 are presented in Tables 1 and 2. In this model, assume that there are three products and three by-products that can be sold: draw textured yarn (X_{11}), greige fabric (X_{21}), finished fabric (X_{22}), scrap yarn (β_1), scrap fabric (β_2) and eco-brick (β_3). The unit prices of products are 97,000, 135,000, 202,500, 570, 2100 and 2300 NTD, respectively, per ton. The main purpose of this model is to accurately estimate production to ascertain cost objectives and operation of production to help corporations control related costs and increase profit.

Table 1. Example data.

Activity Resources			Process 1		Process 2		Process 3		Capacity
			Draw Textured Yarn	Scrap yarn	Greige Fabric	Scrap Fabric	Finished Fabric	Eco-Brick	
Price of per selling unit			$97,000	$570	$135,000	$2,100	$202,500	$2,300	
Production coefficient		e_i	0.96		0.95			0.14	
		$(1 - e_i)$		0.04		0.05			
Direct material cost		M, m_{22d}	$65,000				$7,000		
		m_{22c}, m_{cem}					$1,800	$1,500	
Machine hours constraint	Activity driver								
Machine 1	Machine hrs.	h_i	3						120,000
Machine 2					4				100,000
Machine 3							5		130,000
		l_i	3		2		1		
Direct Labor constraint				*Carbon tax constraint*		q_c	7		
Cost	$L_0 = 47,840,000$		$L_1 = 101,660,000$	Cost		$PC_0 = 0$	$PC_1 = 170,000,000$	$PC_2 = 270,000,000$	
Labor hours	$G_0 = 368,000$		$G_1 = 598,000$	Emission Quantities		$GC_0 = 50,000$	$GC_1 = 170,000$	$GC_2 = 180,000$	
Wage Rate	$w_{r0} = \$130/h$		$w_{r1} = 170/h$	Tax Rate		$T_{r0} = \$0/ton$	$T_{r1} = \$1,000/ton$	$T_{r2} = \$1,500/ton$	

Table 2. Example data for batch-level activities.

Batch-Level Activity		Starting Location (s)	End Location (e)	Draw Textured Yarn	Greige Fabric	Finished Fabric
Handling	σ_{se}, μ_{se}	0	1	5, 2		
		1	2	1, 1		
		1	0	0.5, 1		
		2	3		1, 1	
		2	0		2, 1	
		2	3			3, 2
Set-up	λ, ω					3, 2

4.2. Optimal Solution Analysis

The green production planning model for the example data is shown in Table 3, which is a mixed integer programming (MIP) model, and the optimal solution is shown in Table 4, which is obtained by using Lingo 16.0.

In Table 4, the optimal solution in the model indicates the optimal product portfolio where the profit is 3,361,133,000 NTD from three products and three by-products. The total revenue is (4,371,466,156 NTD) which is comprised of three products: Draw Textured Yarn (1,327,368,370 NTD), Greige Fabric (69,285,713 NTD) and Finished Fabric (2,355,713,870 NTD). Product quantities are (13,684.21/ton, 714.29/ton, 24,285.71/ton) m_{22d} (15,785.71), m_{22c} (2428.57), and M (41,666.67). Besides, the by-product revenue of eco-brick is (3,833,334 NTD) ($\beta_3 e_3 X_{22}$) and the energy recycling cost saving for heat and water is (1,384,286 NTD) ($C_5 RE_h$) and (2,477,143 NTD) ($C_6 RE_w$), respectively.

Three kinds of machine hours relate to false twisting (120,000), weaving (100,000), and dyeing and finishing (121,428.55). Carbon tax costs are (120,000,000 NTD), and CO_2 emission quantities are (1200/ton). Therefore, the mathematical programming in this model combining ABC and TOC, as well as the constraint of carbon emissions, can reduce production costs and enhance profit through the efficient distribution of resources.

4.3. Sensitivity Analysis of the Quantity Discount of Direct Material

Considering the quantity discount of direct material, this paper divided the material purchase discount pricing into high, medium and low degree levels [70,71]. This study used three segments of piecewise linear function, as shown in Figure 5. In Equations (35)–(40), this paper replaces the former material cost (C_1M) to become three segments of piecewise discount ($R_1 \cdot MT_1 + R_2 \cdot MT_2 + R_3 \cdot MT_3$). For example, when the amount of material is more than MQ_1, the material cost would become R_1 and the total material cost would be ($R_2 MT_2$).

Table 3. Green production planning model for the example data.

$$\text{Max } \pi = (97{,}000 \times X_{11} + 135{,}000 \times X_{21} + 202{,}500 \times X_{22} + 570 \times 0.04 \times M + 2100 \times 0.05 \times X_{12} + 2300 \times 0.14 \times X_{22}) - (65{,}000 \times M + 7000 \times m_{22d} + 1800 \times m_{22c} + 1500 \times m_{cem}) - [47{,}840{,}000 + (170 \times Q_1 - 62{,}560{,}000 \times \eta_1)] - (1000 \times TC_1 - 50{,}000{,}000 \times \delta_1 + 1500 \times TC_2 - 75{,}000{,}000 \times \delta_2) + (60 \times RE_h + 120 \times RE_w) - 500{,}000$$

Subject to Input-Output Relationship:

$$X_1 - 0.96 \times M = 0$$
$$X_1 = X_{11} + X_{12}$$
$$X_2 - 0.95 \times X_{12} = 0$$
$$X_2 = X_{21} + X_{22}$$
$$BP_1 = 0.04 \times M$$
$$BP_2 = 0.05 \times X_{12}$$
$$BP_3 = 0.14 \times X_{22}$$
$$m_{22d} = 0.65 \times X_{22}$$
$$m_{22c} = 0.1 \times X_{22}$$
$$m_{cin} = 0.2 \times m_{22c}$$
$$m_{cem} = 6 \times m_{cin}$$
$$RE_h = 0.95 \times X_{22}$$
$$RE_w = 0.85 \times X_{22}$$

Subject to Direct Labor:

$$3 \times X_1 + 2 \times X_2 + 1 \times X_{22} + 2 \times B_{01} + 1 \times B_{12} + 1 \times B_{10} + 1 \times B_{23} + 1 \times B_{20} + 2 \times B_{30} + 2 \times BS - Q_0 - Q_1 \leq 0$$
$$Q_0 \leq \eta_0 \times 368{,}000$$
$$\eta_1 \times 368{,}000 < Q_1$$
$$Q_1 \leq \eta_1 \times 598{,}000$$
$$\eta_0 + \eta_1 = 1$$

Subject to Machine Hour:

$$3 \times X_1 - 120{,}000 \leq 0$$
$$4 \times X_2 - 100{,}000 \leq 0$$
$$5 \times X_{22} - 130{,}000 \leq 0$$

Subject to Batch-Level:

$$M \leq 5 \times B_{01}$$
$$X_{12} \leq 1 \times B_{12}$$
$$X_{11} + 0.04 \times M \leq 0.5 \times B_{10}$$
$$X_{22} \leq 1 \times B_{23}$$
$$X_{21} + 0.05 \times X_{12} \leq 2 \times B_{20}$$
$$X_{22} + 0.14 \times X_{22} \leq 3 \times B_{30}$$
$$X_{22} \leq 3 \times BS$$

Subject to Carbon Tax Function:

$$7 \times X_{22} \leq TC_0 + TC_1 + TC_2$$
$$TC_0 \leq \delta_0 \times 50{,}000$$
$$50{,}000 \times \delta_1 < TC_1$$
$$TC_1 \leq \delta_1 \times 170{,}000$$
$$\delta_2 \times 170{,}000 < TC_2$$
$$TC_2 \leq \delta_2 \times 0.1E + 10$$
$$\delta_0 + \delta_1 + \delta_2 = 1$$

Table 4. The optimal solution in mixed integer programming model (MIP).

$\pi = 3{,}361{,}133{,}000$	$X_1 = 40{,}000$	$X_{11} = 13{,}684.21$	$X_{12} = 26{,}315.79$
$X_2 = 25{,}000$	$X_{21} = 714.2857$	$X_{22} = 24{,}285.71$	$BP_1 = 1666.667$
$BP_2 = 1315.789$	$BP_3 = 2695.714$	$M = 41{,}666.67$	$m_{22d} = 15{,}785.71$
$m_{22c} = 2428.571$	$m_{cin} = 485.7143$	$m_{cem} = 2914.286$	$Q_0 = 0$
$Q_1 = 327{,}455.7$	$\eta_0 = 0.746149$	$\eta_1 = 0.9253851$	$TC_0 = 0$
$TC_1 = 170{,}000$	$TC_2 = 0$	$\delta_0 = 0$	$\delta_1 = 1$
$\delta_2 = 0$	$RE_h = 23{,}071.43$	$RE_w = 20{,}642.86$	$B_{01} = 8334$
$B_{12} = 26{,}316$	$B_{10} = 30{,}702$	$B_{23} = 24{,}286$	$B_{20} = 13{,}158$
$B_{30} = 9229$	$BS = 8334$		

Our corporation in this model obtains the highest discount pricing. It indicates that the material cost would be from R_2 to R_3; the quantities that a plant buys and the total cost of materials are shown as (MT_3R_3):

$$\begin{aligned}\pi = {} & P_1X_{11}+P_2X_{21}+P_3X_{22}+\beta_1(1-e_1)M+\beta_2(1-e_2)X_{12}+\beta_3\cdot e_3X_{22} \\ & -[(R_1\cdot MT_1+R_2\cdot MT_2+R_3\cdot MT_3)+C_2m_{22d}+C_3m_{22c}+C_4m_{22m}] \\ & -[L_O+\eta_1 r_{ot}(Q_1-G_O)]-[\delta_1 r_{c1}(TC_1-GC_O)+\delta_2 r_{c2}(TC_2-GC_O)]+C_5Re-F,\end{aligned} \quad (36)$$

$$M = MT_1+MT_2+MT_3, \quad (37)$$

$$0 \le MT_1 \le \varphi_1 MQ_1, \quad (38)$$

$$\varphi_2 MQ_1 < MT_2 \le \varphi_2 MQ_2, \quad (39)$$

$$\varphi_3 MQ_2 < MT_3, \quad (40)$$

$$\varphi_1+\varphi_2+\varphi_3 = 1 \quad (41)$$

where:

MT_1	The original price of the amount of material
MT_2	The amount of material in the first gradation of the price discount
MT_3	The amount of material in the second gradation of the price discount
MQ_1	The original price of the maximum amount of material
MQ_2	The maximum amount of material in the first gradation of the price discount
φ_1, φ_2, φ_3	Special ordered sets of type 1 (SOS1). The summation of 0/1 variables is 1.

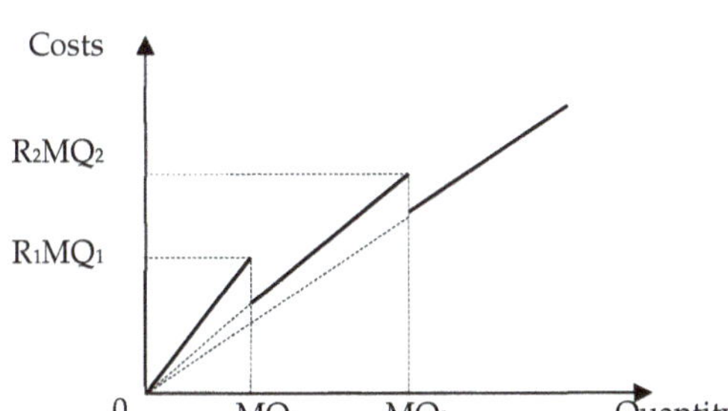

Figure 5. Direct material cost function.

Moreover, this research proposes a sensitivity analysis to measure the cost of materials in the discount pricing function to increase our understanding of material price in terms of how they affect a company's cost and revenue [72]. Sensitivity analysis in this model uses material discount cost and normal material cost to test three kinds of material cost. The interval of cost variation ratio (%) is 5%, from −20% to 20%. The results of the sensitivity analysis of the direct material discount are shown in Table 5.

Table 5. Sensitivity analysis on the material cost.

Cost Variation Ratio (%)	Normal		Price Discount	
	Profit (Thousands)	Profit Variation Ratio (%)	Profit (Thousands)	Profit Variation Ratio (%)
20%	2,819,466	−16.12%	3,144,466	−13.42%
15%	2,954,883	−12.09%	3,266,341	−10.07%
10%	3,090,299	−8.06%	3,388,216	−6.71%
5%	3,225,716	−4.03%	3,510,091	−3.36%
0%	3,361,133	0.00%	3,631,966	0.00%
−5%	3,496,549	4.03%	3,753,841	3.36%
−10%	3,631,966	8.06%	3,875,716	6.71%
−15%	3,767,383	12.09%	3,997,591	10.07%
−20%	3,902,799	16.12%	4,119,466	13.42%

This research adjusted material cost (M) in the sensitivity analysis to estimate the impact on normal material cost with and without discount pricing. The price is altered by 5% from (−20%–20%). If (M)'s cost is raised 5%, the profit will be from (3,361,133 NTD to 3,225,716 NTD), and if the cost that includes discount pricing is changed, that profit would decrease from (3,361,133 NTD) to (3,510,091 NTD). When the cost is decreased by 10%, the profit is (3,496,549 NTD) without discount pricing, and including the discount, it is (3,753,841 NTD).

Even if the change is from 0% to 20% the other products would not change and earnings would be (158,103,200 NTD). However, when price of X_1 is decreased by 12% to (1251), the quantities of X_{22} would become zero; when X_2 is raised to (75399) except for continuing decreased volume of X_1 (99,902) but changed rate to −16%, it is not too different from the original price for the product mix. On the other hand, if a corporation needs to increase profit, it only needs to increase the price of X_1 because the increased price would not reduce the quantities of products. However, when prices are down, this not only cuts down the quantities of all of products, but also reduces a firm's profit.

Therefore, the changed cost of material in the textile industry would not result in a significant change to a corporation's profit even if costs decrease by 20%; however, the profit only increased by $3,902,799. When the profit variation ratio (%) changed to 16.12%, this means that when the material cost changes, the corporation's profit would not increase or decrease significantly; even when considering discount pricing, the result is the same as the normal material cost.

5. Manufacturing Execution System for Production Control under Industry 4.0 in the Textile Industry

In the past, traditional manufacturing produced few types and huge amounts of products, but the Internet has currently changed the situation. Enterprises need to achieve rapid, small quantity and customization production. If an enterprise continues to transport and cooperate production data still in accordance with manual adjustment, then it simply cannot adapt to the demands of the current market [58].

Under Industry 4.0, the Manufacturing Execution System (MES) provides an important bridge between Shop Floor Control System (SFCS) and Enterprise Resource Planning (ERP) [55,57,73]. Compared to the old-fashioned MES, it is not only just scheduling, managing materials and contorting products to increase product yield and quality, but MES can collect data for cloud application and use data to control production process. There is also a need to embrace Corporate Social Responsibility (CSR) to effectively control carbon emissions, room temperature and sewage concentration [74].

5.1. Status Monitoring and Predictive Maintenance

MES can monitor the status of a plant and use the collected data for useful information about status monitoring and predictive maintenance. For successful status monitoring, MES need to meet the following conditions: first, a factory must be digitized and synchronized with networks to respond to actual situation changes in real time [75]. Second, three levels of enterprise management, production management and production must be vertically integrated via automation. Machine and plant control systems (PCS) shift data to react quickly to automation, and vertical integration can ensure that the information to other levels is accurate in the correct real time [76,77]. Third, the ability to exchange consistent data with other applications in production management level is an important attribute of the MES system; this is called horizontal integration. Fourth, the existing MES data are evaluated synthetically by data mining, and identify the quality data and process parameters' they are then adjusted according to a self-optimizing system [48,55]. Fifth, it identifies multiple correlation data, as well as combines the information from MES and plant data to provide meaningful intelligence. Finally, factory machines need preventive maintenance through the data collection and analyzing parameters, in addition to regular maintenance. It is prudent to be warned before a machine malfunctions, thus decreasing the occurrence of unforeseen downtimes [48,78].

5.2. Work-in-Process Tracking

In Industry 4.0, MES can effectively track WIP and reduce losses that WIP causes through the abovementioned features. Work in process tracking is a significant application that aims to control work order, batch and unit production [55,79,80]. Moreover, WIP tracking can efficiently support plants to reduce costly problems such as cancelled orders, assembly errors, line of products downtime, and product recalls. Thus, efficient WIP tracking can offer traceability in all activities, and can effectively increase manufacturing effectiveness. To enable corporations to collect critical data and accurately track progress, the fundamental in WIP tracking is to build on barcode labeling that can accurately identify each item, and to put in place scanning technologies that can scan items as they go through the manufacturing, assembling, maintenance or testing processes [81,82]. These technologies enable the collection of critical data and accurately track progress for full traceability. That is to say, a corporation can more timely track orders to produce related-products through WIP tracking. Moreover, it can reduce the overall cost of production by reducing aspects of inventory. Thus, by effectively controlling cost and reducing WIP, a company can achieve zero inventory that reduces inventory costs, increases turnover and enhances profit to achieve profit maximization [83–85].

5.3. Quality Control

Recently, mass production of customized products in the textile industry has become increasingly important, but pressure is put on quality by the delivery times demanded by customers. In order to control quality, a plant needs to automatically collect data and history information through sensors and IoT technology, and then analyze and optimize process parameters to adjust them with the actual situation, thereby controlling and improving product quality [86,87]. Therefore, the key to Industry 4.0 is real-time communication and machine to machine connection between all systems [88,89]. Machines and equipment exchange data by communicating with each other, and a "smart" plant can record, analyze and make adjustments on its own to enhance the manufacturing process. Due to the need to integrate complex information in real time, achieving stand-alone systems and transparency are the problems on the shop floor that human-machine interaction need to solve [3,38].

6. Discussion

Some issues related to the textile industry and the case company are discussed in this section.

6.1. Energy Saving Method for the Case Company

Equations (18) and (19) are the equations for the energy recycling of heat and water in dyeing factory, respectively. Figure 6 is the heat recovery and recycling system used for the desizing activity in dyeing factory of the case factory. Desizing is the activity of removing the size material from the warp yarns after the textile fabric is woven.

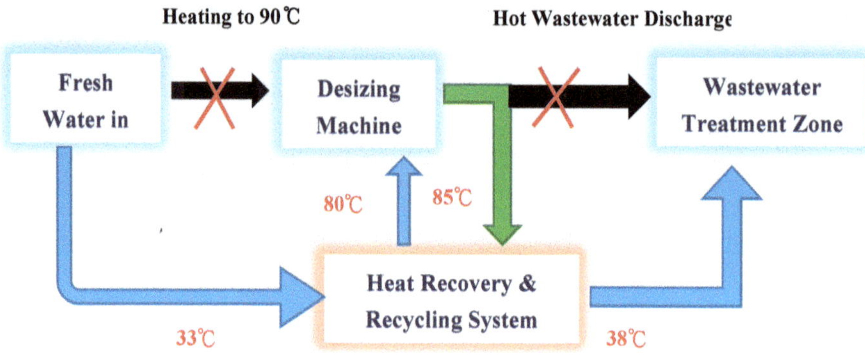

Figure 6. Heat recovery & recycling system in dyeing factory.

Traditionally, the fresh water is heated to 90 °C and put into the desizing machine and the hot wastewater created from the desizing machine is discharged into the wastewater treatment zone. The case company designed a heat recovery and recycling system. The fresh water is only heated to 33 °C and put into the system. On the other hand, the 85 °C water created from the desizing machine is put into the system. These two sources of water are blended in the system. After the processing of the system, 85% of the water with 80 °C returns to the desizing machine and 15% of the wastewater flows into the wastewater treatment zone for further processing. This method will save the energy and water costs and the total cost saving amount is $(C_5RE_h+C_6RE_w)$ in Equation (1). Figure 7 shows the waste-heat recycling system adopted in the case company.

Source: Everest Textile Company

Figure 7. Waste-heat Recycling System.

6.2. From Waste Cinder to Eco-Brick

In addition to the recycling and reuse of wastewater and waste-heat, the case company also reuse the waste cinder produced from coal combustion in coal boiler. There are 10 tons of cinders generated from coal boiler. The cinders are non-toxic, harmless and suitable to be reused and processed into eco-bricks by using a special and precise high pressure forming process. The water permeable turf eco-bricks can be sold in the market or be used to replace the asphalt surface inside the factory as shown in Figure 8. The quantity of eco-bricks in the model is shown in Equation (28), i.e., $BP_3 = e_3 \times X_{22}$, and the revenue of the by-product eco-bricks is $\beta_3 e_3 X_{22}$ in Equation (1).

Source: Everest Textile Company

Figure 8. Eco-brick made from waste cinder and used in the factory.

6.3. Labor Skills and Job Losses Due to the Introduction of Industry 4.0

Some researchers claim that there are four important kinds of skills required for Industry 4.0: (1) Knowledge about Information and Communication Technology such as robots, machine to machine communication, IT security, data protection, and so on; (2) Ability to work with data such as basic statistical knowledge, visual data output and analysis, and so on; (3) Technical know-how about new manufacturing processes and machine maintenance related maintenance related activities; and (4) Person skills such as adaptability and ability to change, working in team, social and communication skills, mindset change for lifelong learning, and so on [89,90].

After introducing Industry 4.0, companies need to face the challenges to find the labor with the skills mentioned above. Besides, the companies may also want to retrain their existing labor by up-skilling the workforce having the retaining jobs equipped with the new technologies or by re-skilling the workforce having the ceasing jobs to new jobs created by Industry 4.0. On the other hand, the employees need to change their mindset for continuously learning new technologies to transit to advanced manufacturing processes [89,91].

The Fourth Industrial Revolution is the annual meeting focus at the World Economic Forum (WEF) held in Davos, Swiss in January, 2016 [92]. The WEF reported that "It sees as many as 7.1 million jobs being lost, mostly in white-collar office and administrative roles, with the creation of 2.1 million new jobs in fields such as computer engineering and mathematics. It leads to a net loss of over 5 million jobs in 15 major developed and emerging economies" [92,93]. After introducing Industry 4.0, the manual labor was replaced by high-performance machines with Programmable logic controller or robotics with the accuracy of 99.9966% [94,95]. Only non-routine manual labor jobs are broadly unaffected [94].

7. Conclusions

This research combined Industry 4.0, Activity-Based Costing (ABC) and the Theory of Constraints (TOC) to achieve profit maximization, and proposes a green production planning model to deal with environmental issues [28,96–99]. The textile industry is one of the world's major sources of industrial pollution; therefore, environmental issues are a major concern. This paper considered carbon emissions, energy recycle and waste reuse. It is found that an Industry 4.0 automated monitoring system will properly maintain and manage all production processes and immediately monitor and record all production process data. In the manufacturing process, it is necessary to effectively control carbon emissions. The carbon taxes must be considered to prompt enterprises to control carbon emissions, and there must be an exploration of the use of renewable energy. This paper also focuses on the recycling of water and heat energy. Energy can be recycled and reused after being processed, not only for a green energy effect, but also to save more costs due to energy saving. Finally, this paper also the reuse of waste cinder. Coal is burned and produces waste cinder in a coal boiler. Cinder is usually regarded as waste. Its transportation cost is very high and burying cinder is very polluting to the environment. Therefore, mixing cinder with cement to form the solid and highly breathable eco-bricks, is not only environmental protective, but can also avoid the high cost of shipping, and so add to green manufacturing.

This research used mathematical programming approach to formulate the decision model with ABC cost data and TOC constraints for the textile companies. This research also clarified the relation between mathematical programming models and Industry 4.0 techniques. Managers in the textile companies can applied this decision model to achieve the optimal product-mix under various constraints and to evaluate the effect on profit of carbon emissions, energy recycling, waste reuse, and material quantity discount.

For the future works, the green production planning model proposed in this paper can be extended to incorporate the cap-and-trade feature [100] the companies can buy carbon emission right if they use up the carbon emission quota, otherwise they also can sell the surplus carbon emission right in the market. Also, this model can be extended to consider different carbon tax cost functions and their effects on profit.

Acknowledgments: The author is extremely grateful to the energies journal editorial team and reviewers who provided valuable comments for improving the quality of this article. The author also would like to thank the Ministry of Science and Technology of Taiwan for financially supporting this research under Grant No. MOST106-2410-H-008-020-MY3.

Conflicts of Interest: The author declares no conflict of interest.

References

1. Choudhury, A.R. Environmental impacts of the textile industry and its assessment through life cycle assessment. In *Roadmap to Sustainable Textiles and Clothing*; Springer: Berlin, Germany, 2014; pp. 1–39.
2. Bullon, J.; González Arrieta, A.; Hernández Encinas, A.; Queiruga Dios, A. Manufacturing processes in the textile industry. Expert Systems for fabrics production. *Adv. Distrib. Comput. Articial Intell. J.* **2017**, *6*, 41–50.
3. Brettel, M.; Friederichsen, N.; Keller, M.; Rosenberg, M. How virtualization, decentralization and network building change the manufacturing landscape: An industry 4.0 perspective. *Int. J. Mech. Ind. Sci. Eng.* **2014**, *8*, 37–44.
4. Rüßmann, M.; Lorenz, M.; Gerbert, P.; Waldner, M.; Justus, J.; Engel, P.; Harnisch, M. *Industry 4.0: The Future of Productivity and Growth in Manufacturing Industries*; Boston Consulting Group: Boston, MA, USA, 2015.
5. Lee, J.; Kao, H.-A.; Yang, S. Service innovation and smart analytics for industry 4.0 and big data environment. *Procedia Cirp* **2014**, *16*, 3–8. [CrossRef]
6. Awad, M.I.; Hassan, N.M. Joint decisions of machining process parameters setting and lot-size determination with environmental and quality cost consideration. *J. Manuf. Syst.* **2018**, *46*, 79–92. [CrossRef]
7. Rojko, A. Industry 4.0 Concept: Background and Overview. *Int. J. Inf. Manag.* **2017**, *11*, 77–90. [CrossRef]
8. Park, S.; Huh, J.-H. Effect of Cooperation on Manufacturing IT Project Development and Test Bed for Successful Industry 4.0 Project: Safety Management for Security. *Processes* **2018**, *6*. [CrossRef]
9. Küsters, D.; Praß, N.; Gloya, Y.-S. Textile up 4.0—Preparing Germany's Textile Industry for the Digital Future. *Procedia Manuf.* **2017**, *9*, 214–221. [CrossRef]

10. Chen, Z.; Xing, M. Upgrading of textile manufacturing based on Industry 4.0. In Proceedings of the 5th International Conference on Advanced Design and Manufacturing Engineering (ICADME 2015), Shenzhen, China, 19–20 September 2015; pp. 2143–2146.
11. Dujin, A.; Blanchet, M.; Rinn, T.; Von Thaden, G.; De Thieullo, G. *Industry 4.0—The New Industrial Revolution: How Europe Will Succeed*; Roland Berger Strategy Consultants: Munich, Germany, 2014.
12. Jerzembeck, J. Industry 4.0 Potential in Textile Production (Dyeing and Finishing). *Melliand Int.* **2016**, *4*, 220–222.
13. RINA Consulting. Industry 4.0: The New Challenge for the ITALIAN TEXTILE MACHINERY INDUSTRY. Executive Summary, Italian Textile Machinery, 2017. Available online: http://www.acimit.it/DOC/Executive17-Ing.pdf (accessed on 20 July 2018).
14. Lee, H.-G.; Huh, J.-H. A Cost-Effective Redundant Digital Excitation Control System and Test Bed Experiment for Safe Power Supply for Process Industry 4.0. *Processes* **2018**, *6*. [CrossRef]
15. Huh, J.-H. *Smart Grid Test Bed Using OPNET and Power Line Communication*; IGI Global: Hershey, PA, USA, 2017.
16. Kamal Abd Rahman, I.; Omar, N.; Zainal Abidin, Z. The applications of management accounting techniques in Malaysian companies: An industrial survey. *J. Financ. Rep. Account.* **2003**, *1*, 1–12. [CrossRef]
17. Holmen, J.S. ABC vs. TOC: It's a matter of time. *Strat. Financ.* **1995**, *76*, 37.
18. Kee, R. Integrating activity-based costing with the theory of constraints to enhance production-related decision-making. *Account. Horiz.* **1995**, *9*, 48–61.
19. Lockhart, J.; Taylor, A. Environmental considerations in product mix decisions using ABC and TOC: As environmental issues increasingly influence corporate performance, they need to be a standard part of management accounting systems. *Manag. Account. Q.* **2007**, *9*, 13.
20. Turney, P.B.B. *Common Cents: The ABC Performance Breakthrough—How to Succeed with Activity-Based Costing*, Revised ed.; McGraw-Hill: New York, NY, USA, 2005.
21. Luebbe, R.; Finch, B. Theory of constraints and linear programming: A comparison. *Int. J. Prod. Res.* **1992**, *30*, 1471–1478. [CrossRef]
22. Plenert, G. Optimizing theory of constraints when multiple constrained resources exist. *Eur. J. Oper. Res.* **1993**, *70*, 126–133. [CrossRef]
23. Tsai, W.-H.; Lin, T.-M. Nonlinear multiproduct CVP analysis with 0–1 mixed integer programming. *Eng. Costs Prod. Econ.* **1990**, *20*, 81–91. [CrossRef]
24. Tsai, W.-H.; Lee, K.-C.; Liu, J.-Y.; Lin, H.-L.; Chou, Y.-W.; Lin, S.-J. A mixed activity-based costing decision model for green airline fleet planning under the constraints of the European Union Emissions Trading Scheme. *Energy* **2012**, *39*, 218–226. [CrossRef]
25. Tsai, W.-H.; Lin, S.-J.; Liu, J.-Y.; Lin, W.-R.; Lee, K.-C. Incorporating life cycle assessments into building project decision-making: An energy consumption and CO_2 emission perspective. *Energy* **2011**, *36*, 3022–3029. [CrossRef]
26. Jaedicke, R.K. Improving breakeven analysis by linear programming techniques. *NAA Bull.* **1961**, 5–12.
27. Entezaminia, A.; Heydari, M.; Rahmani, D. A multi-objective model for multi-product multi-site aggregate production planning in a green supply chain: Considering collection and recycling centers. *J. Manuf. Syst.* **2016**, *40*, 63–75. [CrossRef]
28. Tsai, W.-H.; Yang, C.-H.; Chang, J.-C.; Lee, H.-L. An activity-based costing decision model for life cycle assessment in green building projects. *Eur. J. Oper. Res.* **2014**, *238*, 607–619. [CrossRef]
29. Williams, H.P. *Model Building in Mathematical Programming*; John Wiley & Sons: West Sussex, UK, 2013.
30. Cohon, J.L. *Multiobjective Programming and Planning*; Dover Publications: Mineola, NY, USA, 2003.
31. Hanasusanto, G.A.; Kuhn, D.; Wallace, S.W.; Zymler, S. Distributionally robust multi-item newsvendor problems with multimodal demand distributions. *Math. Program.* **2015**, *152*, 1–32. [CrossRef]
32. Correa, J.; Marchetti-Spaccamela, A.; Matuschke, J.; Stougie, L.; Svensson, O.; Verdugo, V.; Verschae, J. Strong LP formulations for scheduling splittable jobs on unrelated machines. *Math. Program.* **2015**, *154*, 305–328. [CrossRef]
33. Elamvazuthi, I.; Ganesan, T.; Vasant, P.; Webb, J.F. Application of a Fuzzy Programming Technique to Production Planning in the Textile Industry. *Int. J. Comput. Sci. Inf. Secur.* **2009**, *6*, 238–243.
34. Teke, Ç.; Okutkan, C.; Erden, C. Determining the Production Amounts in Textile Industry with Fuzzy Linear Programming. *Int. J. Eng. Technol. Res.* **2017**, *2*, 1–6.

35. Waidner, M.; Kasper, M. Security in industrie 4.0: Challenges and solutions for the fourth industrial revolution. In Proceedings of the 2016 Conference on Design, Automation & Test in Europe, Dresden, Germany, 14–18 March 2016; pp. 1303–1308.
36. Schmidt, R.; Möhring, M.; Härting, R.-C.; Reichstein, C.; Neumaier, P.; Jozinović, P. Industry 4.0—Potentials for creating smart products: Empirical research results. In Proceedings of the International Conference on Business Information Systems, Poznań, Poland, 24–26 June 2015; Springer: Berlin, Germany, 2015; pp. 16–27.
37. Wang, S.; Wan, J.; Zhang, D.; Li, D.; Zhang, C. Towards smart factory for industry 4.0: A self-organized multi-agent system with big data based feedback and coordination. *Comput. Netw.* **2016**, *101*, 158–168. [CrossRef]
38. Lee, J.; Bagheri, B.; Kao, H.-A. A cyber-physical systems architecture for industry 4.0-based manufacturing systems. *Manuf. Lett.* **2015**, *3*, 18–23. [CrossRef]
39. Kagermann, H. Change through digitization—Value creation in the age of Industry 4.0. In *Management of Permanent Change*; Springer: Berlin, Germany, 2015; pp. 23–45.
40. Karre, H.; Hammer, M.; Kleindienst, M.; Ramsauer, C. Transition towards an Industry 4.0 State of the LeanLab at Graz University of Technology. *Procedia Manuf.* **2017**, *9*, 206–213. [CrossRef]
41. Jazdi, N. Cyber physical systems in the context of Industry 4.0. In Proceedings of the 2014 IEEE International Conference on Automation, Quality and Testing, Robotics, Cluj-Napoca, Romania, 22–24 May 2014; pp. 1–4.
42. Schoeberl, P.; Brik, M.; Braun, R.; Fuchs, W. Treatment and recycling of textile wastewater—Case study and development of a recycling concept. *Desalination* **2005**, *171*, 173–183. [CrossRef]
43. Gorecky, D.; Schmitt, M.; Loskyll, M.; Zühlke, D. Human-Machine-Interaction in the industry 4.0 era. In Proceedings of the 2014 12th IEEE International Conference on Industrial Informatics (INDIN), Porto Alegre, Brazil, 27–30 July 2014; pp. 289–294.
44. Lethbridge, T.C. What knowledge is important to a software professional? *Computer* **2000**, *33*, 44–50. [CrossRef]
45. Valdeza, A.C.; Braunera, P.; Schaara, A.K.; Holzingerb, A.; Zieflea, M. Reducing complexity with simplicity-usability methods for industry 4.0. In Proceedings of the 19th Triennial Congress of the IEA, Melbourne, Australia, 9–14 August 2015; pp. 1–8.
46. Lary, H.B. Trade in Labor-Intensive Manufactures. In *Imports of Manufactures from Less Developed Countries*; NBER: Cambridge, MA, USA, 1968; pp. 86–115.
47. Leamer, E.E. Wage inequality from international competition and technological change: Theory and country experience. *Am. Econ. Rev.* **1996**, *86*, 309–314.
48. Kletti, J. *Manufacturing Execution System—MES*; Springer: Berlin, Germany, 2007.
49. Saenz de Ugarte, B.; Artiba, A.; Pellerin, R. Manufacturing execution system—A literature review. *Prod. Plan. Control* **2009**, *20*, 525–539. [CrossRef]
50. Arik Ragowsky, T.M.S. Enterprise resource planning. *J. Manag. Inf. Syst.* **2002**, *19*, 11–15.
51. Leon, A. *Enterprise Resource Planning*; McGraw-Hill Education: New York, NY, USA, 2008.
52. Sumner, M. *Enterprise Resource Planning*; Pearson Education Inc.: New Jersey, NJ, USA, 2005.
53. O'Leary, D.E. *Enterprise Resource Planning Systems: Systems, Life Cycle, Electronic Commerce, and Risk*; Cambridge University Press: Cambridge, UK, 2000.
54. Umble, E.J.; Haft, R.R.; Umble, M.M. Enterprise resource planning: Implementation procedures and critical success factors. *Eur. J. Oper. Res.* **2003**, *146*, 241–257. [CrossRef]
55. Lasi, H.; Fettke, P.; Kemper, H.-G.; Feld, T.; Hoffmann, M. Industry 4.0. *Bus. Inf. Syst. Eng.* **2014**, *6*, 239–242. [CrossRef]
56. Choi, B.K.; Kim, B.H. MES (manufacturing execution system) architecture for FMS compatible to ERP (enterprise resource planning). *Int. J. Comput. Integr. Manuf.* **2002**, *15*, 274–284. [CrossRef]
57. Liu, W.; Chua, T.J.; Larn, J.; Wang, F.-Y.; Yin, X. APS, ERP and MES systems integration for semiconductor backend assembly. In Proceedings of the 7th International Conference on Control, Automation, Robotics and Vision, Singapore, 2–5 December 2002; pp. 1403–1408.
58. Saggiomo, M.; Wischnowski, M.; Winkel, B.; Nierhaus, M.; Gloy, Y.-S.; Gries, T. Industry 4.0 in the field of textile machinery-first steps of implementation. *Melliand Int.* **2015**, *1*, 49–50.
59. Ramanathan, V.; Feng, Y. Air pollution, greenhouse gases and climate change: Global and regional perspectives. *Atmos. Environ.* **2009**, *43*, 37–50. [CrossRef]

60. Dodman, D. Blaming cities for climate change? An analysis of urban greenhouse gas emissions inventories. *Environ. Urban.* **2009**, *21*, 185–201. [CrossRef]
61. Keohane, N.O.; Olmstead, S.M. Introduction. In *Markets and the Environment*; Springer: Berlin, Germany, 2016; pp. 1–10.
62. Davis, S.J.; Caldeira, K. Consumption-based accounting of CO_2 emissions. *Proc. Natl. Acad. Sci. USA* **2010**, *107*, 5687–5692. [CrossRef] [PubMed]
63. Guan, D.; Peters, G.P.; Weber, C.L.; Hubacek, K. Journey to world top emitter: An analysis of the driving forces of China's recent CO_2 emissions surge. *Geophys. Res. Lett.* **2009**, *36*, 1–5. [CrossRef]
64. Steinberger, J.K.; Friot, D.; Jolliet, O.; Erkman, S. A spatially explicit life cycle inventory of the global textile chain. *Int. J. Life Cycle Assess.* **2009**, *14*, 443–455. [CrossRef]
65. Elahee, K. Heat recovery in the textile dyeing and finishing industry: Lessons from developing economies. *J. Energy S. Afr.* **2010**, *21*, 9–15.
66. Oğulata, R.T. Utilization of waste-heat recovery in textile drying. *Appl. Energy* **2004**, *79*, 41–49. [CrossRef]
67. Lopez, A.; Ricco, G.; Ciannarella, R.; Rozzi, A.; Di Pinto, A.; Passino, R. Textile wastewater reuse: Ozonation of membrane concentrated secondary effluent. *Water Sci. Technol.* **1999**, *40*, 99–105. [CrossRef]
68. Chequer, F.M.D.; de Oliveira, G.A.R.; Ferraz, E.R.A.; Cardoso, J.C.; Zanoni, M.V.B.; de Oliveira, D.P. Textile dyes: Dyeing process and environmental impact. In *Eco-Friendly Textile Dyeing and Finishing*; InTech: London, UK, 2013.
69. Li, K.; Zhang, X.; Leung, J.Y.-T.; Yang, S.-L. Parallel machine scheduling problems in green manufacturing industry. *J. Manuf. Syst.* **2016**, *38*, 98–106. [CrossRef]
70. Monahan, J.P. A quantity discount pricing model to increase vendor profits. *Manag. Sci.* **1984**, *30*, 720–726. [CrossRef]
71. Kee, R. The sufficiency of product and variable costs for production-related decisions when economies of scope are present. *Int. J. Prod. Econ.* **2008**, *114*, 682–696. [CrossRef]
72. Sinha, A.; Rämö, J.; Malo, P.; Kallio, M.; Tahvonen, O. Optimal management of naturally regenerating uneven-aged forests. *Eur. J. Oper. Res.* **2017**, *256*, 886–900. [CrossRef]
73. Koh, S.; Saad, S. A holistic approach to diagnose uncertainty in ERP-controlled manufacturing shop floor. *Prod. Plan. Control.* **2003**, *14*, 273–289. [CrossRef]
74. Lin, G.Y.; Solberg, J.J. Integrated shop floor control using autonomous agents. *IIE Trans.* **1992**, *24*, 57–71. [CrossRef]
75. Park, H.-S.; Tran, N.-H. An autonomous manufacturing system for adapting to disturbances. *Int. J. Adv. Manuf. Technol.* **2011**, *56*, 1159–1165. [CrossRef]
76. Zhong, R.Y.; Dai, Q.; Qu, T.; Hu, G.; Huang, G.Q. RFID-Enabled real-time manufacturing execution system for mass-customization production. *Robot. Comput. Integr. Manuf.* **2013**, *29*, 283–292. [CrossRef]
77. Hua, J.; Liang, T.; Lei, Z. Study and design real-time manufacturing execution system based on RFID. In Proceedings of the 2008 Second International Symposium on Intelligent Information Technology Application, Shanghai, China, 20–22 December 2008; pp. 591–594.
78. Almada-Lobo, F. The Industry 4.0 revolution and the future of manufacturing execution systems (MES). *J. Innov. Manag.* **2016**, *3*, 16–21.
79. Wang, W.Y.; Chan, H.K. Virtual organization for supply chain integration: Two cases in the textile and fashion retailing industry. *Int. J. Prod. Econ.* **2010**, *127*, 333–342. [CrossRef]
80. Kuhn, W. Digital factory-simulation enhancing the product and production engineering process. In Proceedings of the 2006 Winter Simulation Conference, Monterey, CA, USA, 3–6 December 2006; pp. 1899–1906.
81. Tajima, M. Strategic value of RFID in supply chain management. *J. Purch. Supply Manag.* **2007**, *13*, 261–273. [CrossRef]
82. Spekman, R.E.; Kamauff, J.W., Jr.; Myhr, N. An empirical investigation into supply chain management: A perspective on partnerships. *Supply Chain Manag. Int. J.* **1998**, *3*, 53–67. [CrossRef]
83. Johnson, H.T. Early cost accounting for internal management control: Lyman Mills in the 1850's. *Bus. Hist. Rev.* **1972**, *46*, 466–474. [CrossRef]
84. Abernathy, F.H.; Dunlop, J.T.; Hammond, J.H.; Weil, D. Globalization in the apparel and textile industries: What is new and what is not? In *Locating Global Advantage: Industry Dynamics in the International Economy*; Stanford University Press: Stanford, CA, USA, 2004.

85. Lummus, R.R.; Vokurka, R.J.; Alber, K.L. Strategic supply chain planning. *Prod. Invent. Manag. J.* **1998**, *39*, 49–58.
86. Montgomery, D.C. *Introduction to Statistical Quality Control*; John Wiley & Sons: Hoboken, NJ, USA, 2007.
87. Stojanovic, R.; Mitropulos, P.; Koulamas, C.; Karayiannis, Y.; Koubias, S.; Papadopoulos, G. Real-Time vision-based system for textile fabric inspection. *Real-Time Imaging* **2001**, *7*, 507–518. [CrossRef]
88. Esmaeilian, B.; Behdad, S.; Wang, B. The evolution and future of manufacturing: A review. *J. Manuf. Syst.* **2016**, *39*, 79–100. [CrossRef]
89. Aulbur, W.; Arvind, C.J.; Bigghe, R. *Whitepaper: Skill Development for Industry 4.0*; BRICS Skill Development Group: Roland Berger, India, 2016.
90. Gehrke, L.; Kühn, A.; Rule, D.; Moore, P.; Bellmann, C.; Siemes, S.; Dawood, D.; Singh, L.; Kulik, J.; Standle, M. *Industry4.0: A Discussion of Qualifications and Skills in the Factory of the Future: A German and American Perspective*; The Association of German Engineers (VDI): Düsseldorf, Germany; American Society of Mechanical Engineers (ASME): Washington, DC, USA, 2015.
91. Hartmann, E.; Bovenschulte, M. Skills Needs Analysis for "Industry 4.0" Based on Roadmaps for Smart Systems. In *SKOLKOVO Moscow School of Management & International Labour Organization: Using Technology Foresights for Identifying Future Skills Needs*; Global Workshop Proceedings, Moscow; Institute for Innovation and Technology: Berlin, Germany, 2013; pp. 24–36.
92. Lowman, S. WEF 2016: 4th Industrial Revolution. 5 mn Jobs, Women in the Firing Line. *BIZNEWS*. 18 January 2016. Available online: https://www.biznews.com/wef/davos-2016/2016/01/18/wef-2016-4th-industrial-revolution-5mn-jobs-women-in-the-firing-line/ (accessed on 20 July 2018).
93. Team, M. Five Million Jobs by 2020: The Real Challenge of the Fourth Industrial Revolution. World Economic Forum, 18 January 2016. Available online: https://www.weforum.org/press/2016/01/five-million-jobs-by-2020-the-real-challenge-of-the-fourth-industrial-revolution/ (accessed on 20 July 2018).
94. Flynn, J.; Dance, S.; Schaefer, D. Industry 4.0 and Its Potential Impact on Employment Demographics in the UK. In *Advanced in Manufacturing Technology XXXI, Proceedings of the 15th International Conference on Manufacturing Research, incorporating the 32nd National Conference on Manufacturing Research, London, UK, 5–7 September 2017*; Gao, J., El Souri, M., Keates, S., Eds.; IOS Press: Amsterdam, The Netherlands, 2017.
95. Matovcikova, D. Industry 4.0 as the Culprit of Unemployment. In Proceedings of the 12th International Workshop on Knowledge Management, Trenčín, Slovakia, 12–13 October 2017.
96. Tsai, W.-H.; Hung, S.-J. A fuzzy goal programming approach for green supply chain optimisation under activity-based costing and performance evaluation with a value-chain structure. *Int. J. Prod. Res.* **2009**, *47*, 4991–5017. [CrossRef]
97. Tsai, W.-H.; Chen, H.-C.; Leu, J.-D.; Chang, Y.-C.; Lin, T.W. A product-mix decision model using green manufacturing technologies under activity-based costing. *J. Clean. Prod.* **2013**, *57*, 178–187. [CrossRef]
98. Tsai, W.-H.; Hung, S.-J. Treatment and recycling system optimisation with activity-based costing in WEEE reverse logistics management: An environmental supply chain perspective. *Int. J. Prod. Res.* **2009**, *47*, 5391–5420. [CrossRef]
99. Tsai, W.-H.; Chen, H.-C.; Liu, J.-Y.; Chen, S.-P.; Shen, Y.-S. Using activity-based costing to evaluate capital investments for green manufacturing systems. *Int. J. Prod. Res.* **2011**, *49*, 7275–7292. [CrossRef]
100. Wang, S.; Wan, L.; Li, T.; Luo, B.; Wang, C. Exploring the effect of cap-and-trade mechanism on firm's production planning and emission reduction strategy. *J. Clean. Prod.* **2018**, *172*, 591–601. [CrossRef]

energies

MDPI

Article

Carbon Taxes and Carbon Right Costs Analysis for the Tire Industry

Wen-Hsien Tsai

Department of Business Administration, National Central University, Jhongli, Taoyuan 32001, Taiwan; whtsai@mgt.ncu.edu.tw; Tel.: +886-3-4267247; Fax: +886-3-4222891

Received: 4 July 2018; Accepted: 9 August 2018; Published: 14 August 2018

Abstract: As enterprises are the major perpetrators of global climate change, concerns about global warming, climate change, and global greenhouse gas emissions continue to attract attention, and have become international concerns. The tire industry, which is a high-pollution, high-carbon emission industry, is facing pressure to reduce its carbon emissions. Thus, carbon prices and carbon trading have become issues of global importance. In order to solve this environmental problem, the purpose of this paper is to combine mathematical programming, Theory of Constraints (TOC), and Activity-Based Costing (ABC) to formulate the green production decision model with carbon taxes and carbon right costs, in order to achieve the optimal product mix decision under various constraints. This study proposes three different scenario models with carbon taxes and carbon right used to evaluate the effect on profit of changes in carbon tax rates.

Keywords: Activity-Based Costing (ABC); carbon emissions; tire industry; carbon trading; mathematical programming

1. Introduction

The international community has paid considerable attention to environmental protection in recent years. Carbon taxes are considered as effective tools to curb greenhouse gas emissions. Analysis of the empirical evidence of carbon tax benefits, including indirect effects on technological developments and other pollutants (i.e., co-benefits) also added suppliers to improve energy price expectations in the future, and empirically propose the first green research contribution [1] and the United Nations has successively enacted relevant laws and regulations, such as the international Greenhouse Gas Emission Act and the 2015 Paris Accord [2], in the hope that the environment we live in will not continue to deteriorate. Due to high technology and industrial development, anthropocentric greenhouse gases are continually generated and accumulate, leading to increased global warming, and deteriorating climate and environmental ecology. In the past decade, it has been well recognized that environmental and economic performance are mutually reinforcing, with improved environmental performance leading to lower costs, increased sales, and greater economic efficiency. That joint effects will help companies to obtain temporary financing and minimize the damage to the environment stemming from carbon emissions [3]. Different perspectives and empirical studies [4] suggest that more attention should be paid to the causal relationship between eco-efficiency and different environmental management approaches, as well as their economic consequences; for example, the implementation of a carbon tax will help to improve the environment, promoting carbon taxes as a policy to improve the cost-effectiveness of carbon emissions. However, in the current practice of environmental policy, only a few countries implement taxation based on carbon content. According to empirical results, the carbon tax may be an interesting policy choice, and the negative impact can be compensated by designing taxes and the fiscal revenue generated by the use [5]. Various environmental policy tools are used in critical assessment criteria, including cost effectiveness, equitable distribution, risk minimization of uncertainty and political feasibility. Additional considerations include carbon taxes, tradable emission allowances, emission reduction subsidies, and research and development subsidies [6]. In the actual situation of the EU, the implementation of the carbon price policy can achieve the goal of emission reduction, and the carbon trading results also show that carbon emissions can be controlled through carbon trading [7]. But when the regulatory company gets compensation, efficiency needs to be paid among the companies to balance the damage caused by the marginal probability. Applying this basic economic logic to the analysis of the compensation rules proposed under the EU Emissions Trading Scheme, emissions permits can be allocated free of charge to carbon-intensive and trade-exposed industries [8]. However, various environmental policy instruments, including carbon emission taxes, tradable emission allowances, emission reduction subsidies, performance standards, technical tasks, and research and development subsidies, can meet major evaluation criteria, including cost-effectiveness, fair distribution, and risk minimization of uncertainty, and political feasibility [9]. Investigating carbon prices in China, the results show that it has important policy implications for regional markets to be included in national carbon trading [10]. The impact of the Emissions Trading Scheme (ETS) is one of the most effective mitigation measures [11]. The tire industry's production process greatly increases atmospheric greenhouse gas concentrations. At present, many countries have adopted a "carbon tax" levy to solve corporate carbon emission problems with the related costs borne by enterprises. Sathre and Gustavsson believed that taxation on environmental issues can reduce pollution [12]. Therefore, the impact of carbon pricing policies on the economy is very significant [13]. In Australia, the carbon premium has become higher and higher after the implementation of the carbon tax, and indicates that a stable long-term policy is needed in the future to enable the carbon pricing mechanism to play its full role [14]. Therefore, the carbon price and carbon trading policy results discussed by the aforementioned research institutes have clearly become the most important issues for governments around the world. In this paper, the carbon taxes and carbon right costs analysis holds only for direct carbon taxes or for cap-and-trade schemes as well through a market determined price for carbon emission allowances.

According to the above research, the effective way to reduce emissions is to implement carbon tax collection and formulate carbon rights trading mechanism. This paper aims to analyze the relevant analysis of carbon emissions in the tire industry. Then the tire industry can employ the Activity-Based Costing (ABC) method and perform related cost calculations. The ABC system provides a means for analyzing non-value-added activity workbooks, such as procedures and production that neglect resource constraints in activity [15,16]. A better solution to the reduction of environmentally harmful products is provided by the Theory of Constraint (TOC) [17,18].

The purpose of this paper is to combine mathematical programming, TOC, and ABC to formulate a green production decision model with carbon taxes and carbon right costs, in order to achieve the optimal product mix decision under various constraints and to analyze the effect on profit of carbon taxes. The tire industry is one of the industries with high pollution and high energy consumption, where its carbon pollution will seriously affect environmental issues [19]. Therefore, further assumes that three different scenario models with carbon taxes and carbon right for tire companies are used for scenario analysis. Hoinka and Ziebik [20] applied mathematical methods to analyze complex energy management systems. In summary this paper will discuss the green ABC production decision-making model. The contribution of this study is that it incorporates the cost of carbon emissions into the mathematical programming, enabling the tire industry to evaluate emissions projects. In recent years, studies on the implementation of carbon trading have provided useful insights and helped to reduce carbon emissions [21–23].

2. Research Background

2.1. Carbon Emissions Cost and Carbon Tax

The increasing attention to global warming and climate change has become a common concern among governments around the world; therefore, it is imperative to adopt effective measures to promote energy conservation and carbon reduction, and implement corporate carbon emissions policies. Carbon pricing is a hot topic in climate policy debates. There are still many problems due to numerous reasons for implementing global carbon prices. Therefore, the main arguments for carbon pricing are proposed for analysis and discussion. The relatively low environmental cost of carbon pricing helps to improve the social acceptability of climate policy. This includes correcting the price of property that stimulates rapid environmental innovation. These arguments are underestimated in open debates, where pricing strategies are often overlooked and innovation policy misconceptions are common. However, carbon pricing and technology policies are largely complementary and therefore necessary for effective climate policy. In addition, other tools are used to complement the carbon pricing, and suggestions for solving these problems are proposed. Whether through carbon taxes or carbon emissions trading, reflections on the implementation of global carbon prices, including energy research and innovation research, go beyond the traditional arguments of environmental economics [24].

Due to the abnormal climate change in the world, the impact of air pollution is becoming more and more serious, and people's health is also seriously threatened. Therefore, effective action must be taken to develop policies to reduce greenhouse gas emissions. People are increasingly concerned about the market instruments in the form of carbon pricing mechanisms. The study found that carbon pricing policies include carbon taxes, limits and transactions, emission reduction credits, clean energy standards and reductions in fossil fuel subsidies. Organize the experience and implementation methods of developing countries. And put forward the carbon pricing applicability for decision-making reference [25]. In South America, a carbon pricing strategy is also proposed for climate issues to cope with environmental pollution problems [26]. Environmental pollution has become a major social and political challenge around the world [27,28], and some studies have yielded positive results. That the application of carbon price is a carbon reduction tool [29]. Because companies with high carbon emissions have higher carbon risks. Recent ex-post evaluations of the European Union (EU) Emissions Trading Scheme (ETS) on regulated industries in the industrial and power industries have found that issues such as carbon dioxide emissions, economic performance and competitiveness, and innovation can all evolve with international developments. Propose more relevant future research directions, continue to expand on these innovative and internationally important issues [30].

In addition, there is an empirical research on the impact of the EU Emissions Trading Scheme on German stock returns. It was found that in the first few years of the program, companies that received national free carbon credits performed significantly better than companies that did not receive tax-free carbon credits. This result is very interesting, if there is a large "carbon premium", the cash flow from the free allocation of carbon allowances increases. Companies with high carbon emissions, higher carbon exposures, show higher expected returns [31]. For example, The Netherlands [32], Denmark, and Sweden have begun implementing carbon tax policies. Carbon and energy taxes will increase fiscal measures to reduce carbon emissions [12,33,34]. The results of these studies show that the implementation of carbon emission cap-and-trade mechanisms can effectively reduce aviation industry carbon emissions [35], and the means used to reduce CO_2 emissions by this policy have been theoretically assessed [36]. And considered the use of two different manufacturing technologies for their carbon emission quota model [37], and the analysis results can reduce the total amount of carbon emissions and pollution of the environment. In considering a mathematical model for carbon emission quota, the results show that carbon emission restrictions will always encourage restricted capital goods manufacturers to produce more remanufactured products to maximize profits and reduce carbon emissions [38]. In addition, the New Zealand Emissions Trading Scheme (NZ ETS) is unique in that it allowed unlimited use of Kyoto grants until 2013. As a result, New Zealand ETS provides an impact on the pricing and limits of carbon pricing for small systems. Using the quota import and export data, empirical analysis, the study found that carbon pricing offset is mainly the price determinant [39]. That the inclusion of flexible pricing mechanisms and the differentiation of tax rates can effectively reduce carbon emissions [29].

Ecological protection has increasingly attracted international attention, and the dramatic increases in various environmentally harmful substances and energy consumption have led to severe air pollution. In view of protecting human health and environmental ecology by preventing the deterioration of air quality, emission reduction policies have become the important issue in the tire industry; in the complex process of manufacturing tires, the carbon emitted cannot be ignored. Higher carbon emissions will lead to higher carbon taxes, which will increase the costs of the tire industry.

The present research, as related to the tire industry, mainly focuses on the analysis of waste tire recycling [40], low carbon or zero carbon production system innovation to reduce carbon dioxide emissions [41]. There is no relevant research analyzing carbon emission costs and carbon trading in combination with mathematical programming models; therefore, the focus of this article is on the costs and expenses of the tire manufacturing process, while adding carbon emission costs to carry out three different scenario models analysis with carbon taxes and carbon right for tire companies. Analysis of the results can be used for decision reference for the tire industry.

2.2. ABC Applied to Green Production and Environmental Protection in the Tire Industry

The ABC method is mainly used for calculating the activity cost of each product, where resources are allocated to activities or activity centers according to resource drivers. Activity costs are allocated to products based on activity drivers, and the resources allocated to activities constitute the cost elements of the activity [42]. This article uses tire companies as the research object. After process differentiation, the cost of each green process project is estimated in order to analyze the total cost; therefore, we consider the green tire manufacturing process as a cost target. In the complex tire industrial process, this article divides cost into direct material cost, direct labor cost, handling cost, carbon tax, and other costs, and also regards the maintenance costs and depreciations of the machines, plants, etc. as the total fixed cost. The ABC method is used to configure the tire industry's manufacturing process and allocate resource costs to corresponding process projects, in order to determine the most suitable green product mix to maximize total profits.

In recent years, due to environmental concerns, new green manufacturing technologies have been more widely applied and explored to improve operation technology [43]; the issue of global warming has drawn ever-greater attention. Economists and international organizations demand the introduction of carbon taxes as a cost-effective way to decrease Greenhouse Gas (GHG) emissions [44]. Moreover, the carbon duty guidelines can promote the growth and application of renewable power sources [45], and assist in the growth of the global economy [46]. With the rapid technological progress, the tire industry has been pushed to recognize the need for environmental protection for the sustainable development of its enterprises regarding indispensable factors, such as raw materials. Processes and products must support the care of the environment and cherish the required resources by embracing the concept of environmental protection, as well as improved product design, equipment, and activities. The tire production process is complicated, and there is little research that analyzes the process costs of this industry. In addition, the tire industry must consider the cost of carbon emissions, in order to accurately predict tire production costs and reduce the impact of environmental pollution. Therefore, this paper proposes three green decision models, including a model without carbon rights trading and two models with two kinds of carbon rights trading costs, in order to analyze the carbon emission costs in the tire industry for managerial decisions.

2.3. Application of TOC Theory of Constraints

This article applies mathematical programming models and analyzes the most profitable tire product mix using the ABC approach. In addition, through fixed operating costs and systematic analysis of overhead costs [41], product mix decisions are evaluated via mathematical programming and TOC. The ABC method can be customized to analyze different types of board decision making, including price and product-mix [47–49], design and development [50], environmental management [51,52], green building project, strategy, and construction method selection [53]. Plenert's economic planning included TOC [54], which is an integer linear programming method, to produce a portfolio with diverse restricted capital and employed TOC-based arithmetic to obtain the best combined product mix resolution [49]. This essay uses the TOC approach to determine production priorities. This approach can be applied to a variety of study themes [55,56] to derive the best product mix.

3. Green Production Decision Model for the Tire Industry

3.1. Tire Manufacturing Process

The manufacturing processes of the tire industry are complex and have six main processes: kneading, pressing out, cutting off, forming, vulcanizing, and inspecting (as shown in Figure 1):

1. *Kneading* (j = 1): The first process mixes all the rubber materials and soot, and transports them to the kneader for kneading to change the strength of the rubber.
2. *Pressing out* (j = 2): The second process uses glue and friction to generate heat, allowing the rubber to mature.
3. *Cutting off* (j = 3): The third process mainly uses a cutting machine to cut off various tire sizes.
4. *Forming* (j = 4): The fourth process is to perform the molding assembly work to assembly all the materials, strips, and steel rings to complete the tire prototype.
5. *Vulcanizing* (j = 5): The fifth process is to perform the vulcanization process. The main purpose of tire vulcanization is to rearrange the molecules of the rubber, meaning to heat and mold the tires through steam. The tires that complete this process will become the finished tires.
6. *Inspecting* (j = 6): Finally, the finished tires are inspected and the tires are moved to the warehouse for shipment.

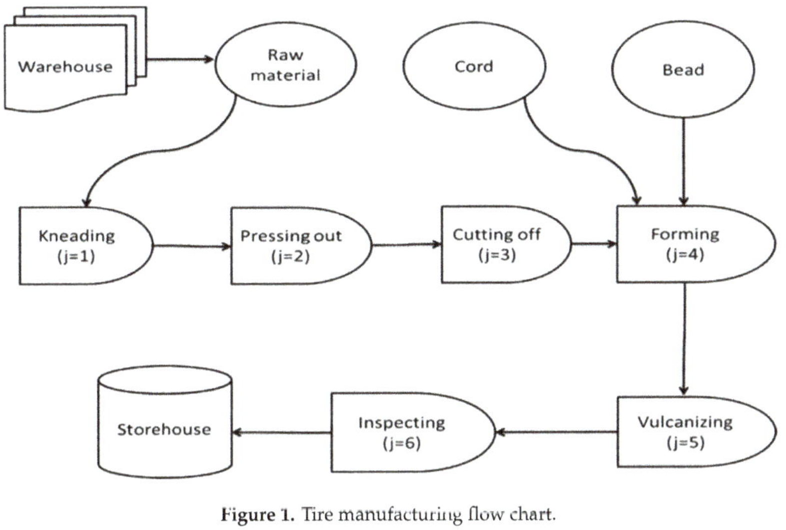

Figure 1. Tire manufacturing flow chart.

3.2. The Model Formulation

From the aforementioned ABC model and the main tire manufacturing process, the corresponding costs are included in the mathematical model. First, this article assumes that the unit product prices are constant within a certain relevant range. Second, the green activity for each production process, as well as its activity driver and related resources, are determined and selected. Third, five materials (natural rubber, soot, synthetic rubber, cord, and bead) used in the production process are the direct material inputs, while other insignificant materials are not included in this study. Fourth, direct labor resources can be increased by additional overtime work with higher wage rates. Fifth, assume that carbon emission costs are considered as variable costs, which depend on carbon emissions quantity and different tax rates. Organic compounds are chemicals that contain carbon and are found in all living things. In addition to carbon, carbon contains other elements, such as hydrogen, oxygen, fluorine, chlorine, bromine, sulfur or nitrogen. Assume that the relationship between carbon tax and carbon emission quantity has been set up by using the concept of carbon equivalency. Sixth, this research is limited by the difficulty of obtaining relevant information, thus, we only consider the direct materials, direct labor, disposal costs, and carbon emissions costs, and assume that all other costs are fixed costs. The data used in this study are in metric tons and U.S. dollars, as shown in Table 1.

Table 1. Data for illustrative example (Company A).

Product					**PCR (i = 1)**	**TBR (i = 2)**	**MC (i = 3)**	**Available Capacity (thousand)**
Maximum demand: (thousand)		f_j	j	B_i	1000	100	1500	
Selling price: (USD)				C_i	310	1010	160	
Direct material:	$k = 1$	21		E_{i1}	4	6	2	$F_1 = 10{,}600$
(USD/ton)	$k = 2$	11		E_{i2}	2	10	1	$F_2 = 8100$
	$k = 3$	6		E_{i3}	5	15	2	$F_3 = 8700$
	$k = 4$	38		E_{i4}	2	6	1	$F_4 = 7100$
	$k = 5$	22		E_{i5}	2	15	1	$F_5 = 7900$
Machine hour constraint:								
Kneading	Machine hours		1	b_{i1}	5	10	1	$H_1 = 13{,}300$
Pressing out	Machine hours		2	b_{i2}	2	4	1	$H_2 = 22{,}100$
Cutting off	Machine hours		3	b_{i3}	2	3	1	$H_3 = 22{,}100$
Forming	Machine hours		4	b_{i4}	3	5	1	$H_4 = 26{,}500$
Vulcanizing	Machine hours		5	b_{i5}	2	3	1	$H_5 = 13{,}300$
Inspecting	Machine hours		6	b_{i6}	3	6	1	$H_6 = 6700$
Material handling constraint:								
Machine hours (h)	$G_1 = 50$		7	b_{i7}	2	3	1	$H_7 = 1860$
Cost (USD)	$G_2 = 150$			a_{i7}	5	10	1	
	$G_3 = 10$							
Direct labor constraint:								
Cost:	$TLC_0 = 7040$	$TLC_1 = 11{,}000$	$TLC_2 = 15{,}840$	d_i	1	1.5	0.5	
Labor hours (h)	$LS_0 = 1760$	$LS_1 = 2200$	$LS_2 = 2640$					
Wage rate (USD/h)	$r_0 = 4$	$r_1 = 9$	$r_2 = 11$					
Carbon emission constraint:								
Cost (USD)	$CT_1 = 20{,}400$	$CT_2 = 46{,}800$	$UCE = 4000$	l_i	0.2	0.1	0.1	
Emission quantities	$CE_1 = 2040$	$CE_2 = 2340$						
Tax rate (USD/ton)	$TR_1 = 10$	$TR_2 = 20$	$TR_3 = 30$					
Total fixed cost: (USD)	22,000							

[Note] Data in Table 1 are used for Model 1, Model 2, and Model 3.

Since the traditional cost system analyzes the cost, most of it is based on direct raw materials, direct labor, and indirect manufacturing costs. The ABC cost system allocates resources to the activity or activity center according to the resource driver, and the activity cost is distributed to the product according to the activity drivers. Through the implementation of the analysis method, the cost of each activity can be correctly calculated. In addition, the TOC theory proposes some standardization methods for defining and eliminating constraints in manufacturing activities. The enterprise is a system, the goal is to obtain more profits, and all the factors that hinder the achievement of the overall goal are constraints, in order to measure the realization. The performance and effectiveness of the target, TOC breaks the traditional concept of accounting costs. In summary, this paper inspires the innovative thinking, applies ABC cost system, TOC theory combined with mathematical planning to establish the model, the objective function and its constraints are described in the sections that follow.

3.2.1. Objective Function

This study establishes the mathematical planning model of the tire industry through the ABC cost system and the TOC theory. Since it is difficult to obtain real cost and expense data, it can be assumed that the factors hindering the target are constraints. Therefore, the following is a discussion of the objective function of the ABC green production model. The objective of the model is to determine the optimal product mix to achieve maximal profit under the constraints of various resources and carbon emissions. The following is the objective of the model:

$$\begin{aligned}\text{Maximize } Z = {} & (\text{Total revenue}) - (\text{Direct material cost}) - (\text{Direct labor cost} \\ & -(\text{Material handing cost}) - (\text{carbon emission cost}) \ (\text{Other fixed cost}) \\ & = \sum_{i=1}^{n} C_i B_i - \sum_{k=1}^{s}\left\{D_k \sum_{i=1}^{n}(E_{ik}B_i)\right\} - (TLC_0 + e_1(TLC_1 - TLC_0) + e_2(TLC_2 - TLC_0)) \\ & -(G_1\beta_1 + G_2\beta_2 + G_3\beta_3) - (TR_1W_1 + TR_2W_2 + TR_3W_3) - F\end{aligned} \tag{1}$$

where B_i is the production quantity of product i; C_i is the unit sales price of product i; D_k is the unit cost of material k; E_{ik} is the requirements of material k for producing a unit of product i; TLC_0 is total direct labor costs at LS_0; TLC_1 is total direct labor costs at LS_1; TLC_2 is total direct labor costs at LS_2; (e_0, e_1, e_2) is a set of non-negative variables of SOS2 (special ordered set of type 2), where at most two adjacent variables may be non-zero in the order of a given set [57,58]; G_i is the other material handling cost, with the exception of machine depreciation per batch of material handling for product i; β_i is the number of batches for material handling of product i; CT_i is the total carbon emission cost in CE_i; F is the fixed cost.

The profit function of Z in the model is shown in Equation (1). Total revenue is $\sum_{i=1}^{n} C_i B_i$ of Equation (1) in the first term. Direct material cost is $\sum_{k=1}^{s}\left\{D_k \sum_{i=1}^{n}(E_{ik}B_i)\right\}$ of Equation (1) in the second term, direct labor cost is $(TLC_0 + e_1(TLC_1 - TLC_0) + e_2(TLC_2 - TLC_0))$ of Equation (1) in the third term, material handing cost is $(G_1\beta_1 + G_2\beta_2 + G_3\beta_3)$ of Equation (1) in the fourth term, carbon emission cost is $(TR_1W_1 + TR_2W_2 + TR_3W_3)$ of Equation (1) in the fifth term, and the other fixed cost is F of Equation (1) in the sixth term. Some other costs are fixed; for example, the depreciations of property, plants, and equipment are fixed, thus, they will have the associated constraints of capacity. While fire insurance charges and fundamental utility fees are also fixed, they will not have the associated constraints of capacity. These fixed costs in total can be expressed as a constant F (USD $22,000 in this example data).

3.2.2. The Associated Constraints

(1) Direct material quantity constraints:

$$\sum_{i=1}^{n}(E_{ik}B_i) \le F_k \ \ k = 1, 2, \ldots, s \tag{2}$$

where F_k is the quantity of material k available for use. Equation (2) represents the constraints for direct materials and the direct material cost is $\sum_{k=1}^{s}\left\{D_k \sum_{i=1}^{n}(E_{ik}B_i)\right\}$ of Equation (1) in the second term.

(2) Direct labor hour constraints:

$$TDL = d_1B_1 + d_2B_2 + d_3B_3 \le LS_0 + e_1(LS_1 - LS_0) + e_2(LS_2 - LS_0) \tag{3}$$

$$e_0 - m_1 \le 0 \tag{4}$$

$$e_1 - m_1 - m_2 \le 0 \tag{5}$$

$$e_2 - m_2 \leq 0 \tag{6}$$

$$e_0 + e_1 + e_2 = 1 \tag{7}$$

$$m_1 + m_2 = 1 \tag{8}$$

$$0 \leq e_0 \leq 1 \tag{9}$$

$$0 \leq e_1 \leq 1 \tag{10}$$

$$0 \leq e_2 \leq 1 \tag{11}$$

where d_i is the requirement of direct labor hours for one unit of product i; B_i is the production quantity of product i; LS_0 is the normal direct labor hours available; LS_1 is the maximal working hours at the first overtime rate plus the normal direct labor hours available; LS_2 is the maximal working hours at the first and second overtime rate plus the normal direct labor hours available. Equations (3)–(11) are the constraints associated with direct labor. TDL in Equation (3) is the total direct labor hours. The direct labor cost is $(TLC_0 + e_1(TLC_1 - TLC_0) + e_2(TLC_2 - TLC_0))$ of Equation (1) in the third term, and the direct labor cost function is shown in Figure 2. (e_0, e_1, e_2) is a set of non-negative variables of SOS2 (special ordered set of type 2), where at most two adjacent variables may be non-zero in the order of a given set [57,58]; (m_1, m_2) is a set of 0–1 variables of SOS1 (special ordered set of type 1), where only one must have one variable that will be non-zero [39,40]. If $m_1 = 1$, then $m_2 = 0$ from Equation (8), $e_2 = 0$ from Equation (6), $e_0, e_1 \leq 1$ from Equations (4) and (5), and $e_0 + e_1 = 1$ from Equation (7). Therefore, total direct labor hours and total labor costs are $LS_0 + e_1(LS_1 - LS_0)$ and $TLC_0 + e_1(TLC_1 - TLC_0)$, respectively, which means that the company works overtime at the first overtime rate. On the other hand, if $m_2 = 1$, then $m_1 = 0$ from Equation (8), $e_0 = 0$ from Equation (4), $e_1, e_2 \leq 1$ from Equations (5) and (6), and $e_1 + e_2 = 1$ from Equation (7). Thus, the total direct labor hours required is $LS_0 + e_1(LS_1 - LS_0) + e_2(LS_2 - LS_0)$ from Equations (3) and the total labor cost is $TLC_0 + e_1(TLC_1 - TLC_0) + e_2(TLC_2 - TLC_0)$, respectively, which means there will be overtime work at the second overtime rate. When $m_1 = 1$ and $e_0 = 1$, then $m_2 = e_1 = e_2 = 0$ from Equations (4)–(9), total direct labor hours = $d_1B_1 + d_2B_2 + d_3B_3 \leq LS_0$, and the total labor cost is TLC_0.

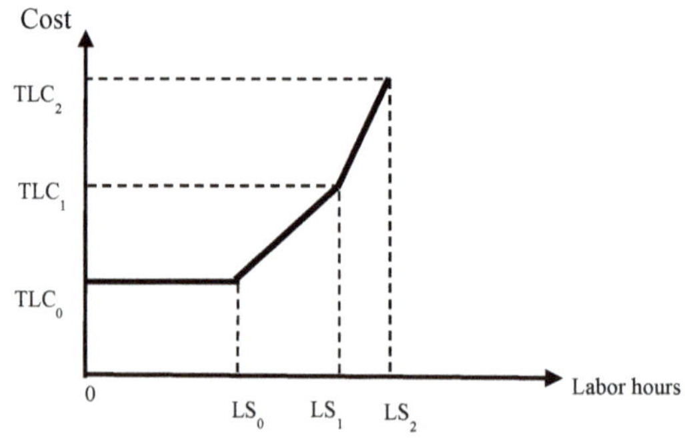

Figure 2. Direct labor cost.

(3) Machine hour constraints:

$$\sum_{i=1}^{n} b_{ij}B_i \leq H_j,\ j = 1, 2, 3, 4, 5, 6 \tag{12}$$

where b_{ij} is the number of machine hours required to produce one unit of product i in activity j; Bi is the production quantity of product i; and H_j is the number of machine hours i available for activity j. Each process requires a different machine for related activities and Equation (12) is the machine hour constraints.

(4) Material handing constraints:

$$(B_i - a_{ij}\beta_i) \le 0,\ i = 1, 2, \ldots, n,\ j = 7 \tag{13}$$

$$\sum_{i=1}^{n} b_{ij}\beta_i \le H_j,\ j = 7 \tag{14}$$

where B_i is the production quantity of product i; β_i is the number of batches for product i; a_{ij} is the quantity of product i in a batch of material handling activity; b_{ij} is the requirement of machine hours for product i in a batch of material handling activity; H_j (j = 7) is the machine hours available for material handling. The material handling activity is a batch-level activity. Equation (13) is the constraints regarding the relationship between the quantity of product i (B_i) and the number of batches for product i; Equation (14) is the resource constraint for the machine hours of material handling activity. In the tire manufacturing process, the relevant materials must be moved from the warehouse to the first process, and then, move the finished tires from the last process to the warehouse. Total material handling cost, with the exception of the depreciation of the material handling machine (it is included in F) is, as follows:

$$\text{Total material handling cos t} = G_1\beta_1 + G_2\beta_2 + G_3\beta_3$$

This is the fourth term of Equation (1), where G_i is the other material handling cost, with the exception of machine depreciation per batch of material handling for product i, and β_i is the number of batches for material handling of product i.

(5) Carbon emissions constraints:

$$GCE = l_1B_1 + l_2B_2 + l_3B_3 = W_1 + W_2 + W_3 \tag{15}$$

$$W_1 + W_2 + W_3 \le UCE \tag{16}$$

$$0 \le W_1 \tag{17}$$

$$W_1 \le CE_1 * n_1 \tag{18}$$

$$CE_1 * n_2 < W_2 \tag{19}$$

$$W_2 \le CE_2 * n_2 \tag{20}$$

$$W_3 > CE_2 * n_3 \tag{21}$$

$$n_1 + n_2 + n_3 = 1 \tag{22}$$

where GCE in the total carbon emission quantity in the tire company; l_i is the carbon emission quantity for one unit of product i; B_i is the production quantity of product i; W_i represents the total carbon emission quantity in the tire company if it falls within the *i*th range of the total carbon emissions in Figure 3; UCE is the upper limit of the carbon emissions, as distributed to the tire company by the government; n_i is the indicator variable and $n_i = 1$ means that the total carbon emission quantity of the tire company falls within the *i*th range of the total carbon emissions in Figure 3; CE_1 and CE_2 are the highest carbon emission quantities for the first and second carbon emission ranges of the carbon emission cost function in Figure 3, respectively.

Levying taxes could increase GDP. Energy taxes will affect the costs of energy-consumption industries due to rising energy costs [59]. On the other hand, energy taxes can help to reduce the concentration of air pollutants, thereby, improving air quality. In addition, calculating capital and operating expenses is based on a simulation of carbon emissions and economic performance under efficiency taxes, as based on the obtained quality and energy balances; the research results of levying carbon taxes show that all of the assessed environmental impact indicators are performing well [60], the new green manufacturing technology (GMT) can reduce carbon emissions [61], which uses a Life Cycle Assessment (LCA) method to measure emissions costs. LCA is a method for assessing the environment over the entire life cycle of a product or service. Based on previous research, we consider all the environmental impacts of carbon emissions in the tire manufacturing process, and reduce carbon footprint through the development of innovative materials and manufacturing sustainability. Other studies have reduced the carbon footprint resistance of trucking by developing innovative low-rolling, reducing truck tires by at least 20% rolling resistance, roughly equivalent to a 5% reduction in fuel consumption and carbon dioxide emissions. Truck tire wear (10% improvement) and safety performance are further improved [43]. According to the method of Ward and Chapman [62], we quantify carbon emissions in Equation (15). The tax rates for carbon emissions differ, and the total emission cost function is a piecewise discontinuity function (as shown in Figure 3). As carbon emissions increase, taxes will increase [63]. That is: TR_1W_1, TR_2W_2, and TR_3W_3 represent the carbon emission cost when total carbon emission falls within the first, the second, and the third range of Figure 3. Total carbon emission costs are, as follows:

$$\text{The total cost of carbon emissions} = TR_1W_1 + TR_2W_2 + TR_3W_3$$

This is the fifth term of Equation (1), i.e., total carbon emissions cost. The related constraints are Equations (15)–(22). In Equations (17)–(22), (n_1, n_2, n_3) is a SOS2 set of 0–1 variables, and only one and just one will be 1. If $n_1 = 1$, then n_2, $n_3 = 0$ from Equation (22), $0 \leq W_1 \leq CE_1$ from Equations (17) and (18), $W_2 = W_3 = 0$ from Equations (19)–(21); that is, the total carbon emission quantity falls within the first range $[0, CE_1]$ of Figure 3, and the total carbon emission cost is TR_1W_1. Similarly, if $n_2 = 1$, then the total carbon emission quantity falls within the first range $(CE_1, CE_2]$ of Figure 3, and the total carbon emission cost is TR_2W_2; if $n_3 = 1$, then the total carbon emission quantity falls within the third range (CE_1, ∞) of Figure 3, and the total carbon emission cost is TR_3W_3. The cost function of Figure 3 is a carbon emission cost function with full progressive tax rates. In summary, the more the tire company generates carbon emissions, the more it should pay for carbon emission taxes.

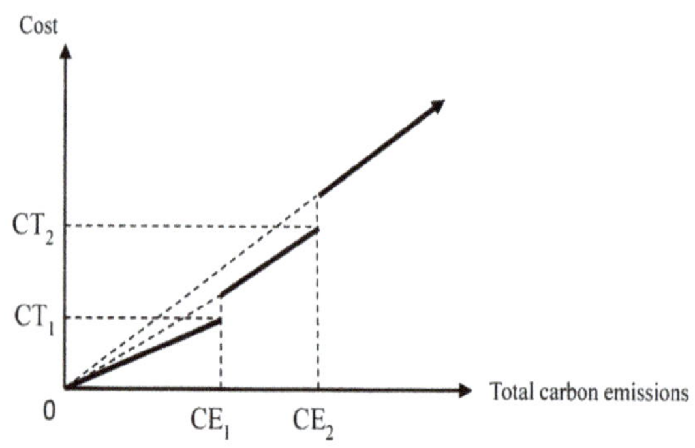

Figure 3. Carbon emission costs (taxes).

3.3. *Summary*

In summary, the above discussion includes the manufacturing process of the tire industry, the application of ABC cost system and TOC theory in mathematical programming, revenue, various costs, and carbon emissions in the objective function and the associated constraints. In the following sections, scenario analysis will be conducted using the assumed three carbon emission cost models. The results of the analysis will be provided for management decision-making.

4. Illustration

The tire company burns exhaust during the production process, and the emission of carbon seriously affects air pollution. This paper proposes a hypothetical discussion of a green decision model for tire Company A. As an illustration, company A in the tire industry has been established in Taiwan for more than 50 years. There are 11 tire manufacturing companies in the world, with more than 24,000 employees worldwide and paid-in capital exceeding 30 billion yuan. The main products are all kinds of tires and rubber products. Company A is used to illustrate how to apply the green decision-making model proposed in this paper. We assume that Company A's costs for producing their products include direct material cost, direct labor cost, kneading activity cost, pressing out activity cost, cutting off activity cost, forming activity cost, vulcanizing activity cost, inspecting activity cost, material handing activity cost, and carbon emission cost (tax).

4.1. Model 1: Green Decision Model without Carbon Rights Trading, but Has a Carbon Cap Allocated for Use by the Government

When using Equations (1)–(22), let Company A consider the production of three main products: Passenger Car Radial (PCR) product (i = 1), Truck & Bus Radial (TBR) product (i = 2), and Motorcycle (MC) product (i = 3), and that these products consume the same direct materials. Then, each product's quantity is shown in thousands, and the costs and input costs are in U.S. dollars. The data for the case are shown in Table 1. Total fixed cost is a constant (USD $22,000 in this example). This study uses the illustrative example data in Table 1 for the three green decision models explored in this paper, including a model without carbon rights trading and two models with two kinds of carbon rights trading costs, in order to analyze the carbon emission costs in the tire industry for managerial decisions.

The green product mix decision model without carbon rights trading for the case data in Table 1 is shown in Table 2. We solved this o-1 mixed integer programming (0–1 MIP) model by using LINGO 16.0 software, and the optimal solution is shown at the bottom area of Table 2.

The total profit Z is USD $53,225,000. It is assumed that the government allocates a carbon emission quantity to Company A, which has been converted to the carbon emission quantity UCE = 4,000,000 metric tons, and the carbon emission cost (tax) is shown in Figure 3. The carbon tax cost function f_z(GCE) are, as follows:

$$f_z\,(\mathrm{GCE}) = \begin{cases} \mathrm{TR}_1 * \mathrm{GCE} \text{ if } 0 \le \mathrm{GCE} \le \mathrm{CE}_1 \\ \mathrm{TR}_2 * \mathrm{GCE} \text{ if } \mathrm{CE}_1 < \mathrm{E} \le \mathrm{CE}_2 \\ \mathrm{TR}_3 * \mathrm{GCE} \text{ if } \mathrm{GCE} > \mathrm{CE}_2 \end{cases}$$

The optimal solution in Table 2 indicates that the total carbon emission quantity is 335,000 metric tons and the carbon emission cost is USD $3,350,000, which falls in the first range of carbon emission quantity in Figure 3. If the remaining 3,665,000 metric tons of carbon emission quantity can be sold to other companies that need to purchase additional carbon credits, according to the EU (2008–2012) under the emissions trading scheme, and is sold at USD $30 per metric ton of carbon dioxide, the tire industry could add an additional profit of USD $109,950,000 to increase its existing maximum gross profit to USD $163,175,000 (=$53,225,000 + $109,950,000).

According to the results of this scenario analysis as shown in Table 2, if the tire industry controls the carbon emission cost according to this mathematical programming model and sells the remaining carbon emissions allocated from the government to the companies that need to purchase carbon emissions. In addition to the original income of the products, the company can also get a considerable income to sell the remaining carbon emissions. It can be seen that carbon emissions are indeed a very important indicator variable affecting the tire company's profit, and the tire industry should carefully control carbon emissions to avoid cost explosion on the premise of maximizing profits.

Table 2. The model and its optimal solutions with the illustrative data for Model 1.

<table>
<tr><td colspan="4">Max Z = $(310^*B_1 + 1010^*B_2 + 160^*B_3) - (21^*4 + 11^*2 + 6^*5 + 38^*2 + 22^*2)^*B_1$
$- (21^*6 + 11^*10 + 6^*15 + 38^*6 + 22^*15)^*B_2\text{-}(21^*2 + 11^*1 + 6^*2 + 38^*1 + 22^*1)^*B_3$
$- 7040 - (11{,}000 - 7040)^*e_1 - (15{,}840 - 7040)^*e_2 - 50^*\beta_1 - 150^*\beta_2 - 10^*\beta_3$
$- 10^*W_1 - 20^*W_2 - 30^*W_3 - F$</td></tr>
<tr><td colspan="2">Subject to direct material;
$4^*B_1 + 6^*B_2 + 2^*B_3 \le 10{,}600;$
$2^*B_1 + 10^*B_2 + 1^*B_3 \le 8100;$
$5^*B_1 + 15^*B_2 + 2^*B_3 \le 8700;$
$2^*B_1 + 6^*B_2 + 1^*B_3 \le 7100;$
$2^*B_1 + 15^*B_2 + 1^*B_3 \le 7900;$</td><td colspan="2">Subject to machine hours;
$5^*B_1 + 10^*B_2 + 1^*B_3 \le 13{,}300;$
$2^*B_1 + 4^*B_2 + 1^*B_3 \le 22{,}100;$
$2^*B_1 + 3^*B_2 + 1^*B_3 \le 22{,}100;$
$3^*B_1 + 5^*B_2 + 1^*B_3 \le 26{,}500;$
$2^*B_1 + 3^*B_2 + 1^*B_3 \le 13{,}300;$
$3^*B_1 + 6^*B_2 + 1^*B_3 \le 6700;$</td></tr>
<tr><td colspan="2">Subject to direct labor;
$1^*B_1 + 1.5^*B_2 + 0.5^*B_3 - 1760\text{-}440^*e_1 - 880^*e_2 \le 0;$
$e_0 - m_1 \le 0;$
$e_1 - m_1 - m_2 \le 0;$
$e_2 - m_2 \le 0;$
$e_0 + e_1 + e_2 = 1;$
$m_1 + m_2 = 1;$
$0 \le e_0 \le 1; 0 \le e_1 \le 1; 0 \le e_2 \le 1; m_1, m_2 = 0{,}1;$</td><td colspan="2">Subject to Material handing;
$B_1 - 5^*\beta_1 = 0;$
$B_2 - 10^*\beta_2 = 0;$
$B_3 - 1^*\beta_3 = 0;$
$2^*\beta_1 + 3^*\beta_2 + 1^*\beta_3 \le 1860;$</td></tr>
<tr><td colspan="4">Subject to carbon emission;
$GCE = 0.2B_1 + 0.1B_2 + 0.1B_3,$
$GCE = W_1 + W_2 + W_3;$
$W_1 + W_2 + W_3 \le 4000;$
$0 \le W_1; W_1 \le 2040n_1;$
$2040^*n_2 < W_2; W_2 \le 2340^*n_2;$
$W_3 > 2340^*n_3;$
$n_1 + n_2 + n_3 = 1.$</td></tr>
<tr><td colspan="4">The optimal solutions for the example data</td></tr>
<tr><td>$B_1 = 895{,}000$
$e_0 = 1$
$\beta_1 = 181{,}000$
$n_1 = 1$
$W_1 = 335{,}000$
$m_1 = 1$</td><td>$B_2 = 85{,}000$
$e_1 = 0$
$\beta_2 = 8000$
$n_2 = 0$
$W_2 = 0$
$m_2 = 0$</td><td>$B_3 = 1{,}475{,}000$
$e_2 = 0$
$\beta_3 = 1{,}475{,}000$
$n_3 = 0$
$W_3 = 0$</td><td>$F = 22{,}000{,}000$
$Z = 53{,}225{,}000$
$GCE = 335{,}000$</td></tr>
</table>

4.2. Model 2: Green Decision Model with Constant Carbon Right Cost

Model 2 assumes that the company can purchase carbon rights with a constant unit cost (R). The carbon right cost function is equal to the $f_4(GCE - UCE)$, as shown in Figure 4, if the company has the need to buy carbon right for additional sales.

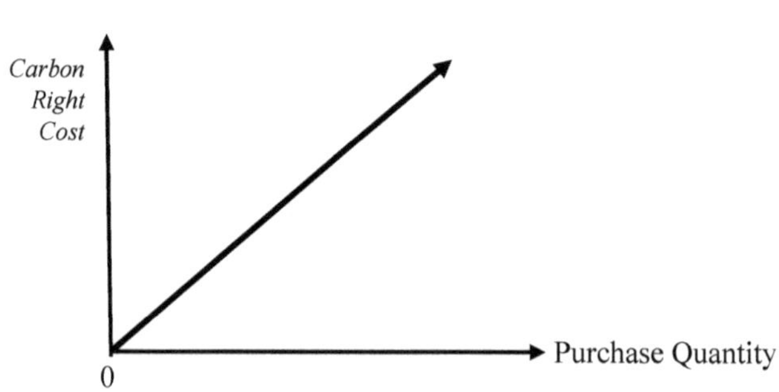

Figure 4. Carbon right cost.

The company's carbon emission cost function is equal to the carbon tax cost plus the carbon right purchase cost. The carbon emission cost function f_2 (GCE) is as follows:

$$f_2\,(\mathrm{GCE}) = \begin{cases} f_z(\mathrm{GCE}) \text{ if } 0 \leq \mathrm{GCE} \leq \mathrm{UCE} \\ f_z(\mathrm{GCE}) + f_4(\mathrm{GCE} - \mathrm{UCE}) \text{ if } \mathrm{GCE} > \mathrm{E} \end{cases}$$

where $f_z(\mathrm{GCE})$ is the carbon tax cost function and $f_4(\mathrm{GCE} - \mathrm{UCE})$ is carbon right cost function. When the company's carbon emission quantity does not exceed the carbon emission quota (UCE), the carbon emission cost only includes the carbon tax cost $f_z(\mathrm{GCE})$, otherwise, the company needs to buy carbon right, and the company's carbon emission cost is $f_z(\mathrm{GCE}) + f_4(\mathrm{GCE} - \mathrm{UCE})$.

The objective for Model 2 is shown as follows:

$$\begin{aligned} \mathrm{Maximize}Z = & \text{(Total revenue)} - \text{(Direct material cost)} - \text{(Direct labor cost} \\ & -\text{(Material handing cost)} - \text{(carbon emission cost) (Other fixed cost)} \\ & = \sum_{i=1}^{n} C_iB_i - \sum_{k=1}^{s}\left\{D_k \sum_{i=1}^{n}(E_{ik}B_i)\right\} - (TLC_0 + e_1(TLC_1 - TLC_0) + e_2(TLC_2 - TLC_0)) \\ & -(G_1\beta_1 + G_2\beta_2 + G_3\beta_3) - \{(TR_1W_1 + TR_2W_2 + TR_3W_3) * \theta_1 \\ & +[(TR_1W_1 + TR_2W_2 + TR_3W_3) + R * (GCE - UCE)] * \theta_2\} - F \end{aligned} \tag{23}$$

The additional associated constraints are, as follows:

$$\mathrm{GCE} = l_1B_1 + l_2B_2 + l_3B_3 = CQ_1 + CQ_2 \tag{24}$$

$$0 \leq CQ_1 \tag{25}$$

$$CQ_1 \leq \mathrm{UCE} * \theta_1 \tag{26}$$

$$\mathrm{UCE} * \theta_2 < CQ_2 \tag{27}$$

$$CQ_2 \leq (\mathrm{UCE} + \mathrm{UCPQ}) * \theta_2 \tag{28}$$

$$\theta_1 + \theta_2 = 1,\ \theta_1,\ \theta_2 = 0,\ 1 \tag{29}$$

where R is the unit cost of carbon right; UCPQ is the maximum quantity of carbon right that the company can purchase. θ_1 and θ_2 are indictor variables of 0–1 variables. If $\theta_1 = 1$, then $\theta_2 = 0$ from Equation (29), $CQ_2 = 0$ from Equations (27) and (28), and $0 \leq \mathrm{GCE} = CQ_1 \leq \mathrm{UCE}$ from Equations (25) and (26). That is, the company need not purchase the carbon rights, and the total carbon emission cost is (TR_1W_1 + TR_2W_2 + TR_3W_3) from Equation (23). On the other hand, if $\theta_2 = 1$, then $\theta_1 = 0$ from Equation (29), $CQ_1 = 0$ from Equations (25) and (26), and $\mathrm{UCE} \leq \mathrm{GCE} = CQ_2 \leq$ (UCE + UCPQ) from Equations (27) and (28). That is, the company needs to purchase carbon rights at the quantity, (GCE − UCE), and the total carbon emission cost is [(TR_1W_1 + TR_2W_2 + TR_3W_3) + R*(GCE − UCE)]. Other variables and parameters are as previously defined.

Equations (2)–(29) form Model 2, as explored in this paper. In Model 2, it is assumed that the carbon rights can be purchased regarding the required carbon emissions, and according to the EU (2008–2012) under the emissions trading scheme, the constant cost per unit tonne of carbon dioxide is USD $30 (R). Assume also that the maximal quantity of carbon rights that the company can purchase (UCPQ) is 160 tonnes of carbon dioxide. Model 2 with R = USD $30, UCPQ = 160, and the illustrative data in Table 1, is as shown in Table 3. We solve this problem and the optimal solution is shown at the bottom area of Table 3. The optimal solution indicates that the required carbon footprint of Company A exceeds the maximum amount of carbon credits, as allocated by the government.

Table 3. The model and its optimal solutions with the illustrative data for Model 2.

<table>
<tr><td colspan="4">Max Z = (310*B₁ + 1010*B₂ + 160*B₃) − (21*4 + 11*2 + 6*5 + 38*2 + 22*2)*B₁
− (21*6 + 11*10 + 6*15 + 38*6 + 22*15)*B₂ − (21*2 + 11*1 + 6*2 + 38*1 + 22*1)*B₃
− 7040 − (11,000 − 7040) *e₁ − (15,840 − 7040) *e₂ − 50*β₁ − 150*β₂ − 10*β₃
− {(10*W₁ − 20*W₂ − 30*W₃)*θ₁ + [(10*W₁ − 20*W₂ − 30*W₃) + 30*(GCE − 4000)]*θ₂} − F</td></tr>
<tr><td colspan="2">Subject to direct material;
4*B₁ + 6* B₂ + 2* B₃ ≤ 10,600;
2*B₁ + 10*B₂ + 1*B₃ ≤ 8100;
5*B₁ + 15*B₂ + 2*B₃ ≤ 8700;
2*B₁ + 6*B₂ + 1*B₃ ≤ 7100;
2*B₁ + 15*B₂ + 1*B₃ ≤ 7900;</td><td colspan="2">Subject to machine hours;
5*B₁ + 10*B₂ + 1*B₃ ≤ 13,300;
2*B₁ + 4*B₂ + 1*B₃ ≤ 22,100;
2*B₁ + 3*B₂ + 1*B₃ ≤ 22,100;
3*B₁ + 5*B₂ + 1*B₃ ≤ 26,500;
2*B₁ + 3*B₂ + 1*B₃ ≤ 13,300;
3*B₁ + 6*B₂ + 1*B₃ ≤ 6700;</td></tr>
<tr><td colspan="2">Subject to direct labor;
1*B₁ + 1.5*B₂ + 0.5*B₃ − 1760 − 440*e₁ − 880*e₂ = 0;
e₀ − m₁ ≤ 0;
e₁ − m₁ − m₂ ≤ 0;
e₂ − m₂ ≤ 0;
e₀ + e₁ + e₂ = 1; m₁ + m₂ = 1;
0 ≤ e₀ ≤ 1; 0 ≤ e₁ ≤ 1; 0 ≤ e₂ ≤ 1; m₁, m₂ = 0,1;</td><td colspan="2">Subject to batch handing;
B₁ − 5*β₁ = 0;
B₂ − 10*β₂ = 0;
B₃ − 1*β₃ = 0;
2*β₁ + 3*β₂ + 1*β₃. ≤ 1860;</td></tr>
<tr><td colspan="4">Subject to carbon emission;</td></tr>
<tr><td colspan="2">GCE = 0.2 ∗ B₁ + 0.1 ∗ B₂ + 0.1 ∗ B₃;
GCE = W₁ + W₂ + W₃;
W₁ + W₂ + W₃ ≤ 4000;</td><td colspan="2">0 ≤ W₁; W₁ ≤ 2040*n₁;
2040*n₂ < W₂; W₂ ≤ 2340*n₂;
W₃ > 2340*n₃;
n₁ + n₂ + n₃ = 1;</td></tr>
<tr><td colspan="4">Subject to carbon emission right purchasing</td></tr>
<tr><td colspan="2">GCE = CQ₁ + CQ₂;
0 ≤ CQ₁;
CQ₁ ≤ 4000*θ₁;</td><td colspan="2">4000*θ₂ < CQ₂;
CQ₂ ≤ (4000 + 160)* θ₂;
θ₁ + θ₂ = 1, θ₁, θ₂ = 0, 1</td></tr>
<tr><td colspan="4">The optimal solutions for the example data</td></tr>
<tr><td>B₁ = 905,000</td><td>B₂ = 80,000</td><td>B₃ = 1,470,000</td><td>F = 22,000,000</td></tr>
<tr><td>e₀ = 1</td><td>e₁ = 0</td><td>e₂ = 0</td><td>Z = 53,050,000</td></tr>
<tr><td>β₁ = 181,000</td><td>β₂ = 8000</td><td>β₃ = 1,470,000</td><td></td></tr>
<tr><td>n₁ = 1</td><td>n₂ = 0</td><td>n₃ = 0</td><td></td></tr>
<tr><td>W₁ = 336,000</td><td>W₂ = 0</td><td>W₃ = 0</td><td></td></tr>
<tr><td>CQ₁ = 336,000</td><td>CQ₂ = 0</td><td>GCE= 336,000</td><td></td></tr>
<tr><td>m₁ = 1</td><td>m₂ = 0</td><td></td><td></td></tr>
<tr><td>θ₁ = 1</td><td>θ₂ = 0</td><td></td><td></td></tr>
</table>

The results of this scenario analysis are compared with scenario 1 and it is found that scenario 2 assumes that carbon rights are purchased, the maximum profit will be reduced by USD $175,000 (from USD $53,225,000 to USD $53,050,000) and carbon emissions will increase by 1000 metric tons. (by 335,000 metric tons increased to 336,000 metric tons). The results of this scenario analysis also prove that carbon emissions are indeed variables that affect the total profitability. It can also provide the tire industry in the future when purchasing carbon rights, we should carefully consider the company's overall revenue, and avoid the purchase and use of carbon rights, so that the cost can be optimally controlled, so that the company will continue to operate.

4.3. Model 3: Green Decision Model with a Minimum Purchase Quantity of Carbon Right

Model 3 assumes that the company should purchase the carbon right with a minimum purchase quantity (LCPQ) that has a constant charge (CF) if the company has the need for additional sales, as shown in Figure 5 and the carbon emission cost function f_3 (GCE) is as follows:

$$f_3\,(\mathrm{GCE}) = \begin{cases} f_z(\mathrm{GCE}) \text{ if } 0 \leq \mathrm{GCE} \leq \mathrm{UCE} \\ f_z(\mathrm{GCE}) + \mathrm{CF} \text{ if } \mathrm{UCE} < \mathrm{GCE} \leq (\mathrm{UCE} + \mathrm{LCPQ}) \\ f_z(\mathrm{GCE}) + \mathrm{CF} + f_4(\mathrm{GCE} - \mathrm{UCE} - \mathrm{LCPQ}) * \mathrm{R} \text{ if } (\mathrm{UCE} + \mathrm{LCPQ}) < \mathrm{GCE} \leq (\mathrm{UCE} + \mathrm{LCPQ}) \end{cases}$$

where, there are three scenarios for the carbon emission cost function: (1) the company's carbon emission quantity does not exceed the carbon emission quota (UCE), (2) the company needs to buy the carbon right whose quantity does not exceed the LCPQ, (3) the company needs to buy the carbon right whose quantity between LCPQ and UCPQ.

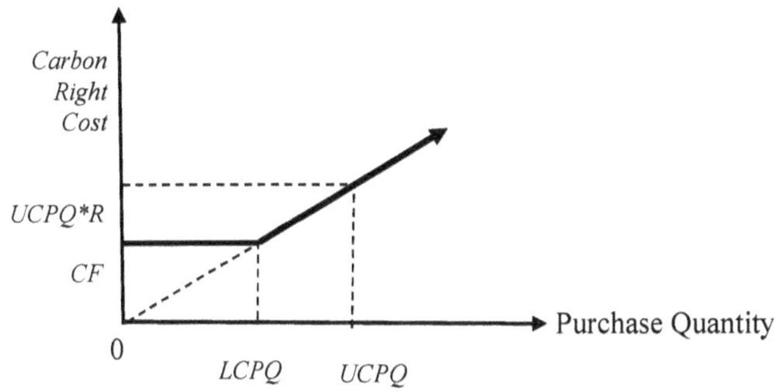

Figure 5. Carbon right cost.

The objective for Model 3 is shown as follows:

$$\begin{aligned}\text{Maximize}Z = & \ (\text{Total revenue}) - (\text{Direct material cost}) - (\text{Direct labor cost} \\ & -(\text{Material handing cost}) - (\text{carbon emission cost}) \ (\text{Other fixed cost}) \\ & = \sum_{i=1}^{n} C_iB_i - \sum_{k=1}^{s}\left\{D_k \sum_{i=1}^{n}(E_{ik}B_i)\right\} - (TLC_0 + e_1(TLC_1 - TLC_0) + e_2(TLC_2 - TLC_0)) \\ & -(G_1\beta_1 + G_2\beta_2 + G_3\beta_3) - \{(TR_1W_1 + TR_2W_2 + TR_3W_3) * \theta_1 \\ & +[(TR_1W_1 + TR_2W_2 + TR_3W_3) + CF] * \theta_2 \\ & +[(TR_1W_1 + TR_2W_2 + TR_3W_3) + CF + R * (GCE - UCE - LCPQ)] * \theta_3\} - F\end{aligned} \quad (30)$$

The additional associated constraints are, as follows:

$$GCE = l_1B_1 + l_2B_2 + l_3B_3 = CQ_1 + CQ_2 + CQ_3 \quad (31)$$

$$0 \leq CQ_1 \quad (32)$$

$$CQ_1 \leq UCE * \theta_1 \quad (33)$$

$$UCE * \theta_2 < CQ_2 \quad (34)$$

$$CQ_2 \leq (UCE + LCPQ) * \theta_2 \quad (35)$$

$$(UCE + LCPQ) * \theta_3 < CQ_3 \quad (36)$$

$$CQ_3 \leq (UCE + UCPQ) * \theta_3 \quad (37)$$

$$\theta_1 + \theta_2 + \theta_3 = 1,\ \theta_1,\ \theta_2,\ \theta_3 = 0,\ 1 \quad (38)$$

where R is the unit cost of carbon right; LCPQ is the minimum purchase quantity; and UCPQ is the maximum quantity of carbon right that the company can purchase. θ_1, θ_2, and θ_3 are indictor variables of 0–1 variables. If $\theta_1 = 1$, then $\theta_2 = \theta_3 = 0$ from Equation (38), $CQ_2 = CQ_3 = 0$ from Equations (34)–(37), and $0 \leq GCE = CQ_1 \leq UCE$ from Equations (32)–(33). That is, the company need not purchase carbon rights, and the total carbon emission cost is (TR_1W_1 + TR_2W_2 + TR_3W_3) from Equation (30). If $\theta_2 = 1$, then $\theta_1 = \theta_3 = 0$ from Equation (38), $CQ_1 = CQ_3 = 0$ from Equations (32)–(33) and (36)–(37), and $UCE \leq GCE = CQ_2 \leq (UCE + LCPQ)$ from Equations (34)–(35). That is, the company needs to purchase carbon rights at the minimum purchase quantity, LCPQ, and the total carbon emission cost is [(TR_1W_1 + TR_2W_2 + TR_3W_3) + CF]. If $\theta_3 = 1$, then $\theta_1 = \theta_2 = 0$. from

Equation (38), $CQ_1 = CQ_2 = 0$ from Equations (32)–(35), and $(UCE + LCPQ) \leq GCE = CQ_3 \leq (UCE + UCPQ)$ from Equations (36)–(37). That is, the company needs to purchase carbon rights exceeding the minimum purchase quantity, LCPQ, and the total carbon emission cost is $[(TR_1W_1 + TR_2W_2 + TR_3W_3) + CF + R^*(GCE - UCE - LCPQ)]$. Other variables and parameters are as previously defined.

Equations (2)–(22) and (30)–(38) form Model 3, as studied in this paper. As in Model 2, it is assumed that R = USD \$30 and UCPQ = 160. Assume also that the minimum purchase of carbon rights is 60,000 metric tonnes, i.e., LCPQ = 60; then, CF = R*LCPQ = \$1800. Model 3, with these data and the illustrative data in Table 1, is shown in Table 4. We solve this problem and the optimal solution is shown at the bottom area of Table 4. The optimal solution indicates that the required carbon footprint of 4,060,000 tonnes for Company A is equal to (UCE + LCPQ). It means that Company A needs to obtain just the minimum purchase quantity of 60,000 metric tonnes of carbon rights in addition to the permissible no need to buy carbon rights emission quantity (UCE), as allocated by the government. As a result, the optimal green portfolio decision and the maximum total profit obtained is USD \$53,050,000.

Table 4. The model and its optimal solutions with the illustrative data for Model 3.

<table>
<tr><td colspan="4">Max Z = $(310^*B_1 + 1010^*B_2 + 160^*B_3) - (21^*4 + 11^*2 + 6^*5 + 38^*2 + 22^*2)^*B_1$
$- (21^*6 + 11^*10 + 6^*15 + 38^*6 + 22^*15)^*B_2 \text{-} (21^*2 + 11^*1 + 6^*2 + 38^*1 + 22^*1)^*B_3$
$- 7040 - (11{,}000 - 7040)\ ^*e_1 - (15{,}840 - 7040)\ ^*e_2 - 50^*\beta_1 - 150^*\beta_2 - 10^*\beta_3$
$- \{(10^*W_1 - 20^*W_2 - 30^*W_3)^*\theta_1 + [(10^*W_1 - 20^*W_2 - 30^*W_3) + 1800]^*\theta_2$
$+ [(10^*W_1 - 20^*W_2 - 30^*W_3) + 1800 + 30^*(GCE - 4000 - 60)]^*\ \theta_3\} - F$</td></tr>
<tr><td colspan="2">Subject to direct material;
$4^*B_1 + 6^*\ B_2 + 2^*\ B_3 \leq 10{,}600$;
$2^*B_1 + 10^*B_2 + 1^*B_3 \leq 8100$;
$5^*B_1 + 15^*B_2 + 2^*B_3 \leq 8700$;
$2^*B_1 + 6^*B_2 + 1^*B_3 \leq 7100$;
$2^*B_1 + 15^*B_2 + 1^*B_3 \leq 7900$;</td><td colspan="2">Subject to machine hours;
$5^*B_1 + 10^*B_2 + 1^*B_3 \leq 13{,}300$;
$2^*B_1 + 4^*B_2 + 1^*B_3 \leq 22{,}100$;
$2^*B_1 + 3^*B_2 + 1^*B_3 \leq 22{,}100$;
$3^*B_1 + 5^*B_2 + 1^*B_3 \leq 26{,}500$;
$2^*B_1 + 3^*B_2 + 1^*B_3 \leq 13{,}300$;
$3^*B_1 + 6^*B_2 + 1^*B_3 \leq 6700$;</td></tr>
<tr><td colspan="2">Subject to direct labor;
$1^*B_1 + 1.5^*B_2 + 0.5^*B_3 - 1760 - 2200^*e_1 - 2640^*e_2 = 0$;
$e_0 - m_1 \leq 0$;
$e_1 - m_1 - m_2 \leq 0; e_2 - m_2 \leq 0$;
$e_0 + e_1 + e_2 = 1; m_1 + m_2 = 1$;
$0 \leq e_0 \leq 1; 0 \leq e_0 \leq 1; 0 \leq e_0 \leq 1; m_1, m_2 = 0{,}1$;</td><td colspan="2">Subject to batch handing;
$B_1 - 5^*\beta_1 = 0$;
$B_2 - 10^*\beta_2 = 0$;
$B_3 - 1^*\beta_3 = 0$;
$2^*\beta_1 + 3^*\beta_2 + 1^*\beta_3 \leq 1860$;</td></tr>
<tr><td colspan="4">Subject to carbon emission;</td></tr>
<tr><td colspan="2">$GCE = 0.2B_1 + 0.1B_2 + 0.1B_3$,
$GCE = W_1 + W_2 + W_3$;
$W_1 + W_2 + W_3 \leq 4000$;</td><td colspan="2">$0 \leq W_1; W_1 \leq 2040^*n_1$;
$2040^*n_2 < W_2; W_2 \leq 2340^*n_2$;
$W_3 > 2340^*n_3$;
$n_1 + n_2 + n_3 = 1$;</td></tr>
<tr><td colspan="4">Subject to carbon emission right purchasing</td></tr>
<tr><td colspan="2">$GCE = CQ_1 + CQ_2 + CQ_3$;
$0 \leq CQ_1; CQ_1 \leq 4000^*\theta_1$;
$4000^*\theta_2 < CQ_2; CQ_2 \leq (4000 + 60)^*\theta_2$;</td><td colspan="2">$(4000 + 60)^*\theta_3 < CQ_3$;
$CQ_3 \leq (4000 + 160)^*\theta_3$;
$\theta_1 + \theta_2 + \theta_3 = 1, \theta_1, \theta_2, \theta_3 = 0, 1$</td></tr>
<tr><td colspan="4">The optimal solutions for the example data</td></tr>
<tr><td>$B_1 = 905{,}000$</td><td>$B_2 = 80{,}000$</td><td>$B_3 = 1{,}470{,}000$</td><td>F = 22,000,000</td></tr>
<tr><td>$e_0 = 1$</td><td>$e_1 = 0$</td><td>$e_2 = 0$</td><td>Z = 53,050,000</td></tr>
<tr><td>$\beta_1 = 181{,}000$</td><td>$\beta_2 = 8000$</td><td>$\beta_3 = 1{,}470{,}000$</td><td></td></tr>
<tr><td>$n_1 = 1$</td><td>$n_2 = 0$</td><td>$n_3 = 0$</td><td></td></tr>
<tr><td>$W_1 = 336{,}000$</td><td>$W_2 = 0$</td><td>$W_3 = 0$</td><td>GCE = 336,000</td></tr>
<tr><td>$CQ_1 = 336{,}000$</td><td>$CQ_2 = 0$</td><td>$CQ_3 = 0$</td><td></td></tr>
<tr><td>$\theta_1 = 1$</td><td>$\theta_2 = 0$</td><td>$\theta_3 = 0$</td><td></td></tr>
<tr><td>$m_1 = 1$</td><td>$m_2 = 0$</td><td></td><td></td></tr>
</table>

The result of this scenario is the same as scenario 2. It means that if the tire industry needs to purchase carbon rights, it will affect the total profit. From the results of the scenario analysis, it is further proved that the purchase of carbon rights will affect the total profit of the tire industry. If carbon tax is levied, in the cost control, special attention should be paid to the quantity of carbon emissions for which there is no need to buy carbon rights to avoid cost explosions.

4.4. Sensitivity Analysis

One of important things for enterprises is the maximization of total profits, thus, this article focuses on high-carbon emission and high-polluting tire industries. With the rise of environmental awareness, corporate carbon emissions policies have been implemented in the branches of other countries. Insufficient carbon emissions due to the allocation of the country will require the additional acquisition of carbon rights, which will result in an additional burden on costly carbon emissions. To further understand the impact of the three key variables of the carbon emissions trading model on total profits in this study, three types of model carbon tax rates were increased or decreased by 5% to 30% for a total of 12 cases for sensitivity analysis (Tables 5–7). The test results are, as follows:

(1) *Model 1*: The original total profit of the company is USD $53,225, and the carbon tax rate is increased by 5%, 10%, 15%, 20%, 25%, and 30%, respectively. The total profit found was reduced by 0.31% (USD $53,057,500), 0.62% (USD $52,890,000), 0.94% (USD $52,772.50), 1.25% (USD $52,555,000), 1.57% (USD) $52,387.50), and 1.88% (USD $52,200), respectively; at the relative decreases of 5%, 10%, 15%, 20%, and 25%, then the total profit was found to increase by 0.31% (USD $53,392.50), 0.62% (USD $53,560), and 0.94% (USD $53,727.50), 1.25% (USD $53,895), 1.57% (USD $54,062.50), and 1.88% (USD $54,230), respectively. Therefore, as the carbon tax rate decreases, the total profit of the enterprise will increase. Conversely, increasing the carbon tax rate will lead to a decrease in total corporate profits.

(2) *Model 2*: The original total profit of the company was USD $53,225, the carbon tax rate was increased by 5%, 10%, 15%, 20%, and 25%, respectively, and the total profit was found to decrease by 0.31% (USD $52,882), 0.63% (USD $52,714), 0.95% (USD $52,546), 1.26% (USD $52,378), 1.58% (USD $52,210), and 1.9% (USD $52,042), respectively; at the relative decrease of 5%, 10%, 15%, 20%, 25%, and 30%, it is found that the total profit increased by 0.31% (USD $53,218), 0.63% (USD $53,386), 0.95% (USD $53,554), 1.26% (USD $53,722), 1.58% (USD $53,890), and 1.9% (USD $54,058), respectively. Therefore, as the carbon tax rate decreases, the total profit of the company will increase. Conversely, increasing the carbon tax rate will lead to a decrease in total corporate profits.

(3) *Model 3*: The original total profit of the company is USD $53,050, and the carbon tax rate is increased by 5%, 10%, 15%, 20%, 25%, and 30% respectively. The total profit found was reduced by 0.31% (USD $52,882), 0.63% (USD $52,714), 0.95% (USD $52,546), 1.26% (USD $52,378), 1.58% (USD $52,210), and 1.9% (USD $52,042), respectively. At the relative decrease of 5%, 10%, 15%, 20%, 25%, and 30%, it is found that the total profit increased by 0.31% (USD $53,218), 0.63% (USD $53,386), 0.95% (USD $53,554), 1.26% (USD $53,722), 1.58% (USD $53,890), and 1.9% (USD $54,058), respectively. Therefore, as the carbon tax rate decreases, the total profit of the company will increase. Conversely, increasing the carbon tax rate will lead to a decrease in total corporate profits.

Table 5. Sensitivity analysis on the carbon tax of model 1.

Carbon Tax Decrease/Increase Ratio (%)	Profit (A)	Increase/Decrease (Compared with the Initial Value) (%) (B) = (D)/(C)	Original Profit (C)	Increase Profit (D) = (A) − (C)
−30%	54,230.00	1.88%	53,225.00	1005.00
−25%	54,062.50	1.57%	53,225.00	837.50
−20%	53,895.00	1.25%	53,225.00	670.00
−15%	53,727.50	0.94%	53,225.00	502.50
−10%	53,560.00	0.62%	53,225.00	335.00
−5%	53,392.50	0.31%	53,225.00	167.50
0%	53,225.00	0.00%	53,225.00	0.00
5%	53,057.50	−0.31%	53,225.00	−167.50
10%	52,890.00	−0.62%	53,225.00	−335.00
15%	52,722.50	−0.94%	53,225.00	−502.50
20%	52,555.00	−1.25%	53,225.00	−670.00
25%	52,387.50	−1.57%	53,225.00	−837.50
30%	52,220.00	−1.88%	53,225.00	−1005.00

Unit: Thousands of USD.

Table 6. Sensitivity analysis on the carbon tax of model 2.

Carbon Tax Decrease/Increase Ratio (%)	Profit (A)	Increase/Decrease (Compared with the Initial Value) (%) (B) = (D)/(C)	Original Profit (C)	Increase Profit (D) = (A) − (C)
−30%	54,058.00	1.90%	53,050.00	1008.00
−25%	53,890.00	1.58%	53,050.00	840.00
−20%	53,722.00	1.26%	53,050.00	672.00
−15%	53,554.00	0.95%	53,050.00	504.00
−10%	53,386.00	0.63%	53,050.00	336.00
−5%	53,218.00	0.31%	53,050.00	168.00
0%	53,050.00	0.00%	53,050.00	0.00
5%	52,882.00	−0.31%	53,050.00	−168.00
10%	52,714.00	−0.63%	53,050.00	−336.00
15%	52,546.00	−0.95%	53,050.00	−504.00
20%	52,378.00	−1.26%	53,050.00	−672.00
25%	52,210.00	−1.58%	53,050.00	−840.00
30%	52,042.00	−1.90%	53,050.00	−1008.00

Unit: Thousands of USD.

Table 7. Sensitivity analysis on the carbon tax of model 3.

Carbon Tax Decrease/Increase Ratio (%)	Profit (A)	Increase/Decrease (Compared with the Initial Value) (%) (B) = (D)/(C)	Original Profit (C)	Increase Profit (D) = (A) − (C)
−30%	54,058.00	1.90%	53,050.00	1008.00
−25%	53,890.00	1.58%	53,050.00	840.00
−20%	53,722.00	1.26%	53,050.00	672.00
−15%	53,554.00	0.95%	53,050.00	504.00
−10%	53,386.00	0.63%	53,050.00	336.00
−5%	53,218.00	0.31%	53,050.00	168.00
0%	53,050.00	0.00%	53,050.00	0.00
5%	52,882.00	−0.31%	53,050.00	−168.00
10%	52,714.00	−0.63%	53,050.00	−336.00
15%	52,546.00	−0.95%	53,050.00	−504.00
20%	52,378.00	−1.26%	53,050.00	−672.00
25%	52,210.00	−1.58%	53,050.00	−840.00
30%	52,042.00	−1.90%	53,050.00	−1008.00

Unit: Thousands of USD.

In summary, three different scenario models with carbon taxes and carbon right are analyzed in this study by sensitivity analysis, and the results show that the increase or decrease of the carbon tax rate is one of the major factors affecting the total profit of the tire industry. However, under the current international focus on greenhouse gas emission reduction and environmental protection, how to reduce carbon emissions has become an important issue. The implementation of the national carbon emission system and carbon rights trading system is the most important issue for international environmental protection today. Therefore, the tire industry should understand the response to the implementation of the carbon tax, as well as the additional costs of carbon emission exchanges, in order to avoid affecting the company's biggest profit decline.

5. Conclusions

The tire industry is a high-carbon and high-pollution industry. As global warming, climate change, and global emissions of greenhouse gases have become international environmental issues of concern, in recent years, some countries have encountered air pollution and environmental problems, and have begun to adopt the carbon price mechanism to implement carbon taxes. Therefore, the carbon tax and the carbon trading mechanism are the main strategies used by various governments around the world to respond to global warming and promote a green economy. Based on this concept, this paper applied a mathematical programming model and put forward three green optimal carbon price models for analysis. In Model 1, the results show that the tire company can sell the remaining 3,664,000 metric tonnes to other companies in need, and thus, earn more money. In addition, based on the assumptions of the two carbon rights purchase cost models of Model 2 and Model 3, analysis shows that the tire companies do not need to purchase additional carbon rights; however, compared with Model 1, if there is a carbon rights transaction, the total profit will decrease. Therefore, in order to reduce the impact of carbon tax and carbon trading on the profits of tire companies, tire companies could reduce carbon emissions as much as possible, and they can significantly reduce costs to create higher profits.

Based on the mathematical programming model, this article developed three different scenario models with carbon tax and carbon right for tire companies and conducted sensitivity analysis. The analysis results can provide references for tire industry management's decision-making. In addition, according to the analysis results, we know that the operating costs will increase if a company wants to purchase additional carbon rights, which may led to a substantial reduction in profits. Therefore, management should pay more attention to the quantity of carbon rights purchased. In addition, the mathematical programming model with carbon tax and carbon rights trading, as developed in this paper, can be extended to other relevant industries, such as construction, aviation, paper, steel, and other high pollution industries with high carbon emissions. In addition, this paper can be extended to the development of a green tax system through the design of a long-term and short-term mixed tax policy [64] to reflect the correct price of carbon tax and to reduce carbon emissions and air pollution. It is also possible to consider the policy of carbon emission and carbon pricing from the perspective of economics. Besides, this research can also be extended to consider the carbon emission permit allocation factors or the game-theoretic problem involving more than one company in a cap-and-trade system [65].

Funding: This research was funded by the Ministry of Science and Technology of Taiwan under Grant No. MOST106-2410-H-008-020-MY3.

Acknowledgments: The author is extremely grateful to the *energies* journal editorial team and reviewers who provided valuable comments for improving the quality of this article. The author also would like to thank the National Science Council of Taiwan for financially supporting this research under Grant No. MOST106-2410-H-008-020-MY3.

Conflicts of Interest: The author declares no conflict of interest.

References

1. Baranzini, A.; Carattini, S. Taxation of Emissions of Greenhouse Gases. In *Global Environmental Change*; Freedman, B., Ed.; Springer: Dordrecht, The Netherlands, 2014; pp. 543–560.
2. Energy Leaders Accelerate Climate Action Global Climate Action at COP22, UNFCCC. 2016. Available online: http://newsroom.unfccc.int/climate-action/energy-leaders-accelerate-climate-action/ (accessed on 3 July 2018).
3. Sarkar, B.; Ahmed, W.; Kim, N. Joint effects of variable carbon emission cost and multi-delay-in-payments under single-setup-multiple-delivery policy in a global sustainable supply chain. *J. Clean. Prod.* **2018**, *185*, 421–445. [CrossRef]
4. Schaltegger, S.; Synnestvedt, T. The link between 'green' and economic success: Environmental management as the crucial trigger between environmental and economic performance. *J. Environ. Manag.* **2002**, *65*, 339–346.

5. Baranzini, A.; Goldemberg, J.; Speck, S. A Future for Carbon Taxes. *Ecol. Econ.* **2000**, *32*, 395–412. [CrossRef]
6. Metcalf, G.E. Designing a Carbon Tax to Reduce U.S. Greenhouse Gas Emissions. *Rev. Environ. Econ. Policy* **2009**, *3*, 63–83. [CrossRef]
7. Fang, G.; Tian, L.; Liu, M.; Fu, M.; Sun, M. How to optimize the development of carbon trading in China—Enlightenment from evolution rules of the EU carbon price. *Appl. Energy* **2018**, *211*, 1039–1049. [CrossRef]
8. Martin, R.; Muûls, M.; De Preux, L.B.; Wagner, U.J. Industry Compensation under Relocation Risk: A Firm-Level Analysis of the EU Emissions Trading Scheme. *Am. Econ. Rev.* **2014**, *104*, 2482–2508. [CrossRef]
9. Goulder, L.H.; Parry, I.W. Instrument Choice in Environmental Policy. *Rev. Environ. Econ. Policy* **2008**, *2*, 152–174. [CrossRef]
10. Fan, J.H.; Todorova, N. Dynamics of China's carbon prices in the pilot trading phase. *Appl. Energy* **2017**, *208*, 1452–1467. [CrossRef]
11. Lin, B.; Jia, Z. The impact of Emission Trading Scheme (ETS) and the choice of coverage industry in ETS: A case study in China. *Appl. Energy* **2017**, *205*, 1512–1527. [CrossRef]
12. Sathre, R.; Gustavsson, L. Effects of energy and carbon taxes on building material competitiveness. *Energy Build.* **2007**, *39*, 488–494. [CrossRef]
13. Basanta, K.; Pradhan, J.G.; Yao, Y.F.; Liang, Q.M. Carbon pricing and terms of trade effects for China and India: A general equilibrium analysis. *Econ. Model.* **2017**, *63*, 60–74.
14. Maryniak, P.; Trück, S.; Weron, R. Carbon pricing and electricity markets—The case of the Australian Clean Energy Bill. *Energy Econ.* **2018**. [CrossRef]
15. Kee, R.; Schmidt, C. A comparative analysis of utilizing activity-based costing and the theory of constraints for making product-mix decisions. *Int. J. Prod. Econ.* **2000**, *63*, 1–17. [CrossRef]
16. Massood, Y.Z. Product-mix decisions under activity-based costing with resource constraints and non-proportional activity costs. *J. Appl. Bus. Res.* **1998**, *14*, 39–46.
17. Lockhart, J.; Taylor, A. Environmental Considerations in Product Mix Decisions Using ABC and TOC. *Manag. Account. Q.* **2007**, *9*, 13–31.
18. Onwubolu, G.C.; Mutingi, M. Optimizing the multiple constrained resources product mix problem using genetic algorithms. *Int. J. Prod. Res.* **2001**, *39*, 1897–1910. [CrossRef]
19. Fernández, Y.F.; Fernández López, M.A.; Blanco, B.O. Innovation for sustainability: The impact of R&D spending on CO_2 emissions. *J. Clean. Prod.* **2018**, *172*, 3459–3467.
20. Hoinka, K.; Ziębik, A. Mathematical model for the choice of an energy management structure of complex buildings. *Energy* **2010**, *35*, 1146–1156. [CrossRef]
21. Chen, X.; Chan, C.K.; Lee, Y.C.E. Responsible production policies with substitution and carbon emissions trading. *J. Clean. Prod.* **2016**, *134*, 642–651. [CrossRef]
22. Chen, X.; Luo, Z.; Wang, X. Impact of efficiency, investment, and competition on low carbon manufacturing. *J. Clean. Prod.* **2017**, *143*, 388–400. [CrossRef]
23. Du, Z.; Lin, B. Analysis of carbon emissions reduction of China's metallurgical industry. *J. Clean. Prod.* **2018**, *176*, 1177–1184. [CrossRef]
24. Baranzini, A.; Van den Bergh, J.C.; Carattini, S.; Howarth, R.B.; Padilla, E.; Roca, J. Carbon Pricing in Climate Policy: Seven Reasons, Complementary Instruments, and Political Economy Considerations. *Wiley Interdiscip. Rev. Clim. Chang.* **2017**, *8*, e462. [CrossRef]
25. Aldy, J.E.; Stavins, R.N. The Promise and Problems of Pricing Carbon: Theory and Experience. *J. Environ. Dev.* **2012**, *21*, 152–180. [CrossRef]
26. CDP. *Putting a Price on Carbon: Integrating Climate Risk into Business Planning*; CDP Report: Carbon Disclosure Project; CDP: London, UK, 2017. Available online: https://b8f65cb373b1b7b15feb-c70d8ead6ced550b4d987d7c03fcdd1d.ssl.cf3.rackcdn.com/cms/reports/documents/000/002/738/original/Putting-a-price-on-carbon-CDP-Report-2017.pdf?1507739326 (accessed on 3 July 2018).
27. Zheng, G.; Jing, Y.; Huang, H.; Zhang, X.; Gao, Y. Application of life cycle assessment (LCA) and extenics theory for building energy conservation assessment. *Energy* **2009**, *34*, 1870–1879. [CrossRef]
28. Fang, Y.P.; Zeng, Y. Balancing energy and environment: The effect and perspective of management instruments in China. *Energy* **2007**, *32*, 2247–2261. [CrossRef]

29. Lee, C.F.; Lin, S.J.; Lewis, C. Analysis of the impacts of combining carbon taxation and emission trading on different industry sectors. *Energy Policy* **2008**, *36*, 722–729. [CrossRef]
30. Martin, R.; Muuls, M.; Wagner, U.J. The Impact of the European Union Emissions Trading Scheme on Regulated Firms: What Is the Evidence after Ten Years? *Rev. Environ. Econ. Policy* **2016**, *10*, 129–148. [CrossRef]
31. Oestreich, A.M.; Tsiakas, I. Carbon Emissions and Stock Returns: Evidence from the EU Emissions Trading Scheme. *J. Bank. Financ.* **2015**, *58*, 294–308. [CrossRef]
32. Herber, B.P.; Raga, J.T. An international carbon tax to combat global warming: An economic and political analysis of the European Union proposal. *Am. J. Econ. Sociol.* **1995**, *54*, 257–267. [CrossRef]
33. Nadel, S. Learning from 19 Carbon Taxes: What Does the Evidence Show? In Proceedings of the 2016 ACEEE Summer Study on Energy Efficiency in Buildings, Pacific Grove, CA, USA, 21–26 August 2016; pp. 9-1–9-7. Available online: https://aceee.org/files/proceedings/2016/data/papers/9_49.pdf (accessed on 3 July 2018).
34. Xu, J.; Qiu, R.; Lv, C. Carbon emission allowance allocation with cap and trade mechanism in air passenger transport. *J. Clean. Prod.* **2016**, *131*, 308–320. [CrossRef]
35. Johansson, B. Climate policy instruments and industry—Effects and potential responses in the Swedish context. *Energy Policy* **2006**, *34*, 2344–2360. [CrossRef]
36. Liu, Y.; Chen, S.; Chen, B.; Yang, W. Analysis of CO_2 emissions embodied in China's bilateral trade: A non-competitive import input–output approach. *J. Clean. Prod.* **2017**, *163*, S410–S419. [CrossRef]
37. Wang, B.; Liu, B.; Niu, H.; Liu, J.; Yao, S. Impact of energy taxation on economy, environmental and public health quality. *J. Environ. Manag.* **2017**, *206*, 85–92. [CrossRef] [PubMed]
38. Cheng, B.; Dai, H.; Wang, P.; Zhao, D.; Masui, T. Impacts of carbon trading scheme on air pollutant emissions in Guangdong Province of China. *Energy Sustain. Dev.* **2015**, *27*, 174–185. [CrossRef]
39. Ivan, D.R.; Tulloch, D.J. Carbon pricing and system linking: Lessons from the New Zealand Emissions Trading Scheme. *Energy Econ.* **2018**, *73*, 66–79.
40. Roy, C.; Labrecque, B.; Caumia, B.D. Recycling of scrap tires to oil and carbon black by vacuum pyrolysis. *Resour. Conserv. Recycl.* **1990**, *4*, 203–213. [CrossRef]
41. Cooper, R.; Kaplan, R.S. *The Design of Cost Management Systems*, 2nd ed.; Prentice-Hall: Upper Saddle River, NJ, USA, 1999.
42. Puurunen, K.; Vasara, P. Opportunities for utillising nanotechnology in reaching near-zero emissions in the paper industry. *J. Clean. Prod.* **2007**, *15*, 1287–1294. [CrossRef]
43. Duez, B. Towards a Substantially Lower Fuel Consumption Freight Transport by the Development of an Innovative Low Rolling Resistance Truck Tyre Concept. *Trans. Res. Proced.* **2016**, *14*, 1051–1060. [CrossRef]
44. Wikipedia, Carbon Tax. Available online: https://en.wikipedia.org/wiki/Carbon_tax (accessed on 3 July 2018).
45. Lin, B.; Li, X. The effect of carbon tax on per capita CO_2 emission. *Energy Policy* **2011**, *39*, 5137–5146. [CrossRef]
46. Conegrey, T.; Gerald, J.D.F.; Valeri, L.M.; Tol, R.S.J. The impact of a carbon tax on economic growth and carbon dioxide emissions in Ireland. *J. Environ. Plan. Manag.* **2013**, *56*, 934–952. [CrossRef]
47. Kee, R. The sufficiency of product and variable costs for production-related decisions when economies of scope are present. *Int. J. Prod. Econ.* **2008**, *114*, 682–696. [CrossRef]
48. Dekker, R.; Bloembof, J.; Mallidis, I. Operations resarch for greenlogistics—An overview of aspects, issues, contributions and challenges. *Eur. J. Oper. Res.* **2012**, *219*, 671–679. [CrossRef]
49. Tsai, W.H.; Lin, W.R.; Fan, Y.W.; Lee, P.L.; Lin, S.J.; Hsu, J.L. Applying A Mathematical Programming Approach for A Green Product Mix Decision. *Int. J. Prod. Res.* **2012**, *50*, 1171–1184. [CrossRef]
50. Qian, L.; David, B.A. Parametric cost estimation based on activity-based costing: A case study for design and development of rotational parts. *Int. J. Prod. Econ.* **2008**, *113*, 805–818. [CrossRef]
51. Tsai, W.H.; Lin, S.J.; Liu, J.Y.; Lin, W.R.; Lee, K.C. Incorporating life cycle assessments into building project decision-making: An energy consumption and CO_2 emission perspective. *Energy* **2011**, *36*, 3022–3029. [CrossRef]
52. Tsai, W.H.; Lin, S.J.; Lee, Y.F.; Chang, Y.C.; Hsu, J.L. Construction method selection for green building projects to improve environmental sustainability by using an MCDM approach. *J. Environ. Plan. Manag.* **2013**, *56*, 1487–1510. [CrossRef]
53. Tsai, W.H.; Yang, C.H.; Huang, C.T.; Wu, Y.Y. The Impact of the Carbon Tax Policy on Green Building Strategy. *J. Environ. Plan. Manag.* **2017**, *60*, 1412–1438. [CrossRef]

54. Plenert, G. Optimizing theory of constraints when multiple constrained resources exist. *Eur. J. Oper. Res.* **1993**, *70*, 126–133. [CrossRef]
55. Tsai, W.H.; Lai, C.W.; Tseng, L.J.; Chou, W.C. Embedding Management Discretionary Power into An ABC Model for A Joint Products Mix Decision. *Int. J. Prod. Econ.* **2008**, *115*, 210–220. [CrossRef]
56. Tsai, W.H.; Kuo, L.; Lin, T.W.; Kuo, Y.C.; Shen, Y.S. Price elasticity of demand and capacity expansion eatufres in an enhanced ABC product-mix Decision Model. *Int. J. Prod. Res.* **2010**, *48*, 6387–6416. [CrossRef]
57. Beale, E.M.L.; Tomlin, J.A. Special facilities in a general mathematical programming system for non-convex problems using ordered sets of variables. *Oper. Res.* **1970**, *69*, 447–454.
58. Williams, H.P. *Model Building in Mathematical Programming*, 2nd ed.; Wiley: New York, NY, USA, 1985; pp. 173–177.
59. Wang, Y.; Chen, W.; Liu, B. Manufacturing/remanufacturing decisions for a capital-constrained manufacturer considering carbon emission cap and trade. *J. Clean. Prod.* **2017**, *140*, 1118–1128. [CrossRef]
60. Igor, L.W.; George, V.B.; Jose, L.D.M.; Ofelia, D.Q.F.A. Carbon dioxide utilization in a microalga-based biorefinery: Efficiency of carbon removal and economic performance under carbon taxation. *J. Environ. Manag.* **2017**, *203*, 988–998.
61. Tsai, W.H.; Chen, H.C.; Liu, J.Y.; Chen, S.P.; Shen, Y.S. Using activity-based costing to evaluate capital investments for green manufacturing technologies. *Int. J. Prod. Res.* **2011**, *49*, 7275–7292. [CrossRef]
62. Ward, S.C.; Chapman, C.B. Risk-management perspective on the project lifecycle. *Int. J. Proj. Manag.* **1995**, *13*, 145–149. [CrossRef]
63. Tsai, W.H.; Shen, Y.S.; Lee, P.L.; Chen, H.C.; Kuo, L.; Huang, C.C. Integrating information about the cost of carbon through activity-based costing. *J. Clean. Prod.* **2012**, *36*, 102–111. [CrossRef]
64. Zetterberg, L.; Wråke, M.; Sterner, T.; Fischer, C.; Burtraw, D. Short-Run Allocation of Emissions Allowances and Long-Term Goals for Climate Policy. *Ambio* **2012**, *41*, 23–32. [CrossRef] [PubMed]
65. Du, S.; Ma, F.; Fu, Z.; Zhu, L.; Zhang, J. Game-theoretic analysis for an emission-dependent supply chain in a 'cap-and-trade' system. *Ann. Oper. Res.* **2015**, *228*, 135–149. [CrossRef]

energies

Article

A Real Options Analysis for Renewable Energy Investment Decisions under China Carbon Trading Market

Yanbin Li, Min Wu * and Zhen Li

School of Economics and Management, North China Electric Power University, Beijing 102202, China; liyb@ncepu.edu.cn (Y.L.); lizhen2020@ncepu.edu.cn (Z.L.)

* Correspondence: wu94090804@163.com

Received: 11 June 2018; Accepted: 6 July 2018; Published: 11 July 2018

Abstract: Under the carbon trading mechanism, renewable energy projects can gain additional benefits through Chinese Certified Emission Reduction transactions. Due to the uncertainty of carbon trading system, carbon prices will fluctuate randomly, which will affect the investment timing of renewable energy projects. Thus, the value of the option will be generated. Therefore, renewable energy power generation project investment has the right of option. However, the traditional investment decision-making method can no longer meet the requirements of renewable energy investment in the current stage. In this paper, a real option model considering carbon price fluctuation is proposed as a tool for renewable energy investment. Considering optimal investment timing and carbon price, the model introduces a carbon price fluctuation as part of the optimization, studies the flexibility of enterprises' delayed investment under the fluctuation of carbon price. A case study is carried out to verify the effectiveness of the proposed real option model by selecting a wind farm in North China. The model is expected to help investors to assess the volatility and risk of renewable energy projects more accurately, and help investors to make a complete plan for the project investment, thus promoting the efficient allocation of resources in the energy industry.

Keywords: carbon price fluctuation; renewable energy; real options analysis; investment under uncertainty

1. Introduction

With the increasingly global climate change, reducing greenhouse gas emissions, optimizing energy structure and decreasing fossil energy consumption have been one of the attentions of the world. The Paris Agreement points out that all sides should strengthen the global response to the threat of climate change. It involves holding on to a clear goal, in this case the 2C limit. China has participated in the cooperation of the climate problem actively, and made solemn commitment to reduce the greenhouse gas emission, that is a decrease in emissions intensity of around 40 percent to 45 percent between 2005 and 2020, at the Copenhagen conference in December 2009. In order to achieve saving energy and reducing emissions, the utilization of renewable energy has great potential for development. Therefore, the 13th Five-Year Plan proposed that the non-fossil energy contributes 15% of primary energy consumption. It is the request of achieving greenhouse gas emission reduction. It means a rare opportunity to invest in renewable energy for power enterprises.

In 2018, China has officially launched the national carbon trading market. As an instrument of policy, carbon trading market can promote the whole society to a low carbon transformation. It also will promote the development of clean energy power generation. Key emission units can reduce carbon emissions by developing clean energy such as photovoltaic and wind power. With the growing size of the market for carbon trading and the development of carbon trading mechanism, more and more enterprises involve in carbon trading. Enterprises will be more active in the use of clean energy, which promotes the development of the renewable energy. Accordingly, the establishment of carbon trading market also brings an opportunity for the investment of renewable energy power generation projects.

Furthermore, with the development of national economy, energy security problem has been an important indicator of future economic development of China. At present, the newly discovered fossil energy resources in China are basically located in remote areas or oceans, and the cost of mining is relatively high. Conversely, renewable energy has no threat of exhaustion, and it is not dependent on import. From the amount of renewable energy reserves in China, renewable energy has good resource conditions, its development potential is huge. Therefore, promoting large-scale investment of renewable energy by power generation enterprises is a feasible and effective measure to ensure China's energy security.

Therefore, this topic is based on the above realistic background. This paper focuses more on the overall impact of the establishment of carbon trading market on renewable energy power generation projects and studies the optimal investment rule of power generation enterprises under carbon price fluctuation to provide useful reference for enterprise decision-makers and thus promoting renewable energy development, achieving national greenhouse gas emission reduction commitments and ensuring energy security supply.

In this paper, the structure of this paper is as follows. Section 1 introduces the background of the research. Section 2 summarizes the current research status at home and abroad. Section 3 constructs the real option model based on the fluctuation of carbon price and carries out the calculation. Section 4 selects the wind power field in North China for an empirical analysis. The research contents of this paper are summarized in Section 5.

2. Literature Review

2.1. The Generation of Real Option Theory

Net present value is a relatively scientific and relatively simple method of investment evaluation. It is suitable for a relatively stable market situation. However, due to the factors that it considers are simple and lack of reflection on future uncertainty, it is likely to underestimate the true value of investment projects. The volatility of carbon prices and the risk of renewable energy investment increased with the gradual maturity of the international carbon trading market and the establishment of China's carbon trading market. The traditional investment decision method cannot comply with the trend. Therefore, we can minimize the risk of renewable energy investment projects by introducing the real option method. Real options method is an extension of financial options theory to real assets. The option pricing model was originally applied to asset appraisal in 1973 [1,2], and then applied to investment projects under uncertainty. Amram and Kulatilaka [3] and Copeland and Antikarov [4] evaluated high risk investment projects by Return on Assets, which including petroleum, aviation, medicine and other fields. Then, due to the investment flexibility of the real option method, it was soon applied to all walks of life [5,6].

2.2. The Application of Real Option Theory

At present, more and more foreign scholars begin to study the volatility of renewable energy investment projects caused by uncertainty. Kyeongseok Kim and Hyoungbae Park [7] put forward the corresponding real option model taking the developing countries as the research object. The model is put forward as a tool to evaluate the investment decision of renewable energy, which help investors to efficiently and accurately invest in projects with high carbon price and high risk. Bøckman et al. [8] selected the hydropower project in Norway as a case. The real option model is constructed, and the optimal scale and time of investment for hydropower projects are finally obtained. Kjærland [9] analyzed the potential project value of Norway under different point and different investment behavior, and quantified the value of the project through the construction of real option model, so as to determine the optimal investment behavior. Abadie and Chamorro [10] designed the real option model under the uncertain conditions and analyzed the influence of the three uncertainties including electricity price, electricity generation, and tariff subsidy policy on the investment value of British wind farm projects. Lee [11] demonstrated the effectiveness of the option model through applying real option model to wind power projects in Taiwan. It is found that the electricity price, the expiration time, the risk free interest rate, and the investment value of the project are proportional to the investment value. Martinez-Cesena and Mutale [12] designed a real option model suitable for small wind power projects. Kumbaroğlu et al. [13] proposed the renewable energy investment strategy based on dynamic programming model considering the uncertainty of renewable energy technology. Lena Kitzing and Nina Juul [14] proposed a real option model of wind energy investment to solve the optimal investment time point and optimal capacity.

At present, the research on renewable energy investment is mainly focused on feasibility evaluation and risk assessment. Mahdi Shahnazari and Adam McHugh [15] studied the comprehensive real option model under the interaction of carbon price and renewable portfolio standard (RPS) instruments. Casals [16] studied the feasibility of the renewable energy power generation project and analyzed its technical feasibility in meeting the stability reliability of the system and regional development planning. Ochoa [17] appraised the economic benefit of the renewable energy generation project, and proposed the evaluation model and multi-objective programming model of the economic benefit of the distributed generation. Rothwell [18] evaluated the risk of developing new nuclear power plants by means of real options, and simulated three uncertain factors: price risk, production risk, and cost risk. The risk premium was derived by Monte Carlo simulation. Fleten et al. [19] founded that the use of a random price real option method to analyze the investment of a power plant required a greater return, not just the return value of the traditional net present value balance point.

Domestic scholars are also concerned about the application of real option theory in renewable energy investment. Han Longxi [20] analyzed the opportunities and challenges facing the renewable energy industry in China through Clean Development Mechanism, and suggested that the renewable energy industry should take the lead in the integration of the domestic greenhouse gas emission reduction trading system, and explore the huge benefits of reducing emissions and increasing revenues by market means. Wang Jianbin [21] fully considered the reaction and influence of market participants to investment decision, established a dynamic model of power investment based on real option theory and game theory, and studied the dynamic decision-making problem of electric power project investors under uncertainty and market competition. Xu nuo [22] proposed two basic assumptions when building the mathematical model of power generation investment based on option game theory, that is, the decision-making behavior of the power generation investors in the market will affect each other. The change of load growth is the most important uncertainty factor in the decision-making of the power generation investors. Ren zhi min and Zeng Ming [23] built a non-cooperative game model of power generation investment decision under the condition of building up carbon price based on bidding principle of free competitive power generation market. Gong Piqin and Li Xinyang [24] studied the investment in renewable energy projects under the fluctuation of carbon price and calculated the NPV of three types of renewable energy project investment and its real option value (ROV). Hou Gang [25] and Zhu Zhenyu [26] studied the risk assessment of renewable energy projects, including the determination of risk factors and risk control strategies.

The main problems involved in the previous research are the application of real options model in different industries, the improvement of the evaluation methods of renewable energy projects and the impact of renewable energy technology on project investment. Through the analysis of previous studies, there is a method of cash flow to assess the impact of carbon prices on renewable energy investment, and also to use capacity constraints as a part of the optimization of renewable energy investment to carry out the real option analysis of carbon prices. However, there is a lack of a real option analysis of the dynamic planning and analysis of carbon prices for China's renewable energy projects. Therefore, this paper attempts to provide an option model for the research of renewable energy investment in China by analyzing the impact of carbon price fluctuation on the investment of renewable energy projects.

On the basis of the above research work, assuming sustainability and scale stability of renewable energy projects the research will introduce the carbon price to build a mathematical model to solve the optimal solution of the carbon price of the renewable energy investment. The study of this paper will fill the gap of optimizing carbon price as an investment value.

3. Methodology

3.1. The Steps of Model for Investment Decisions for Renewable Energy Investment

Although the fixed tariff system provides the guarantee of higher electricity price and power generation capacity, investment in renewable energy generation projects will get CDM support with the fixed price cannot cover the renewable energy project investment and the cost of power generation enterprises. Then, enterprises will obtain carbon emissions subsidies. The uncertainty of carbon price will affect the Investment decision of renewable energy project. It is difficult for investors to accurately evaluate renewable energy investment projects.

In order to solve these difficulties, this paper proposed a real option model for renewable energy projects. The model is divided into three steps, as shown in Figure 1: values confirmation, determination of carbon price threshold, and investment timing determining. Step 1 is the determination of the project value of renewable energy generation. In this step, the random fluctuation of carbon price is considered. Benefits of renewable energy projects include the revenue from certified emission reduction. Step 2 is the determination of carbon price threshold. The Brownian movement of carbon price is an important type of stochastic process. The uncertainty of the price of carbon produces the value of the option. In this step, the dynamic programming method and Bellman equation are introduced, and then Bellman equation and Ito lemma are used to introduce a famous irreversible investment model. Step 3 is to determine whether renewable energy projects will be invested.

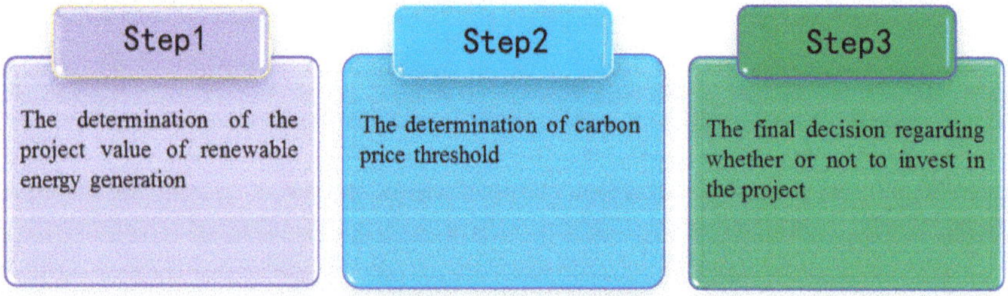

Figure 1. The steps of model for investment decisions for renewable energy projects.

3.2. Model

Considering the volatility of carbon prices and the flexibility of investment time points, the investment problem of renewable energy projects can be considered as a time-lasting real option problem. In regards to uncertainty, the carbon price (p_c) is assumed to be governed by a geometric Brownian motion (GBM) $dp_c = \alpha p_c dt + \sigma p_c dz$, where α is the growth rate of carbon price, σ is the standard deviation of the growth rate, and α and σ are the drift and volatility of the carbon price, and dz is a standard Brownian motion.

The operational gross margin is influenced by the electricity price (p_e), unit generation cost (c), carbon price, and total output of renewable energy generation, all of which are combined in this single measure $\pi(p_c)$.

3.2.1. Value Confirmation

The profit of the renewable energy generation project is composed of two parts of the income of the electricity sales profit and the revenue of the reduced displacement approved by the sales CDM. For given $\pi(p_c)$ [3], the value of the project refers to the expected discount value of gross profit of the project.

$$\pi(p_c) = (p_e + p_c\eta)Q - I \tag{1}$$

where π is the gross profit of the project, and Q is the average production. I denotes investment cost. $I = CQ, C > 0$. C refers to the investment cost of project unit output, which is related to the capacity of the project.

$$\pi(p_c) = (p_e + p_c\eta - C)Q \tag{2}$$

The discounted rate (r) refers to the interest rate that can be invested in a certain investment object without any risk. Where T denotes the life cycle of renewable energy generation projects and η denotes the conversion coefficient of certified emission reduction and generating capacity $\eta = \frac{\text{certified emission reduction}}{\text{generating capacity}}$.

The project value of the renewable energy generation project is equal to the discount value of the project profit, and it is a function of the carbon price. The value of the project is:

$$V(p_c) = \int_0^T e^{-rt} \prod \pi(p_c) dt \tag{3}$$

$$V(p_c) = V_e + V_c = V_e + \theta p_c$$

where V_e denotes the income of the electricity sales profit and V_c denotes the revenue of the reduced displacement approved by the sales CDM, where θ is a compound index of the value of profit with the random fluctuation of carbon price.

3.2.2. Determination of Carbon Price Threshold

$F(p_c)$ represents investment options for renewable energy projects due to the uncertainty of carbon price fluctuation. Bellman equation is [14]:

$$F(p_c) = \max\left\{\max_{0<t<T}\{V(p_c)\}\right\}, \frac{1}{1+rdt} E\left[(dF(p_c + dp_c)|p_c\right] \tag{5}$$

$F(p_c)$ satisfies the dynamic programming equation. The ordinary differential equation is [21]:

$$\frac{1}{2}\sigma^2 p_c^2 F''(p_c) + \mu p_c F'(p_c) - rF(p_c) = 0 \tag{6}$$

$F''(p_c) = \frac{\partial^2 F}{\partial p_c^2}$, $F'(p_c) = \frac{\partial F}{\partial p_c}$. The homogeneous part in the upper form indicates the delayed option value of the power generation enterprise postponing investment. The general solution to the ordinary differential equation is [21] $F(p_c) = A_1 p_c^{\beta_1} + A_2 p_c^{\beta_2}$, where β_1 and β_2 are the two roots of the quadratic equation. Where A_1 and A_2 are undetermined constant determined by boundary condition:

$$k(\beta) = \frac{1}{2}\sigma^2\beta(\beta - 1) + \mu\beta - r = 0 \tag{7}$$

$$\beta_1 = \frac{1}{2} - \frac{\mu}{\sigma^2} + \sqrt{(\frac{\mu}{\sigma^2} - \frac{1}{2})^2 + \frac{2r}{\sigma^2}}(\beta_1 > 1), \quad \beta_2 = \frac{1}{2} - \frac{\mu}{\sigma^2} - \sqrt{(\frac{\mu}{\sigma^2} - \frac{1}{2})^2 + \frac{2r}{\sigma^2}}(\beta_2 < 0)$$

Since $r > 0$, it is that $k(0) = -r < 0$. Also, since $\mu < r$, it is that $k(1) = \mu - r < 0$.

In order to obtain the threshold of carbon price, two conditions must be satisfied. First, if the carbon price is 0, then the option value is 0, then the investment of renewable energy projects will not be optimal. F = 0. $A_2 p_c^{\beta_2} = 0$. It thus follows that $F(p_c) = A_1 p_c^{\beta_1} = A p_c^{\beta}$. Second, to determine p_c^*, the value of the option and the value of the projects are analyzed: They should be equal at p_c^*, moreover, both functions have the same slope (Smooth pasting condition).

$$\begin{cases} F(p_c^*) = V(p_c^*) - I \\ F'(p_c^*) = V'(p_c^*) \end{cases} \tag{8}$$

$$\begin{cases} A_1 p_c^{*\beta_1} = V_e + \theta p_c^* - I \\ \beta_1 A_1 p_c^{*\beta_1 - 1} = \theta \end{cases} \tag{9}$$

Solving upper formula for p_c^*, the investment thresholds are derived:

$$\begin{cases} A_1 = \frac{\theta}{\beta_1} p_c^{*1-\beta_1} \\ p_c^* = \frac{\beta_1}{\theta(\beta_1 - 1)}(I - V_e) \end{cases} \tag{10}$$

The value of option is determined as:

$$F(p_c) = \begin{cases} A_1 p_c^{\beta_1},\ p_c < p_c^* \\ (\theta p_c^* + V_e - I)\left[\frac{p_c}{p_c^*}\right]^{\beta_1}, p_c \geq p_c^* \end{cases} \tag{11}$$

4. Case Study

4.1. Case Description and Sources of Data

In recent years, the power supply tends to be diversified, and the thermal power generation is becoming cleaner. The proportion of renewable energy is increasing obviously. In 2017, renewable energy accounted for 36.6% of the total installed capacity of China's electricity industry and 26.4% of the total generating capacity [20]. Renewable energy investment has accounted for 85% of the total investment in the power sector. According to the "energy production and consumption revolutionary strategy (2016–2030)", by 2030, the proportion of non-fossil energy generated by non-fossil energy, including nuclear energy and renewable energy, will strive to reach 50% of the total power generation [21].

This paper selected the operation of a wind farm in North China and applied the option model, and then solved carbon price threshold, investment time point and investment value by the option model. The source of funds for the project is free financing, and the rest is settled by financing. According to the conditions of wind energy, it is planned to install 33 units of wind turbines with an installed capacity of 49.5 MW, an average annual availability of 2500 h, and a unit generation cost of 0.0728 USD/kwh, with a project period of 20 years. The risk free interest rate R is 0.05 [24]. The electricity price of the Internet is 0.0823 USD/kwh [27]. Table 1 lists the specific data of the case.

Table 1. Summary of input parameters in the case study.

Items	Symbols	Values
Upper generator capacity	q	49.5 MW
Total project cost	I	39.17 million USD
Concession period	T	20 Years
Annual average production	Q	1947.72 USD kwh
Variable investment cost	C	0.0728 USD/kwh

p_c is assumed to follow a GBM, it is that $d\ln(p_c) = \left(\alpha - \frac{1}{2}\sigma^2\right)dt + \sigma dt$. The carbon price is estimated by drift α and volatility σ, and the two parameters are assumed to be constant, indicating that the randomness of carbon price will not change with time.

Carbon emission reduction coefficient is closely related to emission reduction standards, geographical distribution, and technological level. According to the data released by the climate division, the emission reduction coefficient of carbon dioxide in north China in 2015 is 1.0416 t CO_2/MWh [15].

Table 2 summarizes the environmental parameters.

Table 2. Environmental parameters.

Items	Symbols	Values
Risk-free rate interest rate	r	0.05
Electricity tariff	p_e	0.0823 USD/kwh
Conversion coefficient	η	1.0416 t CO_2/MWh
The growth rate of carbon price	α	0.02
The standard deviation of α	σ	0.10

4.2. Real Options Analysis

The seasonal changes were not taken into account in the model construction. It is assumed that renewable energy output is considered constant throughout the life cycle of the project.

According to the correlation coefficient of carbon price in the previous article, the Monte Carlo method is used to simulate the random fluctuation of carbon price. The number of simulation times is 100,000 times, and the carbon price threshold of the power generation enterprises under fixed net price is measured, which is 6.78 dollars/ton.

According to the seven major carbon trading market in China in the past year, the price range of quota price has narrowed year by year in the past two years. At present, the price of carbon emissions in Beijing is 4.5–8 dollars/ton, and Shenzhen's carbon price is 3–5 dollars/ton [28].

According to the above information, the carbon price threshold of the wind power generation project is within the range of domestic carbon price fluctuations, indicating that wind power generation projects can gain revenue through carbon emission reduction transactions. The power generation companies will invest in wind power projects at the level of carbon price higher than 6.78 US dollars/ton.

According to the data of wind power project, the NPV of the wind power project is $330 million. Under the carbon trading mechanism, the investment value of the project with real option value is $460 million because of the uncertainty of carbon price. The total value of the project investment considering the option value is higher than that of the pure present value method. It shows that the renewable energy power generation project can be more comprehensive and flexible to measure the investment value of the project under the carbon trading mechanism.

4.3. Sensitivity Analysis

Sensitivity analysis, that is, using Monte Carlo method to iterate many times, to observe the impact of renewable energy investment options through variable changes. According to the results obtained above, the carbon price threshold of the wind power generation project is of practical significance. Therefore, the sensitivity analysis of carbon price fluctuation of wind power projects is carried out. The effects of carbon price growth rate and volatility of carbon price growth rate on carbon price threshold and investment option value are studied respectively while other parameters remain the same.

(1) The relationship between carbon price growth rate and carbon price threshold and investment options

According to Figure 2, the investment option value of power generation enterprises is positively related to the growth rate of carbon price under the condition that other parameters remain unchanged. The higher the carbon price growth rate, the more inclined the power generation companies to wait for the delayed investment, and when the future carbon price reaches a higher level, the investment will get greater returns. However, with the expansion of the carbon emission market, the rate of carbon price growth will be reduced, and the power generation enterprises will invest earlier in the renewable energy generation projects.

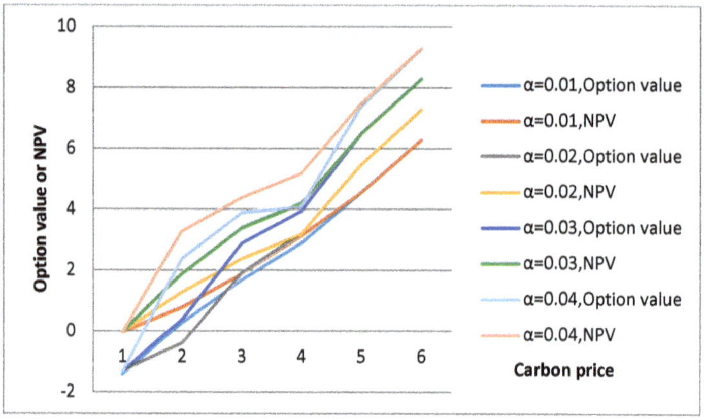

Figure 2. The relationship between F, Net Present Value and α.

(2) The relationship between volatility of carbon price growth rate and carbon price threshold and investment options

As shown in Figure 3, the option value of the power generation enterprise is positively related to the volatility of the carbon price growth rate. This shows that the higher volatility of carbon price increases the value of enterprise investment options, but delays the investment time of enterprises. The reason is that higher volatility shows greater uncertainty, and power generation companies need to wait longer to determine whether the volatility of carbon prices is beneficial to investment. With the continuous improvement of China's carbon trading system, the volatility of carbon prices will be further stabilized, and the volatility will be reduced, promoting the power generation enterprises to invest in renewable energy projects at a lower carbon price level.

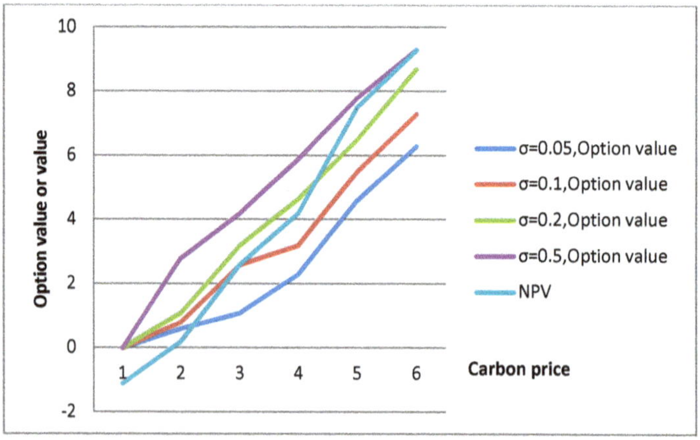

Figure 3. The relationship between F, Net Present Value and σ.

5. Conclusions

The uncertainty of renewable energy investment comes from the volatility of the carbon trading market, the uncertainty of national policies, and the uncertainty of the environment, etc. This paper studies the uncertainty of the stochastic volatility of carbon price to the investment of renewable energy, so as to determine the optimal investment time and option value of renewable energy. The two objectives pursued in this study were: (1) the feasibility of renewable energy investment projects under the uncertainty of carbon price; and (2) the variation of the carbon price threshold was revealed. In order to achieve these two objectives, this paper proposed a real option model considering carbon price fluctuation as a tool to evaluate the investment of renewable energy projects.

A real option model of renewable energy generation investment under the current benchmark electricity price system was constructed. The numerical simulation and sensitivity test of the model were carried out by the Monte-Carlo method. This paper discussed the feasibility of investing in wind power projects under the current net electricity price level in China, analyzed the investment opportunity of the investment in the renewable energy project of wind power generation and revealed the change law of the value of the carbon price gap with the related parameters.

In this paper, a real option model was verified by selecting a case of a wind farm project. The research has somewhat practical significance. Sensitivity analysis was to determine the impact of stochastic characteristics of carbon prices on project value. According to the sensitivity analysis of wind power data, this paper concluded that the higher the rate of carbon price growth and the higher the fluctuation level, the more the power generation enterprises tend to postpone the investment, and these factors have a very significant impact on the return of the enterprise investment.

The future uncertain world carbon trading system will inevitably aggravate the fluctuation of carbon prices. The future research on investment in renewable energy projects will tend to further quantify the uncertainty of projects, which can better control of the high risk of project investment. The uncertainty of the project includes changes in government policies, climate changes, and changes in technology levels. In the construction of real option model, if more scenarios are designed and more detailed data analysis method is adopted, it will be more accurate to analyze the change in investment value of project uncertainty.

Author Contributions: Conceptualization, Methodology: M.W. carried out a real options model in a realistic and easily applicable way as a tool to assess renewable energy investment. Y.L. carried out a general design of the article. Z.L. contributed much in the revised version of our manuscript.

Funding: This research was financed by "Natural Science Foundation of China Project" (Grant No. 71471058), Beijing social science foundation research base project (Grant No. 17JDGLA009).

Acknowledgments: The authors thank the anonymous reviewers for careful reading and many helpful suggestions to improve the presentation of this paper. The authors thanks the anonymous referees for careful reading and many helpful suggestions to improve the presentation of this paper. This paper is supported by "Natural Science Foundation of China Project" (Grant No. 71471058), Beijing social science foundation research base project (Grant No. 17JDGLA009).

Conflicts of Interest: The authors declare that they have no competing interests.

References

1. Black, F.; Scholes, M. The pricing of options and corporate liabilities. *J. Political Econ.* **1973**, *81*, 637–654. [CrossRef]
2. Merton, R. An intertemporal capital asset pricing model. *Econometric* **1973**, *41*, 867–887. [CrossRef]
3. Amram, M.; Kulatilaka, N. *Real Options*; Harvard Business School Press: Boston, MA, USA, 1999.
4. Copeland, T.E.; Antikarov, V. *Real Options: A Practitioner's Guide*; Texere: New York, NY, USA, 2003.
5. Arrow, K.J.; Fisher, A.C. Environmental preservation, uncertainty, and irreversibility. *Q. J. Econ.* **1974**, *88*, 312–319. [CrossRef]
6. Henry, C. Investment decisions under uncertainty: The irreversibility effect. *Am. Econ. Rev.* **1974**, *64*, 1006–1012.

7. Kim, K.; Park, H. Real options analysis for renewable energy investment decisions in developing countries. *Renew. Sustain. Energy Rev.* **2017**, *75*, 918–926. [CrossRef]
8. Bøckman, T.; Fleten, S.-E.; Juliussen, E.; Langhammer, H.J.; Revdal, I. Investment timing and optimal capacity choice for small hydropower projects. *Eur. J. Oper. Res.* **2008**, *190*, 255–267. [CrossRef]
9. Kjærland, F. A real option analysis of investments in hydropower—The case of Norway. *Energy Policy* **2007**, *35*, 5901–5908. [CrossRef]
10. Abadie, L.M.; Chamorro, J.M. Valuation of wind energy projects: A real options approach. *Energies* **2014**, *7*, 3218–3255. [CrossRef]
11. Lee, S.-C. Using real option analysis for highly uncertain technology investments: The case of wind energy technology. *Renew. Sustain. Energy Rev.* **2011**, *15*, 4443–4450. [CrossRef]
12. Martinez-Cesena, E.A.; Mutale, J. Wind power projects planning considering real options for the wind resource assessment. *IEEE Trans. Sustain. Energy* **2012**, *3*, 158–166. [CrossRef]
13. Venetsanos, K.; Angelopoulou, P.; Tsoutsos, T. Renewable energy sources project appraisal under uncertainty: The case of wind energy exploitation within a changing energy market environment. *Energy Policy* **2002**, *30*, 293–307. [CrossRef]
14. Kitzing, L.; Juul, N. A real options approach to analyze wind energy investments under different support schemes. *Appl. Energy* **2017**, *188*, 83–96. [CrossRef]
15. Shahnazari, M.; McHugh, A. Overlapping carbon pricing and renewable support schemes under political uncertainty: Global lessons from an Australian case study. *Appl. Energy* **2017**, *200*, 237–248. [CrossRef]
16. Casals, X. Technical and economic analysis on the introduction of a high percentage of renewable energy in the Spanish energy system. In Proceedings of the ISES Solar World Congress 2007: Solar Energy and Human Settlement, Beijing, China, 18–21 September 2007.
17. Ochoa, L. Evaluating distributed generation impacts with a multiobjective index. *IEEE Trans. Power Deliv.* **2006**, *21*, 1452–1458. [CrossRef]
18. Rothwell, G.A. A real options approach to evaluating new nuclear power plants. *Energy J.* **2006**, *27*, 37–53. [CrossRef]
19. Fleten, S.; Maribu, K.; Wangensteen, I. Optimal investment strategies in decentralized renewable power generation under uncertainty. *Energy* **2007**, *32*, 803–815. [CrossRef]
20. Han, L.; Zhai, J.; Zhang, L. The challenges and opportunities of China's greenhouse gas emission trading on renewable energy. *Ecol. Econ.* **2015**, *31*, 31–35.
21. Wang, J. A dynamic modeling of investments in power generation based on real option theory and game theory. *Trans. China Electrotech. Soc.* **2007**, *22*, 163–167.
22. Xu, N.; Wen, F.; Huang, M. An option-game based approach for generation investment decision-making. *Autom. Electr. Power Syst.* **2007**, *31*, 25–30.
23. Ren, Z.; Zeng, M.; Li, C. Investment decision-making of renewable energy generation with increasing of carbon price. *Water Resour. Power* **2015**, *33*, 211–215.
24. Gong, P.; Li, X. Study on the investment value and investment opportunity of renewable energies under the carbon trading system. *China Popul. Resour. Environ.* **2017**, *27*, 22–29. [CrossRef]
25. Hou, G. *Research on the Risk Evaluation and Management of Biomass Energy in China*; Northwest Agriculture & Forestry University: Yangling, China, 2009; pp. 43–76.
26. Zhu, Z. *Decision on Investment Risks of solar Photovoltaic Industry in China*; China University of Geosciences: Beijing, China, 2010; pp. 83–96.
27. Polaris Power Network Database. 2017. Available online: http://news.bjx.com.cn/html/20170420/821314.shtml (accessed on 5 May 2018).
28. Carbon Emissions Trading Database. 2017. Available online: http://www.tanpaifang.com/ (accessed on 5 May 2018).

energies

MDPI

Article

Life Cycle Assessment of a Buoy-Rope-Drum Wave Energy Converter

Qiang Zhai *, Linsen Zhu * and Shizhou Lu

Department of Mechanical Engineering, Shandong University at Weihai, Weihai 264209, China; lushizhou@sdu.edu.cn

* Correspondence: zhaiqiang@sdu.edu.cn (Q.Z.); zlinsen@sdu.edu.cn (L.Z.)

Received: 7 August 2018; Accepted: 10 September 2018; Published: 13 September 2018

Abstract: This study presents a life cycle assessment (LCA) study for a buoy-rope-drum (BRD) wave energy converter (WEC), so as to understand the environmental performance of the BRD WEC by eco-labeling its life cycle stages and processes. The BRD WEC was developed by a research group at Shandong University (Weihai). The WEC consists of three main functional modules including buoy, generator and mooring modules. The designed rated power capacity is 10 kW. The LCA modeling is based on data collected from actual design, prototype manufacturing, installation and onsite sea test. Life cycle inventory (LCI) analysis and life cycle impact analysis (LCIA) were conducted. The analyses show that the most significant environmental impact contributor is identified to be the manufacturing stage of the BRD WEC due to consumption of energy and materials. Potential improvement approaches are proposed in the discussion. The LCI and LCIA assessment results are then benchmarked with results from reported LCA studies of other WECs, tidal energy converters, as well as offshore wind and solar PV systems. This study presents the energy and carbon intensities and paybacks with 387 kJ/kWh, 89 gCO_2/kWh, 26 months and 23 months respectively. The results show that the energy and carbon intensities of the BRD WEC are slightly larger than, however comparable, in comparison with the referenced WECs, tidal, offshore wind and solar PV systems. A sensitivity analysis was carried out by varying the capacity factor from 20–50%. The energy and carbon intensities could reach as much as 968 kJ/kWh and 222 gCO_2/kWh respectively while the capacity factor decreasing to 20%. Limitations for this study and scope of future work are discussed in the conclusion.

Keywords: wave energy converter; life cycle assessment; energy intensity; carbon intensity

1. Introduction

Carbon dioxide emissions from electricity generation are a major cause of anthropogenic climate change [1]. Around 30% of the world's population is currently exposed to climatic conditions exceeding a deadly threshold for at least 20 days a year. By 2100, this percentage is projected to increase to 48% and 74% under scenarios with drastic reductions and no considerable reductions of GHG emissions [2]. Overtaking the United States, China has become the world's biggest carbon emission contributor since 2007 due to its reliance on fossil fuel. In 2009, China committed to reduce its carbon intensity (CO_2 per unit of GDP) by 40% to 45% percent below 2005 levels by 2020. Achieving this ambitious 2020 goal, China will have to further expand its renewable energies' portion in its power supply structure [3].

Among various renewable energies, marine energy has been believed as one to partially replace conventional fossil fuel energy due to its carbon free electricity production mechanism. In recent years the world has witnessed the emergence of marine energy conversion technologies. Research and development programmes on wave energy extraction have been carried out in many countries, such as USA, Australia, UK, Portugal, Norway and Japan. Thousands of patents were granted on how to absorb wave energy and convert it to electricity or other forms of energy [4]. Since the beginning of the century, tidal stream technology and wave energy technology development have started to ramp up [5]. Although there no large-scale WEC installation has been reported in the U.S., the U.S. Department of Energy sponsored the Wave Energy Prize in 2015, an 18-month public design-build-test competition to increase the diversity of organizations involved in WEC technology development, while motivating and inspiring existing stakeholders [6]. The UK and Portugal have the vast majority of wave energy deployments in Europe [5]. The Carbon Trust estimates that a contribution of up to 20% of total UK energy generation could be provided by marine energy by 2050 [7,8]. China possesses a total marine area of 4,700,000 km^2 [9]. With annual mean wave power of up to 7.73 kW/m wave front, China has a wave power potential of 128.5 GW, nearly half of the electricity production of China [10,11]. However, the majority of China's wave energies are without any exploitation and thus far, most of the WECs in China are still in concept design or pre-commercial stages. Aiming to stimulate and encourage the marine renewable energy technology research and development, China National Oceanic Administration set up the Specific Fund for Marine Renewable Energies in 2010. Sponsored by this specific funding, quite a few new WECs have been invented, prototyped and sea tested. Representative systems of these WECs include 100 kW Sharp Eagle by Guangzhou Institute of Energy Conversion, 120 kW oscillating Buoy WEC by Shandong University, 300 kW floating hydromantic WEC device Hailong I, etc. Most of these Chinese WECs are still in their early stages under research and development and are not ready for commercial scale production.

On the other hand, marine energy extraction technology is still at its infancy stage as wave energy conversion has been proven to be difficult [12]. In spite of decades of research and thousands of patents there is still no consensus on the best way to harness the energy of ocean waves [13]. In this context, efforts for current wave conversion technologies' research and development are still focused on the concept optimization and efficiency improvement, as well as reliability enhancement. These technologies have not emphasized much on the WECs' life cycle environmental performance improvement. Although new technologies of marine energy conversion are constantly emerging, only few full LCAs have been carried out and reported [14]. None of these studies are based on devices developed in China.

Up to 2016, only a few LCA case studies reported the energy and carbon intensities of wave and tidal energy devices. Most among those studies have looked only at energy and carbon as impact categories [8,15,16]. Douglas et al. in the LCA study of Seagen, the world's first commercial-scale grid-connected tidal current energy installation, reported the energy and carbon intensities with 214 kJ/kWh and 15 gCO_2/kWh. The energy payback period is approximately 14 months and the CO_2 payback is around 8 months [15]. LCA studies for Pelamis WEC conducted by Parker et al. in 2007 and Thomson et al. in 2011 demonstrate the Pelamis' energy and carbon intensities with 293 kJ/kWh and 22.8 g CO_2/kWh [8]. Thomson et al. expanded this analysis by covering more environmental impacts, demonstrating the same energy intensity of 293 kJ/kWh, and a justified carbon intensity of 23 gCO_2/kWh [8,14]. The estimated energy and carbon payback periods are approximately 20 months and 13 months, respectively [8,14]. Walker et al. and Douziech et al. in their studies compared the life cycle environmental performance of wave and tidal energy plants [16,17]. The studied WEC, Oyster, was reported to have energy and carbon intensities of 236 kJ/kWh and 25 gCO_2/kWh, with energy and carbon payback periods of 12 months and 8 months. Claimed to be the first LCA study for tidal current energy devices taking impacts beyond climate change into account, Douziech et al. demonstrated the WEC Oyster800 has on average 3.5 times higher impact than offshore wind power, but eight times lower impact than electricity generated from coal power [17]. Uihlein et al. presented a LCA study to assess the environmental impacts of eight types of WECs and seven types of tidal energy devices. The reported total greenhouse gas emissions of ocean energy devices range from about 15 to 105 g$CO_{2\text{-eq}}$/kWh. Average global warming potential for all device types is 53 $\pm$ 29 g$CO_{2\text{-eq}}$/kWh [18].

These pioneering in-depth LCA studies indicate that wave energy conversion have the advantage and potential to become a promising contributor to achieve reduction of energy and carbon intensities from energy sector, by comparison with the high carbon intensity level of typical fossil fuel energy ranging from 400–1000 gCO_2/kWh [15,16]. These studies compare wave energy favorably against current renewable energy sources such as wind (8–12 gCO_2/kWh), solar PV (~30 gCO_2/kWh) and nuclear (~70 gCO_2/kWh) [16]. As per these studies, the WEC environmental impacts are closely linked to material inputs and are caused mainly by structural components and mooring foundations; while impacts from assembly, installation and use are insignificant for all device types [8,13,18]. Possible improvement to reduce WECs' entire life cycle environmental impact, especially energy and carbon intensity, can be achieved by using more environmentally sustainable materials for the structural components, utilizing more fuel-efficient transportation, as well as reducing the shipping associated with maintenance [8,13].

The purpose of this work is to increase the knowledge of life cycle environmental performance of wave energy conversion technologies and identify high environmental impact stages and processes for future system development and improvement. While marine energy sources are carbon-free, there are indirect GHG emissions related to the manufacturing, operation, and maintenance as well as decommissioning of generators and other system components. Detail understanding of the life cycle impacts of WECs is important for future developments of the WEC system and environmental impact evaluation for technology maturating and commercializing. In this study, the life cycle of a BRD WEC is analytically modeled using SimaPro v 8.3.0.0, a widely used LCA software developed by PRé Consultants (Amersfoort, The Netherlands). The life cycle of the BRD WEC consists of stages such as materials extraction, manufacture and assembly, installation, operations and maintenance (O&M), decommissioning and recycling. The LCI and LCIA analyses are performed for the BRD WEC device to identify significant energy and carbon intensity and environmental impact contributing stages and processes. In order to comprehensively understand its environmental impacts beyond energy and carbon, the ReCiPe Midpoint method is applied to cover as wide range of environmental impacts. The LCI and LCIA results are benchmarked with reported pioneering LCA studies of other WECs, tidal, offshore wind and solar PV systems.

2. Method

2.1. BRD WEC Structural Breakdown

The studied BRD WEC system was designed by a research group of Shandong University (Weihai). The prototype device has been manufactured, deployed offshore and sea tested. The first stage of the project has been fully conducted and completed. With additional financial support by National Marine Renewable Energy Specific Fund, the research group has been focusing on the efficiency improvement and partial redesign of its structural components. As illustrated in Figure 1, the BRD WEC consists of such modules as buoy, generator and mooring modules. A rope with its lower end connected to a gravity anchor on the seabed is attached and wound around the drum of the generator casing through the rope guide. The wave energy extraction and energy conversion mechanism consists of two primary strokes. As shown in Figure 1a, buoy heaves upward due to coming wave's pushing; the drum rotates by pulling force from the rope; the coaxial permanent magnet rotates around the stator winding to generate alternating current which will supply power for the resistive load. During this stroke, wave energy is extracted and converted to elastic potential energy and stored by the coiled spring and mechanical kinetic energy of the drum, and then converted to electrical energy generated through the rotation of the coaxial rotator.

Figure 1b shows that, when the buoy heaves downward when waves coming off, the built in coiled spring will reposition the drum by releasing its elastic potential energy and winding the rope back to its initial position. During this stroke, with an overrunning clutch installed between the drum and magnetic steel casing, the magnetic steel casing does not rotate when the drum rotates backward, and the buoyant hull does not generate power during falling [19]. Figure 1c,d show the deployment and in situ status of the prototype BRD WEC at offshore.

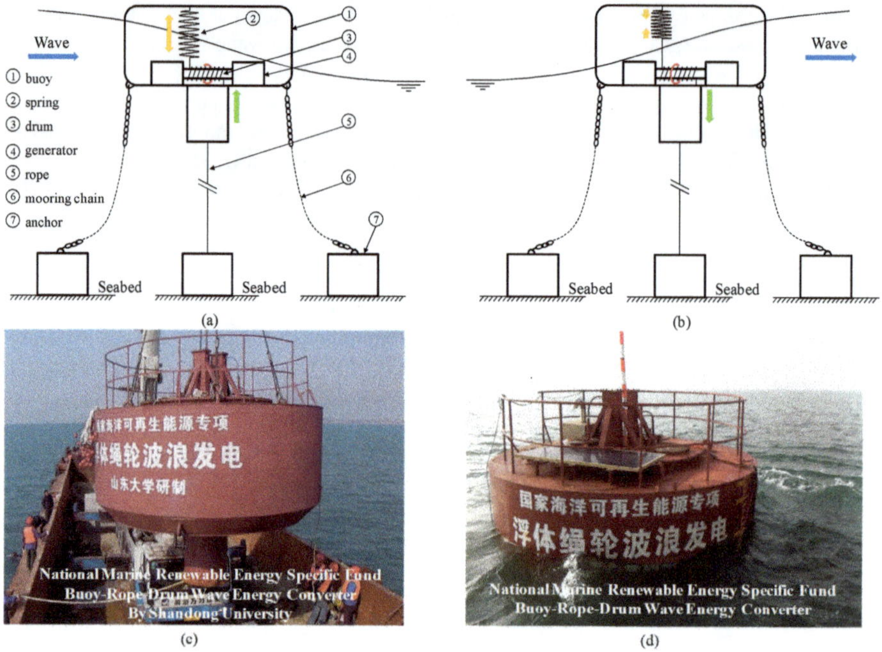

Figure 1. (**a**) Wave energy extraction stroke; (**b**) Drum repositioning stroke; (**c**) BRD WEC transportation; (**d**) BRD WEC installation [19].

2.2. LCA Goal and Scope for BRD WEC

The aim of this LCA study is to analyze the environmental performance of the BRD WEC over its entire life cycle, identify stages and processes with significant environmental impacts, suggest environmental improvement approaches, as well as benchmark its environmental performance with reported LCA studies of other renewable energy systems.

Functional unit is normally defined as the marine energy system itself in existing LCA studies for marine tidal and wave energy systems [16,20,21]. Nevertheless, it is sometimes defined as 1 kWh produced electricity delivered to grid by the marine energy power system, so as to compare its environmental impacts with other renewable energies [13,16–18,21]. As the designed BRD WEC will be primarily used for electricity supply for navigation buoys and small islands rather than delivering electricity to the grid on mainland, the functional unit of this LCA study is thus defined as the entire BRD WEC system during its service lifespan of 20 years.

As illustrated in Figure 2, the system boundary encompasses all the materials and processes through the BRD WEC life cycle, including raw materials extraction and manufacturing, component manufacturing, module production, system assembly, installation, O&M, decommissioning and recycling. Physically, the system boundary excludes small mechanical components used for upstream module and system assembly as well as downstream maintenance (repairs and replacement), such as bolts, nuts and studs, which account for negligible portion of weight and minimal environmental impacts. For this LCA model, materials include the entire raw material extraction and processing. Manufacturing and assembly include materials and processes for the components manufacturing, module assembly through to the WEC system assembly. Installation primarily includes the transportation to ship the assembled system to the installation location at sea, as well as energy and materials consumption. O&M includes energy and materials consumption and component replacement associated with materials and energy used within the O&M during the service life span of the WEC. O&M also includes the transportation of replacement material and disposal of demolished materials.

Decommissioning and recycling includes energy and emissions related to the disassembly process as well as recycling or disposal of materials. Related transportation at the end of the life cycle is included as well. The end of life (EoL) includes materials to be recycled and reused to avoid raw materials extraction. Due to the fact that there has no industrial practice of WEC system decommissioning and recycling or disposal [8,18], here it is assumed that majority of metals will be recycled; the gravity anchors and chains will remain in the seabed and reused for subsequent WEC anchoring; and other materials will be sent to either landfill or incineration.

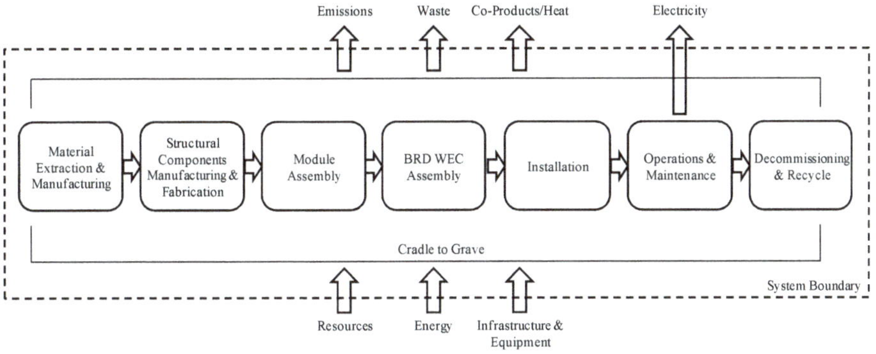

Figure 2. Life cycle of BRD WEC and system boundary.

The ReCiPe method is applied for the LCIA at a midpoint level, so as to examine as much wide range of environmental impacts of the BRD WEC beyond energy and carbon intensities. The environmental impacts of materials, energy consumptions as well as emissions are evaluated for each life cycle stage.

2.3. Inventory Analysis of BRD WEC

2.3.1. Inventory Analysis of Raw Material Extractions

A LCI is conducted by following the flow chart in Figure 2, according to the system boundary definition, including all stages associated with primary materials and unit processes. Embodied energy and emissions of raw materials acquisition and manufacturing are based on the normalized values by Ecoinvent 3 database and published LCA studies [8,21,22]. The materials and processes inputs are listed in Table 1.

2.3.2. Inventory Analysis of Manufacturing, Assembly and Installation

The buoy structure is designed to be made of 10,467 kg of carbon steel, while 640 kg of polyurethane foam equipped in void spaces in the buoy structure to provide protection by absorbing and reservoiring flooded sea water in case of local structural damage and failure. A total amount of 63 kg of alkyd paint is applied to both inside and outside surfaces of the buoy structure so as to provide antifouling and as anticorrosive protection. Zinc blocks are applied as a cathodic protection mechanism to provide additional anticorrosive protection for the submerged steel structural components. Primary processes include arc welding, painting and transportation in between manufacturing plants in different locations in Shandong to the installation location 2 km off the shore at Weihai, Shandong.

The generator module contains two identical 5 kW generators made of 2056 kg of electrical steel, 171 kg of copper windings and a 75 kg neodymium magnet. The primary process is the machining and arc welding for the generator manufacturing. Other processes such as wire drawing and transportation from manufacturing plant in Tai'an to the assembly plant in Zibo are also taken into consideration.

Since the input data of electrical steel is not available in the Ecoinvent database, here the data from [21] is used for the inventory input. Input data of machining is referenced from [8,22].

Table 1. Primary materials and processes for BRD WEC manufacturing, assembly and installation.

Material/Process	Weight (kg)	Source	Comments
Buoy Structure			
Steel, general	10,467	Ecoinvent	Structural components of buoy
Polyurethane	640	Ecoinvent	Provides protection by absorbing flooded sea water
Alkyd paint	63	Ecoinvent	Antifouling and anticorrosive protection coating
Zinc	86	Ecoinvent	Used as sacrificial anode (cathodic protection)
Arc welding	-	Ecoinvent	Primary process for steel structural construction
Casting	-	Ecoinvent	Primary process for Zinc block manufacturing
Painting	-	Ecoinvent	-
Transportation	-	Ecoinvent	-
Generator			
Electrical Steel	2056	[21]	Stator/Translator
Copper winding	171	Ecoinvent	Stator/Translator
Permanent magnet	75	Ecoinvent	Rotator
Wire drawing, copper	-	Ecoinvent	Copper winding manufacturing approximation
Machining	-	[8,22]	Generator manufacturing process approximation
Arc welding	-	Ecoinvent	Generator manufacturing process approximation
Transportation	-	Ecoinvent	-
Mooring			
Concrete	120,000	Ecoinvent	Gravity anchor
Steel, low-alloyed	4698	Ecoinvent	Material for mooring chain
Transportation	-	Ecoinvent	Shipping for anchors and chains
Rope			
Fleece	30	Ecoinvent	Approximating UHMWPE rope, due to unavailability of data
Polar fleece production	-	Ecoinvent	Rope production
Transportation	-	Ecoinvent	Shipping for rope from manufacturing plant to installation

The mooring module contains four gravity anchors and three steel mooring chains. Each mooring chain connects the buoy structure with one gravity anchor sitting in the seabed touchdown site. The lower end of the rope is fixed on the 4th gravity anchor and the upper end is connected to the drum. The gravity anchors are made of square shaped concrete block in a local concrete factory. The mooring chains are made of steel and manufactured in a mooring manufacturing facility in Qingdao. The process of transportation from Qingdao to the WEC installation location is included in the analysis. Another included process is the steel casting of chain links. The distances taken into account in this study are listed in Table 2.

2.3.3. Inventory Analysis of O&M

The BRD WEC does not consume energy and produces negligible amount of emissions during its electricity production. The use and maintenance connected with energy and emissions are mainly from replacements and routine inspection caused material consumption and transportations. Here two assumptions are made: one is that routine inspections are planned at an annual basis; the other is that the entire BRD WEC system's structural integrity is well maintained with no need of major repairs and replacements. In this case, only the transportations are taken into consideration for the inventory analysis of the use and maintenance phase. The transportation is modeled as workboat commuting in between the BRD WEC offshore installation site and the onshore maintenance workshop.

Table 2. Distances taken into account for BRD WEC LCA.

Location	Distance (km)	Comments
Zibo, Shandong, China	20	Manufacturing/Assembly location of BRD WEC structure; distance from Zibo to Weihai
Qingdao, Shandong, China	270	Manufacturing location of rope; distance from Qingdao to Weihai
Qingdao, Shandong, China	270	Manufacturing location of mooring chains; distance from Qingdao to Weihai
Weihai, Shandong, China	20	Manufacturing location of mooring module; distance from local plant to BRD WEC assembly location at Weihai
Tai'an, Shandong, China	150	Manufacturing location of generators; distance from Tai'an to Zibo
Xiaoshi Island, Weihai, Shandong, China	2	Offshore Weihai, Installation location of the BRD WEC; Operation and Maintenance location

2.3.4. Inventory Analysis of EoL and Waste Scenarios

Up to date of the conduction of this study, there has no reported or published disposal or recycling practice of WECs, thus reasonable assumptions have to be made in order for the LCA to be convincing. According to ISO 14040, a credit is given to reduce the total energy and emissions, as the recycling of certain components avoid the extraction of raw materials [8]. Given the level of uncertainty, different recycling rates are taken for modeling by currently existing LCA studies. A recycling rate of 90% was applied for steel recycling [8]. In order to investigate the impacts of including and excluding recycling activities at EoL, a 100% recycling rate for metals and concretes was applied by [13] for the purpose of sensitivity analysis. Another study excluded the disposal or recycling stage because of the unavailability of information [21]. A comprehensive list of recycling or disposal rates regarding the material categories are given in [18].

Here in this study, four EoL routes—recycling, incineration, landfill and reuse—are assumed for the EoL modeling. For offshore steel facilities, current industrial practice suggests a corrosion rate 0.2–0.4 mm/year for offshore steel structures [23]. With the assumption of good protection by the built in zinc sacrificial anodes and anticorrosive coating, here the lower corrosion rate value of 0.2 mm/year was applied for the BRD WEC steel structure corrosion estimation, which results in a 0.75% annual corrosion rate, in terms of steel weight. Due to lack of data for the design life of actual WEC installations, the service life of WECs is usually assumed to be 20–25 years [8,13,15,18,20], in order to set up the temporal coverage for LCA studies. Here in this work a service life of 20 years is assumed considering the structural similarity of this BRD WEC with other WECs and other types of offshore energy systems. The life time corrosion rate of the steel is then calculated to be 15%, which indicates that a remaining 85% of the structural steel is available for recycling at its EoL. As for the mooring module, with the assumption that both the gravity anchors and mooring chains can be well maintained during the WEC's lifetime, the 85% of mooring module can be left in seabed for reuse for subsequent WEC deployments and fixation. The waste routes at EoL for this study are shown in Table 3.

Table 3. Assumed waste routes at EoL.

Material	Recycling %	Incineration %	Landfill %	Reuse %	Comments
Ferrous metals (buoy) [a]	85	0	0	0	15% of buoy ferrous metals assumed to be corroded
Non-ferrous metals [b]	95	0	5	0	Data source [18,24]
Plastics [b]	80	20	0	0	Data source [18,24]
Composites [b]	0	100	0	0	Data source [18,24]
Concrete (mooring) [c]	0	0	0	85	15% of mooring concrete assumed to be deteriorated
Ferrous metals (other) [a]	90	0	10	0	Data source [18,24]

Source: [a] Calculation; [b] [18,24]; [c] Assumption.

2.3.5. Uncertainties

There are a few uncertainties for this LCA study due to the lack of available data such as geographical difference, specific technological parameters, alternative materials, as well as assumptions made in this study. Here the geographical difference refers to the actual embodied energy and emissions from materials and manufacturing processes in China versus Europe, as most of the data for input come from Europe cases based database by Ecoinvent. Technological parameters are another group of factors that can have significant impact on the LCA results. For instance, the capacity factor, the generator efficiency, as well as the system reliability among others, are directly related to the input of for the analysis, thus the change of these parameters will directly result in difference of the environmental impacts. As discussed in the LCI and LCIA analysis, consumption of materials is the most significant contributor to the environmental impact of the entire life cycle of the WEC system. Applying alternative materials in any components can have expected impact on the LCA results due to the embodied energy and emission differences in raw material acquisition and manufacturing processes. Assumptions made for this LCA study introduce uncertainties to the analysis results. For instance, the service lifetime, the installation location at sea, the O&M requirements, the recycling rate of wasted materials at EoL, among others, are all important input data which can have impact on the analysis results.

Numerically quantifying of the impacts of these uncertainties on the LCA results can be investigated through the calculation of those interested indicators. Here an example is given by looking into the following calculation formula of embodied energy or carbon intensity as shown below:

$$I_{em.} = \frac{\sum_{i=j=1}^{n}\left(E_{ij} \times A_{ij}\right)}{P_{ins.} \times CF_{Act.} \times \eta_{sys.} \times T_{lc}}$$

where $I_{em.}$ represents the embodied energy or carbon intensity; E_{ij} is the embodied energy or carbon of the jth material or process in the ith life cycle stage for its unit amount of consumption; A_{ij} is the amount of the jth material or process in the ith life cycle stage; $P_{ins.}$ represents the installed capacity; $CF_{act.}$ is the actual capacity factor; $\eta_{sys.}$ is the system efficiency; T_{lc} is the design service life of the entire system. From this mathematical expression it can be indicated that any specific parameter as an input for the calculation is not necessarily linearly related with the embodied intensity calculation result. For instance, adopting an alternative material with lower unit embodied energy or emission is not necessarily expected to reduce the system embodied energy or emission intensity because it is possible that at the same time the total amount of material consumption could also be changed, not to mention that the system specs such as capacity factor, efficiency and design life could also be changed due to the new material adoption. In this regard, the final embodied energy or emission is not necessarily reduced with the application of such material with low embodied energy or emission per unit amount. Assuming another case that the O&M related activities are planned with shortened time intervals, the associated with energy and emissions are expected to increase, on the other hand, this enhanced maintenance plan may also lead to the system's functional life extension beyond its design life, which provides more electricity production. Taking these possibilities into consideration, the above formula gives possibilities for either reduction or increase of energy or carbon intensities.

3. LCA Results

3.1. *Life Cycle Inventory of the BRD WEC*

The inventory analysis results show that the most significant contribution to the life cycle embodied energy comes from the manufacturing stage of the BRD WEC system, with buoy structure contributing 52.97%, generator contributing 26.77% and mooring contributing 13.12%. It is noted that the contribution of mooring manufacturing is compensated by the reuse of the mooring module after its design life. The EoL and O&M contribute much less to the total energy consumption compared with the manufacturing stage. For the embodied CO_2, the largest contributor is also the WEC system manufacturing, with 62.22% of the total embodied CO_2. The manufacturing of buoy structure contributes the most with 35.31% which mainly comes from the structural steel consumption, followed by the manufacturing of generator with 21.45% of the total CO_2. The embodied CO_2 from the generator module manufacturing primarily comes from the consumption of electrical steel and permanent magnet. The EoL disposal of buoy structure contributes 37.42% of total embodied CO_2, which is primarily due to the emissions from treatment of the waste polyurethane. The O&M contributes only 0.13% of total embodied energy and 0.15% of total embodied CO_2 coming from the transportation for annual inspection activities to maintain the structural and functional integrity of the WEC system during its service life time.

3.2. Life Cycle Impact Assessment Results

3.2.1. Life Cycle Impact Analysis Results

With application of ReCiPe midpoint method, a sum of 18 environmental impact categories are analyzed, including climate change, ozone depletion, terrestrial acidification, freshwater eutrophication, marine eutrophication, human toxicity, photochemical oxidant formation, particulate matter formation, terrestrial ecotoxicity, freshwater ecotoxicity, marine ecotoxicity, ionising radiation, agricultural and occupation, urban land occupation, natural land transformation, water depletion, metal depletion and fossil depletion. The total characterized environmental impacts are presented in Figure 3. Here a few results are highlighted: (1) the manufacturing stage of the WEC system is the most significant contributor to all categories of environmental impact except for marine eutrophication, freshwater ecotoxicity and marine ecotoxicity; (2) the O&M stage contributes relatively little to the total environmental impact; (3) manufacturing of the buoy structure by consumption of steel is the largest contributor to 14 out of 18 categories of impacts including climate change (with steel contributing 31.3%), Ozone depletion (with a steel contribution of 42.3%), terrestrial acidification (with steel contributing 50.6%), freshwater eutrophication (with steel contributing 65.5%), human toxicity (with steel contributing 36.7%), photochemical oxidant formation (with steel contributing 53.7%), particulate matter formation (with steel contributing 64.9%), ionising radiation (with concrete contributing 42.3%), agricultural land occupation (with concrete contributing 33.8%), urban land occupation (with concrete contributing 54.2%), natural land transformation (with concrete contributing 43.2%), water depletion (with concrete contributing 36.3%), metal depletion (with steel contributing 77%) and fossil depletion (with steel contributing 48.1%); (4) manufacturing of generator is the most significant contributor to terrestrial ecotoxicity (with permanent magnet contributing 50.2%); (5) disposal of buoy structure contributes the most to marine eutrophication (with treatment of waste polyurethane contributing 86.3%), freshwater ecotoxicity (with treatment of waste polyurethane contributing 57.4%) and marine ecotoxicity (with treatment of waste polyurethane contributing 49.2%).

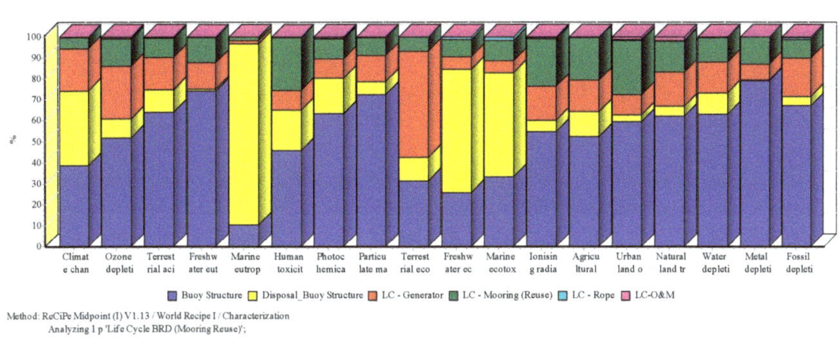

Figure 3. Life cycle impact of BRD WEC.

3.2.2. Embodied Energy and Carbon: Intensities and Paybacks

In order to quantitatively ecolabel the high environmental impact associated with life cycle stages, annual electricity production has to be estimated. Here for the calculation of annual electricity output for the BRD WEC, a few assumed parameters are taken into calculation including the total wave energy conversion efficiency and the capacity factor of the BRD WEC system. Since the BRD WEC is in its development and test stages, the actual capacity factor has not been specified yet for use. Assumed capacity factor values 20–50% are suggested by other WEC LCA studies [15,18,21,25] for the purpose of calculating annual electricity productions. Here in this study, the capacity factor is assumed to be 50%, since this prototype WEC is still in its early stage of development and technologically immature. Assuming that the total design generator efficiency 80% [26] is taken into the calculation and given the design power capacity of 10 kW, the annual electricity output is then calculated to be about 35 MWh. Within a service lifetime of 20 years, the total electricity production is calculated to be 700 MWh.

The energy intensity is calculated by dividing the life cycle embodied energy with total life time energy production; likewise the carbon intensity is calculated by dividing the life cycle embodied carbon with total life time energy production [27]. The energy payback period is an important environmental indicator for renewable energies as it demonstrates the period of time for the renewable power system to produce as much amount of energy as the energy consumption to produce the energy system itself. Here the impact assessment method Cumulative Energy Demand [27–29] is applied for the calculation of the payback time of the BRD WEC system. The results show that the cumulative energy demand of the BRD WEC is 0.108 $MJ_{eq}/MJ_{electricity}$, which gives the energy payback time is 26 month. The calculated energy intensity of the BRD WEC is 387 kJ/kWh. The carbon payback is another important environmental indicator for renewable energies, which is calculated by dividing the life cycle embodied carbon with annual avoided carbon [27]. Here the annual avoided carbon is calculated by taking the difference of the BRD WEC carbon intensity and carbon intensity of local conventional fossil fueled electricity. The fossil fueled electricity emission factor is taken as 1004 gCO_2/kWh [30], which is the latest emission factor published in 2014 for Shandong, China. The calculated results of the carbon intensity and carbon payback are 89 gCO_2/kWh and 23 months, respectively.

3.3. Improvement Potentials

The above analyses show that the consumption of materials is the most significant contributor of the BRD WEC's life cycle inventory and life cycle impacts. However, reduction of the overall energy and carbon intensities should be comprehensively strategized. Here a few approaches are proposed so as to achieve overall life cycle environmental improvement: (1) to increase the system specific mass, according to the inherent connection between the specific mass with energy and carbon intensities; (2) to improve the system reliability and energy conversion so as to achieve system life time extension which results in more electricity production; (3) to utilize alternative materials with lower embodied energy and emissions per unit mass; (4) to involve more environmentally friendly transportations; among others.

3.4. Comparison with Other Renewable

In order to benchmark the environmental performance of the BRD WDC with other WECs as well as other renewables, a comparison of energy and carbon intensities is made with two other types of WECs including Pelamis and Oyster. The comparison is also included five types of tidal energy converters including Seagen, TGL, Openhydro, Flumill and ScoRenewables. Comparison with other renewable such as offshore wind and polycrystalline silicone photovoltaic systems is also conducted.

As shown in Figure 4, the energy intensity of BRD WEC is slightly larger than Pelamis and Oyster but still within the range of 200–400 kJ/kWh. These energy intensities of WECs are comparable to tidal energy converters. The carbon intensity of BRD WEC is approximately three times larger than Pelamis' and Oyster's, which is also larger than the referenced tidal, offshore wind and PV systems. Energy and carbon intensities of WECs are dependent on many factors, including geographical, technological, and temporal and among others. Specific mass in terms of kilograms per kilowatt nominal power capacity was introduced for comparison of different types of tidal and wave energy devices [18]. Here in this work the approximate linear dependence with specific mass is analytically illustrated in Figure 4. Comparison with Pelamis, Seagen and Oyster shows that BRD has much larger specific mass, which is most likely the inherit reason for its relatively larger energy and carbon intensities.

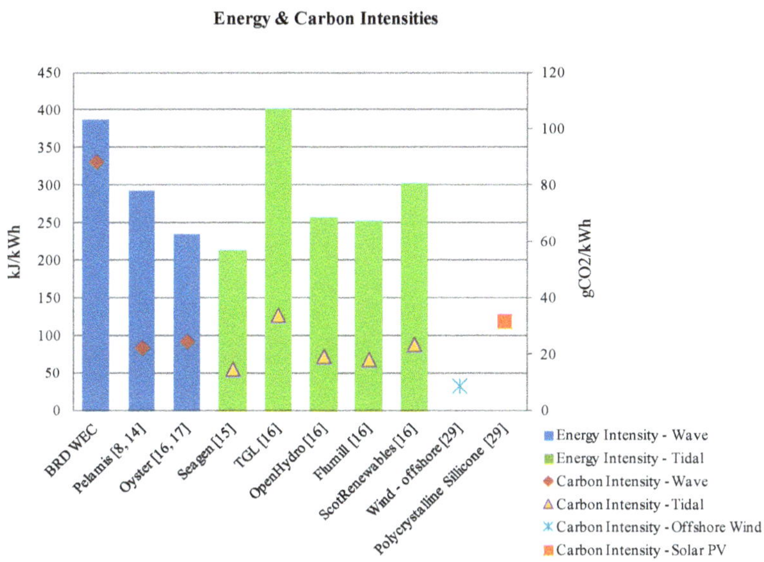

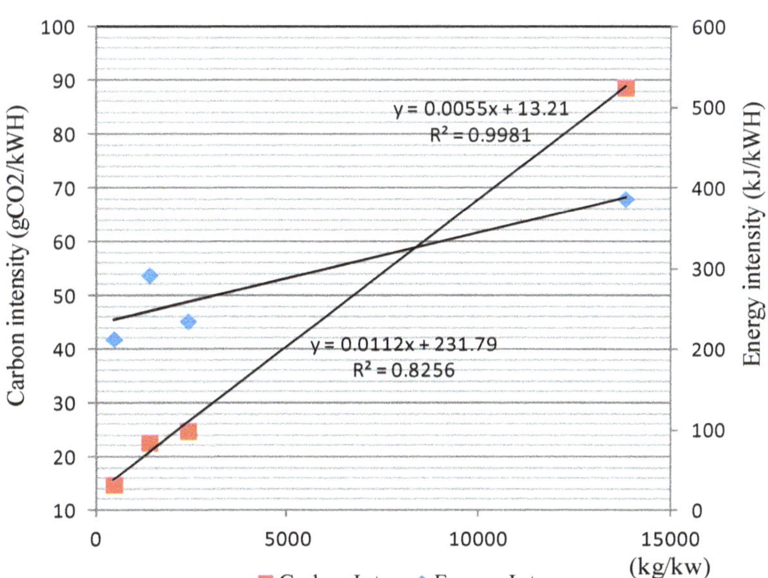

Figure 4. Energy and carbon intensities of WECs, comparison and vs. specific mass.

Figure 5 demonstrates the comparison of energy and carbon paybacks with other WECs including Pelamis and Seabased Case Norway, as well as tidal energy converters including Seagen, TGL, Openhydro, Flumill and ScoRenewables. It is indicated that the energy and carbon paybacks of BRD WEC are comparable with other WECs. Secondly, all the studied WECs have relatively less payback of energy and carbon than tidal energy converters, excluding Seagen. Further, the carbon payback of all WECs is expected to be shorter than tidal energy converters. This further indicates that application of wave energy conversion can achieve compensation for the system manufacturing related energy consumption and carbon emission within shorter period of time.

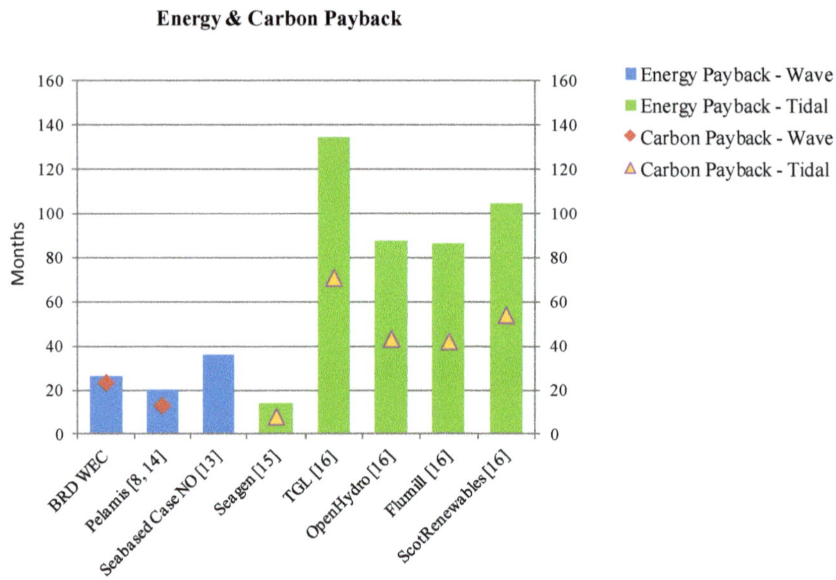

Figure 5. Comparison of energy and carbon paybacks of wave and tidal energy converters.

4. Sensitivities

Among the parameters that might have a significant influence on the environmental impact, capacity factor is one of the most interested ones for this study due to its uncertainty, since the system is still immature. Hence, a sensitivity analysis was carried out for the wave energy system by varying the capacity factor values in order to investigate its influence on the impact results. Here for the capacity factor values, 20–50% was taken, as discussed in Section 3.2.2. Figure 6 indicates that both carbon intensity and energy intensity decrease with the increase of capacity factor. For carbon intensity, a value of 222 gCO_2/kWh is achieved, when the capacity factor decreases to 20%. Energy intensity could be as much as 968 kJ/kWh with a capacity factor 20%. Here for the sensitivity analysis, more specific examples or scenarios are not exhaustively discussed since the BRD WEC is still at its development and optimization stage. However, the above discussion should give the designers and interested stakeholders a sense of complexity to reduce the system's energy and carbon intensities so as to improve its overall environmental performance. A tradeoff has to be taken into consideration among optimization of different technological parameters for future development and improvement.

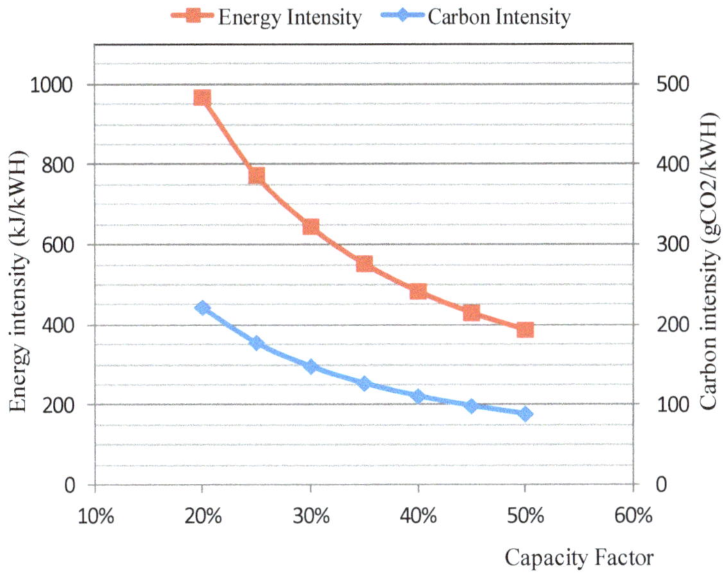

Figure 6. Carbon and energy intensities depending on capacity factors.

5. Discussion and Conclusions

In this study, a full life cycle is analytically modeled for the BRD WEC. The conducted LCI and LCIA show that manufacturing is the most significant contributor to energy consumption, carbon emissions as well as interested environmental impacts. The energy and carbon intensities are calculated to be 387 kJ/kWh and 89 gCO_2/kWh. The energy and carbon paybacks are calculated to be 26 and 23 months. Analytical comparisons show that the energy and carbon intensities and paybacks are comparable to reported LCA studies of other WECs and other renewables such as tidal, offshore wind and solar PV. Improvements can be achieved through optimization of the system's specific mass, utilization of alternative materials which contains low unit embodied energy and carbon. Shipping of the structural components and modules through more environmentally friendly transportations with relatively low environmental impacts could also reduce the overall life cycle environmental impacts. Since the life cycle impacts cover a wide range of categories of indicators, a tradeoff is to be comprehensively and systematically made so as to achieve overall environmental performance improvement.

Novelty of this study consists of but not limited to: (1) this study is the first LCA case study for a WEC developed in China; (2) this study is one of the few LCAs for WECs based on an actual system installation; (3) this study originally suggests taking corrosion rate of offshore steel structures into account for wave energy converters at EoL stage for LCA; (4) this study is one of the few fully LCA studies taking into account not only energy and carbon, but also other environmental impacts.

Limitations of this study include: (1) lack of available data such as geographical difference, specific technological parameters such as input data for operation and maintenance; (2) system immaturity caused uncertainties such as materials, system efficiency and capacity factor; (3) accuracy of other assumptions made for the analysis such as the service lifetime, the O&M requirements, the recycling rate of wasted materials at EoL, and among others.

According to the above discussion, our future research will focus on acquisition of more available data, data update with the system improvement as well as consideration of wider marine environmental effects such as marine life. Last but not least, we realized that for any product or service, regular LCA practice does not take into account economic and social factors. In this context, researchers have been dedicated to the integration of economic models as well as ecological and social theories [31–34]. Nowadays life cycle sustainability assessment (LSCA) has been developed rapidly because of its comprehensiveness. Although this is currently beyond scope of our study, economic and social factors could be taken into consideration for our future research in order to comprehensively understand the system's sustainability performance.

Author Contributions: Data collection, Q.Z. and L.Z.; Analysis and manuscript drafting, Q.Z.; Review, S.L.

Funding: This research was funded by National Marine Renewable Energy Specific Fund (GHME2011BL02). The APC was funded by Shandong University (Weihai) Scientific Research Funding (1050517300003) and the National Natural Science Foundation of China (No. 51605261).

Conflicts of Interest: The authors declare no conflict of interest.

References

1. MacDonald, A.E.; Clack, C.T.M.; Alexander, A.; Dunbar, A.; Wilczak, J.; Xie, Y. Future cost-competitive electricity systems and their impact on US CO_2 emissions. *Nat. Clim. Chang.* **2016**, *6*, 526–531. [CrossRef]
2. Mora, C.; Dousset, B.; Caldwell, I.R.; Powell, F.E.; Geronimo, R.C.; Bielecki, C.R.; Counsell, C.W.W.; Dietrich, B.S.; Johnston, E.T.; Louis, L.V.; et al. Global risk of deadly heat. *Nat. Clim. Chang.* **2017**, *7*, 501–506. [CrossRef]
3. Zhou, N.; Fridley, D.; McNeil, M.; Zheng, N.; Ke, J.; Levine, M. *China's Energy and Carbon Emissions Outlook to 2050*; Lawrence Berkeley National Laboratory: Foshan, China, 2011.
4. Sheng, S.; Wang, K.; Lin, H.; Zhang, Y.; You, Y.; Wang, Z.; Chen, A.; Jiang, J.; Wang, W.; Ye, Y. Model research and open sea tests of 100 kW wave energy converter Sharp Eagle Wanshan. *Renew. Energy* **2017**, *113*, 587–595. [CrossRef]
5. Liu, Y.; Li, Y.; He, F.; Wang, H. Comparison study of tidal stream and wave energy technology development between China and some Western Countries. *Renew. Sustain. Energy Rev.* **2017**, *76*, 701–716. [CrossRef]
6. Office of Energy Efficiency & Renewable Energy, U.S. Department of Energy. What Is the Wave Energy Prize? Available online: https://waveenergyprize.org/about (accessed on 26 October 2017).
7. Carbon Trust. *Future Marine Energy*; Carbon Trust: London, UK, 2006; Available online: http://ukerc.rl.ac.uk/Roadmaps/Marine/Marine_Roadmaps_MEC_v2.pdf (accessed on 11 September 2018).
8. Parker, R.P.M.; Harrison, G.; Chick, J. Energy and carbon audit of an offshore wave energy converter. *Proc. Inst. Mech. Eng. Part A J. Power Energy* **2007**, *221*, 1119–1130. [CrossRef]
9. Map of People's Republic of China. Available online: http://www.gov.cn/test/2005-06/15/content_18252.htm (accessed on 26 October 2017).
10. Gao, H.; Zhang, A. *Technology and Application of Renewable Energy*; National Defense Industry Press: Beijing, China, 2007.
11. Zhang, D.; Li, W.; Lin, Y. Wave energy in China: Current status and perspectives. *Renew. Energy* **2009**, *34*, 2089–2092. [CrossRef]
12. Amponsah, N.; Troldborg, M.; Kington, B.; Aalders, I.; Hough, R.L. Greenhouse gas emissions from renewable energy sources: A review of lifecycle considerations. *Renew. Sustain. Energy Rev.* **2014**, *39*, 461–475. [CrossRef]
13. Dahlsten, H. *Life Cycle Assessment of Electricity from Wave Power*; Swedish University of Agricultural Sciences: Uppsala, Sweden, 2009.
14. Thomson, R.C.; Harrison, G.P.; Chick, J.P. Life Cycle Assessment in the Marine Renewable Energy Sector. In *LCA Report—Instruments for Green Futures Markets, Proceedings of the LCA XI International Conference, Chicago, IL, USA, 4–6 October 2011*; American Center for Life Cycle Assessment: Chicago, IL, USA, 2011; pp. 120–125.
15. Douglas, C.A.; Harrison, G.; Chick, J. Life cycle assessment of the Seagen marine current turbine. *Proc. Inst. Mech. Eng. Part M J. Eng. Marit. Environ.* **2008**, *222*, 1–12. [CrossRef]
16. Walker, S.R.; Howell, R.; Hodgson, P.; Griffin, A. Tidal energy machines: A comparative life cycle assessment study. *Proc. Inst. Mech. Eng. Part M J. Eng. Marit. Environ.* **2015**, *229*, 124–140. [CrossRef]
17. Douziech, M.; Hellweg, S.; Verones, F. Are wave and tidal energy plants new green technologies. *Environ. Sci. Technol.* **2016**, *50*, 7870–7878. [CrossRef] [PubMed]

18. Uihlein, A. Life cycle assessment of ocean energy technologies. *Int. J. Life Cycle Assess.* **2016**, *21*, 1425–1437. [CrossRef]
19. Zhu, L.; Wang, Y.; Yang, Z.; Qu, Y. Buoy-Rope-Drum Wave Power System. *Adv. Mech. Eng.* **2013**, *5*, 609464. [CrossRef]
20. Cavallaro, F.; Coiro, D. Life Cycle Assessment (LCA) of a marine current turbine for cleaner energy production. In Proceedings of the 3rd International Conference on Life Cycle Management, Zurich, Switzerland, 27–29 August 2007.
21. Zepeda, L.G. *Life Cycle Assessment of a Wave Energy Converter*; Degree Project in Mechanical Engineering; KTH Royal Institute of Technology: Stockholm, Sweden, 2017.
22. Kalpakjian, S. *Manufacturing Processes for Engineering Materials*; Addison-Wesley Publishing Company: Reading, MA, USA, 1992.
23. Corrosion Protection of Ships. Available online: https://rules.dnvgl.com/docs/pdf/DNVGL/CG/2017-05/DNVGL-CG-0288.pdf (accessed on 11 September 2018).
24. Zimmermann, T. Parameterized tool for site specific LCAs of wind energy converters. *Int. J. Life Cycle Assess.* **2012**, *18*, 49–60. [CrossRef]
25. Banerjee, S.; Duckers, L.; Blanchard, R.E. An overview on green house gas emission characteristics and energy evaluation of ocean energy systems from life cycle assessment and energy accounting studies. *J. Appl. Nat. Sci.* **2013**, *5*, 535–540. [CrossRef]
26. Ju, Y. Research on the Buoy-Rope-Drum Wave Energy Collection Device. Master's Thesis, Shandong University, Weihai, China, 2016.
27. Dalton, G.; Madden, D.; Daly, M.C. Life cycle assessment of the Wavestar. In Proceedings of the 2014 Ninth International Conference on Ecological Vehicles and Renewable Energies (EVER), Monte Carlo, Monaco, 25–27 March 2014; pp. 1–9.
28. Weinzettel, J.; Reenaas, M.; Solli, C.; Hertwich, E.G. Life cycle assessment of a floating offshore wind turbine. *Renew. Energy* **2009**, *34*, 742–747. [CrossRef]
29. Hischier, R.; Weidema, B.; Althaus, H.-J.; Bauer, C.; Doka, G.; Dones, R.; Frischknecht, R.; Hellweg, S.; Humbert, S.; Jungbluth, N.; et al. *Implementation of Life Cycle Impact Assessment Methods*; Final Report ecoinvent 2000; Swiss Centre for LCI: Duebendorf, Switzerland, 2003.
30. Ma, C.; Ge, Q. Method for calculating CO_2 emissions from the power sector at the provincial level in China. *Adv. Clim. Chang. Res.* **2014**, *5*, 92–99.
31. Curran, M.A. Life cycle assessment: A review of the methodology and its application to sustainability. *Curr. Opin. Chem. Eng.* **2013**, *2*, 273–277. [CrossRef]
32. Falcone, P.M.; Imbert, E. Social Life Cycle Approach as a Tool for Promoting the Market Uptake of Bio-Based Products from a Consumer Perspective. *Sustainability* **2018**, *10*, 1031. [CrossRef]
33. Schaubroeck, T.; Rugani, B. A revision of what life cycle sustainability assessment should entail: Towards modeling the Net Impact on Human Well-Being. *J. Ind. Ecol.* **2017**, *21*, 1464–1477. [CrossRef]
34. Weidema, B. The integration of economic and social aspects in life cycle impact assessment. *Int. J. Life Cycle Assess.* **2006**, *11*, 89–96. [CrossRef]

energies

Article

Ethylene Supply in a Fluid Context: Implications of Shale Gas and Climate Change

Gillian Foster

Institute for Ecological Economics, Department of Socioeconomics, Vienna University of Economics and Business, Welthandelsplatz 1, 1020 Vienna, Austria; gillian.foster@wu.ac.at; Tel.: +43-1-313-36-5521

Received: 19 September 2018; Accepted: 27 October 2018; Published: 1 November 2018

Abstract: The recent advent of shale gas in the U.S. has redefined the economics of ethylene manufacturing globally, causing a shift towards low-cost U.S. production due to natural gas feedstock, while reinforcing the industry's reliance on fossil fuels. At the same time, the global climate change crisis compels a transition to a low-carbon economy. These two influencing factors are complex, contested, and uncertain. This paper projects the United States' (U.S.) future ethylene supply in the context of two megatrends: the natural gas surge and global climate change. The analysis models the future U.S. supply of ethylene in 2050 based on plausible socio-economic scenarios in response to climate change mitigation and adaptation pathways as well as a range of natural gas feedstock prices. This Vector Error Correction Model explores the relationships between these variables. The results show that ethylene supply increased in nearly all modeled scenarios. A combination of lower population growth, lower consumption, and higher natural gas prices reduced ethylene supply by 2050. In most cases, forecasted CO_2 emissions from ethylene production rose. This is the first study to project future ethylene supply to go beyond the price of feedstocks and include socio-economic variables relevant to climate change mitigation and adaptation.

Keywords: ethylene supply; shale gas; non-energy uses of fossil fuels; socio-economic scenarios; climate change

1. Introduction

The chemical ethylene (C_2H_4) is the star in an unfolding drama of two defining and contentious megatrends influencing today's energy landscape in the U.S. and globally. First, the increased supply of shale gas due to unconventional hydraulic fracturing (fracking) in the United States. Second, the crisis of global climate change compelling a transition to a low-carbon economy. These two megatrends influencing ethylene supply in the U.S. are fluid because they are complex, contested, and uncertain. Therefore, this analysis developed medium- (2035) and long-term (2050) projections for the future U.S. supply of ethylene using the Shared Socioeconomic Pathways (SSPs) scenario framework for future climate change mitigation and adaptation and a range of natural gas prices. Additionally, this paper estimates the CO_2 equivalent emissions of greenhouse gas associated with future ethylene supply.

Ethylene is an important commodity in today's global economy because it is widely produced and used. It is used primarily to make plastics for the automotive, building, and packaging industries around the world. It is the key ingredient in about half of all plastics [1]. Sixty percent of ethylene produced worldwide is used to make polyethylene terephthalate (PET) [2]. Products for which ethylene is the basic building block are common, including plastic sheeting, wire coatings, antifreeze, solvents, golf balls, food packaging, and bottles. Most humans use ethylene every day.

Ethylene represents a class of materials that are made from fossil fuel feedstock. These materials are ubiquitous in modern life, including chemicals, plastics, solvents, fertilizers, synthetic fiber, asphalt, and lubricants. They are referred to as "non-energy" uses of fossil fuels. According to the U.S.

Environmental Protection Agency (EPA), non-energy uses contribute 126 million metric tons CO_2 equivalent (MMT CO_2 eq.) or 2 percent of overall fossil fuel emissions [3]. U.S. ethylene production in 2015 contributed 20 MMT of CO_2, which is 18 percent of the industrial non-energy use of fuel category [3]. In 2015, the ethylene supply was 391.4 million barrels [4]. U.S. ethylene is produced at about 40 factories in five states [5]. With only 40 factories responsible for 20 MMT of CO_2, the ethylene industry is a large greenhouse gas emitter with significant impact for its size.

The research and policy context of this analysis of future ethylene supply is described by three dynamics. First, the U.S. is one of the top manufacturers of ethylene in a growing market [6]. Chemical sales are growing worldwide, increasing by 2.2 times in value between 2005 and 2015 [7]. Second, the recent advent of shale gas in the U.S. has redefined the economics of ethylene manufacturing globally. North American producers of ethylene have a price advantage due to the availability of relatively low-cost natural gas-derived feedstocks [6,8]. The price of ethylene is heavily influenced by the cost of its feedstock, which can be up to 60 percent of its market price [9]. Consequently, ethylene production has shifted to lower-cost U.S. production due to ethane, propane, and butane, collectively called "light" feedstocks, derived from shale gas production [10]. Fracking produces both natural gas used for heating and electricity, and the ethane that is used primarily for ethylene production. Investments in U.S. ethylene production with ethane feedstock are drawing investment from other regions, such as Europe [11]. The increased availability of this low-cost light feedstock encouraged U.S. producers to switch from naphtha and crude oil "heavy" feedstocks to light feedstocks from natural gas. Third, the risks of global climate change and the need to reduce CO_2 emissions due to fossil fuel use worldwide are pressing concerns that require analysis of widely-used materials that are currently produced from fossil fuels. The U.S. production of ethylene is completely dependent on non-renewable fossil fuels [12]. The goal of replacing fossil fuel as the main feedstock for materials such as ethylene is founded, not only on the need to reduce global output of carbon dioxide and greenhouse gases [13], but the broader concept of sustainable consumption and production. The present case study on ethylene in the U.S. is representative of and applicable to the discussion of decarbonization of non-energy uses of fossil fuels, their future supply in the context of climate change and natural gas industry developments. Research on the future U.S. ethylene supply in light of societal approaches to climate change and sustainable consumption has national and global relevance for policymaking today.

Although the future supply of ethylene and similar materials is vital for business decisions, policymaking, and estimating environmental impacts, there are few analyses of ethylene supply in the literature. The 1978 article "Ethylene Economics and Production Forecasting in a Changing Environment" is notable as a historical reference because it includes a detailed discussion of feedstock trends at the time. The article documents the industry's shift from natural gas in the 1960s to crude oil and naphtha [14]. Now, because of shale gas, the industry is reverting back to natural gas. Recent studies focus on competitiveness and market share of the U.S. ethylene industry without regard to environmental impact. The present analysis takes a different approach to estimating future supply of ethylene because the primary goal is not estimating competitiveness, but rather future supply and the potential environmental impact measured as greenhouse gas equivalent CO_2.

The literature review identified the corpus of relevant peer-reviewed articles, which are described as follows. This article contributes to a small group of studies that forecast basic chemicals and/or ethylene supply for estimating environmental impacts, e.g., References [15–18]. Dornburg et al. created future scenarios for petrochemical demand using production volumes from the year 2000 and applied growth rates between 0% and 3% per year based on interviews with experts [15]. Broeren and Patel forecasted the basic chemical industry by applying technology and policy scenarios to production capacity estimates [16]. Herman and Patel projected annual greenhouse gas savings from ethylene from technology improvements by 2030 but kept the supply of ethylene constant [17]. Ruth et al. developed a scenario model with policy and technology assumptions with the age of production facilities to include replacement of capital stocks [18]. Unlike the models mentioned, the present analytical model is driven by new feedstocks and socioeconomic developments that embody climate change mitigation and adaptation narratives as described in Section 2 rather than technological and specific policy change estimates.

The article also contributes to articles that examine the dynamics of ethylene markets including feedstocks using similar econometric methods. The literature review identified only one directly applicable paper, in which Masih et al. researched the drivers of ethylene price by modeling crude oil price (WTI), a feedstock for ethylene, and ethylene prices in three regions using a vector error correction model (VECM) [19]. They found no other study with a similar approach to olefin price (ethylene price) and crude oil prices. The present study also applied a VECM model and used crude oil price (WTI). Similar to the present study, Masih et al. found that crude oil prices and ethylene prices were co-integrated. However, the current paper is more in depth because it investigated the relationships between more than two commodities and included more feedstocks than the previous contribution.

This paper presents a VECM of ethylene supply using historical data to project future scenarios for ethylene. Additionally, this paper estimates the CO_2 equivalent emissions of greenhouse gas associated with future ethylene supply scenarios. The scenarios were derived from: (1) historical data (1986–2014); (2) plausible socio-economic scenarios in response to climate change mitigation and adaptation pathways; and (3) a range of gas feedstock prices. The results of the model for the years 2014, 2035, and 2050 are shown. The results show that ethylene supply increases in nearly all modeled scenarios. Ethylene supply is projected to grow by 18 percent by 2035 and 28 percent by 2050 using historical data. Only one of the scenarios resulted in ethylene supply reductions and corresponding greenhouse gas reductions by 2050.

The paper proceeds as follows. Section 2 describes the materials and methods and details the scenarios. Section 3 presents the empirical results. Section 4 interprets the results. Section 5 concludes with policy implications and opportunities for future research.

2. Materials and Methods

The goal of this research was a scenario-based assessment of future climate change impacts of ethylene manufacture in the U.S. accounting for variations in natural gas price. This section details the model estimation, reviews the data, and describes the SSPs and how they were used innovatively to create the scenarios. The model was built in three steps.

- The first step was to create a business-as-usual "base" model that projects future ethylene supply from production and socio-economic data. This historical data describes the system of ethylene supply in the U.S. without regard to climate change implications. This projection was used as a baseline in this analysis. To do this, an econometric VECM of the U.S. ethylene supply was developed with time series data (1986–2014). A VECM is an autoregressive model designed to account for co-integration amongst the variables. See Model Estimation below.
- The second step created the climate change relevant scenarios by applying the SSP socioeconomic drivers in the VECM. The SSPs were recently developed by a consortium of climate change researchers to "serve as a framework for systematic future research of climate change mitigation, climate impacts and adaptation as well as broader sustainability issues aiming to integrate studies from a great diversity of research fields" [20].
- A third step built a range of natural gas prices into the model. Finally, the greenhouse gas emissions of the quantities of ethylene supplied under each scenario were estimated.

Model Estimation—A multivariate autoregressive model rather than a multiple linear regression model (MLR) was chosen for this time series analysis because it better reflects the complexity of interactions amongst variables and to avoid two analytical limitations of MLRs. In contrast to an MLR, an autoregressive model is able to model ongoing relationships between same-time variables in the future, rather than only continuing historical trends, which is relevant in this case. Additionally, an autoregressive model can show the effect of a one-time shock of each variable on ethylene supply, which leads to better understanding of the system described as a whole. In the real world, ethylene supply is in flux and long-term trends have recently changed. Therefore, an autoregressive model is appropriate for gaining insight into the real system of ethylene supply mimicked by the model.

The preliminary analysis began with a vector auto regressive model (VAR). A VAR describes relationships between all variables in concert at each period in a time series. This type of model does not single out a dependent variable, but is a series of equations in which all variables interact, all are endogenous, and every variable is forecasted in relation to the other variables. A VAR is a standard time series modeling equation. This analysis used the statistical software (language and computing environment) R [21] and the packages VARS and URCA to carry out the modeling [22,23]. The VAR's general form is:

$$X_t = \prod_1 X_{t-1} + \ldots + \prod_k X_{t-k} + \mu + \Phi D_{t+} \varepsilon_t, (t = 1, \ldots, T)$$

Standard tests showed that the variables in this analysis were co-integrated, meaning that the variables exhibited common stochastic trends, which means spurious regression is possible. A common method for avoiding spurious regression in a multivariate time series is to transform a VAR into a VECM. A VECM is a co-integrated vector autoregressive model that is statistically adjusted to eradicate spurious regression [24]. The analysis applied the Engle–Granger two-step procedure for VECM model building, whereby each variable was tested for stationarity (presence of a unit root), followed by VECM estimation using the lagged residuals [25]. A VECM has the advantage that its linear equations express the long-term relationships between variables. Furthermore, the error-correction term is the short-run adjustment to the long run relationships (see Supplementary Materials). The VECM is written as follows [23]:

$$\Delta X_t = \Gamma_1 \Delta X_{t-1} + \ldots + \Gamma_{k-1} \Delta X_{t-k+1} + \prod X_{t-k} + \mu + \Phi D_{t+} \varepsilon_t$$

if

$$\Gamma i = -(I - \prod_1 - \ldots - \prod_i), (i = 1, \ldots, k-1),$$

and

$$\prod = -(I - \prod_1 - \ldots - \prod_k)$$

In step 1 of the Engle–Granger procedure, the augmented Dickey–Fuller unit-root tests and the Philips–Peron unit-root tests for time series data indicated that several of the variables were nonstationary. Various model runs were tested with the goal of determining the number of co-integrating vectors [26,27]. The final model was estimated using a lag (k) of 3 guided by the final prediction error (FPE) information criterion rather than the Akaike's information criterion (AIC), or the Bayesian information criterion (BIC). The goal of the model was to optimize projections by reducing the mean square error (MSE), thus the FPE was most appropriate [24]. The second step applied the Johansen maximum likelihood procedure to test for co-integration resulting in the co-integration rank of 2 [23,28]. The resulting restricted VEC was converted to a level VAR for further structural analysis of Granger causality, orthogonal impulse response functions (OIRF), and future projections of ethylene supply as shown in the Results section.

The accuracy of the model was evaluated with an ex-post sample of actual data. The mean absolute percentage error (MAPE) was calculated for ex-post forecast accuracy for the ethylene supply variable. The predicted values for 24 periods, January 2015 through to December 2016, were compared to actual data. The MAPE was 4.5 percent, which is an acceptable range of forecast accuracy for a statistical model. Once the "base" VECM was made stable and reliable for forecasting, it was further modified to implement the scenarios. Three variables were made exogenous in the outyears: Gross Domestic Product (GDP)/Personal Consumption Expenditures (PCE), resident population, and gas price. The next section describes the data and how the scenarios were structured.

Data—The historical dataset (1986–2014) included monthly data for seven variables from several U.S. government agencies. The final list of selected variables was chosen based on the Granger-causation principle. Causal inference with the Granger causality tests determined which variables would be useful for forecasting other variables and should be retained. The variables were

U.S. ethylene quantities, feedstock prices, and socio-economic data. The data from the U.S. Energy Information Administration (EIA) are as follows: (1) ethane/ethylene supply, which is a calculated total for refinery, blender, and gas plants in thousand barrels; (2) stocks of ethane/ethylene stock in thousand barrels; (3) gas plant production of natural gas liquids and liquid refinery gases supply in thousand barrels; (4) crude oil (Cushing OK, WTI spot price dollars per barrel); and (5) industrial natural gas for feedstock price (City Gate Price) in dollars per thousand cubic feet. The data was downloaded from the EIA's publicly available database, the "Total Energy Browser" (Available at https://www.eia.gov/totalenergy/data/browser/). WTI spot price for crude oil was selected rather than Europe Brent crude oil spot price assuming that ethylene producers would choose the lowest cost feedstocks. According to EIA, WTI and Brent were almost equally priced throughout the study period, with Brent marginally lower until 2011. Thereafter, "WTI crude oil has priced at a persistent discount to Brent crude" [29]. WTI was the lowest cost crude oil option for producers in comparison to Brent crude because of its lower spot price and lower transportation costs. Almost all ethane is used for ethylene production in the U.S., so the EIA combines them into one category named "ethane/ethylene". The historical socio-economic variables were as follows: (6) GDP/PCE is personal consumption expenditure on goods and is per capita in this analysis; and (7) resident population. These were downloaded from the U.S. Bureau of Economic Analysis (BEA) and the U.S. Census Bureau, respectively [30,31]. The PCE data was adjusted to 2009 dollars using BEA deflators [32]. All data in the models are in log. These are U.S. official data sources and are considered high quality and accurate. See Table 1 for summary statistics of the dataset. This data was used to develop the VECM and baseline projections of ethylene supply in step one.

Climate Change Relevant Scenarios with SSPs—A description of the SSP framework is required to understand the scenarios. The SSPs consist of five narratives that envision possible futures by the degree of climate change adaptation and climate change mitigation challenges defined as follows. Also see Figure 1.

- Socioeconomic challenges to mitigation—"(1) factors that tend to lead to high reference emissions in the absence of climate policy because, all else equal, higher reference emissions makes that mitigation task larger; and (2) factors that would tend to reduce the inherent mitigative capacity of a society" [33].
- Challenges to adaptation—"a function of the socioeconomic determinants of exposure to climate change hazards, sensitivity to these hazards, and the adaptive capacity to deploy coping measures" [33].

The second element of the SSP narratives is the quantified data for socio-economic drivers (population, gross domestic product (GDP), and urbanization) that illustrate each of the scenarios for each country. The future ethylene supply model uses three SSP narratives and their population and GDP trajectories to define its scenarios. Urbanization is not included because there is no difference in any SSP's urbanization estimates for the U.S. [34]. The SSPs applied are as follows:

- SSP1 "Sustainability": low challenges to adaptation and mitigation (progress towards a sustainable low carbon economy) [35];
- SSP3 "Regional Rivalry": high challenges to adaptation and mitigation (heavy fossil fuel use, low global cooperation on environmental issues, low economic growth rates, and low investment in education with high birth rates in some countries and low birth rates in the U.S.) [35]; and
- SSP5 "Fossil-Fueled Development": high challenges to mitigation and low challenges to adaptation resulting in heavy fossil fuel use [35].

Table 1. Summary Statistics.

Variable	Obs	Mean	Standard Deviation	Median	Trimmed Mean	Median Absolute Deviation	Minimum	Maximum	Range	Skewness	Kurtosis	Standard Error
Ethane/Ethylene Supply (Thousand Barrels)	348	9.93	0.23	9.93	9.93	0.23	9.43	10.46	1.04	0.06	−0.66	0.01
Cushing, OK WTI Crude Oil Price (Dollars per Barrel)	348	3.51	0.68	3.27	3.48	0.62	2.43	4.9	2.47	0.47	−1.28	0.04
Natural Gas Liquids Quantity (Thousand Barrels)	348	10.93	0.16	10.91	10.91	0.11	10.64	11.48	0.85	1.33	1.92	0.01
Ethane/Ethylene Stocks (Thousand Barrels)	348	9.97	0.24	9.94	9.96	0.23	9.47	10.62	1.16	0.52	−0.21	0.01
GDP/PCE per Capita (in 2009 $U.S.)	348	8.95	0.36	9.02	8.97	0.43	8.21	9.5	1.29	−0.25	−1.07	0.02
Resident Population	348	19.45	0.09	19.46	19.45	0.11	19.29	19.58	0.29	−0.19	−1.23	0
Natural Gas Price (Dollars per Thousand Cubic Feet)	348	1.47	0.4	1.35	1.44	0.44	0.96	2.52	1.56	0.55	−0.91	0.02

"Middle-of-the-Road Development" (SSP2) and "Inequality" (SSP4) are not included in this paper for several reasons. First, the author wanted to reflect a wide-range of scenarios. Second, SSP2 "Middle-of-the-Road" is not needed because the base model herein is also a "business-as-usual" scenario that models ongoing relationships found in the historical data specifically for ethylene supply. Third, the SSPs reflect expert judgement regarding future pathways that assumes similar trends for SSP1, SSP2, and SSP4 for several highly developed countries, including the U.S., because rates of population growth and other trends are unlikely to change up to 2035. For example, SSP4 and SSP1 presume functioning international institutions working cooperatively and the integration of low carbon technologies. These two factors already exist for highly developed countries, for example, the members of the Group of Seven (G-7) and the Organization for Economic Cooperation and Development (OECD). On the hand, the data for SSP1, SSP3, and SSP5 diverge in the U.S. case, resulting in a wider range of scenarios in comparison to the base "business-as-usual case".

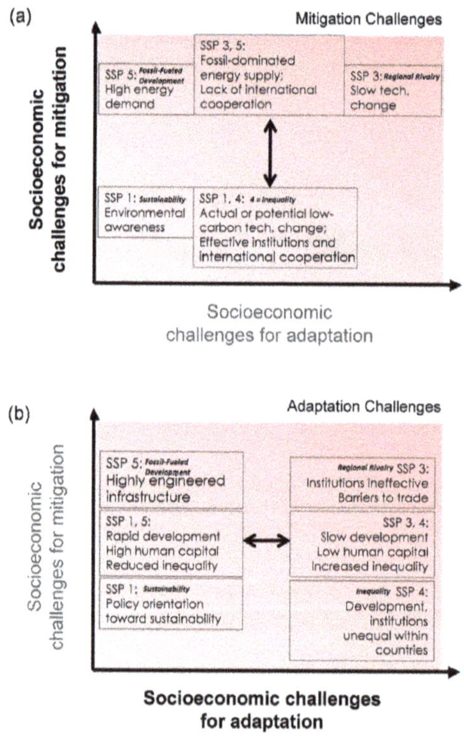

Figure 1. SSP Scenarios from O'Neill et al. [35] (Reprinted with permission. Bold italics added to figure). (**a**) Socioeconomic challenges for mitigation (**b**) Socioeconomic challenges for adaptation.

Modeling the Scenarios—As discussed above, the "base" VECM model of historical data (1986–2014) projected a baseline for future ethylene supply through to 2050 without considering climate change as the first step of the analysis. Second, climate change-relevant scenarios were developed using the SSPs as exogenous variables. Third, these were combined with natural gas scenarios (also exogenous) to produce the nine-scenario matrix for 2015–2050 shown in Table 2. This section briefly explains how the scenario data was developed.

Table 2. Overview of Future Ethylene Supply Scenarios 2015–2050.

	Natural Gas Feedstock Price Scenarios		
SSP Scenarios	SSP1 Sustainability EIA Ref. Gas Price	SSP1 Sustainability Low Gas Price	SSP1 Sustainability High Gas Price
	SSP3 Regional Rivalry EIA Ref. Gas Price	SSP3 Regional Rivalry Low Gas Price	SSP3 Regional Rivalry High Gas Price
	SSP5 Fossil-Fueled Dev. EIA Ref. Gas Price	SSP5 Fossil-Fueled Dev. Low Gas Price	SSP5 Fossil-Fueled Dev. High Gas Price

The time series projections of the U.S. socioeconomic drivers were downloaded from the SSP Scenario Database—Version 1.1, which may be found at the website https://secure.iiasa.ac.at/web-apps/ene/SspDb/dsd?Action=htmlpage&page=about. GDP and population data were in five-year increments. A monthly data set was created using the cubic splines method in R [21]. The population estimates were derived from "fertility, mortality, migration and educational transitions" [36]. The SSP GDP estimates were "based on a convergence process and places emphasis on the key drivers of economic growth in the long run: population, total factor productivity, physical capital, employment and human capital, and energy and fossil fuel resources (specifically oil and gas)" [37]. The growth rate of GDP from the SSPs was applied to GDP/PCE to capture the consumption of ethylene in products as GDP/PCE excludes services. "The measurement of GDP captures the value of products that are consumed and not used in a later stage of production, those that are sold, given away, or otherwise transferred to foreign residents, those that are used to produce other goods and that last more than a year, and those that may be inventoried for future consumption" [38]. To justify this technique, a strong positive correlation between the percentage change in PCE and the percentage change in GDP over the research period 1986–2014 was shown. The Pearson product-moment correlation coefficient is 0.522 [$r(115) = 0.52$, $p = 0.05$]. Each SSP's growth rates for the period 2015–2050 were used for GDP/PCE and population.

The natural gas feedstock price scenario data was developed from the EIA Annual Energy Outlook 2014 data [8], which contains historical data and projections for natural gas price by sector. The author used the reference case data in "6. Industrial Sector Key Indicators and Consumption," "Natural Gas Feedstock" (nominal dollars per million Btu) [8]. The EIA reference case projection, which includes existing regulations and policies to 2040 was extended to 2050. A low gas price case that was 10 percent below the reference case and a high gas price case that was 10 percent above the reference case were developed. The average annual change in natural gas price during the sample period was 7 percent. This reflects the range of uncertainties inherent in estimating future industry conditions that lead to natural gas price change. Price changes can be driven by policy change, technological change, and other factors. The range was commensurate with the wide range of estimates for future gas price developed by modelers [39].

In summary, the future ethylene supply model design is innovative in three ways. First, the SSPs are often used in highly complex integrated assessment models (IAM) that are global in scope for the entire world economy or macro-level sectors [40]. This future ethylene supply model is at a finer analytical level as it focuses on one commodity and uses SSP data for one country. Second, the research applies GDP/PCE rather than GDP to exclude services and focus on the consumption of materials, which is also unusual for an SSP analysis. Third, it is a novel approach because it is a straightforward exploratory econometric model using R, an open-source software environment for statistical analysis, instead of a proprietary IAM. This is the first study to project future ethylene supply to go beyond the price of feedstocks to include socio-economic variables relevant to climate change mitigation and adaptation. All data is available upon request from the author.

3. Results

The objectives of the analysis were to shed light on the relationships between the selected variables using the "base" VECM to find the direction (increasing or decreasing) and long- and short-run behaviors of the variables (Section 3.1). The VECM was used to find the volume of ethylene supplied in each scenario in 2035 and 2050 (Section 3.2). In addition, the volume of greenhouse gasses associated with ethylene production in 2035 and 2050 were calculated using 2015 emissions ratios, which reflected the shift from heavy to light feedstocks (Section 3.3). This section describes these results.

3.1. Relationships between Variables

The "base" VECM was used to better understand the interrelationships between variables. Three methods were used. The deterministic coefficients measure the contribution of each variable in predicting ethylene supply (see Table 3). For details on the model please view Supplementary Materials. The orthogonal impulse response functions (OIRFs) show the effect of a one-time shock of each variable on ethylene supply. In addition, the forecast error variance decomposition (FEVD) compares how much of the variance is due to shocks to the other variables.

The OIRF results suggest that several variables have long-term effects (greater than ten periods) on ethylene supply. The OIRF was calculated using the Wold moving average for a VAR (p)-process. Notably, gas price and Natural Gas Liquid (NGL) quantity are in line with the theoretical ethylene supply models used by some industry analyses [16,41]. These variables behaved as expected, causing long-term, non-transitory changes to ethylene supply. The results show that the impact of natural gas price is negative, meaning that as natural gas price goes up, ethylene supply decreases. This is in-keeping with real-world conditions because light feeds (ethane, propane, and butane) are co-produced with shale gas. The variable NGL quantity is a proxy measure for shale gas quantity in the market. As shown in Figure 2a, an innovation in NGL has a lasting negative effect on gas price and crude oil price. Last, ethylene supply responds positively to ethylene supply, as this is an autoregressive model. These results reflect expectations; however, some results were surprising.

The relationship between crude oil price and ethylene supply was surprising. Although natural gas is an important and growing ethylene feedstock, crude oil price exerts a significantly stronger influence on ethylene supply. This finding may be explained by the fact that heavier feedstocks such as crude oil and naphtha remain dominant outside the U.S. and ethylene-derived products are globally traded; therefore, the economics of global ethylene supply are contingent on crude oil price. The global market gives U.S. ethylene manufacturing an advantage in its production cost compared to other countries because of the low price of U.S. ethane feedstock derived from recently exploited shale gas [12].

Another interesting result of the OIRF is that an innovation in GDP/PCE has a slight long-term influence on ethylene supply in the model, whereas population has a long-term, robust, and positive impact. These are two of the variables that are made exogenous to represent the SSPs. The OIRF from GDP/PCE is shown in Figure 2b. The OIRF from the population is shown in Figure 2c. The relative forecast error variance decomposition (FEVD) of these two variables is shown in Figure 3. The FEVD is related to the OIRF. It provides the contribution of each variable to the forecast error variance. The contribution of population and crude oil price to the forecast error variance of ethylene supply increases over time, but GDP/PCE does not. This result raised the question: "Is the level of U.S. GDP/PCE per capita leveling off?" The answer is yes. The average share of PCE on goods of GDP over the sample period, 1986 to 2014, is 23.8 percent. This percentage has varied little since 1986 (−0.7% to +0.7%). There is recent literature investigating downward consumption trends in wealthy countries [42,43]. The trends that have driven ethylene consumption have levelled off in the U.S. because it is a mature and wealthy market, but this may not be the case for other countries, which consume products derived from U.S. ethylene.

Table 3. Matrix of Deterministic Coefficients for the Base VECM.

Variables with Significance Codes	Constant
Ethane/Ethylene Supply (Thousand Barrels)	−23.6
Cushing, OK WTI Crude Oil Price (Dollars per Barrel)	5.53
Natural Gas Liquids Quantity (Thousand Barrels) ***	−1.65
Ethane/Ethylene Stocks (Thousand Barrels) ***	14.00
GDP/PCE per Capita (in 2009 $U.S.) *	1.36
Resident Population ***	0.02
Natural Gas Price (Dollars per Thousand Cubic Feet)	2.50
-OLS regression of the unrestricted VECM (lags 1–3)	

Significance codes: 0 '***' 0.001 '**' 0.01 '*' 0.05 '.' 0.1 ' '.

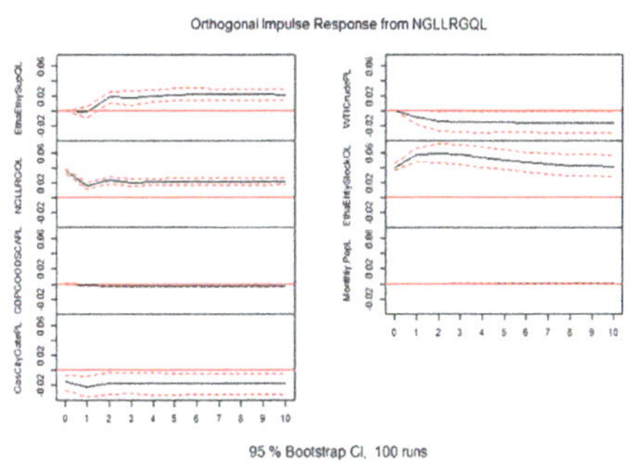

(**a**)

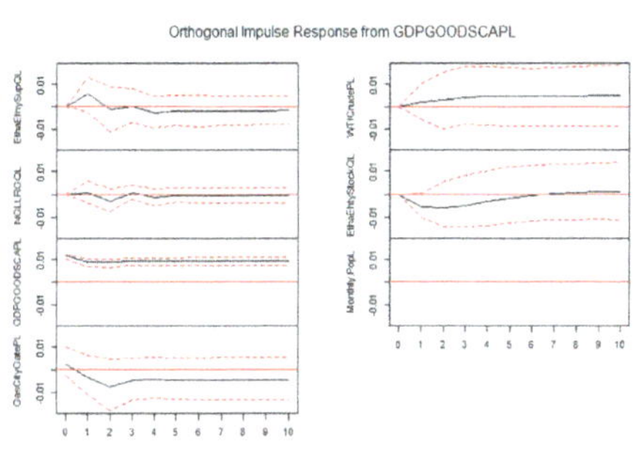

(**b**)

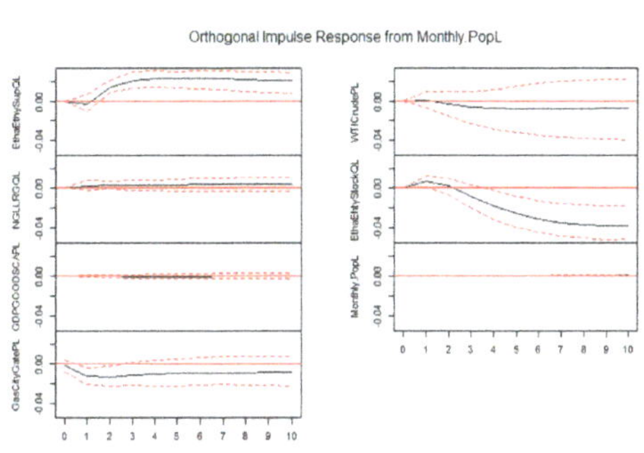

(**c**)

Figure 2. (**a**) OIRF for natural gas liquids, (**b**) OIRF for GDP/PCE, and (**c**) OIRF for population.

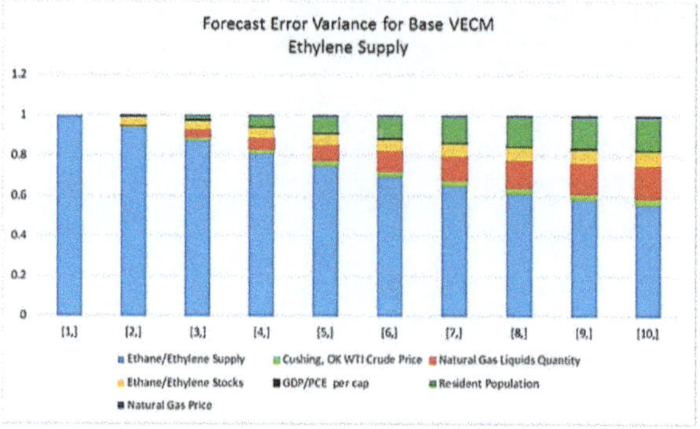

Figure 3. Forecast Error Variance Decomposition.

3.2. *The Future of Ethylene Supply*

This section presents the modeled volumes of ethylene for 2035 and 2050 (see Table 4). The projections were made using a confidence interval of 0.95. The scenarios are represented graphically in Figure 4. For example, Table 4 shows that in 2035, SSP3 with a high gas price reduced supply by 10 percent. Alternatively, SSP3 with low gas price, increased supply by 2 percent. The SSP3 long-term results in 2050 showed reductions between 26 and 37 percent, irrespective of gas price. These dynamic estimates answer the question: "What is the future U.S. supply of ethylene considering new shale gas exploitation and socio-economic development pathways?"

Table 4. Model Results for Future Ethylene Supply (thousand barrels per year).

Years and % Change	"Base" VECM	SSP1 Sustainability EIA Ref. Gas Price	SSP1 Sustainability Low Gas Price	SSP1 Sustainability High Gas Price
2014 (actual)	375,309	375,309	375,309	375,309
2035 (projection)	441,771	633,244	655,684	579,193
2050 (projection)	481,183	716,908	752,334	633,876
% Change from 2014–2035	18%	69%	75%	54%
% Change from 2014–2050	28%	91%	100%	69%
% Change from Base 2035		43%	48%	31%
% Change from Base 2050		49%	56%	32%
Years and % Change	**"Base" VECM**	**SSP3 Regional Rivalry EIA Ref. Gas Price**	**SSP3 Regional Rivalry Low Gas Price**	**SSP3 Regional Rivalry High Gas Price**
2014 (actual)	375,309	375,309	375,309	375,309
2035 (projection)	441,771	435,267	450,691	398,117
2050 (projection)	481,183	340,796	357,636	301,327
% Change from 2014–2035	18%	16%	20%	6%
% Change from 2014–2050	28%	−9%	−5%	−20%
% Change from Base 2035		−1%	2%	−10%
% Change from Base 2050		−29%	−26%	−37%
Years and % Change	**"Base" VECM**	**SSP5 Fossil-Fueled Dev. EIA Ref. Gas Price**	**SSP5 Fossil-Fueled Dev. Low Gas Price**	**SSP5 Fossil-Fueled Dev. High Gas Price**
2014 (actual)	375,309	375,309	375,309	375,309
2035 (projection)	441,771	830,595	860,029	759,698
2050 (projection)	481,183	1,212,416	1,272,330	1,071,991
% Change from 2014–2035	18%	121%	129%	102%
% Change from 2014–2050	28%	223%	239%	186%
% Change from Base 2035		88%	95%	72%
% Change from Base 2050		152%	164%	123%

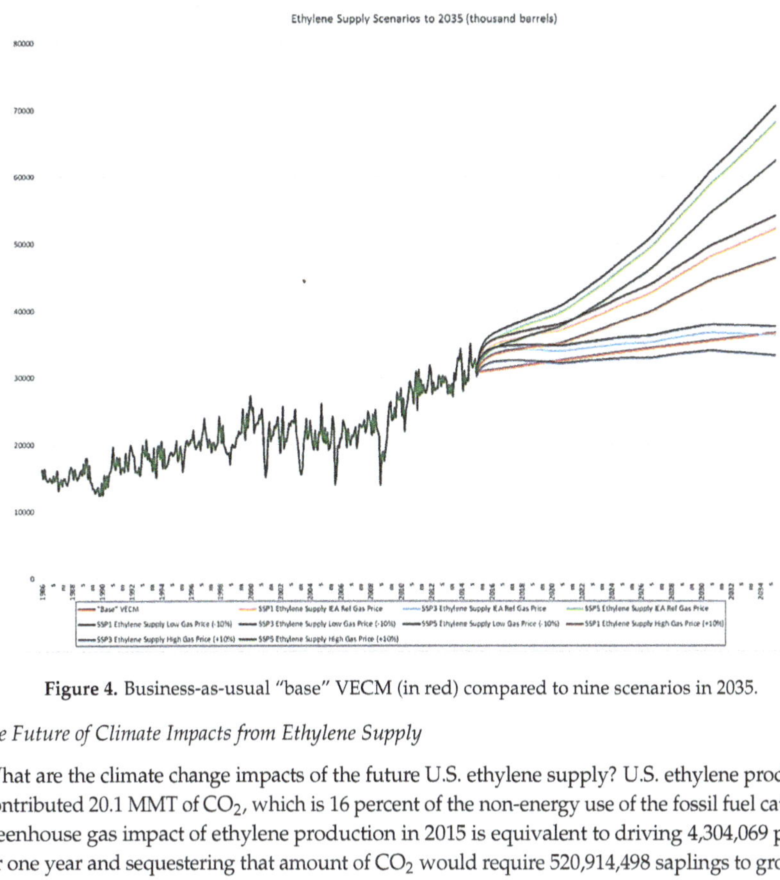

Figure 4. Business-as-usual "base" VECM (in red) compared to nine scenarios in 2035.

3.3. *The Future of Climate Impacts from Ethylene Supply*

What are the climate change impacts of the future U.S. ethylene supply? U.S. ethylene production in 2015 contributed 20.1 MMT of CO_2, which is 16 percent of the non-energy use of the fossil fuel category [3]. The greenhouse gas impact of ethylene production in 2015 is equivalent to driving 4,304,069 passenger cars for one year and sequestering that amount of CO_2 would require 520,914,498 saplings to grow for ten years [44]. Of available estimates, the EPA's "Inventory of U.S. Greenhouse Gases and Sinks: 1990–2015" estimate was chosen for this analysis (0.78 metric tons of CO_2 per metric ton of ethylene) [3]. This ratio, developed from 2010–2015 production data, reflects recent feedstock trends and is from a reliable source. The analysis assumes that the CO_2 to ethylene ratio applies in the short term (2035) and long term (2050). This means that no lower CO_2 emitting options have come online and that the general feedstock mix of majority ethane from shale gas is unchanged in the short and long term. As discussed in Section 2, the SSPs imply futures that are more or less amenable to climate change mitigation and adaptation policies in general. The range of gas prices used in the analysis reflects the uncertainties in the field including climate change policies that would affect natural gas price. The results show that these factors affect projections of CO_2 emissions associated with ethylene. For example, the SSP3 with low gas price scenario was −5% lower in the long-term (2050) when compared to 2014 CO_2 emissions. The SSP3 high gas price scenario CO_2 emissions was (−19%) lower than in 2014.

4. Discussion

Overall, the U.S. ethylene supply grew considerably over the medium and long term in most scenarios. Ethylene supply increased by 18 percent in 2035 and 28 percent in 2050 in the business-as-usual case. Only the SSP3 "Regional Rivalry" showed significant reductions in ethylene supply in the long term with a −5% to −20% change from 2014 actual, and −26% to −37% change from the business-as-usual estimate. SSP3 presumes a fragmented global economy with less international trade. The pathway indicated for the U.S. is low GDP growth and low population growth. As seen in the OIRF and FEVD results, ethylene supply is more sensitive to population rather than GDP/PCE. Therefore, SSP3's sharp downturn in population over time contributes to the downturn in ethylene supply. A combination of all three exogenous variables—lower consumption, lower population, and higher prices—was needed to slow and reduce ethylene supply.

The upward trend in most scenarios means that ethylene's growth was largely inelastic and reflects its integral role in modern consumption. Notably, consumption per capita alone was not a substantial driver in the U.S. case. Ethylene increased, not because of the consumption of goods per capita, but because of increased population. Population was a more important lever than consumption. This may be based on the overall maturity/saturation of the U.S. market, leading to a flat consumption per capita trend. Additionally, the inflexibility of ethylene supply in response to consumption drivers suggests a lack of sufficient alternative products and path dependency. It can be inferred that a country with rapidly rising personal consumption per capita would have a higher rate of ethylene supply growth than the U.S. and China for example. Consequently, country-level consumption trends, rather than global or regional trends, are pertinent for analyzing ethylene and non-energy uses of fossil fuels. The results also emphasize the importance of country-level feedstock trends.

The results indicated that the availability of shale gas in the U.S. and low-priced feedstocks from natural gas relative to crude oil were key factors influencing ethylene supply. Unexpectedly, crude oil price exerted a stronger influence on the U.S. ethylene supply than natural gas price. Therefore, future research is needed on the price dynamics of competing feedstocks in the U.S. ethylene market and emphasis is placed on the importance of analysis at regional and global levels to better understand future changes in ethylene supply.

In general, the scenario results were in line with the SSP narratives; however, there were some unexpected results. Surprisingly, all SSP1 scenarios, "Sustainability—Taking the Green Road," resulted in increasing supply by at least 54 percent. This outcome was unexpected because the narrative and quantification of SSP1 included lower consumption. "Consumption is oriented toward low material growth and lower resource and energy intensity" [35]. The SSP5 "Fossil-Fueled Development" scenario of high challenges to climate change mitigation and low challenges to adaptation resulted in the highest levels of ethylene supply. The SSP5 low gas price scenario projects 121% growth over 2014 by 2035 and 223% by 2050. This outcome was expected because, in SSP5, policy would encourage rather than curtail fossil fuel use [33]. The starkest differences in outcome were between SSP3 and SSP5 at all gas prices, which reflects the SSP narratives and range of natural gas prices.

This scenario analysis with econometric modeling created a new method for using the SSPs. It showed that it is possible to use the SSPs to create scenarios without an intricate IAM model. This method is not argued to be more robust than an IAM, which is by definition complex. It is a statistically valid alternative. Additionally, the SSPs may be down-sized to successfully study one commodity in one country. These are two positive outcomes for researchers that need credible vetted climate change-relevant scenarios, but do not have the ability to convene multidisciplinary teams of experts and/or access to large-scale IAMs.

The future ethylene supply model has some limitations and criticisms. For one, it does not distinguish the effects of discrete policies. Second, the macro-level assumptions on social and technological change in the SSP narratives are not precise enough to capture effects on ethylene supply. Third, the Lucas critique states that the impact of future policies on a phenomenon cannot be econometrically modeled from historical data, which is an overarching criticism of all econometric forecasting [45] and IAMs. To address these criticisms, the author notes that modeling short- and long-run relationships is informative for decision making, but not infallible. In the SSP framework, technological change is included in the GDP estimates and is modeled as a consequence of the total factor productivity frontier growth, convergence speed, and openness [37]. Certainly, technological change is also driven by policy. Underlying policy and institutional approaches are implicit in the SSP data and in the range of natural gas price data. Future research can apply explicit policies to this case using the output of this analysis.

Although explicit technology and explicit policy scenario data were not included in the model during the phase of the research represented in this paper, these are important variables for future research. For example, technological advances could be captured by data on the availability and price of new biomass-based feedstocks for ethylene, e.g., ethanol, and policy measures that would promote them. Public engagement campaigns and regulations lowering consumption of plastic products that contain ethylene may also impact ethylene supply in future. Also, governmental climate policies that limit fossil fuel use and carbon emissions by increasing or decreasing subsidies and imposing or reducing taxes could impact ethylene supply. These variables will be evaluated in future uses of the model.

The new method for using the SSPs also points to a need for further development of the SSP framework. The SSP1 results did not project a lower consumption pathway as expected based on the SSP 1 narrative. A pathway is needed that embodies socio-technical transition to a low-carbon, low-consumption sustainable economy that represents deep decarbonization and de-growth concepts in its narrative and quantification of drivers for population, GDP, urbanization, etc. The SSP framework, in particular SSP1, may not be progressive enough to develop scenarios that reduce consumption, which is the key to sustainability policies.

5. Conclusions

In summary, the results showed that ethylene is a significant and rising source of CO_2 emissions and it is difficult to reverse this trend. These results have broad policy implications because ethylene is indicative of non-energy uses of fossil fuels in general. Given that global climate change concerns compel a transition to a low-carbon economy with less reliance on fossil fuels, there are three policy implications of these findings:

- Lifecycle perspectives are needed to inspire alternative low-carbon feedstocks for ethylene and its uses.
- Policies that target reducing the consumption of ethylene-based products, such as plastics, are needed.
- Better recovery and reuse of ethylene-based products is needed with the aim of reducing consumption.

This is the first study to project future ethylene supply to go beyond the price of feedstocks and include socio-economic variables relevant to climate change mitigation and adaptation. The scenario projections may be used in future research on transition to low-carbon production of important petrochemicals such as ethylene. These projections are needed to estimate environmental benefits and economic impacts. In addition, the scenario analysis with econometric modeling methods applied in this paper can be applied to other important commodities derived from fossil fuels, such as fertilizers. In addition, the method may be applied to other countries or regions. The research findings and methods are relevant to the community of scientists, manufacturers, and policymakers that are concerned about the future availability of industrial chemicals, the impact of various feedstocks on supply, and speeding up the transition to a low carbon economy.

Supplementary Materials: The following are available online at http://www.mdpi.com/1996-1073/11/11/2967/s1.

Funding: This research received no external funding.

Acknowledgments: This research benefited from the input and guidance of researchers at the International Institute for Applied Systems Analysis during the author's participation in the Young Scientists Summer Program.

Conflicts of Interest: The author declares no conflict of interest.

References

1. Broeren, M. Production of Bio-Ethylene—Technology Brief. Available online: IRENA-ETSAPTechBriefI13Production_of_Bio-ethylene.pdf (accessed on 18 September 2018).
2. Plotkin, J.S. Beyond the Ethylene Steam Cracker. Available online: https://www.acs.org/content/acs/en/pressroom/cutting-edge-chemistry/beyond-the-ethylene-steam-cracker.html (accessed on 30 August 2018).
3. Inventory of U.S. Greenhouse Gas Emissions and Sinks: 1990–2015. Available online: https://www.epa.gov/ghgemissions/inventory-us-greenhouse-gas-emissions-and-sinks-1990-2015 (accessed on 18 September 2018).
4. U.S. Product Supplied of Ethane-Ethylene (Thousand Barrels). Available online: https://www.eia.gov/dnav/pet/hist/LeafHandler.ashx?n=pet&s=metupus1&f=m (accessed on 18 September 2018).
5. Koottungal, L. International survey of ethylene from steam crackers. *Oil Gas J.* **2012**, *110*, 85.
6. True, W.R. Global ethylene capacity poised for major expansion. *Oil Gas J.* **2013**, *111*, 90–95.
7. Facts and Figures of the European Chemicals Industry 2016. Available online: http://www.cefic.org/Facts-and-Figures/ (accessed on 18 September 2018).
8. *Annual Energy Outlook 2014: With Projections to 2040*; US Energy Information Administration: Washington, DC, UK, 2014.
9. True, W.R. Global ethylene capacity continues advance in 2011. *Oil Gas J.* **2012**, *110*, 78–84.
10. Lippe, D. 2013 ethylene output rises; growth to continue in early 2014. *Oil Gas J.* **2014**, *112*, 84.
11. Facts and Figures of the European Chemicals Industry 2014. Available online: https://www.google.com.tw/url?sa=t&rct=j&q=&esrc=s&source=web&cd=2&ved=2ahUKEwjmvcyLmK3eAhWMabwKHYCSDwsQFjABegQIABAC&url=http%3A%2F%2Fwww.cefic.org%2FDocuments%2FFactsAndFigures%2F2014%2FFacts%2520and%2520Figures%25202014%2520-%2520The%2520Brochure.pdf&usg=AOvVaw3v6YC2aFFHO8gK5KOC3o2C (accessed on 18 September 2018).
12. Fattouh, B.; Brown, C. US NGLs Production and Steam Cracker Substitution: What Will the Spillover Effects Be in Global Petrochemical Markets. Available online: https://www.oxfordenergy.org/publications/us-ngls-production-and-steam-cracker-substitution-what-will-the-spillover-effects-be-in-a-global-petrochemicals-market/ (accessed on 18 September 2018).
13. Liverman, D. From uncertain to unequivocal. *Environ.Sci. Policy Sustain. Dev.* **2007**, *49*, 28–32. [CrossRef]
14. Remer, D.S.; Jorgens, C. Ethylene economics and production forecasting in a changing environment. *Eng. Process Econ.* **1978**, *3*, 267–278. [CrossRef]
15. Dornburg, V.; Hermann, B.G.; Patel, M.K. Scenario projections for future market potentials of biobased bulk chemicals. *Environ. Sci. Technol.* **2008**, *42*, 2261–2267. [CrossRef] [PubMed]
16. Broeren, M.; Saygin, D.; Patel, M.K. Forecasting global developments in the basic chemical industry for environmental policy analysis. *Energy Policy* **2014**, *64*, 273–287. [CrossRef]
17. Hermann, B.G.; Blok, K.; Patel, M.K. Producing Bio-Based Bulk Chemicals Using Industrial Biotechnology Saves Energy and Combats Climate Change. *Environ. Sci. Technol.* **2007**, *41*, 7915–7921. [CrossRef] [PubMed]
18. Ruth, M.; Amato, A.D.; Davidsdottir, B. Carbon emissions from us ethylene production under climate change policies. *Environ. Sci. Technol.* **2002**, *36*, 119–124. [CrossRef] [PubMed]
19. Masih, M.; Algahtani, I.; De Mello, L. Price dynamics of crude oil and the regional ethylene markets. *Energy Econ.* **2010**, *32*, 1435–1444. [CrossRef]
20. Bauer, N.; Calvin, K.; Emmerling, J.; Fricko, O.; Fujimori, S.; Hilaire, J.; Eom, J.; Krey, V.; Kriegler, E.; Mouratiadou, I.; et al. Shared socio-economic pathways of the energy sector–quantifying the narratives. *Glob. Environ. Chang.* **2016**, *42*, 316–330. [CrossRef]
21. R Development Core Team. R: A Language and Environment for Statistical Computing. Available online: https://scholar.google.com.tw/scholar?hl=zh-TW&as_sdt=0%2C5&q=R%3A+A+language+and+environment+for+statistical+computing.&btnG=19 (accessed on 18 September 2018).
22. Pfaff, B. VAR, Svar and Svec models: Implementation within R package vars. *J. Stat. Softw.* **2008**, *27*, 1–32. [CrossRef]
23. Pfaff, B.; Stigler, M.; Pfaff, M.B. Package 'Urca'. Available online: https://www.google.com.tw/url?sa=t&rct=j&q=&esrc=s&source=web&cd=1&ved=2ahUKEwjL1Oruma3eAhXGjLwKHcorCYMQFjAAegQIARAC&url=https%3A%2F%2Fcran.r-project.org%2Fweb%2Fpackages%2Furca%2Furca.pdf&usg=AOvVaw0SmP8V1aQQoA0bKmW44LTk (accessed on 18 September 2018).

24. Lütkepohl, H. *New Introduction to Multiple Time Series Analysis*; Springer Science and Business Media: New York, NY, USA, 2005.
25. Engle, R.F.; Granger, C.W. Co-integration and error correction: representation, estimation, and testing. *J. Econ. Soc.* **1987**, *55*, 251–276. [CrossRef]
26. Johansen, S. Statistical analysis of cointegration vectors. *J. Econ. Dyn. Control* **1988**, *12*, 231–254. [CrossRef]
27. Johansen, S. Interpretation of cointegrating coefficients in the cointegrated vector autoregressive model. *Oxf. Bull. Econ. Stat.* **2005**, *67*, 93–104. [CrossRef]
28. Pfaff, B. *Analysis of Integrated and Cointegrated Time Series with R*; Springer Science & Business Media: New York, NY, USA, 2008.
29. U.S. Energy Information Administration. Price Difference between Brent and WTI Crude Oil Narrowing. Available online: https://www.eia.gov/todayinenergy/detail.php?id=11891 (accessed on 18 September 2018).
30. Table 2.8.5. Personal Consumption Expenditures by Major Type of Product, Monthly. Available online: https://www.google.com.tw/url?sa=t&rct=j&q=&esrc=s&source=web&cd=1&cad=rja&uact=8&ved=2ahUKEwiNrOH-mq3eAhWCXbwKHY_jDTUQFjAAegQICRAB&url=https%3A%2F%2Ffred.stlouisfed.org%2Frelease%2Ftables%3Feid%3D3220%26rid%3D54&usg=AOvVaw2RlNWPcF64Hj-wKor5AzK- (accessed on 18 September 2018).
31. U.S. Census Bureau, P.D. Historical Population US Census. Available online: https://www.google.com.tw/url?sa=t&rct=j&q=&esrc=s&source=web&cd=3&cad=rja&uact=8&ved=2ahUKEwi0_6-im63eAhVIvLwKHRPNA-EQFjACegQICBAB&url=https%3A%2F%2Fwww.census.gov%2Fhistory%2Fwww%2Freference%2Fpublications%2Fdemographic_programs_1.html&usg=AOvVaw3gwcYffZ_XjLxbQfegneat (accessed on 18 September 2018).
32. Table 2.8.4. Price Indexes for Personal Consumption Expenditures by Major Type of Product, Monthly. Available online: https://www.google.com.tw/url?sa=t&rct=j&q=&esrc=s&source=web&cd=1&cad=rja&uact=8&ved=2ahUKEwihqOaLna3eAhWCWLwKHRykCAsQFjAAegQICRAB&url=https%3A%2F%2Ffred.stlouisfed.org%2Frelease%2Ftables%3Feid%3D3208%26rid%3D54&usg=AOvVaw3ZKqLdUe8CC33ltJGDYCt6 (accessed on 18 September 2018).
33. O'Neill, B.C.; Kriegler, E.; Riahi, K.; Ebi, K.L.; Hallegatte, S.; Carter, T.R.; Mathur, R.; van Vuuren, D.P. A new scenario framework for climate change research: the concept of shared socioeconomic pathways. *Clim. Chang.* **2014**, *122*, 387. [CrossRef]
34. Jiang, L.; O'Neill, B.C. Global urbanization projections for the Shared Socioeconomic Pathways. *Glob. Environ. Chang.* **2015**, *42*, 193–199. [CrossRef]
35. O'Neill, B.C.; Kriegler, E.; Ebi, K.L.; Kemp-Benedict, E.; Riahi, K.; Rothman, D.S.; van Ruijven, B.J.; van Vuuren, D.P.; Birkmann, J.; Kok, K.; et al. The roads ahead: Narratives for shared socioeconomic pathways describing world futures in the 21st century. *Glob. Environ. Chang.* **2017**, *42*, 169–180.
36. Samir, K.; Lutz, W. The human core of the shared socioeconomic pathways: Population scenarios by age, sex and level of education for all countries to 2100. *Glob. Environ. Chang.* **2014**, *42*, 181–192.
37. Dellink, R.; Chateau, J.; Lanzi, E.; Magné, B. Long-term economic growth projections in the Shared Socioeconomic Pathways. *Glob. Environ. Chang.* **2017**, *42*, 200–214. [CrossRef]
38. Measuring the Economy: A Primer on GDP and the National Income and Product Accounts. Available online: https://www.bea.gov/resources/methodologies/measuring-the-economy (accessed on 18 September 2018).
39. Changing the Game? Emissions and Market Implications of New Natural Gas Supplies. Available online: https://www.osti.gov/biblio/1411245 (accessed on 18 September 2018).
40. Riahi, K.; van Vuuren, D.P.; Kriegler, E.; Edmonds, J.; O'Neill, B.C.; Fujimori, S.; Bauer, N.; Calvin, K.; Dellink, R.; Fricko, O. The shared socioeconomic pathways and their energy, land use, and greenhouse gas emissions implications: An overview. *Glob. Environ. Chang.* **2017**, *42*, 153–168. [CrossRef]
41. Armor, J.N. Emerging importance of shale gas to both the energy & chemicals landscape. *J. Energy Chem.* **2013**, *22*, 21–26.
42. Pearce, F. Peak planet: Are we starting to consume less? *New Sci.* **2012**, *214*, 38–43. [CrossRef]
43. Cohen, M.J. Collective dissonance and the transition to post-consumerism. *Futures* **2013**, *52*, 42–51. [CrossRef]

44. U.S. EPA. Greenhouse Gas Equivalencies Calculator. September 2017 edition. Available online: https://www.epa.gov/energy/greenhouse-gas-equivalencies-calculator (accessed on 28 August 2018).
45. Lucas, R.E. Econometric policy evaluation: A critique. In *Carnegie-Rochester Conference Series on Public Policy*; Elsevier: Amsterdam, The Netherlands, 1976; pp. 19–46.

energies

MDPI

Article

The Efficiency of Long-Term Forecasting Model on Final Energy Consumption in Thailand's Petroleum Industries Sector: Enriching the LT-ARIMAXS Model under a Sustainability Policy

Pruethsan Sutthichaimethee * and Kuskana Kubaha

Division of Energy Management Technology, School of Energy, Environment and Materials, King Mongkut's University of Technology Thonburi, 126 Pracha Uthit Road., Bang Mod, Thung Khru, Bangkok 10140, Thailand; kuskana.kub@kmutt.ac.th

* Correspondence: pruethsan.sut@gmail.com; Tel.: +66-639-645-195

Received: 11 July 2018; Accepted: 6 August 2018; Published: 8 August 2018

Abstract: Presently, Thailand runs various sustainable development-based policies to boost the growth in economy, society, and environment. In this study, the economic and social growth was found to continuously increase and negatively deteriorate the environment at the same time due to a more massive final energy consumption in the petroleum industries sector than any other sectors. Therefore, it is necessary to establish national planning and it requires an effective forecasting model to support Thailand's policy-making. This study aimed to construct a forecasting model for a final energy consumption prediction in Thailand's petroleum industry sector for a longer-term (2018–2037) at a maximum efficiency from a certain class of methods. The Long Term-Autoregressive Integrated Moving Average with Exogeneous variables and Error Correction Mechanism model (LT-ARIMAXS model) (p, d, q, Xi, $ECT_{(t-1)}$) was adapted from the autoregressive and moving average model incorporating influential variables together in both long-term relationships to produce the best model for prediction performance. All relevant variables in the model are stationary at Level I(0) or Level I(1). In terms of the extraneous variables, they consist of per capita GDP, population growth, oil price, energy intensity, urbanization rate, industrial structure, and net exports. The study found that the variables used are the causal factors and stationary at the first difference as well as co-integrated. With such features, it reflects that the variables are influential over the final energy consumption. The LT-ARIMAXS model (2,1,2) determined a proper period ($t - i$) through a white noise process with the Q test statistical method. It shows that the LT-ARIMAXS model (2,1,2) does not generate the issues of heteroskedasticity, multicollinearity, and autocorrelation. The performance of LT-ARIMAXS model (2,1,2) was tested based on the mean absolute percentage error (MAPE) and the root mean square error (RMSE). The LT-ARIMAXS model (2,1,2) can predict the final energy consumption based on the Sustainable Development Plan for the 20 years from 2018 to 2037. The results showed that the final energy consumption continues to increase steadily by 121,461 ktoe in 2037. Furthermore, the findings present that the growth rate (2037/2017) increases by 109.8%, which is not in line with Thailand's reduction policy. In this study, the MAPE was valued at 0.97% and RMSE was valued at 2.12% when compared to the other old models. Therefore, the LT-ARIMAXS model (2,1,2) can be useful and appropriate for policy-making to achieve sustainability.

Keywords: long-term; final energy consumption; LT-ARIMAXS model; sustainable development; economic growth and the environment; error correction mechanism model

1. Introduction

Thailand has continuously put efforts into accelerating the economic development of the country by focusing on widening urbanization. In parallel, the government is trying its best to encourage both domestic and international private investment. This is to ensure that the industrial structure is broadened. At the same time, Thailand is also focusing on export activities where Thailand is to be a production base, so that Thailand's market share will continue to expand. Additionally, there are also policies designed to increase spending, attract more foreign tourists and increase the minimum wage rate, resulting in the increments of both local and foreign labors. Therefore, these policies have supported the Thai economy to grow with a 4.3% growth rate in 2016/2017 [1], and a 2.5% population growth rate (2016/2017) [2]. However, the economic and population growth in Thailand has continuously caused the environment to deteriorate. In 2017, CO_2 emissions from energy consumption increased by 1.3% when compared to 2016 [3]. These CO_2 emissions are highly contributed by the petroleum industries sector, accounting for 50.1% of the final energy consumption (2017). In fact, the final energy consumption has resulted in continuous economic growth, and that growth has also been affected by inflation due to the constant increase of world oil prices [1,3]. In addition, 89% of carbon dioxide is released by the energy sector with a growth rate of 10.3% (2016/2017). The petroleum industries sector produces more CO_2 due to its maximal power consumption. This reflects the fact that the above sector releases the most greenhouse gas. Emissions are expressed in the form of CO_2 (with the highest emissions) as well as other gases including methane (CH_4), nitrous oxide (N_2O), hydrofluorocarbons (HFC), perfluorocarbons (PFC), sulfur hexafluoride (SF_6), and nitrogen trifluoride (NF_3) [4,5].

The sustainable development policy is the future policy that Thailand aims to achieve. The focus of the policy covers three main areas: economic growth, social growth, and environmental growth. The policy is achieved when those three areas are simultaneously developed. For Thailand, both short-term (five years) and long-term (20 years) plans have been set [1]. Nonetheless, the implementation of Thailand's sustainable development policy results in growth in both the economy and population. This also affects the increment of energy consumption. Thus, Thailand has set a long-term reduction goal of 20 years (2018–2037) in the final energy consumption based on the petroleum industries sector not exceeding 90,000 ktoe [3]. This is because the petroleum industries sector accounts for highest energy consumption (50.1%) and produces most of the greenhouse gases [5]. Therefore, the most important tool in effective policy planning for sustainability is to forecast the future possibility [3,5].

However, the best forecasting model on energy consumption must also be able to support sustainable development policy planning. From the various relevant studies that have been reviewed, there are different models and forecasting techniques optimized for different forecasting timelines, be it short-term or long-term. Therefore, it is necessary to examine what has been done in this area to increase the quality of the proposed model. In fact, there have been few stream studies exploring total energy consumption. For instance, Zhao, Zhao, and Guo [6] started to estimate the electricity consumption of Inner Mongolia by deploying gray model (GM(1,1) model) optimized by moth-flame optimization (MFO) with a rolling mechanism from 2010 to 2014. Their study indicated which model could improve the forecasting performance of annual electricity consumption significantly. Li and Li [7] also initiated a comparative study by using the autoregressive integrated moving average model (ARIMA model), GM(1,1) model, and ARIMA–GM model to forecast energy consumption in Shandong, China from 2016 to 2020. Their prediction results showed that the energy demand of Shandong Province between those years would increase at an average annual rate of 3.9%. Similarly, Xiong, Dang, Yao and Wang [8] proposed a novel GM(1,1) model based on optimizing the initial condition in accordance with the new information priority principle to predict China's energy consumption and production from 2013 to 2017. The study produced findings indicating that China's energy consumption and production will keep increasing, as will the gap between them. Furthermore, Panklib, Prakasvudhisarn, and Khummongkol [9] attempted to forecast electricity consumption in Thailand by using an artificial neural network and multiple linear regression model (MLR model) for the years 2010, 2015, and 2020. Their estimation revealed that the electricity consumption of Thailand in 2010, 2015, and 2020, retrieved from the regression, would reach 160,136, 188,552, and 216,986 GWh, respectively, whereas 155,917, 174,394, and 188,137 GWh were the results obtained from the artificial neural network model (ANN model). Additionally, an ANN integrated with genetic algorithm was also presented by Azadeh, Ghaderi, Tarverdian, and Saberi [10] to estimate the electricity consumption in the Iranian agriculture sector in 2008. They observed that the integrated genetic algorithm (GA) and ANN model dominated the time series approach, yielding less mean absolute percentage error.

By incorporating values of socio-economic indicators and climatic conditions, Günay [11] modeled artificial neural networks with the use of predicted values of socio-economic indicators and climatic conditions to predict the annual gross electricity demand of Turkey in 2028, which produced a result where the demand would double, accounting for 460 TW in 2028, when compared to the years 2007 to 2013. Dai, Niu and Li [12] explored energy consumption forecasting in China from 2018 until 2022 by adopting a model of ensemble empirical mode decomposition and least squares support vector machine with the technology of the improved shuffled frog leaping algorithm. Their results showed China's energy consumption to have a significant growth trend. Based on Wang and Li [13], they tried to find whether China's coal consumption during 2016 to 2020 would be higher or lower than the level of 2014. Here, they optimized a time series model with a comprehensive analysis of data reliability. According to the analysis, it indicated that the annual Chinese coal consumption during 2016–2020 would be lower than the level of 2014 provided the annual average GDP growth rate was less than 8.2% per year. Suganthi and Samuel [14] developed econometric models to study the influence of the socioeconomic variables on energy consumption in India from 2030 to 2031 and found that the electricity demand depended on the Gross National Product (GNP) and electricity price, and the total energy requirement was found to be 22.944×10^{15} kJ.

In addition, Xu et al. [15] analyzed the change of energy consumption and CO_2 emissions in China's cement industry and its driving factors over the period between 1990 to 2009 by applying a log-mean Divisia index (LMDI) method. With such analysis, the study reveals that, by applying the best available technology, an additional energy saving potential of 26% and a CO_2 mitigation potential of 33% can be gained when compared with 2009. Kishita, Yamaguchi, and Umeda [16] tried to analyze electricity consumption in the telecommunications industry in 2030 by deploying an electricity demand model for the telecommunications industry (EDMoTI). The prediction results pointed out that electricity consumption in 2030 would be 0.7–1.6 times larger than the level of 2012 (10.7 TWh per year). For a shorter time of prediction, Zhao, Wang and Lu [17] conducted a study to forecast the monthly electricity consumption in China by proposing a time-varying-weight combining method: the High-order Markov chain based time-varying weighted average (HM–TWA) method. Their forecasting performance evaluation showed that the HM–TWA produced a better outcome for the component models and traditional combining methods.

Nonetheless, several studies have examined the total energy demands and its consumption for a longer term of forecasting. For instance, Hamzacebi and Es [18] implemented optimized grey modeling to forecast the total electric energy demand of Turkey from 2013 to 2025. Their prediction reflected that the direct forecasting approach resulted in better predictions than the iterative forecasting approach in estimating the electricity consumption in Turkey. An Improved Gray Forecast Model was also drawn by Mu et al. [19] to predict CO_2 emissions, energy consumption, and economic growth in China from 2011 and 2020 by using an improved grey model. Based on their prediction results, China's compound annual emissions, energy consumption, and real GDP growth for the predicted years was found to be 4.47–0.06% and 6.67%, respectively. Furthermore, Zeng, Zhou, and Zhang [20] proposed a Homologous Grey Prediction Model to predict the energy consumption of China's manufacturing from 2018 to 2024 where their study revealed that the total energy consumption in China's manufacturing was slowing down, however, the amount was still too large. Additionally, Jiang, Yang and Li [21] adapted a metabolic grey model (MGM), ARIMA model, MGM–ARIMA model, and back propagation neural network (BP) to forecast energy demand from 2017 to 2030. From their estimation, it showed that India's energy consumption would increase by 4.75% a year in the next 14 years at a 5% growth rate. By using the same, but improved, forecasting model, Ediger and Akar [22] analyzed the primary energy demand by fuel in Turkey from 2005 to 2020 using the ARIMA model and seasonal ARIMA (SARIMA) methods to estimate the above demand, and showed that the average annual growth rates of individual energy sources and total primary energy would decrease in all cases, except wood, and the animal–plant went negative.

Furthermore, Ekonomou [23] developed an artificial neural network to estimate the Greek long-term energy consumption from 2005 to 2008, 2010, 2012, and 2015. Overall, the study has constituted an accurate tool for the forecasting problem in Greek long-term energy consumption. In addition, Ardakani and Ardehali [24] utilized an IPSO (improved particle swarm optimization)–ANN model to forecast EEC (electrical energy consumption) for Iran and the U.S. from 2010 to 2030, which resulted in the mean absolute percentage error of 1.94% and 1.51% for Iran and the U.S., respectively. In the context of Thailand, a study of characteristics and factors towards energy consumption was conducted by Supasa et al. [25], who explored five household group energy consumption characteristics and seven driving forces of growth in residential energy consumption from 2000 to 2010 by applying the energy input–output method. Their calculations indicated that about 70% of total residential energy consumption was indirect energy consumption from consuming products and services. Seung et al. [26] predicted the future electricity demand for cooling in the building sectors in Singapore from 2014 to 2030 by applying a MLR model. Their study revealed that the electricity demands accounted for 31 $\pm$ 2% of the total electricity consumption in Singapore. Additionally, Wang et al. [27] attempted to estimate the total industrial energy consumption and energy-related carbon emissions in Tianjin from 2003 to 2012 by using an energy decomposition analysis. From their evaluation, energy efficiency could be enhanced by energy-saving efforts and the optimization of the industrial structure.

In fact, Zou, Liu, and Tang [28] analyzed the factors that contributed towards the changes in energy consumption in Tangshan city from 2007 to 2012 by applying the logarithmic mean Divisia index. Their findings showed that the technical effect played a vital role in reducing energy consumption in most sectors. Another investigation of the impacts of urban land use on energy consumption in China from 2000 to 2010 was undertaken by Zhao, Thinh, and Li [29]. They used a panel data analysis with nighttime light (NTL) data estimation. Their study on sigh has shown that an increase in the irregularity of urban land forms and the expansion of urban land will accelerate energy consumption, which indicates the relationship between urban growth and energy consumption. Similarly, Tian, Xiong, and Ma [30] evaluated the potential impacts of China's industrial structure on energy consumption by deploying a fuzzy multi-objective optimization model based on the input–output model from 2015 to 2020. From their analysis, they concluded that the industrial structure adjustment had great potential in energy conservation, and such an adjustment could save energy by 19% (1129.17 Mtce) at the average annual growth rate of 7% GDP. Ayvaz and Kusakci [31] employed a nonhomogeneous discrete grey model (NDGM) to forecast electricity consumption from 2014 to 2030. In their findings, they proved that the grey model proposed produced a better forecasting performance.

Previous studies have used varied methodologies and analyses, while the forecasting timeline includes short-term (1–5 years), mid-term (6–10 years), and long-term (11–20 years). From this point of view, only few studies have been conducted for long-term forecasting, accounting for about 28% out of the reviewed research. Moreover, the long-term forecasting studies are very limited, and that limitation may result in lower quality when compared to short-term and mid-term studies. From the study of related research on prediction models, we have found some shortcomings in long-term forecasting including a lack of true variable selection for a causality based on context and study interest, a lack of co-integration test and the error correction mechanism test, and a lack of a spurious test. In addition, those models did not identify the problems of heteroskedasticity, multicollinearity, and autocorrelation. In the context of Thailand, in the past, most energy consumption forecasting models used were of those models adapted from traditional approaches such as the Ordinary Least Square (OLS) model, the Autoregressive Moving Average (ARMA) model, the ARIMA model, and the ANN model. In fact, the above models were for forecasting with potentially high errors. They did not consider the causal variables in the real context of Thailand. Therefore, the influence of the factors towards dependent variables were unknown. When the output was used in national policy-making, this would negatively affect the country at large. However, the models are for short-term forecasting [5]. These models cannot be used for national long-term policy-making. As a result, the country has failed to head in the right direction for achieving the reduction goal and sustainable development.

Hence, we considered the above gap as an important issue that has to be addressed. Simultaneously, we developed a forecasting model for final energy consumption by adapting various theoretical concepts, conceptual frameworks, research, variable selection, and the implementation of the heteroskedasticity test, multicollinearity test, and autocorrelation for spurious check. Additionally, the co-integration model was optimized by incorporating an error correction mechanism test to differentiate this model from the other existing models. This newly developed model comes under the name of the Long Term-Autoregressive Integrated Moving Average with Exogeneous variables and Error Correction Mechanism model (LT-ARIMAXS model). However, we developed the LT-ARIMAXS model to differentiate from other models and to fill the recent gap existing in old models that was found in the research review. The existing models include the MLR model, ANN model, BP model, GM(1,1) model, ARMA model, and ARIMA model, among others. The LT-ARIMAXS model is a forecasting model that aims to create an effectiveness in long-term forecasting to support long-term policy planning. Hence, the findings of this study become useful and applicable in both Thailand's context and other contexts. The research's flow chart is illustrated in Figure 1 and determines all the relevant variables for the final energy consumption forecasting model, whose characteristics fall under the long-term sustainable development policy of 20 years (2018–2037), with the Augment Dickey Fuller theory only at the same level by using data from 1985 to 2017. Moreover, only crucial and influential variables are used in the forecasting model.

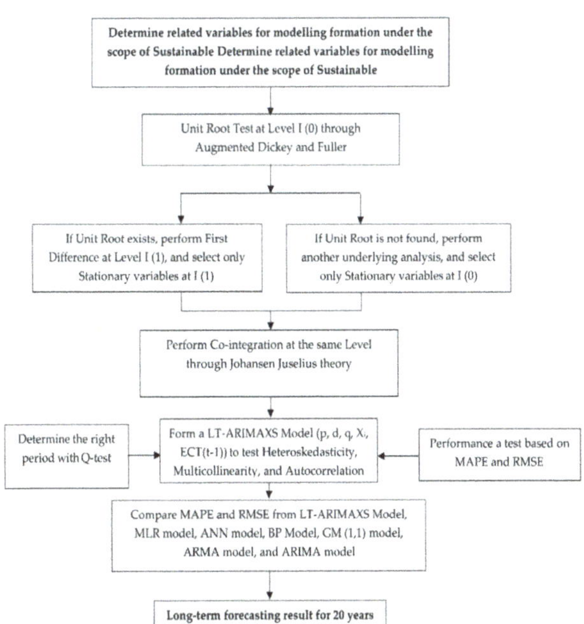

Figure 1. The flowchart of the LT-ARIMAXS model.

The remainder of this paper is as follows: Section 2 discusses the materials and methods. Section 3 shows the results. Section 4 summarizes the discussion. Section 5 presents the conclusion.

2. Materials and Methods

2.1. Autoregressive Model and Moving Average Model

The autoregressive model and moving average model or Box–Jenkins are two models that emphasize only the stationary data [32,33], described as follows.

In the case of the random seasonal process, it is an uncertain or specific seasonal process. For instance, a country encounters a political conflict for the past many years. At the same time, a demonstration occurs at the second quarter. This situation causes a sales drop. However, the political conflicts may seem stable this year in the same quarter. Thus, the sale is consistent. Here, the seasonal process that took place last year at the second quarter temporarily affects this year's second quarter, which can be called the stationary seasonal process. In this case, it is not necessary to drop off the season, but it can be incorporated into the model. This can be called the seasonal autoregressive moving average or (seasonal ARMA). The model is explained as below [34,35]:

X_t is a quarter time series and falls under a stationary seasonal process. This time series X_t can be written as:

$$X_t = A_1 X_{t-4} + v_t, \quad |A_1| < 1 \tag{1}$$

where v_t is a random error variable, which is a white noise. The above equation is a AR(4) model where the coefficient of X_{t-1}, X_{t-2} and X_{t-3} is 0, and $|A_1| < 1$ is the condition indicating the stationary seasonal process in time series X_t. If this X_t is brought to find an average value, a variance of the Theoretical Autocorrelation Function (TAC) and Theoretical Partial Autocorrelation Function (TPAC) is computed through the following equation:

Let $\mu = 0$, Variance $\gamma_0 = \frac{\sigma^2}{1-A_1^2}$, the TAC is pointed in Equation (2) and TPAC is drawn in Equation (3).

$$\rho_k = \begin{cases} (A_1)^{\frac{k}{4}}, & k = 0, 4, 8, \ldots \text{ when it is other case} \\ 0, & \end{cases} \tag{2}$$

$$\phi_{kk} = \begin{cases} \rho_4 & k = 4 \text{when it is other case} \\ 0, & \end{cases} \tag{3}$$

Since $|A_1| < 1$, when considering Equation (2), it can be concluded that, if $0 < A_1 < 1$, TAC will exponentially reduce at time 4, 8, 12, . . . ; if, as time slowly passes, TAC will be exponentially up and down at time 4, 8, 12, . . . ; if $|A_1|$ is closely approaching 1, a seasonal pattern will be clearer and last longer; and if $|A_1|$ is close to 0, the pattern will disappear. Equation (3) shows that the TPAC is not equivalent to 0.

Based on Equation (1), it reflects only on the impact of season in AR, but the time series can be as the ARMA in practice, and this can be written as below:

$$A(L^s)X_t = B(L^s)v_t \tag{4}$$

where s is the time duration of season.

$$A(L^s) = 1 - A_1 L^s - A_2 L^{2s} - \ldots A_p L^{ps} \tag{5}$$

$$B(L^s) = 1 - B_1 L^s - B_2 L^{2s} - \ldots B_Q L^{Qs} \tag{6}$$

We consider Equation (4) as the pure seasonal $ARMA$ model at $(P, Q)_s$. In practice, it is possible that X_t is in AR(1), together influencing the season as the equation below:

$$X_t = A_1 X_{t-1} + A_1 X_{t-4} + v_t, \quad |\alpha_1| < 1 \text{ and } |A_1| < 1 \tag{7}$$

Equation (7) indicates the influence of seasonal process when $s = 4$, and time series X_t in quarter 2 is related to quarter 1, while quarter 2 of this year is related to quarter 2 last year.

Meanwhile, Equation (4) shows X_t in some seasons of the $ARMA$ model at $(P,Q)_s$ or $ARMA$ (p,q) together, and v_t in $ARMA$ (p,q) is as follows:

$$v_t = \frac{\beta(L)}{\alpha(L)}\varepsilon_t \tag{8}$$

where ε_t is the random error variable with white noise $\alpha(L) = 1 - \alpha_1 L - \alpha_2 L^2 - \ldots \alpha_p L^p$ and $\beta(L) = 1 - \beta_1 L - \beta_2 L^2 - \ldots \beta_q L^q$. Therefore, Equation (4) can be drawn as below:

$$A(L^s)\alpha(L)X_t = B(L^s)\beta(L)\varepsilon_t \tag{9}$$

Equation (9) is called the multiplicative seasonal $ARMA$ model at $(p,q) \times (P,Q)_s$, and can be denoted as $ARMA$ $(p,q)(P,Q)_s$ or $ARMA$ $(p,q) \times (P,Q)_s$.

2.2. LT-ARIMAXS Model

In the construction of the LT-ARIMAXS model for forecasting, the autoregressive model (AR) and moving average model (MA) were basically integrated to first structure an ARIMA model. Once the ARIMA model was obtained, it was then applied to generate the LT-ARIMAXS model together with a co-integration test at the same level for every variable in the equations. In addition, there was also an adaptation of an error correction mechanism test in this particular model, as discussed below.

2.2.1. A Forecasting Model with ARIMA Model

It is a notion that differentiating at d with a particular time series will make a non-stationary time series a stationary time series. With such differentiation applied in the Box–Jenkins model, it can become known as $ARIMA(p,d,q)$ [34,35].

For better understanding, X_t is denoted as the non-stationary time series, where $Z_t = \Delta X_t = X_t - X_{t-1}$ is the stationary time series. Here, a proper model for this time series X_t is $ARIMA(1,1,0)$ and it can be written as:

$$Z_t = \alpha_0 + \alpha_1 Z_{t-1} + \varepsilon_t \quad \text{where } t = 1,2,\ldots,T \tag{10}$$

If time at T is taken into account, the $ARIMA(1,1,0)$ becomes:

$$Z_T = \alpha_0 + \alpha_1 Z_{T-1} + \varepsilon_T \tag{11}$$

and $X_1, X_2, \ldots X_T$ (or denoted as I_T) is now known for their value.

When using Equation (11), we forecast $\hat{Z}_{T+1}, \hat{Z}_{T+2}, \hat{Z}_{T+3}$ from the following equation.

$$\left.\begin{array}{lllll} \hat{Z}_{T+1} & = & \alpha_0 & + & \alpha_1 \Delta X_T \\ \hat{Z}_{T+2} & = & \alpha_0 & + & \alpha_1 \Delta \hat{X}_{T+1} \\ \hat{Z}_{T+3} & = & \alpha_0 & + & \alpha_1 \Delta \hat{X}_{T+2} \\ & & \vdots & & \\ \hat{Z}_{T+j} & = & \alpha_0 & + & \alpha_1 \Delta \hat{X}_{T+(j-1)} \end{array}\right\} \tag{12}$$

From Equation (12), it can be seen that $\hat{X}_T$ may not need forecasting. This is because the true information is known, which is X_T; hence, the forecasting result of $\hat{X}_{T+1}$ can be computed from.

$$\hat{X}_{T+1} = X_T + \hat{Z}_{T+1} \tag{13}$$

while the forecasting result of $\hat{X}_{T+2}, \hat{X}_{T+3}, \ldots, \hat{X}_{T+j}$ can be calculated as follows:

$$\hat{X}_{T+j} = X_T + \sum_{k=1}^{j} \hat{Z}_{T+k} \tag{14}$$

As for forecasting with $ARIMA(p,1,q)$, it can be applied by Equation (14), but the equation transformation is complicated.

Assuming the $ARIMA(p,1,q)$ model is written as below:

$$\begin{aligned} X_t &= \varphi_0 + \varphi_1 X_{t-1} + \varphi_2 X_{t-2} + \varphi_3 X_{t-3} + \ldots + \varphi_p X_{t-p} \\ &+ \varphi_{p+1} X_{t-p-1} + \varepsilon_t - \beta_1 \varepsilon_{t-1} - \beta_2 \varepsilon_{t-2} - \ldots - \beta_1 \varepsilon_{t-q} \end{aligned} \tag{15}$$

where $\varphi_0 = \alpha_{0,} \varphi_1 = \alpha_1 + 1, \varphi_j = \alpha_j - \alpha_{j-1}$ and $\varphi_{p+1} = -\alpha_p$.

When the ARIMA model is obtained, it can then be used to construct the LT-ARIMAXS model. The construction is explained below.

2.2.2. A Forecasting Model with LT-ARIMAXS Model

For the LT-ARIMAXS model, we have adapted the concept from the basic models including ARIMA models [32,33], co-integration and error correction mechanism model [36,37]. This LT-ARIMAXS model was examined for the unit root test and variable selection for stationary into this model formation. We have determined the Level (I(0)) or first difference (I(1)) to analyze co-integration [36]. This point of analysis must reflect the relationship at the same level. However, this LT-ARIMAXS model must consist of the co-integration and error correction mechanism test (ECT) [37] to increase efficiency and the zero error in the model. In addition, the LT-ARIMAXS model comes with the suitability of future application in different areas in line with the policy of a particular country. This is due to the difference of the LT-ARIMAXS model with other models so that the ARIMA model focuses on the variables of Autoregressive (AR), Integrated (I), and Moving Average (MA) only at time $t-i$ especially in past data. In this paper, the LT-ARIMAXS model differs from other old models due to the emphasis of Exogeneous Variables ($\sum_{i=1}^{p} Y_{t-i}$), which is believed to be an important yet appropriate variable in the study. As for the reason, it is the influential variable that can affect the dependent variable. Additionally, the LT-ARIMAXS model uses Autoregressive (AR), Integrated (I), and Moving Average (MA) during time $t-i$ in the study's model. The LT-ARIMAXS model utilizes the co-integration and error correction mechanism test from the theory of Johansen and Juselius to increase the effectiveness of the model [36,37]. The co-integration model and error correction mechanism model can be explained below.

This model applies the Johansen co-integration test to examine a pattern called multivariate co-integration, which is the method proposed by Johansen and Juselius [36]; it is used to examine the long-term relationship between the variables. The essence of cointegration is that the linear combination of variables is stationary. Cointegration tests also require that all variables are integrated in the same order [37]. We can use the following formula to conduct the cointegration test.

$$X_t = \alpha_0 + \sum_{i=1}^{p} \alpha_i X_{t-i} + \varepsilon_t \tag{16}$$

where X_t is a $(n \times 1)$ vector of variables, α_0 is a vector of constants, α_i is a $(n \times n)$ matrix of parameters, and ε_t is a $(n \times 1)$ vector of error term. Subtracting X_{t-1} from each side of Equation (16) and letting I be an $(n \times n)$ identity matrix, it can be rewritten as follows.

$$\Delta X_t = \alpha_0 + \pi X_{t-1} + \sum_{i=1}^{p} \prod_i \Delta X_{t-i} + \varepsilon_t \tag{17}$$

where $\prod_i$ and π are the coefficient matrix and πX_{t-1} is the error correction term, while the coefficient matrix π provides information about the long term relationships among the variables. The number of the co-integration vectors can be determined by using the trace test and maximum eigenvalue test suggested by Johansen [37], as demonstrated in Equation (18) and (19).

$$\lambda_{trace} = -T \sum_{i=r+1}^{n} \ln(1-\hat{\lambda}_1) \tag{18}$$

$$\lambda_{\max}(r, r+1) = -T \sum_{i=r+1}^{n} \ln(1-\hat{\lambda}_{r+1}) \tag{19}$$

where T is the sample size and λ is the eigenvalue. Based on Equations (17) and (18), if the null hypothesis is rejected, it shows the testing variables consist of co-integration. On the other hand, if the null hypothesis is accepted, there is no co-integration.

After performing co-integration test, another important test must be carried out, which is error correction mechanism test [37]. We can find that the change of X_t not only depends on the change of Y_t but also depends on the change of the last period Y_{t-1} and X_{t-1}. Considering the non-stationarity, the OLS test cannot be used to perform the regression. Therefore, Equation (32) can deform to the equation below.

$$\Delta X_t = \beta_1 \Delta Y_t - (1-\delta)(X_{t-1} - \frac{\beta_0}{1-\delta} - \frac{\beta_1+\beta_2}{1-\delta} Y_{t-1}) + \varepsilon_t \tag{20}$$

Hence, the LT-ARIMAXS model can be written below.

$$\begin{aligned} X_t &= \varphi_0 + \varphi_1 X_{t-1} + \varphi_2 X_{t-2} + \varphi_3 X_{t-3} + \ldots + \varphi_p X_{t-p} \\ &+ \varphi_{p+1} X_{t-p-1} + \varepsilon_t - \beta_1 \varepsilon_{t-1} - \beta_2 \varepsilon_{t-2} - \ldots - \beta_1 \varepsilon_{t-q} \\ &+ \sum_{i=1}^{p} Y_{t-i} + \sum_{i=1}^{p} ECT_{t-i} \end{aligned} \tag{21}$$

Let $\varphi_0 = \alpha_0, \varphi_1 = \alpha_1 + 1, \varphi_j = \alpha_{j-1}, \varphi_{p+1} = -\alpha_p, \sum_{i=1}^{p} Y_{t-i}$ = exogeneous variables, which are stationary at the level and $\sum_{i=1}^{p} ECT_{t-i}$ = the error correction mechanism test.

Equation (21) indicates the components of the LT-ARIMAXS model comprised of: (1) Autoregressive variables (AR); (2) Moving Average (MA); (3) exogenous variables (); and (4) error correction mechanism $\sum_{i=1}^{p} ECT_{t-i}$. The LT-ARIMAXS model is built and developed with the assurance of being Heteroskedasticity, Multicollinearity, and Autocorrelation free. There is also an analysis of period identification with the Q-statistics test as to ensure that the model is not spurious while it becomes efficient in the forecasting with fewer errors. The model is then able to be applied in a different context and management policy.

2.2.3. Measurement of the Forecasting Performance

To evaluate the forecasting effect of each model, we employed the mean absolute percentage error (MAPE) and the root mean square error (RMSE) to compare the forecasting accuracy of each model. The calculated equations are shown below [35,38].

$$\text{MAPE} = \frac{1}{n} \sum_{i=1}^{n} \left| \frac{\hat{y}_i - y_i}{y_i} \right| \tag{22}$$

$$\text{RMSE} = \sqrt{\frac{1}{n} \sum_{i=1}^{n} (\hat{y}_i - y_i)^2} \tag{23}$$

The LT-ARIMAXS model is a newly developed method completed by adapting various concepts from the general models including autoregressive (AR), integrate (I), and moving average (MA). For the variable selection criterion, the variables must only be causal factors or stationary at the same level. The stationary level I(0) or first difference I(1) are used to test the unit root test. With the right variables, we fulfill the criterion. Those variables are put forth for a co-integration test. When they are found to be co-integrated, they are then used in structuring the LT-ARIMAXS model (p, d, q, Xi, $ECT_{(t-1)}$) with an appropriateness check of period ($t - i$) through the implementation of a white noise process by the Q test statistic method. In this paper, the LT-ARIMAXS model (p, d, q, Xi, $ECT_{(t-1)}$) must not be free from heteroskedasticity, multicollinearity, and autocorrelation. Testing the LT-ARIMAXS model (p, d, q, Xi, $ECT_{(t-1)}$) can be done based with MAPE and RMSE, and comparing those two values with existing models. Once the model is obtained, forecasting the future is the next essential step. We have combined the dataset using Microsoft Office Excel. In addition, EViews 9.5 software is deployed to implement the model, and it flows as below.

(1) Place the stationary variables at the same level in the analysis of the long-term relationship based on the Johansen and Juselius concept.
(2) Create a forecasting model by adapting the advance statistics of the so-called LT-ARIMAXS model with full consideration of the relationship of all causal variables in terms of both the error correction mechanism test and the co-integration test.
(3) Examine the goodness of fit in two aspects: (1) appropriateness check of period ($t - i$) through the implementation of a white noise process by the Q test statistic method; and (2) performance test for the LT-ARIMAXS model based on MAPE and RMSE. Compare those two values derived from the LT-ARIMAXS model with the existing model, including MLR model, the ANN model, BP Model, GM(1,1) model, ARMA model, and ARIMA model.
(4) Forecast final energy consumption by using the LT-ARIMAXS model for the period from 2018 to 2037, totaling 20 years. The flowchart of the LT-ARIMAXS model is shown in Figure 1.

3. Results

3.1. Screening of Influencing Factors for Model Input

We tested the factors in the context of Thailand's sustainable development policy. Here, we deployed the time series data of the period 1987–2017. The tested factors consisted of eight variables, namely final energy consumption (ln(EC)), per capita GDP (ln(GDP)), population growth (ln(Population)), oil price (ln(OP)), energy intensity (ln(EI)), urbanization rate (ln(UG)), industrial structure (ln(IS)), and net exports (ln(X − E)). The test was conducted based on the Augment Dickey Fuller theory at Level I(0) and first difference I(1), as illustrated in Table 1.

Table 1. Unit root test at Level I(0) and First Difference I(1).

ADF Test at Level I(0)		ADF Test at First Difference I(1)		MacKinnon Critical Value		
Variables	Value	Variables	Value	1%	5%	10%
ln(EC)	−2.85	Δ ln(EC)	−4.56 ***	−4.22	−3.53	−3.20
ln(GDP)	−2.24	Δ ln(GDP)	−5.26 ***	−4.22	−3.53	−3.20
ln(Population)	−2.75	Δ ln(Population)	−4.25 ***	−4.22	−3.53	−3.20
ln(OP)	−3.05	Δ ln(OP)	−5.11 ***	−4.22	−3.53	−3.20
ln(EI)	−3.11	Δ ln(EI)	−4.95 ***	−4.22	−3.53	−3.20
ln(UG)	−2.39	Δ ln(UG)	−4.77 ***	−4.22	−3.53	−3.20
ln(IS)	−3.12	Δ ln(IS)	−5.71 ***	−4.22	−3.53	−3.20
ln(X − E)	−3.60	Δ ln(X − E)	−4.78 ***	−4.22	−3.53	−3.20

Note: EC is the final energy consumption; GDP is the per capita GDP; Population is the population growth; OP is the oil price; EI is the energy intensity; UG is the urbanization rate; IS is the industrial structure; X − E is the net export, *** denotes a significance, α = 0.01, compared to the Tau test with the MacKinnon Critical Value, Δ is the first difference, and ln is the natural logarithm.

Table 1 presents the testing result of the Tau test compared to the MacKinnon critical value. The result showed that all variables had the unit root or were found to be non-stationary at Level I(0), and explained the insignificance at 1%, 5%, and 10%. Therefore, carrying out the First Difference I(1) was required. When testing the unit root at Level I(1), it was found that all variables were stationary at Level I(1) with the significance level of 1%, 5%, and 10%. Later, all stationary variables are taken for the co-integration test by Johansen Juselius to analyze the long-term relationship of every variable at the same level as shown in Table 2.

Table 2. Co-integration test by Johansen Juselius.

Variables	Hypothesized No of CE(S)	Trace Statistic Test	Max-Eigen Statistic Test	MacKinnon Critical Value		Status
				1%	5%	
$\Delta\ln(EC)$, $\Delta\ln(GDP)$, $\Delta\ln(Population)$, $\Delta\ln(OP)$, $\Delta\ln(EI)$, $\Delta\ln(UG)$, $\Delta\ln(IS)$, $\Delta\ln(X-E)$	None ***	231.15	165.85	18.75	15.40	I(1)
	At Most 1 ***	79.41	81.45	5.50	3.12	I(1)

*** denotes significance $\alpha = 0.01$.

3.2. *Analysis of Co-Integration*

Table 2 shows that all variables had a long-term relationship (co-integration) because the results of the trace test were 231.15 and 79.41, which were higher than the MacKinnon critical values at significance levels of 1% and 5%. The maximum eigenvalue test results were 165.85 and 81.45, which were higher than the MacKinnon critical values at the same significance levels. Consequently, those variables were used to form a forecasting model by adapting the LT-ARIMAXS model and applying short- and long-term relationships into the model.

3.3. *Formation of Analysis Modeling with the LT-ARIMAXS Model*

All stationary variables at the first difference are tested for the co-integration at the same level to construct the LT-ARIMAXS Model at time (1,1,1). All exogenous variables at time $t-1$ and ECT($t-1$) are not proper as evaluated by the Q-statistic. However, we have started to build the Best model named as the LT-ARIMAXS model at period ($t-i$) of p,d,q, and the good fit of period ($t-i$) falls at LT-ARIMAXS (2,1,2), which is shown in Figure 2 and Table 3.

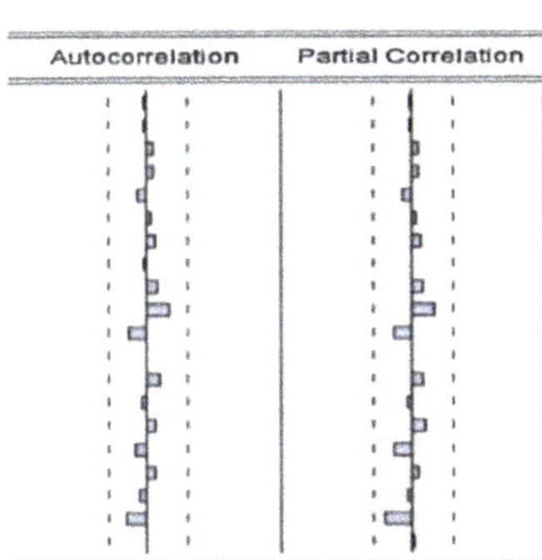

	AC	PAC	Q-Stat	Prob
1	-0.024	-0.024	0.0244	0.876
2	-0.015	-0.016	0.0347	0.983
3	0.057	0.056	0.1853	0.980
4	0.049	0.052	0.3015	0.990
5	-0.074	-0.070	0.5675	0.989
6	0.039	0.034	0.6437	0.996
7	0.080	0.075	0.9778	0.995
8	-0.012	-0.003	0.9858	0.998
9	0.087	0.092	1.3987	0.998
10	0.187	0.179	3.3925	0.971
11	-0.147	-0.145	4.6666	0.946
12	-0.002	-0.003	4.6668	0.968
13	0.112	0.086	5.4509	0.964
14	-0.026	-0.026	5.4953	0.978
15	0.069	0.112	5.8220	0.983
16	-0.090	-0.150	6.3974	0.983
17	0.076	0.051	6.8212	0.986
18	-0.057	-0.035	7.0708	0.990
19	-0.156	-0.227	9.0261	0.973
20	0.001	0.013	9.0262	0.983

Figure 2. The correlogram of the residual error of the LT-ARIMAXS model (2,1,2). Note: Columns 1 and 2 show the velocity trend of the correlation coefficient that shrank to two times the standard deviation (obtained by EViews). AC is the value of the autocorrelation coefficient. PAC is the value of the partial correlation coefficient. Q-stat denotes the Q test statistic method at time $t-i$.

Table 3. The result of the LT-ARIMAXS model (2,1,2).

Independent Variables	Dependent Variable $\Delta\ln(\mathrm{EC})_t$
AR(1)	2.79 ***
AR(2)	3.15 ***
MA(1)	2.99 ***
MA(2)	2.41 **
$\Delta\ln(\mathrm{GDP})_{t-1}$	7.06 ***
$\Delta\ln(\mathrm{Population})_{t-1}$	2.45 **
$\Delta\ln(\mathrm{OP})_{t-2}$	6.15 ***
$\Delta\ln(\mathrm{EI})_{t-1}$	3.42 ***
$\Delta\ln(\mathrm{UG})_{t-1}$	5.26 ***
$\Delta\ln(\mathrm{IS})_{t-1}$	7.34 ***
$\Delta\ln(\mathrm{X}-\mathrm{E})_{t-2}$	5.01 ***
ECT_{t-1}	−2.15 **

Note: AR is Autoregressive model, MA is Moving Average model, *** denotes significance $\alpha = 0.01$, ** denotes significance $\alpha = 0.05$, R-squared is 0.91, adjusted R-squared is 0.90, the Durbin–Watson statistic is 2.01, the F-statistic is 229.25 (probability is 0.00), the ARCH test is 35.01 (probability is 0.1), the LM test is 1.65 (probability is 0.10), and the response test (χ^2 > critical) represents the significance.

Figure 2 reflects that the LT-ARIMAXS (2,1,2) model became the best forecasting model as all values of the Q test statistic at time $(t-i)$ were in the criteria and met all conditions, or the insignificance fell as follows: $\alpha = 0.01$, $\alpha = 0.05$, and $\alpha = 0.1$. Therefore, this model can be used to forecast the final energy consumption. However, we have discovered the best model at time LT-ARIMAXS (2,1,2), and this allowed us to understand the influence in changes or elasticity of all independent variables causing the changes over the final energy consumption at time $(t-i)$, as illustrated in Table 3.

Table 3 illustrates the parameters of the LT-ARIMAXS (2,1,2) model at a statistically significant level of 1% and 5%. The findings illustrated that, when per capita GDP ($\Delta\ln(\mathrm{GDP})_{t-1}$at time $(t-1)$ changed about 1%, it changed the final energy consumption ($\Delta\ln(\mathrm{EC})_t$) equivalent to the elasticity coefficient of 7.06% at the significance level of 1%, which went in the same direction. When population growth ($\Delta\ln(\mathrm{Population})_{t-1}$at time $(t-1)$ changed about 1%, it showed influence over the final energy consumption ($\Delta\ln(\mathrm{EC})_t$) equivalent to the elasticity coefficient of 2.45% at the significance level of 5%, which also went in the same direction. When oil price ($\Delta\ln(\mathrm{OP})_{t-2}$) at time $(t-2)$ changed about 1%, the final energy consumption ($\Delta\ln(\mathrm{EC})_t$) was affected to change by the elasticity coefficient of 6.15% at the significance level of 1%, whose change was in the same direction. When energy intensity ($\Delta\ln(\mathrm{EI})_{t-1}$) at time $(t-1)$ changed about 1%, the final energy consumption ($\Delta\ln(\mathrm{EC})_t$) was also changed by the elasticity coefficient of 3.42% at the significance level of 1%, which went in the same direction. When the change in urbanization rate ($\ln(\mathrm{UG})_{t-1}$) at time $(t-1)$ accounted for 1%, the final energy consumption ($\Delta\ln(\mathrm{EC})_t$) was also affected to change by the elasticity coefficient of 5.26% at the significance level of 1%, which went in the same direction. When industrial structure ($\ln(\mathrm{IS})_{t-1}$) at time $(t-1)$ changed about 1%, it affected the final energy consumption ($\Delta\ln(\mathrm{EC})_t$) equivalent to the elasticity coefficient of 7.34% at the significance level of 1%, whose direction went in the same direction. Furthermore, when net exports ($\ln(\mathrm{X}-\mathrm{E})_{t-2}$) at time $(t-2)$ changed about 1%, the final energy consumption ($\Delta\ln(\mathrm{EC})_t$) was changed equally with the elasticity coefficient of 5.01% at the significance level of 1% in the same direction.

However, from the analysis of the LT-ARIMAXS model (2,12), it was found that ECT_{t-1} was equal to −2.15% at the significance level of 5%. This shows that the error-correction mechanism (ECT_{t-i}) can precisely explain the fluctuations and the adjustment. Specifically, ECT_{t-1} denotes that, under the impacts of controlled variables, when short-term fluctuations deviate from the long-term equilibrium, the changes of final energy consumption in t time can eliminate the non-equilibrium error of the $t-1$ time by 2.15% and make a reverse adjustment to bring the non-equilibrium point back to the equilibrium point. Furthermore, the LT-ARIMAXS model (2,1,2) is free from the issue of heteroskedasticity, multicollinearity, and autocorrelation.

In addition, we compared some selected forecasting models in terms of their effectiveness with MAPE and RMSE, which is indicated in Table 4. LT-ARIMAXS model was compared with other models: the MLR model, the BP model, the ANN model, the ARMA model, the GM(1,1) model, and the ARIMA model shown below.

Table 4. The performance monitoring of the forecasting model (%).

Forecasting Model	MAPE	RMSE
MLR model	19.76	20.76
BP model	10.67	14.63
ANN model	8.55	9.95
ARMA model	8.51	9.17
GM(1,1) model	6.69	8.52
ARIMA model	5.75	6.41
LT-ARIMAXS model (2,1,2)	0.97	2.12

Table 4 shows that LT-ARIMAXS model (2,1,2) with analytical data for 1985–2017 consisting of independent variables in the model including final energy consumption at the time period $t-1$ $(\ln(EC)_{t-1})$ or AR(1), final energy consumption at time period $t-2$ $(\ln(EC)_{t-2})$ or AR(2), Moving Average (1) or MA(1), Moving Average (2) or MA(2), per capita GDP at time period $t-1$ $(\ln(GDP)_{t-1})$, population growth at time period $t-1$ $(\ln(Population)_{t-1})$,oil price at time period $t-2$ $(\ln(OP)_{t-2})$, energy intensity at time period $t-1$ $(\ln(EI)_{t-1})$, urbanization rate at time period $t-1$ $(\ln(UG)_{t-1})$, industrial structure at time period $t-1$ $(\ln(IS)_{t-1})$, net exports at time period $t-2$ $(\ln(X-E)_{t-2})$, and the error-correction mechanism at time period $t-1$ (ECT_{t-1}) provided the lowest MAPE value at 0.97% and RMSE value at 2.12%. The ARIMA model, the GM(1,1) model, the ARMA model, the ANN model, the BP model, and the MLR model had MAPE values of 5.75%, 6.69%, 8.51%, 8.55%, 10.67%, and 19.76%, respectively, and RMSE values of 6.41%, 8.52%, 9.17, 9.95%, 14.63%, and 20.76%, respectively. The findings show that the LT-ARIMAXS model (2,1,2) is most effective. This was observed as MAPE and RMSE being the lowest when compared to the old models. Furthermore, it is very useful for long-term forecasting and national policy-making and planning in boosting sustainable development in the long-run. Therefore, the LT-ARIMAXS model (2,1,2) was used to forecast CO_2 emissions in the following step.

3.4. *Final Energy Consumption Forecasting Based on the LT-ARIMAXS Model*

Table 4 shows that the LT-ARIMAXS Model comes with highest efficiency by looking at the lowest value of MAPE and RMSE compared with past forecasting models. Therefore, we chose the LT-ARIMAXS Model for long-term forecasting (2018–2037). Once we attained the best model, i.e., the LT-ARIMAXS Model (2,1,2), the long-term forecasting on the final energy consumption in Thailand's petroleum industries sector for 20 years (2018–2037) was conducted, as shown in Figure 3.

Figure 3 shows that final energy consumption from 2018 to 2037 in Thailand constantly increases where the 2037 rate was found to be 109.8% higher than 2017. At the same time, it presents that the 2037 final energy consumption would be equivalent to 121,461 ktoe, which is higher than the government's reduction goal, i.e., the final energy consumption in Thailand's petroleum industries sector should not exceed more than 90,000 ktoe. As for the study, it reflects that the final energy consumption does not go along with the national policy effectively, and has negative effects on Thailand's sustainable development in the long-run.

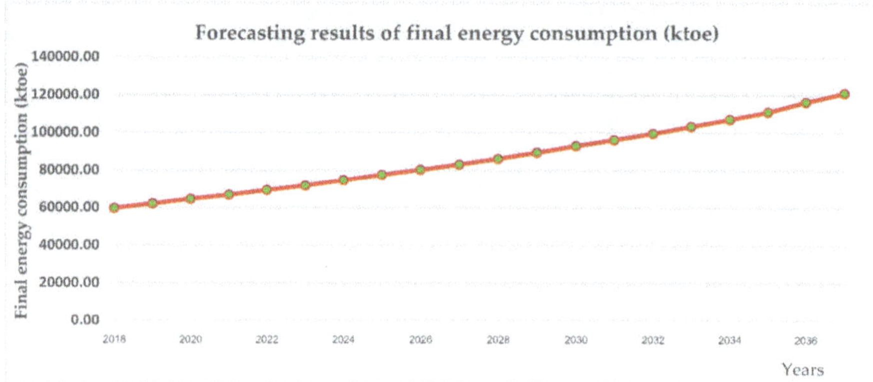

Figure 3. The forecasting results of final energy consumption from 2018 to 2037 in Thailand.

4. Discussion

This research differs from other previous studies, as this LT-ARIMAXS model has a higher effectiveness and better long-term forecasting, while producing fewer discrepancies in prediction. Based on the many relevant studies reviewed, most existing models were only made available for short-term forecasting capability ranging from one to five years. Zhao, Zhao, and Guo [6] applied the GM model optimized by MFO with rolling mechanism in the forecasting for the period 2010–2014. Li and Li [7] used the ARIMA model, GM model, and ARIMA-GM model to forecast energy consumption in Shandong, China from 2016 until 2020. Xiong, Dang, Yao, and Wang [8] utilized the GM(1,1) model based on optimizing the initial condition in accordance with the new information priority principle in the prediction from 2013 to 2017. Panklib, Prakasvudhisarn and Khummongkol [9] chose ANN model and MLR model in the forecasting for 2010, 2015, and 2020. In addition, Azadeh, Ghaderi, Tarverdian, and Saberi [10] incorporated ANN integrated with the genetic algorithm in a one-year prediction. Günay [11] implemented Artificial Neural Networks using predicted values of socio-economic indicators and climatic conditions for the same one-year coverage of forecasting. Dai, Niu, and Li [12] applied ensemble empirical mode decomposition and least squares support vector machine based on an improved shuffled frog leaping algorithm for 2018–2022 forecasting. Suganthi and Samuel [14] used the econometrics model for 2030–2031 forecasting. Additionally, there have also been some studies that have attempted to forecast for six years but not 20 years. Hamzacebi and Es [18] implemented an optimized Grey Modeling for 2013–2025 forecasting. Mu et al. [19] used improved grey model for the 2011–2020 prediction. Zeng, Zhou, and Zhang [20] developed a homologous grey prediction model for 2018–2024 forecasting. Furthermore, Jiang, Yang, and Li [21] used MGM model, ARIMA model, MGM–ARIMA model, and BP model for the forecasting period of 2017–2030. Ediger and Akar [22] deployed ARIMA model and SARIMA model for 2005–2020 forecasting. Ekonomou [23] applied ANN model for 2005–2008, 2010, 2012, and 2015 prediction. Seung et al. [26] also deployed a MLR model for 2014–2030 forecasting. Ayvaz and Kusakci [31] chose to apply a NDGM model in prediction for the period 2014–2030. However, by reviewing various studies, it was found that long-term forecasting has become a popular topic most researchers have chosen to study, and various different methodologies have been implemented. In this study, we selected the most appropriate model for long-term forecasting. The LT-ARIMAXS model functioned better and was more efficient, with less erro, when compared to other models. With the output of the model, it is very useful and fits in Thailand's policy-making and planning. Furthermore, it can become the best guideline for any interested researchers to further develop and explore.

Nonetheless, the limitation of this research lies upon the diesel price as the government controls the price by using the Diesel Oil Support Fund. As a result, it does not reflect the real economy and energy demand due to this government intervention, which may result in inaccurate forecasting. If the government allows the oil price to move with the world market, we would be able to obtain the true influence of the diesel price over the change in final energy consumption. In addition, the government's policy does not specifically define the government expenditure, especially on mega projects the government has invested in, which heavily effects the economy, society, and environment. If such variable can be utilized and considered in policy-making, we wouls also be able to see the influence affecting the change in energy consumption. However, we highly expect that the LT-ARIMAXS model is applied for a formulation of sustainable development-based policy and future research. In addition, it is used to forecast greenhouse gases as determined by Thailand in all terms of durations including short-term (1–5 years), mid-term (6–10 years), and long-term (11–20 year). However, the variables must be contributed as the causal factors, especially a global diesel price and government expenditure. Nonetheless, all factors are tested for the stationary, co-integration, and the error correction mechanism. Most importantly, these two factors of global diesel price and government expenditure are used in direct and indirect relationship analysis to ensure the real influence.

A government policy formulation requires a number of factors. Apart from variables used in this study, the other true influential factors must be considered. Those factors are those truly affecting the change in final energy consumption especially in long-term forecasting. The true and complete factors must be emphasized and applied in future policy-making since their characteristics qualify and fulfill the criterion as complete factors. Increasing the numbers of foreign tourists, the increment of foreign workers, and carbon emission intensity are some factors to consider. Since the existing models do not produce a precise output and are less efficient in forecasting, the LT-ARIMAXS model has, therefore, been designed to fill this gap. Additionally, the LT-ARIMAXS model is structured with detailed research methodology together with a fine selection of variables emphasizing on the influential factors over the changes in the final energy consumption. At the same time, both co-integration and an error correction mechanism test are carried out to ensure zero heteroscedasticity, multicollinearity, and autocorrelation. The proper period is specified based on the Q-statistic test for constructing the LT-ARIMAXS model (2,1,2). With all processes put together, the model is completely ready for long-term prediction (2018–2037) and analysis of MAPE and RMSE for further comparison with the MLR model, the BP model, the ANN model, the ARMA Model, the GM(1,1) model, and the ARIMA model. The study reflected that the LT-ARIMAXS model (2,1,2) has a higher effectiveness with less error based on the evaluation from MAPE and RMSE.

However, the LT-ARIMAXS model is unique in terms of its variables used, as only causal factors affecting future forecasting are considered. This means all irrelevant factors are removed from the modeling. This study is very instrumental not only for Thailand but also other countries. To ensure its structure, all variables are rightly deployed, according to the proper context by analyzing the co-integration and carrying out an error correction test together by specifying the right period for the right sectors. The LT-ARIMAXS model does not limit the casual factors, but the users must give details in all processes for a better performance of the model. We used various software for our analysis by collecting data via Excel and constructing the LT-ARIMAXS model with EViews 9.5 due to its capability in precise and long-term analysis and supportive to window (64 bit). Nonetheless, for those individuals interested in using the LT-ARIMAXS model, they can download the EViews 9.5 student version for Windows (Windows 10, Windows 8, Windows 7, and Windows Vista) for free, but it only allows free usage for two years. A Pentium or better CPU with 512 MB Memory and 270 MB Disk Space is required.

However, from this study, it can be concluded that the final energy consumption in the long-run (2018–2037) by using the LT-ARIMAXS model will be higher than the targeted amount. In addition, this model differs from any existing models used on the prediction within Thailand since the model's prediction result can really support Thailand's national policy-making at a higher efficiency. Since the LT-ARIMAXS model is developed, only true influential variables over the final energy consumption in particular sectors are deployed and become applicable at a wider scope of prediction compared to past predictions. If the government decides to utilize the LT-ARIMAXS model as part of obtaining sustainability, all influences produced from the LT-ARIMAXS model (2,1,2) must be considered to attain a proper management of changes in all casual factors. One of the major key variables in the LT-ARIMAXS model is the error-correction mechanism (ECT). This is because the implication of the model is very beneficial for a long-term prediction for Thailand and other countries. When there are changes, shocks, or variations in the variables deviating from the equilibrium, the ECT is the value telling that those variables will adjust to the equilibrium in the next period $(t - i)$. However, the ECT parameter will indicate the capability range of the adaptation to the equilibrium and it will guide the government to set a clearer path in policy-planning. In addition, the focus in modeling should be emphasized while other untaken variables are used for consideration by specifying a proper period of application to maximize the model's use.

5. Conclusions

This study has formed another area to explore and acted as a guideline for future research. This has made the model standout from other past models. At the same time, it has helped to narrow the gap and strengthen the existing weaknesses in previous studies, reducing potential discrepancies. Therefore, this study was necessary as it is beneficial and instrumental for both academia and strengthening future sustainable development policy. This LT-ARIMAXS model has been structured based on previous models, and has become the first model to optimize the advance statistic. Simultaneously, it was designed to fill the gap of forecasting capability, especially long-term forecasting, which is important and necessary to develop to reduce any potential residual discrepancies. The reason for this assurance is that it allows us to improve policy formulation in the right direction in the most efficient and effective manner. We have developed this LT-ARIMAXS model by commencing with variable selection. This selection uses only influential variables, which have an impact on the change in final energy consumption, and must be felt within the sustainable development concept in both the long-term and short-term. When the right variables are obtained, they are used for the unit root test to identify the stationary at the same level. If any variables are found to be stationary yet at different levels, they are immediately eliminated. In this study, it was found that all involved variables were stationary at first difference, and they were used for the co-integration test to evaluate the long-term relationship. This test showed that all variables were co-integrated at Level I(1). After all those processes, the LT-ARIMAXS model (2,1,2) was structured consisting of the autoregressive model (AR), moving average model (MA), exogeneous variable, and error correction mechanism test (ECT_{t-1}). However, the LT-ARIMAXS model (2,1,2) was improved for the right period $(t - i)$ as shown in the correlogram of the residual error with the use of the Q statistic test. With this test, we were able to identify the period for this modeling to be the most effective. In addition, we tested the effectiveness of the LT-ARIMAXS model (2,1,2) by MAPE and RMSE, whose values were later found to be lowest, equivalent to 0.97% and 2.12% when compared to the ARIMA model, GM(1,1) model, ARMA model, ANN model, BP model, and MLR model. Thus, the LT-ARIMAXS model (2,1,2) was used to forecast the final energy consumption in the petroleum industries sector in Thailand for 20 years (2018–2037). As a result, the model produced outcomes where the rate 2037 was 109.8% higher than 2017. Additionally, the final energy consumption was found to be 121,461 ktoe by 2037, and this exceeds the government limit of 90,000 ktoe. Hence, the above output can be applied in policy-making and planning in the future to ensure that the right policy for the right direction is established, unlike any other previous years (1985–2017).

Author Contributions: P.S. and K.K. were involved in the data collection and preprocessing phase, model constructing, empirical research, results analysis and discussion, and manuscript preparation. All authors have approved the submitted manuscript.

Funding: This research received no external funding.

Acknowledgments: This work was performed with the approval of King Mongkut's University of Technology Thonburi and the Office of the National Economic and Social Development Board.

Conflicts of Interest: The authors declare no conflict of interest.

References

1. Office of the National Economic and Social Development Board (NESDB). Available online: http://www.nesdb.go.th/nesdb_en/more_news.php?cid=154&filename=index (accessed on 27 June 2018).
2. National Statistic Office Ministry of Information and Communication Technology. Available online: http://web.nso.go.th/index.htm (accessed on 28 June 2018).
3. Department of Alternative Energy Development and Efficiency. Available online: http://www.dede.go.th/ewtadmin/ewt/dede_web/ewt_news.php?nid=47140 (accessed on 29 June 2018).
4. Achawangkul, Y. Thailand's Alternative Energy Development Plan. Available online: http://www.unescap.org/sites/default/files/MoE%20_%20AE%20policies.pdf (accessed on 29 June 2018).
5. Thailand Greenhouse Gas Management Organization (Public Organization). Available online: http://www.tgo.or.th/2015/thai/content.php?s1=7&s2=16&sub3=sub3 (accessed on 29 June 2018).
6. Zhao, H.R.; Zhao, H.R.; Guo, S. Using GM(1,1) Optimized by MFO with Rolling Mechanism to Forecast the Electricity Consumption of Inner Mongolia. *Appl. Sci.* **2016**, *6*, 20. [CrossRef]
7. Li, S.; Li, R. Comparison of forecasting energy consumption in Shandong, China Using the ARIMA model, GM model, and ARIMA-GM model. *Sustainability* **2017**, *9*, 1181.
8. Xiong, P.P.; Dang, Y.G.; Yao, T.X.; Wang, Z.X. Optimal modeling and forecasting of the energy consumption and production in China. *Energy* **2014**, *77*, 623–634. [CrossRef]
9. Panklib, K.; Prakasvudhisarn, C.; Khummongkol, D. Electricity Consumption Forecasting in Thailand Using an Artificial Neural Network and Multiple Linear Regression. *Energy Sources Part B Econ. Plan. Policy* **2015**, *10*, 427–434. [CrossRef]
10. Azadeh, A.; Ghaderi, S.F.; Tarverdian, S.; Saberi, M. Integration of artificial neural networks and genetic algorithm to predict electrical energy consumption. *Appl. Math. Comput.* **2007**, *186*, 1731–1741. [CrossRef]
11. Günay, M.E. Forecasting annual gross electricity demand by artificial neural networks using predicted values of socio-economic indicators and climatic conditions: Case of Turkey. *Energy Policy* **2016**, *90*, 92–101. [CrossRef]
12. Dai, S.; Niu, D.; Li, Y. Forecasting of Energy Consumption in China Based on Ensemble Empirical Mode Decomposition and Least Squares Support Vector Machine Optimized by Improved Shuffled Frog Leaping Algorithm. *Appl. Sci.* **2018**, *8*, 678. [CrossRef]
13. Wang, Q.; Li, R. Decline in China's coal consumption: An evidence of peak coal or a temporary blip? *Energy Policy* **2017**, *108*, 696–701. [CrossRef]
14. Suganthi, L.; Samuel, A.A. Modelling and forecasting energy consumption in INDIA: Influence of socioeconomic variables. *Energy Sources Part B Econ. Plan. Policy* **2016**, *11*, 404–411. [CrossRef]
15. Xu, J.; Fleiter, T.; Eichhammer, W.; Fan, Y. Energy consumption and CO2 emissions in China's cement industry: A perspective from LMDI decomposition analysis. *Energy Policy* **2012**, *50*, 821–832. [CrossRef]
16. Kishita, Y.; Yamaguchi, Y.; Umeda, Y. Describing Long-Term Electricity Demand Scenarios in the Telecommunications Industry: A Case Study of Japan. *Sustainability* **2016**, *8*, 52. [CrossRef]
17. Zhao, W.; Wang, J.; Lu, H. Combining forecasts of electricity consumption in China with time-varying weights updated by a high-order Markov chain model. *Omega* **2014**, *45*, 80–91. [CrossRef]
18. Hamzacebi, C.; Es, H.A. Forecasting the annual electricity consumption of Turkey using an optimized grey model. *Energy* **2014**, *70*, 165–171. [CrossRef]
19. Mu, H.; Dong, X.; Wang, W.; Ning, Y.; Zhou, W. Improved Gray Forecast Models for China's Energy Consumption and CO_2 Emission. *J. Desert Res.* **2002**, *22*, 142–149.
20. Zeng, B.; Zhou, M.; Zhang, J. Forecasting the Energy Consumption of China's Manufacturing Using a Homologous Grey Prediction Model. *Sustainability* **2017**, *9*, 1975. [CrossRef]
21. Jiang, F.; Yang, X.; Li, S. Comparison of Forecasting India's Energy Demand Using an MGM, ARIMA Model, MGM-ARIMA Model, and BP Neural Network Model. *Sustainability* **2018**, *10*, 2225. [CrossRef]
22. Ediger, V.S.; Akar, S. ARIMA forecasting of primary energy demand by fuel in Turkey. *Energy Policy* **2007**, *35*, 1701–1708. [CrossRef]
23. Ekonomou, L. Greek long-term energy consumption prediction using artificial neural networks. *Energy* **2010**, *35*, 512–517. [CrossRef]

24. Ardakani, F.J.; Ardehali, M.M. Long-term electrical energy consumption forecasting for developing and developed economies based on different optimized models and historical data types. *Energy* **2014**, *65*, 452–461. [CrossRef]
25. Supasa, T.; Hsiau, S.S.; Lin, S.M.; Wongsapai, W.; Wu, J.C. Household Energy Consumption Behaviour for Different Demographic Regions in Thailand from 2000 to 2010. *Sustainability* **2017**, *9*, 2328. [CrossRef]
26. Seung, J.O.; Kim, C.N.; Kyaw, T.; Wongee, C.; Kian, J.E.C. Forecasting Long-term Electricity Demand for Cooling of Singapore's Buildings Incorporating an Innovative Air-conditioning Technology. *Energy Build.* **2016**, *127*, 183–193.
27. Wang, Y.; Ge, X.-L.; Liu, J.-L.; Ding, Z. Study and analysis of energy consumption and energy-related carbon emission of industrial in Tianjin, China. *Energy Strategy Rev.* **2016**, *10*, 18–28. [CrossRef]
28. Zou, J.; Liu, W.; Tang, Z. Analysis of Factors Contributing to Changes in Energy Consumption in Tangshan City between 2007 and 2012. *Sustainability* **2017**, *9*, 452. [CrossRef]
29. Zhao, J.; Thinh, N.X.; Li, C. Investigation of the Impacts of Urban Land Use Patterns on Energy Consumption in China: A Case Study of 20 Provincial Capital Cities. *Sustainability* **2017**, *9*, 1383. [CrossRef]
30. Tian, Y.; Xiong, S.; Ma, X. Analysis of the Potential Impacts on China's Industrial Structure in Energy Consumption. *Sustainability* **2017**, *9*, 2284. [CrossRef]
31. Ayvaz, B.; Kusakci, A.O. Electricity consumption forecasting for Turkey with nonhomogeneous discrete grey model. *Energy Sources Part B Econ. Plan. Policy* **2017**, *12*, 260–267. [CrossRef]
32. Dickey, D.A.; Fuller, W.A. Likelihood ratio statistics for autoregressive time series with a unit root. *Econometrica* **1981**, *49*, 1057–1072. [CrossRef]
33. Enders, W. *Applied Econometrics Time Series*; Wiley Series in Probability and Statistics; University of Alabama: Tuscaloosa, AL, USA, 2010.
34. MacKinnon, J. Critical Values for Cointegration Tests. In *Long-Run Economic Relationships*; Engle, R., Granger, C., Eds.; Oxford University Press: Oxford, UK, 1991.
35. Harvey, A.C. *Forecasting, Structural Time Series Models and the Kalman Filter*; Cambridge University Press: Cambridge, UK, 1989.
36. Johansen, S.; Juselius, K. Maximum likelihood estimation and inference on cointegration with applications to the demand for money. *Oxf. Bull. Econ. Stat.* **1990**, *52*, 169–210. [CrossRef]
37. Johansen, S. *Likelihood-Based Inference in Cointegrated Vector Autoregressive Models*; Oxford University Press: New York, NY, USA, 1995.
38. Cryer, J.D.; Chan, K. *Time Series Analysis with Applications in R*, 2nd ed.; Springer: New York, NY, USA, 2008.

MDPI
St. Alban-Anlage 66
4052 Basel
Switzerland
Tel. +41 61 683 77 34
Fax +41 61 302 89 18
www.mdpi.com

Energies Editorial Office
E-mail: energies@mdpi.com
www.mdpi.com/journal/energies

www.ingramcontent.com/pod-product-compliance
Lightning Source LLC
Chambersburg PA
CBHW041750220926
51847CB00014B/159
* 9 7 8 3 0 3 9 2 1 3 1 1 5 *